FIELD EXPERIMENTS

FIELD EXPERIMENTS

Design, Analysis, and Interpretation

Alan S. Gerber YALE UNIVERSITY

Donald P. Green COLUMBIA UNIVERSITY

W. W. NORTON & COMPANY NEW YORK • LONDON

W. W. Norton & Company has been independent since its founding in 1923, when William Warder Norton and Mary D. Herter Norton first published lectures delivered at the People's Institute, the adult education division of New York City's Cooper Union. The firm soon expanded its program beyond the Institute, publishing books by celebrated academics from America and abroad. By midcentury, the two major pillars of Norton's publishing program—trade books and college texts—were firmly established. In the 1950s, the Norton family transferred control of the company to its employees, and today—with a staff of four hundred and a comparable number of trade, college, and professional titles published each year—W. W. Norton & Company stands as the largest and oldest publishing house owned wholly by its employees.

Editor: Ann Shin
Associate Editor: Jake Schindel
Project Editor: Jack Borrebach
Marketing Manager, political science: Sasha Levitt
Production Manager: Eric Pier-Hocking
Text Design: Joan Greenfield / Gooddesign Resource
Design Director: Hope Miller Goodell
Composition by Jouve International—Brattleboro, VT
Manufacturing by the Maple Press—York, PA

Printed in the United States of America.
First edition.

Library of Congress Cataloging-in-Publication Data

Gerber, Alan S.
Field experiments : design, analysis, and interpretation / Alan S. Gerber, Donald P. Green. — 1st ed.
p. cm.
Includes bibliographical references and index.
ISBN 978-0-393-97995-4 (pbk.)
1. Political science—Research—Methodology. 2. Social science—Research—Methodology.
3. Political science—Study and teaching (Higher) 4. Social science—Study and teaching (Higher)
I. Green, Donald P., 1961-II. Title.
JA86.G36 2012
001.4'34—dc23

2011052337

W. W. Norton & Company, Inc., 500 Fifth Avenue, New York, NY 10110-0017
wwnorton.com
W. W. Norton & Company Ltd., Castle House, 75/76 Wells Street, London W1T 3QT

1 2 3 4 5 6 7 8 9 0

THIS BOOK IS DEDICATED TO OUR PARENTS, WHO HELPED INSTILL IN US A LOVE OF SCIENCE.

CONTENTS

PREFACE

For more than a decade, we have taught a one-semester course on experimental research methods to undergraduate and graduate students in the social sciences. Although readings and discussion sometimes address experiments conducted in the lab, the course focuses primarily on "field" experiments, studies conducted in natural settings in which subjects are allocated randomly to treatment and control groups. Students read research articles that illustrate key principles of experimental design or analysis, and class time is devoted to explaining these principles. Students often find the material engaging and even inspiring, but the fact that they read selections from a broad research literature rather than a textbook means that even very talented students frequently fail to assimilate important terms, concepts, and techniques.

Our aim in writing this book is to provide a systematic introduction to experimentation that also conveys the excitement of encountering and conducting primary research. Each chapter weaves abstract principles together with examples drawn from a wide range of social science disciplines: criminology, economics, education, political science, social psychology, and sociology. The exercises at the end of each chapter invite students to reflect on abstract problems of research design and to analyze data from (or inspired by) important experiments. Our aim is to alert readers to the vast range of experimental applications and opportunities for future investigation.

Developing expertise as an experimental researcher is part technical training and part apprenticeship. The former requires the reader to think about experimentation in abstract terms. What inferences can be drawn from an experiment, and under what conditions might these inferences be jeopardized? Any explanation of abstract principles must inevitably invoke statistical terminology, because the language of statistics brings precision and generality. The presentation in this book presupposes that the reader has at some point taken a one- or two-semester introduction to statistical inference and regression. Recognizing that the reader's memory of statistical principles may be hazy, the book continually defines, explains, and illustrates. In an effort to make the presentation accessible, we have freely renamed arcane terms of art.[1]

1 Our aim throughout is to use naming conventions that convey the intuition behind the idea or procedure. The term "external validity, " for example, is replaced by "generalizability. " We also depart from the academic convention of using scholars' names to refer to ideas or procedures. The term "extreme value bounds," for example, replaces the term "Manski bounds." References are provided so that originators of key ideas receive appropriate credit.

We have also sidestepped most of the standard formulas used to conduct hypothesis tests and to construct confidence intervals in favor of a unified framework that relies on statistical simulation. This framework not only makes the presentation more systematic, it also makes the book more concise—from a few core principles, one can deduce a large number of design recommendations that would otherwise require hundreds of pages to explicate.

Our years in the classroom suggest that presentation of abstract principles needs to be reinforced by instructive examples and hands-on experience. It is one thing to memorize key assumptions and quite another to be able to recognize which assumptions come into play in a given application. In an effort to develop this skill, the exercises at the end of each chapter introduce a wide array of experiments and invite readers to reflect on issues of design, analysis, and interpretation. Chapter 12 further illustrates principles laid out in earlier chapters by offering a close reading of several important field experiments.

The text and exercises are designed to prepare readers for the challenge of developing and implementing their own experimental projects. In our courses, we require students to conduct their own field experiments because the experience forces them to link statistical concepts to the specifics of their application, imparts valuable lessons about planning and implementation, and makes them more perceptive readers of other researchers' work. We urge instructors to assign a small-scale field experiment in order to solidify students' understanding of how to frame a testable hypothesis, allocate subjects to experimental conditions, and contend with complications such as attrition or noncompliance. Although field experiments are sometimes dismissed as prohibitively expensive, difficult, or ethically encumbered, experience shows that a wide variety of field experimental studies may be conducted with limited resources and minimal risk to human subjects. Sample topics include applying for jobs, searching for apartments, asking for assistance, fundraising, tipping, tutoring, dieting, petitioning, advertising, and exercising. In Appendix B we suggest field experiments (and accompanying readings) that students may use as inspiration for term papers or capstone projects.

Although designed for a stand-alone course on experimental research, this book may also be used as a supplementary text for courses on research design, causal inference, or applied statistics. Chapters 1 through 4 provide a concise introduction to core concepts, such as potential outcomes, sampling distributions, and statistical inference. Chapters 5 through 11 cover more advanced topics, such as noncompliance, attrition, interference, mediation, and meta-analysis. In an effort to make the book accessible, each chapter supplies plenty of worked examples; for those seeking additional technical details, we furnish a list of suggested readings at the end of each chapter. Supplementary materials at http://isps.research.yale.edu/FEDAI provide readers with data and computer code to perform all analyses and simulations. The code for all of the book's examples has been written using the free software package R, so that readers from all over the world can use the statistical procedures we demonstrate at no cost. Readers are encouraged to visit the Web site for supplementary materials, updates, and errata.

This book is in many ways a collective undertaking. The data for the examples and exercises were furnished by an extraordinary array of scholars: Kevin Arceneaux, Julia Azari, Marianne Bertrand, Rikhil Bhavnani, Elizabeth Campbell, David Clingingsmith, Sarah Cotterill, Ruth Ditlmann, Pascaline Dupas, Leslie Hough, Peter John, Asim Ijaz Khwaja, Michael Kremer, Paul Lagunes, Sendhil Mullainathan, Karthik Muralidharan, David Nickerson, Ben Olken, Jeffrey Rosen, Venkatesh Sundararaman, Rocio Titiunik, and Ebonya Washington. We thank David Torgerson and Iain Chalmers for suggesting examples from the history of medicine and randomized trials. Several of the chapters draw on our collaborations with colleagues: Christopher Larimer (Chapters 6 and 10),[2] Betsy Sinclair and Margaret McConnell (Chapter 8),[3] John Bullock and Shang Ha (Chapter 10),[4] and Edward Kaplan and Holger Kern (Chapters 5, 9 and 11).[5] For comments on chapter drafts, we are grateful to Josh Angrist, Kevin Arceneaux, Noah Buckley, John Bullock, Daniel Butler, Ana De La O, Thad Dunning, Brian Fried, Grant Gordon, Justine Hastings, Susan Hyde, Macartan Humphreys, Edward Kaplan, Jordan Kyle, Paul Lagunes, Malte Lierl, Jason Lyall, Neil Malhotra, David Nickerson, Laura Paler, Elizabeth Levy Paluck, Ben Pasquale, Limor Peer, Kenneth Scheve, Betsy Sinclair, Pavithra Suryanarayan, David Szakonyi, and Lauren Young. Special thanks go to Cyrus Samii and Rocío Titiunik, who provided valuable comments on the entire manuscript.

The authors wish to thank the talented team of researchers who assisted with the preparation of the statistical examples and exercises. Peter Aronow and Holger Kern made important technical and substantive contributions to every chapter. Peter Aronow, Cyrus Samii, and Neelan Sircar developed the R package that we use throughout the book to conduct randomization inference. Bibliographic research and manuscript preparation benefited enormously from the work of Mary McGrath and Josh Kalla, as well as from Lucas Leemann, Malte Lierl, Arjun Shenoy, and John Williams. Allison Sovey and Paolo Spada assisted in preparing the exercises and solutions. We are grateful to Limor Peer and Alissa Stollwerk, who archived the data and programs featured in the book. The Institution for Social and Policy Studies at Yale University provided generous support for research and data preservation. We thank the editorial team at W. W. Norton, Ann Shin, Jack Borrebach, and Jake Schindel, for their outstanding work. The authors take full credit for errors and oversights.

Finally, we owe a special debt of gratitude to our families, for their support, encouragement, and patience during the long process of writing, re-writing, and re-re-writing.

---

2 Gerber, Green, and Larimer 2008.
3 Sinclair, McConnell, and Green 2012.
4 Bullock, Green, and Ha 2010; Green, Ha, and Bullock 2010.
5 Gerber, Green, and Kaplan 2004; Green and Kern 2011; Gerber, Green, Kaplan, and Kern 2010.

CHAPTER 1

Introduction

Daily life continually presents us with questions of cause and effect. Will eating more vegetables make me healthier? If I drive a bit faster than the law allows, will the police pull me over for a speeding ticket? Will dragging my reluctant children to museums make them one day more interested in art and history? Even actions as banal as scheduling a dental exam or choosing an efficient path to work draw on cause-and-effect reasoning.

Organizations, too, grapple with causal puzzles. Charities try to figure out which fundraising appeals work best. Marketing agencies look for ways to boost sales. Churches strive to attract congregants on Sundays. Political parties maneuver to win elections. Interest groups attempt to influence legislation. Whether their aim is to boost donations, sales, attendance, or political influence, organizations make decisions based (at least in part) on their understanding of cause and effect. In some cases, the survival of an organization depends on the skill with which it addresses the causal questions that it confronts.

Of special interest to academic researchers are the causal questions that confront governments and policy makers. What are the economic and social effects of raising the minimum wage? Would allowing parents to pay for private school using publicly funded vouchers make the educational system more effective and cost-efficient? Would legal limits on how much candidates can spend when running for office affect the competitiveness of elections? In the interest of preventing bloodshed, should international peacekeeping troops be deployed with or without heavy weapons? Would mandating harsher punishments for violent offenders deter crime? A list of policy-relevant causal questions would itself fill a book.

An even larger tome would be needed to catalog the many theoretical questions that are inspired by causal claims. For example, when asked to contribute to a collective cause, such as cutting down on carbon emissions in order to prevent global climate change, to what extent are people responsive to appeals based on social norms or ideology? Prominent scholars have argued that collective action will founder unless individuals are given some sort of reward for their participation; according to this argument, simply telling people that they ought to contribute to a collective cause will not work.[1] If this underlying causal claim is true, the consequences for policymaking are profound: tax credits may work, but declaring a national Climate Change Awareness Day will not.

Whether because of their practical, policy, or theoretical significance—or simply because they transport us to a different time and place—causal claims spark the imagination. How does the pilgrimage to Mecca affect the religious, social, and political attitudes of Muslims?[2] Do high school dropout rates in low-income areas improve when children are given monetary rewards for academic performance?[3] Are Mexican police more likely to demand bribes from upper- or lower-class drivers who are pulled aside for traffic infractions?[4] Does your race affect whether employers call you for a job interview?[5] In the context of a civil war, do civilians become more supportive of the government when local economic conditions improve?[6] Does artillery bombardment directed against villages suspected of harboring insurgent guerrillas increase or decrease the likelihood of subsequent insurgent attacks from those villages?[7]

In short, the world is brimming over with causal questions. How might one go about answering them in a convincing manner? What methods for answering causal questions should be viewed with skepticism?

1.1 Drawing Inferences from Intuitions, Anecdotes, and Correlations

One common way of addressing causal questions is to draw on intuition and anecdotes. In the aforementioned case of artillery directed at insurgent villages, a scholar might reason that firing on these villages could galvanize support for the rebels, leading to more insurgent attacks in the future. Bombardment might also prompt the rebels to demonstrate to villagers their determination to fight on by escalating their insurgent activities. In support of this hypothesis, one might point out that the anti-Nazi insurgency in Soviet Russia in 1941 became more determined after occupation forces stepped up their military suppression. One problem with building causal arguments

1 Olson 1965.
2 Clingingsmith, Khwaja, and Kremer 2009.
3 Angrist and Lavy 2009; see also Fryer 2010.
4 Fried, Lagunes, and Venkataramani 2010.
5 Bertrand and Mullainathan 2004.
6 Beath, Christia, and Enikolopov 2011.
7 Lyall 2009.

around intuitions and anecdotes, however, is that such arguments can often be adduced for both sides of a causal claim. In the case of firing on insurgents, another researcher could argue that insurgents depend on the goodwill of villagers; once a village is fired upon, villagers have a greater incentive to expel the rebels in order to prevent future attacks. Supplies dry up, and informants disclose rebel hideouts to government forces. This researcher could defend the argument by describing the government suppression of the Sanusi uprising in Libya, which seemed to deal a lasting blow to these rebels' ability to carry out insurgent attacks.[8] Debates based on intuition and anecdotes frequently result in stalemate.

A critique of anecdote and intuition can be taken a step further. The method is susceptible to error even when intuition and anecdotes seem to favor just one side of an argument. The history of medicine, which is instructive because it tends to provide clearer answers to causal questions than research in social science, is replete with examples of well-reasoned hypotheses that later proved to be false when tested experimentally. Consider the case of aortic arrhythmia (irregular heartbeat), which is often associated with heart attacks. A well-regarded theory held that arrhythmia was a precursor to heart attack. Several drugs were developed to suppress arrhythmia, and early clinical reports seemed to suggest the benefits of restoring a regular heartbeat. The Cardiac Arrhythmia Suppression Trial, a large randomized experiment, was launched in the hope of finding which of three suppression drugs worked best, only to discover that two of the three drugs produced a significant *increase* in death and heart attacks, while the third had negative but seemingly less fatal consequences.[9] The broader point is that well-regarded theories are fallible. This concern is particularly acute in the social sciences, where intuitions are rarely uncontroversial, and controversial intuitions are rarely backed up by conclusive evidence.

Another common research strategy is to assemble statistical evidence showing that an outcome becomes more likely when a certain cause is present. Researchers sometimes go to great lengths to assemble large datasets that allow them to track the correlation between putative causes and effects. These data might be used to learn about the following statistical relationship: to what extent do villages that come under attack by government forces tend to have more or less subsequent insurgent activity? Sometimes these analyses turn up robust correlations between interventions and outcomes. The problem is that correlations can be a misleading guide to causation. Suppose, for example, that the correlation between government bombardment and subsequent insurgent activity were found to be strongly positive: the more shelling, the more subsequent insurgent activity. If interpreted causally, this correlation would indicate that shelling prompted insurgents to step up their attacks. Other

8 See Lyall 2009 for a discussion of these debates and historical episodes.
9 Cardiac Arrhythmia Suppression Trial II Investigators 1992.

interpretations, however, are possible. It could be that government forces received intelligence about an escalation of insurgent activity in certain villages and directed their artillery there. Shelling, in other words, could be a marker for an uptick in insurgent activity. Under this scenario, we would observe a positive correlation between shelling and subsequent insurgent attacks even if shelling per se had no effect.

The basic problem with using correlations as a guide to causality is that correlations may arise for reasons that have nothing to do with the causal process under investigation. Do SAT preparation courses improve SAT scores? Suppose there were a strong positive correlation here: people who took a prep class on average got higher SAT scores than those who did not take the prep class. Does this correlation reflect the course-induced improvement in scores, or rather the fact that students with the money and motivation to take a prep course tend to score higher than their less affluent or less motivated counterparts? If the latter were true, we might see a strong association even if the prep course had no effect on scores. A common error is to reason that where there's smoke, there's fire: correlations at least hint at the existence of a causal relationship, right? Not necessarily. Basketball players tend to be taller than other people, but you cannot grow taller by joining the basketball team.

The distinction between correlation and causation seems so fundamental that one might wonder why social scientists rely on correlations when making causal arguments. The answer is that the dominant methodological practice is to transform raw correlations into more refined correlations. After noticing a correlation that might have a causal interpretation, researchers attempt to make this causal interpretation more convincing by limiting the comparison to observations that have similar background attributes. For example, a researcher seeking to isolate the effects of the SAT preparatory course might restrict attention to people with the same gender, age, race, grade point average, and socioeconomic status. The problem is that this method remains vulnerable to *unobserved* factors that predict SAT scores and are correlated with taking a prep course. By restricting attention to people with the same socio-demographic characteristics, a researcher makes the people who took the course comparable to those who did not in terms of *observed* attributes, but these groups may nevertheless differ in ways that are unobserved. In some cases, a researcher may fail to consider some of the factors that contribute to SAT scores. In other cases, a researcher may think of relevant factors but fail to measure them adequately. For example, people who take the prep course may, on average, be more motivated to do well on the test. If we fail to measure motivation (or fail to measure it accurately), it will be one of the unmeasured attributes that might cause us to draw mistaken inferences. These unmeasured attributes are sometimes called *confounders* or *lurking variables* or *unobserved heterogeneity*. When interpreting correlations, researchers must always be alert to the distorting influence of unmeasured attributes. The fact that someone chooses to take the prep course may reveal something about how they are likely to perform on the test. Even if the course truly has no effect, people with the same age, gender, and affluence may seem to do better when they take the course.

Whether the problem of unobserved confounders is severe or innocuous will depend on the causal question at hand and the manner in which background attributes are measured. Consider the so-called "broken windows" theory, which suggests that crime increases when blighted areas appear to be abandoned by property owners and unsupervised by police.[10] The causal question is whether one could reduce crime in such areas by picking up trash, removing graffiti, and repairing broken windows. A weak study might compare crime rates on streets with varying levels of property disrepair. A more convincing study might compare crime rates on streets that currently experience different levels of blight but in the past had similar rates of disrepair and crime. But even the latter study may still be unconvincing because unmeasured factors, such as the closing of a large local business, may have caused some streets to deteriorate physically and coincided with an upsurge in crime.[11]

Determined to conquer the problem of unobserved confounders, one could set out to measure each and every one of the unmeasured factors. The intrepid researcher who embarks on this daunting task confronts a fundamental problem: no one can be sure what the set of unmeasured factors comprises. The list of all potential confounders is essentially a bottomless pit, and the search has no well-defined stopping rule. In the social sciences, research literatures routinely become mired in disputes about unobserved confounders and what to do about them.

1.2 Experiments as a Solution to the Problem of Unobserved Confounders

The challenge for those who seek to answer causal questions in a convincing fashion is to come up with a research strategy that does not require them to identify, let alone measure, all potential confounders. Gradually, over the course of centuries, researchers developed procedures designed to sever the statistical relationship between the treatment and all variables that predict outcomes. The earliest experiments, such as Lind's study of scurvy in the 1750s and Watson's study of smallpox in the 1760s, introduced the method of systematically tracking the effects of a researcher-induced intervention by comparing outcomes in the treatment group to outcomes in one or more control groups.[12] One important limitation of these early studies is that they assumed that their subjects were identical in terms of their medical trajectories. What if this assumption

10 Wilson and Kelling 1982.

11 See Keizer, Lindenberg, and Steg 2008, but note that this study does not employ random assignment. For a randomized field experiment see Mazerolle, Price, and Roehl 2000.

12 Hughes 1975; Boylston 2008.

were false, and treatments tended to be administered to patients with the best chances of recovery? Concerned that the apparent effects of their intervention might be attributable to extraneous factors, researchers placed increasing emphasis on the procedure by which treatments were assigned to subjects. Many pathbreaking studies of the nineteenth century assigned subjects alternately to treatment and control in an effort to make the experimental groups comparable. In 1809, a Scottish medical student described research conducted in Portugal in which army surgeons treated 366 sick soldiers alternately with bloodletting and other palliatives.[13] In the 1880s, Louis Pasteur tested his anthrax vaccine on animals by alternately exposing treatment and control groups to the bacteria. In 1898, Johannes Fibiger assigned an experimental treatment to diphtheria patients admitted to a hospital in Copenhagen on alternate days.[14] Alternating designs were common in early agricultural studies and investigations of clairvoyance, although researchers gradually came to recognize potential pitfalls of alternation.[15] One problem with alternating designs is that they cannot definitively rule out confounding factors, such as sicker diphtheria patients coming to the hospital on certain days of the week. The first to recognize the full significance of this point was the agricultural statistician R. A. Fisher, who in the mid-1920s argued vigorously for the advantages of assigning observations at random to treatment and control conditions.[16]

This insight represents a watershed moment in the history of science. Recognizing that no planned design, no matter how elaborate, could fend off every possible systematic difference between the treatment and control groups, Fisher laid out a general procedure for eliminating systematic differences between treatment and control groups: random assignment. When we speak of experiments in this volume, we refer to studies in which some kind of random procedure, such as a coin flip, determines whether a subject receives a treatment.

One remarkable aspect of the history of randomized experimentation is that the idea of random assignment occurred to several ingenious people centuries before it was introduced into modern scientific practice. For example, the notion that one could use random assignment to form comparable experimental groups seems to have been apparent to the Flemish physician Jan Baptist Van Helmont, whose 1648 manuscript "Origin of Medicine" challenged the proponents of bloodletting to perform the following randomized experiment:

> Let us take out of the hospitals . . . 200 or 500 poor people, that have fevers, pleurisies. Let us divide them into halves, let us cast lots, that one halfe of them may fall to

13 Chalmers 2001.

14 Hróbjartsson, Gøtzsche, and Gluud 1998.

15 Merrill 2010. For further reading on the history of experimentation, see Cochran 1976; Forsetlund, Chalmers, and Bjørndal 2007; Hacking 1990; and Salsburg 2001. See Greenberg and Shroder 2004 on social experiments and Green and Gerber 2003 on the history of experiments in political science.

16 Box 1980, p. 3.

> my share, and the other to yours; I will cure them without bloodletting and sensible evacuation; but you do, as ye know . . . We shall see how many funerals both of us shall have.[17]

Unfortunately for those whose physicians prescribed bloodletting in the centuries following Van Helmont, he never conducted his proposed experiment. One can find similar references to hypothetical experiments dating back to medieval times, but no indication that any were actually put into practice. Until the advent of modern statistical theory in the early twentieth century, the properties of random assignment were not fully appreciated, nor were they discussed in a systematic manner that would have allowed one generation to recommend the idea to the next.

Even after Fisher's ideas became widely known in the wake of his 1935 book *The Design of Experiments*, randomized designs met resistance from medical researchers until the 1950s, and randomized experiments did not catch on in the social sciences until the 1960s.[18] In the class of brilliant twentieth-century discoveries, the idea of randomization contrasts sharply with the idea of relativity, which lay completely hidden until uncovered by genius. Randomization was more akin to crude oil, something that periodically bubbled to the surface but remained untapped for centuries until its extraordinary practical value came to be appreciated.

1.3 Experiments as Fair Tests

In the contentious world of causal claims, randomized experimentation represents an evenhanded method for assessing what works. The procedure of assigning treatments at random ensures that there is no systematic tendency for either the treatment or control group to have an advantage. If subjects were assigned to treatment and control groups and no treatment were actually administered, there would be no reason to expect that one group would outperform the other. In other words, random

17 Chalmers 2001, p. 1157.

18 The advent of randomized experimentation in social and medical research took roughly a quarter century. Shortly after laying the statistical foundations for random assignment and the analysis of experimental data, Fisher collaborated on the first randomized agricultural experiment (Eden and Fisher 1927). Within a few years, Amberson, McMahon, and Pinner (1931) performed what appears to be the first randomized medical experiment, in which tuberculosis patients were assigned to clinical trials based on a coin flip. The large-scale studies of tuberculosis conducted during the 1940s brought randomized clinical trials to the forefront of medicine. Shortly afterward, the primacy of this methodology in medicine was cemented by a series of essays by Hill (1951, 1952) and subsequent acclaim of the polio vaccine trials of the 1950s (Tanur 1989). Randomized clinical trials gradually came to be heralded as the gold standard by which medical claims were to be judged. By 1952, books such as Kempthorne's *Design and Analysis of Experiments* (pp. 125–126) declared that "only when the treatments in the experiment are applied by the experimenter using the full randomization procedure is the chain of inductive inference sound."

assignment implies that the observed *and unobserved* factors that affect outcomes are equally likely to be present in the treatment and control groups. Any given experiment may overestimate or underestimate the effect of the treatment, but if the experiment were conducted repeatedly under similar conditions, the average experimental result would accurately reflect the true treatment effect. In Chapter 2, we will spell out this feature of randomized experiments in greater detail when we discuss the concept of unbiased estimation.

Experiments are fair in another sense: they involve transparent, reproducible procedures. The steps used to conduct a randomized experiment may be carried out by any research group. A random procedure such as a coin flip may be used to allocate observations to treatment or control, and observers can monitor the random assignment process to make sure that it is followed faithfully. Because the allocation process precedes the measurement of outcomes, it is also possible to spell out beforehand the way in which the data will be analyzed. By automating the process of data analysis, one limits the role of discretion that could compromise the fairness of a test.

Random allocation is the dividing line that separates experimental from nonexperimental research in the social sciences. When working with nonexperimental data, one cannot be sure whether the treatment and control groups are comparable because no one knows precisely why some subjects and not others came to receive the treatment. A researcher may be prepared to assume that the two groups are comparable, but assumptions that seem plausible to one researcher may strike another as far-fetched.

This is not to say that experiments are free from problems. Indeed, this book would be rather brief were it not for the many complications that may arise in the course of conducting, analyzing, and interpreting experiments. Entire chapters are devoted to problems of noncompliance (subjects who receive a treatment other than the one to which they were randomly assigned), attrition (observations for which outcome measurements are unavailable), and interference between units (observations influenced by the experimental conditions to which other observations are assigned). The threat of bias remains a constant concern even when conducting experiments, which is why it is so important to design and analyze them with an eye toward maintaining symmetry between treatment and control groups and, more generally, to embed the experimental enterprise in institutions that facilitate proper reporting and accumulation of experimental results.

1.4 Field Experiments

Experiments are used for a wide array of different purposes. Sometimes the aim of an experiment is to assess a theoretical claim by testing an implied causal relationship. Game theorists, for example, use laboratory experiments to show how the introduction

BOX 1.1

Experiments in the Natural Sciences

Readers with a background in the natural sciences may find it surprising that random assignment is an integral part of the definition of a social science experiment. Why is random assignment often unnecessary in experiments in, for example, physics? Part of the answer is that the "subjects" in these experiments—e.g., electrons—are more or less interchangeable, and so the method used to assign subjects to treatment is inconsequential. Another part of the answer is that lab conditions neutralize all forces other than the treatment.

In the life sciences, subjects are often different from one another, and eliminating unmeasured disturbances can be difficult even under carefully controlled conditions. An instructive example may be found in a study by Crabbe, Wahlsten, and Dudek (1999), who performed a series of experiments on mouse behavior in three different science labs. As Lehrer (2010) explains:

> Before [Crabbe] conducted the experiments, he tried to standardize every variable he could think of. The same strains of mice were used in each lab, shipped on the same day from the same supplier. The animals were raised in the same kind of enclosure, with the same brand of sawdust bedding. They had been exposed to the same amount of incandescent light, were living with the same number of littermates, and were fed the exact same type of chow pellets. When the mice were handled, it was with the same kind of surgical glove, and when they were tested it was on the same equipment, at the same time in the morning.

Nevertheless, experimental interventions produced markedly different results across mice and research sites.

of uncertainty or the opportunity to exchange information prior to negotiating affects the bargains that participants strike with one another.[19] Such experiments are often couched in very abstract terms, with rules that stylize the features of an auction, legislative session, or international dispute. The participants are typically ordinary people (often members of the university community), not traders, legislators, or diplomats, and the laboratory environment makes them keenly aware that they are participating in a research study.

At the other end of the spectrum are experiments that strive to be as realistic and unobtrusive as possible in an effort to test more context-specific hypotheses.

19 See Davis and Holt 1993; Kagel and Roth 1995; Guala 2005.

Quite often this type of research is inspired by a mixture of theoretical and practical concerns. For example, to what extent and under what conditions does preschool improve subsequent educational outcomes? Experiments that address this question shed light on theories about childhood development while at the same time informing policy debates about whether and how to allocate resources to early childhood education in specific communities.

The push for realism and unobtrusiveness stems from the concern that unless one conducts experiments in a naturalistic setting and manner, some aspect of the experimental design may generate results that are idiosyncratic or misleading. If subjects know that they are being studied or if they sense that the treatment they received is supposed to elicit a certain kind of response, they may express the opinions or report the behavior they believe the experimenter wants to hear. A treatment may seem effective until a more unobtrusive experiment proves otherwise.[20] Conducting research in naturalistic settings may be viewed as a hedge against unforeseen threats to inference that arise when drawing generalizations from results obtained in laboratory settings. Just as experiments are designed to test causal claims with minimal reliance on assumptions, experiments conducted in real-world settings are designed to make generalizations less dependent on assumptions.

Randomized studies that are conducted in real-world settings are often called *field experiments*, a term that calls to mind early agricultural experiments that were literally conducted in fields. The problem with the term is that the word *field* refers to the setting, but the setting is just one aspect of an experiment. One should invoke not one but several criteria: whether the treatment used in the study resembles the intervention of interest in the world, whether the participants resemble the actors who ordinarily encounter these interventions, whether the context within which subjects

20 Whether this concern is justified is an empirical question, and the answer may well depend on the setting, context, and subjects. Unfortunately, the research literature on this topic remains underdeveloped. Few studies have attempted to estimate treatment effects in both lab and field contexts. Gneezy, Haruvy, and Yafe (2004), for example, use field and lab studies to test the hypothesis that the quantity of food consumed depends on whether each diner pays for his or her own food or whether they all split the bill. When this experiment is conducted in an actual cafeteria, splitting the bill leads to significantly more food consumption; when the equivalent game is played in abstract form (with monetary payoffs) in a nearby lab, the average effect is weak and not statistically distinguishable from zero. Jerit, Barabus, and Clifford (2012) compare the effects of exposure to a local newspaper on political knowledge and opinions. In the field, free Sunday newspapers were randomly distributed to households over the course of one month; in the lab, subjects from the same population were invited to a university setting, where they were presented with the four most prominent political news stories airing during the same month. For the 17 outcome measures, estimated treatment effects in the lab and field are found to be weakly correlated (Table 2). See also Rondeau and List (2008), who compare the effectiveness of different fundraising appeals on behalf of the Sierra Club directed at 3,000 past donors, as measured by actual donations. The fundraising appeals, which involve various combinations of matching funds, thresholds, and money-back guarantees, are then presented in abstract form in a lab setting with monetary payoffs. The correspondence between lab and field results was relatively weak, with average contributions in the lab predicting about 5% of the variance in average contributions in the field across the four conditions.

receive the treatment resembles the context of interest, and whether the outcome measures resemble the actual outcomes of theoretical or practical interest.

For example, suppose one were interested in the extent to which financial contributions to incumbent legislators' reelection campaigns buy donors access to the legislators, a topic of great interest to those concerned that the access accorded to wealthy donors undermines democratic representation. The hypothesis is that the more a donor contributes, the more likely the legislator is to grant a meeting to discuss the donor's policy prescriptions. One possible design is to recruit students to play the part of legislative schedulers and present them with a list of requests for meetings from an assortment of constituents and donors in order to test whether people described as potential donors receive priority. Another design involves the same exercise, but this time the subjects are actual legislative schedulers.[21] The latter design would seem to provide more convincing evidence about the relationship between donations and access in actual legislative settings, but the degree of experimental realism remains ambiguous. The treatments in this case are realistic in the sense that they resemble what an actual scheduler might confront, but the subjects are aware that they are participating in a simulation exercise. Under scrutiny by researchers, legislative schedulers might try to appear indifferent to fundraising considerations; in an actual legislative setting where principals provide feedback to schedulers, donors might receive special consideration. More realistic, then, would be an experiment in which one or more donors contribute randomly assigned sums of money to various legislators and request meetings to discuss a policy or administrative concern. In this design, the subjects are actual schedulers, the treatment is a campaign donation, the treatment and request for a meeting are authentic, and the outcome is whether a real request is granted in a timely fashion.

Because the degree of "fieldness" may be gauged along four different dimensions (authenticity of treatments, participants, contexts, and outcome measures), a proper classification scheme would involve at least sixteen categories, a taxonomy that far exceeds anyone's interest or patience. Suffice it to say that field experiments take many forms. Some experiments seem naturalistic on all dimensions. Sherman et al. worked with the Kansas City police department in order to test the effectiveness of police raids on locations where drug dealing was suspected.[22] The treatments were raids by teams of uniformed police directed at 104 randomly chosen sites among the 207 locations for which warrants had been issued. Outcomes were crime rates in nearby areas. Karlan and List collaborated with a charity in order to test the effectiveness of alternative fundraising appeals.[23] The treatments were fundraising letters; the experiment was unobtrusive in the sense that recipients of the fundraising appeals were

21 See Chin, Bond, and Geva 2000.

22 Sherman et al. 1995.

23 Karlan and List 2007.

unaware that an experiment was being conducted; and the outcomes were financial donations. Bergan teamed up with a grassroots lobbying organization in order to test whether constituents' e-mail to state representatives influences roll call voting.[24] The lobbying organization allowed Bergan to extract a random control group from its list of targeted legislators; otherwise, its lobbying campaign was conducted in the usual way, and outcomes were assessed based on the legislators' floor votes.

Many field experiments are less naturalistic, and generalizations drawn from them are more dependent on assumptions. Sometimes the interventions deployed in the field are designed by researchers rather than practitioners. Eldersveld, for example, fashioned his own get-out-the-vote campaigns in order to test whether mobilization activities cause registered voters to cast ballots.[25] Much may be learned when researchers craft their own treatments—indeed, the development of theoretically inspired interventions is an important way in which researchers may contribute to theoretical and policy debates. However, if the aim of an experiment is to gauge the effectiveness of typical candidate- or party-led voter mobilization campaigns, researcher-led campaigns may be unrepresentative in terms of the messages used or the manner in which they are communicated. Suppose that the researcher's intervention were to prove ineffective. This finding alone would not establish that a typical campaign's interventions are ineffective, although this interpretation could be bolstered by a series of follow-up experiments that test different types of campaign communication.[26] Sometimes treatments are administered and outcomes are measured in a way that notifies participants that they are being studied, as in Paluck's experimental investigation of intergroup prejudice in Rwanda.[27] Her study enlisted groups of Rwandan villagers to listen to recordings of radio programs on a monthly basis for a period of one year, at which point outcomes were measured using surveys and role-playing exercises. Finally, experimental studies with relatively little field content are those in which actual interventions are delivered in artificial settings to subjects who are aware that they are part of a study. Examples of this type of research may be found in the domain of commercial advertising, where subjects are shown different types of ads either in the context of an Internet survey or in a lab located in a shopping center.[28]

Whether a given study is regarded as a field experiment is partly a matter of perspective. Ordinarily, experiments that take place on college campuses are consid-

24 Bergan 2009.

25 Eldersveld 1956.

26 For example, in an effort to test whether voter mobilization phone calls conducted by call centers are typically ineffective, Panagopoulos (2009) compares partisan and nonpartisan scripts, Nickerson (2007) assesses whether effectiveness varies depending on the quality of the calling center, and other scholars have conducted studies in various electoral environments. See Green and Gerber 2008 for a review of this literature.

27 Paluck 2009.

28 See, for example, Clinton and Lapinski 2004; Kohn, Smart, and Ogborne 1984.

ered lab studies, but some experiments on cheating involve realistic opportunities for students to copy answers or misreport their own performance on self-graded tests.[29] An experimental study that examines the deterrent effect of exam proctoring would amount to a field experiment if one's aim were to understand the conditions under which students cheat in school. This example serves as a reminder that what constitutes a field experiment depends on how "the field" is defined.

1.5 Advantages and Disadvantages of Experimenting in Real-World Settings

Many field experiments take the form of "program evaluations" designed to gauge the extent to which resources are deployed effectively. For example, in order to test whether a political candidate's TV advertising campaign increases her popularity, a field experiment might randomize the geographic areas in which the ads are deployed and measure differences in voter support between treatment and control regions. From the standpoint of program evaluation, this type of experiment is arguably superior to a laboratory study in which voters are randomly shown the candidate's ads and later asked their views about the candidate. The field experiment tests the effects of deploying the ads and allows for the possibility that some voters in targeted areas will miss the ad, watch it inattentively, or forget its message amid life's other distractions. Interpretation of the lab experiment's results is complicated by the fact that subjects in lab settings may respond differently to the ads than the average voter outside the lab. In this application, preliminary lab research might be useful insofar as it suggests which messages are most likely to work in field settings, but only a field experiment allows the researcher to reliably gauge the extent to which an actual ad campaign changed votes and to express this outcome in relation to the resources spent on the campaign.

As we move from program evaluation to tests of theoretical propositions, the relative merits of field and lab settings become less clear-cut. A practical advantage of delivering treatments under controlled laboratory conditions is that one can more easily administer multiple variations of a treatment to test fine-grained theoretical propositions. Field interventions are often more cumbersome: in the case of political advertisements, it may be logistically challenging or politically risky to air multiple advertisements in different media markets. On the other hand, field experiments are sometimes able to achieve a high level of theoretical nuance when a wide array of treatments can be distributed across a large pool of subjects. Field experiments that deploy multiple versions of a treatment are common, for example, in research

29 Canning 1956; Nowell and Laufer 1997.

on discrimination, where researchers vary ethnicity, social class, and a host of other characteristics to better understand the conditions under which discrimination occurs.[30]

Even when limited to a single, relatively blunt intervention, a researcher may still have reason to conduct experiments in the field. Advertising research in field settings is often unobtrusive in the sense that subjects are not viewing the ad at the behest of a researcher, and outcomes are measured in a way that does not alert subjects to the fact that they are being studied.[31] Whereas outcomes in lab settings are often attitudes and behaviors that can be measured in the space of one sitting,[32] field studies tend to monitor behaviors over extended periods of time. The importance of ongoing outcome measurement is illustrated by experiments that find strong instantaneous effects of political advertising that decay rapidly over time.[33]

Perhaps the biggest disadvantage of conducting experiments in the field is that they are often challenging to implement. In contrast to the lab, where researchers can make unilateral decisions about what treatments to deploy, field experiments are often the product of coordination between researchers and those who actually carry out the interventions or furnish data on subjects' outcomes. Orr[34] and Gueron[35] offer helpful descriptions of how these partnerships are formed and nurtured over the course of a collaborative research project. Both authors stress the importance of building consensus about the use of random assignment. Research partners and funders sometimes balk at the idea of randomly allocating treatments, preferring instead to treat everyone or a hand-picked selection of subjects. The researcher must be prepared to formulate a palatable experimental design and to argue convincingly that the proposed use of random assignment is both feasible and ethical. The authors also stress that successful implementation of the agreed-upon experimental design—the allocation of subjects, the administration of treatments, and the measurement of outcomes—requires planning, pilot testing, and constant supervision.

Managing research collaboration with schools, police departments, retail firms, or political campaigns sounds difficult and often is. Nevertheless, field experimentation is a rapidly growing form of social science research, encompassing hundreds of

---

30 See Doleac and Stein 2010 for a study of racial discrimination by bidders on Internet auctions or Pager, Western, and Bonikowski 2009 for a study of labor market discrimination. We discuss discrimination experiments in Chapters 9 and 12.

31 In cases where surveys are used to assess outcomes, measurement may be unobtrusive in the more limited but nevertheless important sense that subjects are unaware that the survey aims to gauge the effects of the intervention.

32 Orchestrating return visits to the lab often presents logistical challenges, and failure to attract all subjects back to the lab may introduce bias (see Chapter 7).

33 See, for example, Gerber, Gimpel, Green, and Shaw 2011. See also the discussion of outcome measurement in Chapter 12.

34 Orr 1999, Chapter 5.

35 Gueron 2002.

studies on topics like education, crime, employment, savings, discrimination, charitable giving, conservation, and political participation.[36] The set of noteworthy and influential studies includes experiments of every possible description: small-scale interventions designed and implemented by researchers; collaborations between researchers and firms, schools, police agencies, or political campaigns; and massive government-funded studies of income taxes, health insurance, schooling, and public housing.[37]

Time and again, researchers overcome practical hurdles, and the boundaries of what is possible seem to be continually expanding. Consider, for example, research on how to promote government accountability. Until the 1990s, research in this domain was almost exclusively nonexperimental, but a series of pathbreaking studies have shown that one can use experiments to investigate the effects of government audits and community forums on accounting irregularities among public works programs,[38] the effects of grassroots monitoring efforts on the performance of legislators,[39] and the effects of information about constituents' preferences on legislators' roll call votes.[40] Field experiments are sometimes faulted for their inability to address big questions, such as the effects of culture, wars, or constitutions, but researchers have grown increasingly adept at designing experiments that test the effects of mechanisms that are thought to transmit the effects of the hard-to-manipulate variables.[41] Given the rapid pace of innovation, the potential for experimental inquiry remains an open question.

1.6 Naturally Occurring Experiments and Quasi-Experiments

Another way to expand the domain of what may be studied experimentally is to seize on *naturally occurring experiments*. Experimental research opportunities arise when interventions are assigned by a government or institution.[42] For example, the

36 Michalopoulos 2005; Green and Gerber 2008.

37 See, e.g., Robins 1985 on income taxes; Newhouse 1989 on health insurance; Krueger and Whitmore 2001 and U.S. Department of Health and Human Services 2010 on schooling. On public housing, see Sanbonmatsu et al. 2006; Harcourt and Ludwig 2006; and Kling, Liebman, and Katz 2007.

38 Olken 2007.

39 Humphreys and Weinstein 2010; Grose 2009.

40 Butler and Nickerson 2011.

41 Ludwig, Kling, and Mullainathan 2011; Card, Della Vigna, and Malmendier 2011.

42 Unfortunately, the term "natural experiment" is sometimes used quite loosely, encompassing not only naturally occurring randomized experiments but also any observational study in which the method of assignment is haphazard or inscrutable. We categorize studies that use near-random or arguably random assignment as quasi-experiments. For definitions of the term *natural experiment* that do not require random assignment, see Dunning 2012 and Shadish, Cook, and Campbell 2002, p. 17.

Vietnam draft lottery,[43] the random assignment of defendants to judges,[44] the random audit of local municipalities in Brazil,[45] lotteries that assign parents the opportunity to place their children in different public schools,[46] the assignment of Indian local governments to be headed by women or members of scheduled castes,[47] the allocation of visas to those seeking to immigrate,[48] and legislative lotteries to determine which representative will be allowed to propose legislation[49] are a few examples where randomization procedures have been employed by government, setting the stage for an experimental analysis. Researchers have also seized on natural experiments conducted by nongovernmental institutions. Universities, for example, occasionally randomize the pairing of roommates, allocation of instructors, and composition of tenure review committees.[50] Sports of all kinds use coin flips and lotteries to assign everything from the sequence of play to the colors worn by the contestants.[51] This list of naturally occurring experimental opportunities might also include revisiting random allocations conducted for other research purposes. A *downstream experiment* refers to a study whose intervention affects not only the proximal outcome of interest but, in so doing, potentially influences other outcomes as well (see Chapter 6). For example, a researcher might revisit an experiment that induced an increase in high school graduation rates in order to assess whether this randomly induced change in educational attainment in turn caused an increase in voter turnout.[52] In this book, we scarcely distinguish between field experiments and naturally occurring experiments, except to note that extra effort is sometimes required in order to verify that draft boards, court systems, or school districts implemented random assignment.

Quite different are *quasi-experiments*, in which near-random processes cause places, groups, or individuals to receive different treatments. Since the mid-1990s, a growing number of scholars have studied instances where institutional rules cause near-random treatment assignments to be allocated among those who fall just short of or just beyond a cutoff, creating a discontinuity. One of the most famous examples of this research design is a study of U.S. congressional districts in which one party's candidate narrowly wins a plurality of votes.[53] The small shift in votes that separates a narrow victory from a narrow defeat produces a treatment—winning the seat in the House of Representatives—that might be construed as random. One

43 Angrist 1991.
44 Kling 2006; Green and Winik 2010.
45 Ferraz and Finan 2008.
46 Hastings, Kane, Staiger, and Weinstein 2007.
47 Beaman et al. 2009; Chattopadhyay and Duflo 2004.
48 Gibson, McKenzie, and Stillman 2011.
49 Loewen, Koop, Settle, and Fowler 2010.
50 Sacerdote 2001; Carrell and West 2010; De Paola 2009; Zinovyeva and Bagues 2010.
51 Hill and Barton 2005; see also Rowe, Harris, and Roberts 2005 for a response to Hill and Barton.
52 Sondheimer and Green 2009.
53 Lee 2008.

could compare near-winners to near-losers in order to assess the effect of a narrow victory on the probability that the winning party wins reelection in the district two years later.

Because quasi-experiments do not involve an explicit random assignment procedure, the causal inferences they support are subject to greater uncertainty. Although the researcher may have good reason to believe that observations on opposite sides of an arbitrary threshold are comparable, there is always some risk that the observations may have "sorted" themselves so as to receive or avoid the treatment. Critics who have looked closely at the pool of congressional candidates who narrowly win or lose have pointed out that there appear to be systematic differences between near-winners and near-losers in terms of their political resources.[54]

The same concerns apply to a wide array of quasi-experiments that take weather patterns, natural disasters, colonial settlement patterns, national boundaries, election cycles, assassinations and so forth to be near-random "treatments." In the absence of random assignment, there is always some uncertainty about how nearly random these treatments are. Although these studies are similar in spirit to field experimentation insofar as they strive to illuminate causal effects in real-world settings, they fall outside the scope of this book because they rely on argumentation rather than randomization procedures. In order to present a single, coherent perspective on experimental design and analysis, this book confines its attention to randomized experiments.

1.7 Plan of the Book

This chapter has introduced a variety of important concepts without pausing for rigorous definitions or proofs. Chapter 2 delves more deeply into the properties of experiments, describing in detail the underlying assumptions that must be met for experiments to be informative. Chapter 3 introduces the concept of sampling variability, the statistical uncertainty introduced whenever subjects are randomly allocated to treatment and control groups. Chapter 4 focuses on how covariates, variables that are measured prior to the administration of the treatment, may be used in

54 Grimmer et al. 2011; Caughey and Sekhon 2011. In addition, regression discontinuity analyses often confront the following conundrum: the causal effect of the treatment is identified at the point of discontinuity, but data are sparse in the close vicinity of the boundary. One may expand the comparison to include observations farther from the boundary, but doing so jeopardizes the comparability of groups that do or do not receive the treatment. In an effort to correct for unmeasured differences between the groups, researchers typically use regression to control for trends on either side of the boundary, a method that introduces a variety of modeling decisions and attendant uncertainty. See Imbens and Lemieux 2008 and Green et al. 2009.

experimental design and analysis. Chapters 5 and 6 discuss the complications that arise when subjects are assigned one treatment but receive another. The so-called *noncompliance* or *failure-to-treat* problem is sufficiently common and conceptually challenging to warrant two chapters. Chapter 7 addresses the problem of attrition, or the failure to obtain outcome measurements for every subject. Because field experiments are frequently conducted in settings where subjects communicate, compare, or remember treatments, Chapter 8 considers the complications associated with interference between experimental units. Because researchers are often interested in learning about the conditions under which treatment effects are especially strong or weak, Chapter 9 discusses the detection of heterogeneous treatment effects. Chapter 10 considers the challenge of studying the causal pathways by which an experimental effect is transmitted. Chapter 11 discusses how one might draw generalizations that go beyond the average treatment effect observed in a particular sample and apply them to the average treatment effect in a broader population. The chapter provides a brief introduction to meta-analysis, a statistical technique that pools data from multiple experiments in order to summarize the findings of a research literature. Chapter 12 discusses a series of noteworthy experiments in order to highlight important principles introduced in previous chapters. Chapter 13 guides the reader through the composition of an experimental research report, providing a checklist of key aspects of any experiment that must be described in detail. Appendix A discusses regulations that apply to research involving human subjects. In order to encourage you to put the book's ideas to work, Appendix B suggests several experimental projects that involve low cost and minimal risk to human subjects.

SUGGESTED READINGS

Accessible introductions to experimental design in real-world settings can be found in Shadish, Cook, and Campbell 2002 and Torgerson and Torgerson 2008. For a discussion of the limitations of field experimentation, see Heckman and Smith 1995. Morgan and Winship (2007), Angrist and Pischke (2009), and Rosenbaum (2010) discuss the challenges of extracting causal inferences from nonexperimental data. Imbens and Lemieux (2008) provide a useful introduction to regression-discontinuity designs.

EXERCISES: CHAPTER 1

1. Core concepts:
   (a) What is an experiment, and how does it differ from an observational study?
   (b) What is "unobserved heterogeneity," and what are its consequences for the interpretation of correlations?
2. Would you classify the study described in the following abstract as a field experiment, a naturally occurring experiment, a quasi-experiment, or none of the above? Why?

   "This study seeks to estimate the health effects of sanitary drinking water among low-income villages in Guatemala. A random sample of all villages with fewer than 2,000 inhabitants was selected for analysis. Of the 250 villages sampled, 110 were found to have unsanitary drinking water. In these 110 villages, infant mortality rates were, on average, 25 deaths per 1,000 live births, as compared to 5 deaths per 1,000 live births in the 140 villages with sanitary drinking water. Unsanitary drinking water appears to be a major contributor to infant mortality."

3. Based on what you are able to infer from the following abstract, to what extent does the study described seem to fulfill the criteria for a field experiment?

   "We study the demand for household water connections in urban Morocco, and the effect of such connections on household welfare. In the northern city of Tangiers, among homeowners without a private connection to the city's water grid, a random subset was offered a simplified procedure to purchase a household connection on credit (at a zero percent interest rate). Take-up was high, at 69%. Because all households in our sample had access to the water grid through free public taps . . . household connections did not lead to any improvement in the quality of the water households consumed; and despite a significant increase in the quantity of water consumed, we find no change in the incidence of waterborne illnesses. Nevertheless, we find that households are willing to pay a substantial amount of money to have a private tap at home. Being connected generates important time gains, which are used for leisure and social activities, rather than productive activities."[55]

4. A parody appearing in the *British Medical Journal* questioned whether parachutes are in fact effective in preventing death when skydivers are presented with severe "gravitational challenge."[56] The authors point out that no randomized trials have assigned parachutes to skydivers. Why is it reasonable to believe that parachutes are effective even in the absence of randomized experiments that establish their efficacy?

55 Devoto et al. 2011.
56 Smith and Pell 2003.

CHAPTER 2

Causal Inference and Experimentation

Although the logic of experimentation is for the most part intuitive, researchers can run into trouble if they lack a firm grasp of the key assumptions that must be met in order for experiments to provide reliable assessments of cause and effect. This point applies in particular to field experimental researchers, who must frequently make real-time decisions about research design. Failure to understand core statistical principles and their practical implications may cause researchers to squander resources and experimental opportunities. It is wise, therefore, to invest time studying the formal statistical properties of experiments before launching a research project.

This chapter introduces a system of notation that will be used throughout the book. By depicting the outcomes that potentially manifest themselves depending on whether the treatment is administered to each unit, the notation clarifies a number of key concepts, such as the idea of a treatment effect. This notational system is then used to shed light on the conditions under which experiments provide persuasive evidence about cause and effect. The chapter culminates with a list of core assumptions and what they imply for experimental design. The advantage of working methodically from core principles is that a long list of design-related admonitions flows from a relatively compact set of ideas that can be stored in working memory.

2.1 Potential Outcomes

Suppose we seek to gauge the causal effect of a treatment. For concreteness, suppose we wish to study the budgetary consequences of having women, rather than men, head Indian village councils, which govern rural areas in West Bengal and Rajasthan.[1]

1 See Chattopadhyay and Duflo 2004.

What you will learn from this chapter:

1. The system of notation used to describe potential outcomes.
2. Definitions of core terms: average treatment effect, expectation, random assignment, and unbiasedness.
3. Assumptions that must be met in order for experiments to produce unbiased estimates of the average treatment effect.

Students of legislative politics have argued that women bring different policy priorities to the budgetary process in developing countries, emphasizing health issues such as providing clean drinking water. Leave aside for the time being the question of how this topic might be studied using randomly assigned treatments. For the moment, simply assume that each village either receives the treatment (a woman serves as village council head) or remains untreated (with its village council headed by a man). For each village, we also observe the share of the local council budget that is allocated to providing clean drinking water. To summarize, we observe the treatment (whether the village head is a woman or not) and the outcome (what share of the budget goes to a policy issue of special importance to women).

What we do not observe is how the budget in each village headed by a man would have been allocated if it had been headed by a woman, and vice versa. Although we do not observe these counterfactual outcomes, we can nevertheless imagine them. Taking this mental exercise one step further, we might imagine that each village has two *potential outcomes*: the budget it would enact if headed by a woman and the budget it would enact if headed by a man. The gender of the village head determines which potential budget we observe. The other budget remains imaginary or counterfactual.

Table 2.1 provides a stylized example of seven villages in order to introduce the notation that we will use throughout the book. The villages constitute the subjects in this experiment. Each subject is identified by a subscript i, which ranges from 1 to 7. The third village on the list, for example, would be designated as $i = 3$. The table imagines what would happen under two different scenarios. Let $Y_i(1)$ be the outcome if village i is exposed to the treatment (a woman as village head), and let $Y_i(0)$ be the outcome if this village is not exposed to the treatment. For example, Village 3 allocates 30% of its budget to water sanitation if headed by a woman but only 20% if headed by a man, so, $Y_3(1) = 30\%$, and $Y_3(0) = 20\%$. These are called potential outcomes because they describe what would happen if a treatment were or were not administered.

For purposes of this example, we assume that each village has just two potential outcomes, depending on whether it receives the treatment; villages are assumed to be unaffected by the treatments that other villages receive. In section 2.7, we spell out

TABLE 2.1
Illustration of potential outcomes for local budgets when village council heads are women or men. (Entries are shares of local budgets allocated to water sanitation.)

Village i	$Y_i(0)$ *Budget share if village head is male*	$Y_i(1)$ *Budget share if village head is female*	τ_i *Treatment effect*
Village 1	10	15	5
Village 2	15	15	0
Village 3	20	30	10
Village 4	20	15	−5
Village 5	10	20	10
Village 6	15	15	0
Village 7	15	30	15
Average	**15**	**20**	**5**

more precisely the assumptions that underlie the model of potential outcomes and discuss complications that arise when subjects are affected by the treatments that other subjects receive.

2.2 Average Treatment Effects

For each village, the causal effect of the treatment (τ_i) is defined as the difference between two potential outcomes:

$$\tau_i \equiv Y_i(1) - Y_i(0). \tag{2.1}$$

In other words, the treatment effect for each village is the difference between two potential states of the world, one in which the village receives the treatment and another in which it does not. For Village 3, this causal effect is $30 - 20 = 10$.

The empirical challenge that researchers typically face when observing outcomes is that at any given time one can observe $Y_i(1)$ or $Y_i(0)$ but not both. (Bear in mind that the only reason we are able to see both potential outcomes for each village in Table 2.1 is that this is a hypothetical example!) Building on the notational system introduced above, we define Y_i as the observed outcome in each village and d_i as the observed treatment that is delivered in each village. In this case, Y_i is the observed share of the budget allocated to water sanitation, and d_i equals 1 when a woman is village head and 0 otherwise.

BOX 2.1

Potential Outcomes Notation

In this system of notation, the subscript i refers to subjects 1 through N.

The variable d_i indicates whether the ith subject is treated: $d_i = 1$ means the ith subject receives the treatment, and $d_i = 0$ means the ith subject does not receive the treatment. It is assumed that d_i is observed for every subject.

$Y_i(1)$ is the potential outcome if the ith subject were treated. $Y_i(0)$ is the potential outcome if the ith subject were not treated. In general, potential outcomes may be written $Y_i(d)$, where d indexes the treatment. These potential outcomes are fixed attributes of each subject and represent the outcome that would be observed hypothetically if that subject were treated or untreated.

A schedule of potential outcomes refers to a comprehensive list of potential outcomes for all subjects. The rows of this schedule are indexed by i, and the columns are indexed by d. For example, in Table 2.1 the $Y_i(0)$ and $Y_i(1)$ potential outcomes for the fifth subject may be found in adjacent columns of the fifth row.

The connection between the observed outcome Y_i and the underlying potential outcomes is given by the equation $Y_i = d_iY_i(1) + (1 - d_i)Y_i(0)$. This equation indicates that the $Y_i(1)$ are observed for subjects who are treated, and the $Y_i(0)$ are observed for subjects who are not treated. For any given subject, we observe either $Y_i(1)$ or $Y_i(0)$, never both.

It is sometimes useful to refer to potential outcomes for a subset of all subjects. Expressions of the form $Y_i(\cdot) | X = x$ denote potential outcomes when the condition $X = x$ holds. For example, $Y_i(0) | d_i = 1$ refers to the untreated potential outcome for a subject who actually receives the treatment.

Because we often want to know about the statistical properties of a hypothetical random assignment, we distinguish between d_i, the treatment that a given subject receives (a variable that one observes in an actual dataset), and D_i, the treatment that could be administered hypothetically. D_i is a random variable, and the ith subject might be treated in one hypothetical study and not in another. For example, $Y_i(1) | D_i = 1$ refers to the treated potential outcome for a subject who would be treated under some hypothetical allocation of treatments.

The budget that we observe in each village may be summarized using the following expression:

$$Y_i = d_i Y_i(1) + (1 - d_i) Y_i(0). \quad (2.2)$$

Because d_i is either 0 or 1, one of the terms on the right side of the equals sign will always be zero. We observe the potential outcome that results from treatment, $Y_i(1)$, if the treatment is administered ($d_i = 1$). If the treatment is not administered ($d_i = 0$), we observe the potential outcome that results when no treatment occurs, $Y_i(0)$.

The average treatment effect, or ATE, is defined as the sum of the τ_i divided by N, the number of subjects:

$$\text{ATE} \equiv \frac{1}{N}\sum_{i=1}^{N} \tau_i. \quad (2.3)$$

An equivalent way to obtain the average treatment effect is to subtract the average value of $Y_i(0)$ from the average value of $Y_i(1)$:

$$\frac{1}{N}\sum_{i=1}^{N} Y_i(1) - \frac{1}{N}\sum_{i=1}^{N} Y_i(0) = \frac{1}{N}\sum_{i=1}^{N}(Y_i(1) - Y_i(0)) = \frac{1}{N}\sum_{i=1}^{N} \tau_i. \quad (2.4)$$

The average treatment effect is an extremely important concept. Villages may have different τ_i, but the ATE indicates how outcomes would change on average if every village were to go from untreated (male village council head) to treated (female village council head).

From the rightmost column of Table 2.1, we can calculate the ATE for the seven villages. The average treatment effect in this example is 5 percentage points: if all villages were headed by men, they would on average spend 15% of their budgets on water sanitation, whereas if all villages were headed by women, this figure would rise to 20%.

BOX 2.2

Definition: Average Treatment Effect

The average treatment effect (ATE) is the sum of the subject-level treatment effects, $Y_i(1) - Y_i(0)$, divided by the total number of subjects. An equivalent way to express the ATE is to say that it equals $\mu_{Y(1)} - \mu_{Y(0)}$, where $\mu_{Y(1)}$ is the average value of $Y_i(1)$ for all subjects and $\mu_{Y(0)}$ is the average value of $Y_i(0)$ for all subjects.

2.3 Random Sampling and Expectations

Suppose that instead of calculating the average potential outcome for all villages, we drew a random sample of villages and calculated the average among the villages we sampled. By *random sample*, we mean a selection procedure in which v villages are selected from the list of N villages, and every possible set of v villages is equally likely to be selected. For example, if we select one village at random from a list of seven villages, seven possible samples are equally likely. If we select three villages at random from a list of seven villages,

$$\frac{N!}{v!(N-v)!} = \frac{7!}{3!4!} = \frac{7\times6\times5\times4\times3\times2\times1}{(3\times2\times1)(4\times3\times2\times1)} = 35 \tag{2.5}$$

possible samples are equally likely. If potential outcomes vary from one village to the next, the average potential outcome in the villages we sample will vary, depending on which of the possible samples we happen to select. The sample average may be characterized as a *random variable*, a quantity that varies from sample to sample.

The term *expected value* refers to the average outcome of a random variable. (See Box 2.3.) In our example, the random variable is the number we obtain when we sample villages at random and calculate their average outcome. Recall from introductory statistics that under random sampling, the expected value of a sample average is equal to the average of the population from which the sample is drawn.[2] This principle may be illustrated using the population of villages depicted in Table 2.1. Recall that the average value of $Y_i(0)$ among all villages in Table 2.1 is 15. Suppose we sample two villages at random from the list of seven villages and calculate the average value of $Y_i(0)$ for the two selected villages. There are

$$\frac{N!}{v!(N-v)!} = \frac{7!}{2!5!} = 21 \tag{2.6}$$

possible ways of sampling two villages at random from a list of seven, and each sample is equally likely to be drawn. Any given sample of two villages might contain an average value of $Y_i(0)$ that is higher or lower than the true average of 15, but the expected value refers to what we would obtain on average if we were to examine all 21 possible samples, for each one calculating the average value of $Y_i(0)$:

$$\{10, 12.5, 12.5, 12.5, 12.5, 12.5, 12.5, 15, 15, 15, 15, 15, 15, 15, 17.5, 17.5, 17.5, 17.5, 17.5, 17.5, 20\}. \tag{2.7}$$

2 The easiest way to see the intuition behind this principle is to consider the case in which we randomly sample just one village. Each village is equally likely to be sampled. The average over all seven possible samples is identical to the average for the entire population of seven villages. This logic generalizes to samples where $v > 1$ because each village appears in exactly $v/7$ of all possible samples.

BOX 2.3

The expectation of a discrete random variable X is defined as

$$E[X] = \Sigma x \Pr[X = x],$$

where $\Pr[X = x]$ denotes the probability that X takes on the value x, and where the summation is taken over all possible values of x.

For example, what is the expected value of a randomly selected value of τ_i from Table 2.1?

$$\begin{aligned} E[\tau_i] &= \Sigma \tau \Pr[\tau_i = \tau] \\ &= (-5)\left(\frac{1}{7}\right) + (0)\left(\frac{2}{7}\right) + (5)\left(\frac{1}{7}\right) + (10)\left(\frac{2}{7}\right) + (15)\left(\frac{1}{7}\right) = 5. \end{aligned}$$

Properties of Expectations

The expectation of the constant α is itself: $E[\alpha] = \alpha$.

For a random variable X and constants α and β, $E[\alpha + \beta X] = \alpha + \beta E[X]$.

The expectation of a sum of two random variables, X and Y, is the sum of their expectations: $E[X + Y] = E[X] + E[Y]$.

The expectation of the product of two random variables, X and Y, is the product of their expectations plus the covariance between them: $E[XY] = E[X]E[Y] + E[(X - E[X])(Y - E[Y])]$.

The average of these 21 numbers is 15. In other words, the expected value of the average $Y_i(0)$ obtained from a random sample of two villages is 15.

The concept of expectations plays an important role in the discussion that follows. Because we will refer to expectations so often, a bit more notation is helpful. The notation $E[X]$ refers to the expectation of a random variable X. (See Box 2.3.) The expression "the expected value of $Y_i(0)$ when one subject is sampled at random" will be written compactly as $E[Y_i(0)]$. When a term like $Y_i(0)$ appears in conjunction with an expectations operator, it should be read not as the value of $Y_i(0)$ for subject i but instead as a random variable that is equal to the value of $Y_i(0)$ for a randomly selected subject. When the expression $E[Y_i(0)]$ is applied to values in Table 2.1, the random variable is the random selection of a $Y_i(0)$ from the list of all $Y_i(0)$; since there are seven possible random selections, the average of which is 15, it follows that $E[Y_i(0)] = 15$.

Sometimes attention is focused on the expected value of a random variable within a subgroup. *Conditional expectations* refer to subgroup averages. In terms of notation, the logical conditions following the | symbol indicate the criteria that define the subgroup. For example, the expression "the expectation of $Y_i(1)$ when one village is selected at random from those villages that were treated" is written $E[Y_i(1) \mid d_i = 1]$. The idea of a conditional expectation is straightforward when working with quantities that are in principle observable. More mind-bending are expressions like $E[Y_i(1) \mid d_i = 0]$, which denotes "the expectation of $Y_i(1)$ when one village is selected at random from those villages that were not treated." In the course of conducting research, we will never actually see $Y_i(1)$ for an untreated village, nor will we see $Y_i(0)$ for a treated village. These potential outcomes can be imagined but not observed.

One special type of conditional expectation arises when the subgroup is defined by the outcome of a random process. In that case, the conditional expectation may vary depending on which subjects happened to meet the condition in any particular realization of the random process. For example, suppose that a random process, such as a coin flip, determines which subjects are treated. For a given treatment assignment d_i, we could calculate $E[Y_i(1) \mid d_i = 0]$, but this expectation might have been different had the coin flips come out differently. Suppose we want to know the expected conditional expectation, or how the conditional expectation would come out, on average, across all possible ways that d_i could have been allocated. Let D_i be a random variable that indicates whether each subject would be treated in a hypothetical experiment. The conditional expectation $E[Y_i(1) \mid D_i = 0]$ is calculated by considering all possible realizations of D_i (all the possible ways that N coins could have been flipped) in order to form the joint probability distribution function for $Y_i(1)$ and D_i. As long as we know the joint probability of observing each paired set of values $\{Y(1), D\}$, we can calculate the conditional expectation using the formula in Box 2.4.[3]

With this basic system of notation in place, we may now describe the connection between expected potential outcomes and the average treatment effect (ATE):

$$\begin{aligned} E[Y_i(1) - Y_i(0)] &= E[Y_i(1)] - E[Y_i(0)] \\ &= \frac{1}{N}\sum_{i=1}^{N} Y_i(1) - \frac{1}{N}\sum_{i=1}^{N} Y_i(0) \\ &= \frac{1}{N}\sum_{i=1}^{N}[Y_i(1) - Y_i(0)] \equiv \text{ATE}. \end{aligned} \tag{2.8}$$

3 The notation $E[Y_i(1) \mid D_i = 0]$ may be regarded as shorthand for $E[E[Y_i(1) \mid d_i = 0, \boldsymbol{d}]]$, where $\boldsymbol{d}$ refers to a vector of treatment assignments and d_i refers its ith element. Given $\boldsymbol{d}$, we may calculate the probability distribution function for all $\{Y(1), d\}$ pairs and the expectation given this set of assignments. Then we may take the expectation of this expected value by summing over all possible $\boldsymbol{d}$ vectors.

BOX 2.4

Definition: Conditional Expectation

For discrete random variables Y and X, the conditional expectation of Y given that X takes on the value x is

$$E[Y|X = x] = \sum y \Pr[Y = y|X = x] = \sum y \frac{\Pr[Y = y, X = x]}{\Pr[X = x]},$$

where $\Pr[Y = y, X = x]$ denotes the joint probability of $Y = y$ and $X = x$, and where the summation is taken over all possible values of y.

For example, in Table 2.1 what is the conditional expectation of a randomly selected value of τ_i, for villages where $Y_i(0) > 10$? This question requires us to describe the joint probability distribution function for the variables τ_i and $Y_i(0)$ so that we can calculate $\Pr[\tau_i = \tau, Y_i(0) > 10]$. Table 2.1 indicates that the $\{\tau, Y(0)\}$ pair $\{0, 15\}$ occurs with probability $2/7$, while the other pairs $\{5, 10\}$, $\{10, 20\}$, $\{-5, 20\}$, $\{10, 10\}$, and $\{15, 15\}$ each occur with probability $1/7$. The marginal distribution of $Y_i(0)$ reveals that 5 of the 7 $Y_i(0)$ are greater than 10, so $\Pr[Y_i(0) > 10] = 5/7$.

$$E[\tau_i | Y_i(0) > 10] = \sum \tau \frac{\Pr[\tau_i = \tau, Y_i(0)>10]}{\Pr[Y_i(0)>10]}$$

$$= (-5)\frac{\frac{1}{7}}{\frac{5}{7}} + (0)\frac{\frac{2}{7}}{\frac{5}{7}} + (5)\frac{0}{\frac{5}{7}} + (10)\frac{\frac{1}{7}}{\frac{5}{7}} + (15)\frac{\frac{1}{7}}{\frac{5}{7}} = 4.$$

In order to illustrate the idea of a conditional expectation when conditioning on the outcome of a random process, suppose we randomly assign one of the observations in Table 2.1 to treatment ($D_i = 1$) and the remaining six observations to control ($D_i = 0$). If each of the seven possible assignments occurs with probability $1/7$, what is the expected value of a randomly selected τ_i given that $D_i = 1$? Again, we start with the joint probability density function for τ_i and D_i and consider all possible pairings of these two variables' values. The $\{\tau, D\}$ pairings $\{-5, 1\}$, $\{5, 1\}$, and $\{15, 1\}$ occur with probability $1/49$, while the pairings $\{0, 1\}$ and $\{10, 1\}$ occur with probability $2/49$; the remaining $\{\tau, D\}$ pairings are instances in which τ is paired with 0. The marginal distribution $\Pr[D_i = 1] = 3(1/49) + 2(2/49) = 1/7$.

$$E[\tau_i | D_i = 1] = \sum \tau \frac{\Pr[\tau_i = \tau, D_i = 1]}{\Pr[D_i = 1]}$$

$$= (-5)\frac{\frac{1}{49}}{\frac{1}{7}} + (0)\frac{\frac{2}{49}}{\frac{1}{7}} + (5)\frac{\frac{1}{49}}{\frac{1}{7}} + (10)\frac{\frac{2}{49}}{\frac{1}{7}} + (15)\frac{\frac{1}{49}}{\frac{1}{7}} = 5.$$

The first line of equation (2.8) expresses the fact that when a village is selected at random from the list of villages, its expected treatment effect is equal to the difference between the expected value of a randomly selected treated potential outcome and the expected value of a randomly selected untreated potential outcome. The second equality in equation (2.8) indicates that the expected value of a randomly selected $Y_i(1)$ equals the average of all $Y_i(1)$ values, and that the expected value of a randomly selected $Y_i(0)$ equals the average of all $Y_i(0)$ values. The third equality reflects the fact that the difference between the two averages in the second line of equation (2.8) can be expressed as the average difference in potential outcomes. The final equality notes that the average difference in potential outcomes is the definition of the average treatment effect. In sum, the difference in expectations equals the difference in average potential outcomes for the entire list of villages, or the ATE.[4]

This relationship is apparent from the schedule of potential outcomes in Table 2.1. The column of numbers representing the treatment effect (τ_i) is, on average, 5. If we were to select villages at random from this list, we would expect their average treatment effect to be 5. We get the same result if we subtract the expected value of a randomly selected $Y_i(0)$ from the expected value of a randomly selected $Y_i(1)$.

2.4 Random Assignment and Unbiased Inference

The challenge of estimating the average treatment effect is that at a given point in time each village is either treated or not: either $Y_i(1)$ or $Y_i(0)$ is observed, but not both. To illustrate the problem, Table 2.2 shows what outcomes would be observed if Village 1 and Village 7 were treated, while the remaining villages were not. We observe $Y_i(1)$ for Villages 1 and 7 but not $Y_i(0)$. For Villages 2, 3, 4, 5, and 6, we observe $Y_i(0)$ but not $Y_i(1)$. The unobserved or "missing" values in Table 2.2 are indicated with a "?".

4 The notation used here is just one way to explicate the link between expectations and the ATE. Samii and Aronow (2012) suggest an alternative formalization. Their model envisions a finite population U consisting of units j in 1, 2, . . . , N, each of which has an associated triple $(y_j(1), y_j(0), D_j')$ such that $y_j(1)$ and $y_j(0)$ are fixed potential outcomes and D_j' is a random variable indicating the treatment status of unit j. Reassign a random index ordering i in 1, 2, . . . , N. Then, for an arbitrary unit i, there exists an associated triple of random variables $(Y_i(1), Y_i(0), D_i)$ such that the random variable $Y_i = D_iY_i(1) + (1 - D_i)Y_i(0)$. It follows that for equation (2.8):

$$E[Y_i(1)] - E[Y_i(0)] = \frac{1}{N}\sum_{j=1}^{N} y_j(1) - \frac{1}{N}\sum_{j=1}^{N} y_j(0) = \text{ATE}.$$

Statistical operators such as expectations or independence refer to random variables associated with an arbitrary index i. Looking ahead to later chapters, one might expand this system to include other unit-level attributes, such as covariates or missingness, by attaching them to the triple indexed by j before reassigning the ordering.

TABLE 2.2
Illustration of observed outcomes for local budgets when two village councils are headed by women.

Village i	$Y_i(0)$ *Budget share if village head is male*	$Y_i(1)$ *Budget share if village head is female*	τ_i *Treatment effect*
Village 1	?	15	?
Village 2	15	?	?
Village 3	20	?	?
Village 4	20	?	?
Village 5	10	?	?
Village 6	15	?	?
Village 7	?	30	?
Estimated average based on observed data	**16**	**22.5**	**6.5**

Note: The observed outcomes in this table are based on the potential outcomes listed in Table 2.1.

Random assignment addresses the "missing data" problem by creating two groups of observations that are, in expectation, identical prior to application of the treatment. When treatments are allocated randomly, the treatment group is a random sample of all villages, and therefore the expected potential outcomes among villages in the treatment group are identical to the average potential outcomes among all villages. The same is true for villages in the control group. The control group's expected potential outcomes are also identical to the average potential outcomes among all villages. Therefore, in expectation, the treatment group's potential outcomes are the same as the control group's. Although any given random allocation of villages to treatment and control groups may produce groups of villages that have different average potential outcomes, this procedure is fair in the sense that it does not tend to give one group a higher set of potential outcomes than the other.

As Chattopadhyay and Duflo point out, random assignment is in fact used in rural India to assign women to head one-third of the local village councils.[5] Ordinarily, men would head the village councils, but Indian law mandates that selected

5 Chattopadhyay and Duflo 2004. A lottery is used to assign council positions to women in Rajasthan. In West Bengal, a near-random assignment procedure is used whereby villagers are assigned according to their serial numbers.

villages install a female representative as head of the council. For purposes of illustration, suppose that our collection of seven villages were subject to this law, and that two villages will be randomly assigned female council heads. Consider the statistical implications of this arrangement. This random assignment procedure implies that every village has the same probability of receiving the treatment; assignment bears no systematic relationship to villages' observed or unobserved attributes.

Let's take a closer look at the formal implications of this form of random assignment. When villages are assigned such that every village has the same probability of receiving the treatment, the villages that are randomly chosen for treatment are a random subset of the entire set of villages. Therefore, the expected $Y_i(1)$ potential outcome among treated villages is the same as the expected $Y_i(1)$ potential outcome for the entire set of villages:

$$E[Y_i(1) | D_i = 1] = E[Y_i(1)]. \quad (2.9)$$

BOX 2.5

Two Commonly Used Forms of Random Assignment

Random assignment refers to a procedure that allocates treatments with known probabilities that are greater than zero and less than one.

The most basic forms of random assignment allocate treatments such that every subject has the same probability of being treated. Let N be the number of subjects, and let m be the number of subjects who are assigned to the treatment group. Assume that N and m are integers such that $0 < m < N$. Simple random assignment refers to a procedure whereby each subject is allocated to the treatment group with probability m/N. Complete random assignment refers to a procedure that allocates exactly m units to treatment.

Under simple or complete random assignment, the probability of being assigned to the treatment group is identical for all subjects; therefore treatment status is statistically independent of the subjects' potential outcomes and their background attributes ($\boldsymbol{X}$):

$$Y_i(0), Y_i(1), \boldsymbol{X} \perp\!\!\!\perp D_i,$$

where the symbol $\perp\!\!\!\perp$ means "is independent of." For example, if a die roll is used to assign subjects to treatment with probability $1/6$, knowing whether a subject is treated provides no information about the subject's potential outcomes or background attributes. Therefore, the expected value of $Y_i(0)$, $Y_i(1)$, and X_i is the same in treatment and control groups.

When we randomly select villages into the treatment group, the villages we leave behind for the control group are also a random sample of all villages. The expected $Y_i(1)$ in the control group ($D_i = 0$) is therefore equal to the expected $Y_i(1)$ for the entire set of villages:

$$E[Y_i(1) \mid D_i = 0] = E[Y_i(1)]. \tag{2.10}$$

Putting equations (2.9) and (2.10) together, we see that under random assignment the treatment and control groups have the same expected potential outcome:

$$E[Y_i(1) \mid D_i = 1] = E[Y_i(1) \mid D_i = 0]. \tag{2.11}$$

Equation (2.11) also underscores the distinction between realized and unrealized potential outcomes. On the left side of the equation is the expected treated potential outcome among villages that receive the treatment. The treatment causes this potential outcome to become observable. On the right side of the equation is the expected treated potential outcome among villages that do not receive the treatment. Here, the lack of treatment means that the treated potential outcome remains unobserved for these subjects.

The same logic applies to the control group. Villages that do not receive the treatment ($D_i = 0$) have the same expected untreated potential outcome $Y_i(0)$ that the treatment group ($D_i = 1$) would have if it were untreated:

$$E[Y_i(0) \mid D_i = 0] = E[Y_i(0) \mid D_i = 1] = E[Y_i(0)]. \tag{2.12}$$

Equations (2.11) and (2.12) follow from random assignment: D_i conveys no information whatsoever about the potential values of $Y_i(1)$ or $Y_i(0)$. The randomly assigned values of D_i determine which value of Y_i we actually *observe*, but they are nevertheless statistically independent of the *potential* outcomes $Y_i(1)$ and $Y_i(0)$. (See Box 2.5 for discussion of the term *independence*.)

When treatments are assigned randomly, we may rearrange equations (2.8), (2.11), and (2.12) in order to express the average treatment effect as

$$\text{ATE} = E[Y_i(1) \mid D_i = 1] - E[Y_i(0) \mid D_i = 0]. \tag{2.13}$$

This equation suggests an empirical strategy for estimating the average treatment effect. The terms $E[Y_i(1) \mid (D_i = 1)]$ and $E[Y_i(0) \mid (D_i = 0)]$ may be estimated using experimental data. We do not observe the $Y_i(1)$ potential outcomes for all observations, but we do observe them for the random sample of observations that receive the treatment. Similarly, we do not observe the $Y_i(0)$ potential outcomes for all observations, but we do observe them for the random sample of observations in the control group. If we want to estimate the average treatment effect, equation (2.13) suggests that we should take the difference between two sample means: the average outcome in the treatment group minus the average outcome in the control group. Ideas that enable researchers to use observable quantities (e.g., sample averages) to reveal parameters of interest (e.g., average treatment effects) are termed *identification* strategies.

Statistical procedures used to make guesses about parameters such as the average treatment effect are called *estimators*. In this example, the estimator is very simple, just a difference between two sample averages. Before applying an estimator to actual data, a researcher should reflect on its statistical properties. One especially important property is *unbiasedness*. An estimator is unbiased if it generates the right answer, on average. In other words, if the experiment were replicated an infinite number of times under identical conditions, the average *estimate* would equal the true parameter. Some guesses may be too high and others too low, but the average guess will be correct. In practice, we will not be able to perform an infinite number of experiments. In fact, we might just perform one experiment and leave it at that. Nevertheless, in theory we can analyze the properties of our estimation procedure to see whether, on average, it recovers the right answer. (In the next chapter, we consider another property of estimators: how precisely they estimate the parameter of interest.)

In sum, when treatments are administered using a procedure that gives every subject the same probability of being treated, potential outcomes are independent of the treatments that subjects receive. This property suggests an identification strategy for estimating average treatment effects using experimental data.

The remaining task is to demonstrate that the proposed estimator—the difference between the average outcome in the treatment group and the average outcome in the control group—is an unbiased estimator of the ATE when all subjects have the same probability of being treated. The proof is straightforward. Because the units assigned to the control group are a random sample of all units, the average of the control group outcomes is an unbiased estimator of the average value of $Y_i(0)$

BOX 2.6

Definition: Estimator and Estimate

An estimator is a procedure or formula for generating guesses about parameters such as the average treatment effect. The guess that an estimator generates based on a particular experiment is called an estimate. Estimates are denoted using a "hat" notation. The estimate of the parameter θ is written $\hat{\theta}$.

among all units. The same goes for the treatment group: the average outcome among units that receive the treatment is an unbiased estimator of the average value of $Y_i(1)$ among all units. Formally, if we randomly shuffle the villages and place the first m subjects in the treatment group and the remaining $N - m$ subjects in the control group, we can analyze the expected, or average, outcome over all possible random assignments:

$$E\left[\frac{\sum_1^m Y_i}{m} - \frac{\sum_{m+1}^N Y_i}{N-m}\right] = \overbrace{E\left[\frac{\sum_1^m Y_i}{m}\right]}^{\text{Average outcome among treated units}} - \overbrace{E\left[\frac{\sum_{m+1}^N Y_i}{N-m}\right]}^{\text{Average outcome among untreated units}}$$

$$= \frac{E[Y_1] + E[Y_2] + \cdots + E[Y_m]}{m} - \frac{E[Y_{m+1}] + E[Y_{m+2}] + \cdots + E[Y_N]}{N-m}$$

$$= E[Y_i(1) \mid D_i = 1] - E[Y_i(0) \mid D_i = 0]$$

$$= E[Y_i(1)] - E[Y_i(0)] = E[\tau_i] = \text{ATE}. \quad (2.14)$$

Equation (2.14) conveys a simple but extremely useful idea. When units are randomly assigned, a comparison of average outcomes in treatment and control groups (the so-called *difference-in-means estimator*) is an unbiased estimator of the average treatment effect.

BOX 2.7

Definition: Unbiased Estimator

An estimator is unbiased if the expected value of the estimates it produces is equal to the true parameter of interest. Call θ the parameter we seek to estimate, such as the ATE. Let $\hat{\theta}$ represent an estimator, or procedure for generating estimates. For example, $\hat{\theta}$ may represent the difference in average outcomes between treatment and control groups. The expected value of this estimator is the average estimate we would obtain if we apply this estimator to all possible realizations of a given experiment or observational study. We say that $\hat{\theta}$ is unbiased if $E(\hat{\theta}) = \theta$; in words, the estimator $\hat{\theta}$ is unbiased if the expected value of this estimator is θ, the parameter of interest. Although unbiasedness is a property of estimators and not estimates, we refer to the estimates generated by an unbiased estimator as "unbiased estimates."

2.5 The Mechanics of Random Assignment

The result in equation (2.14) hinges on random assignment, and so it is important to be clear about what constitutes random assignment. *Simple random assignment* is a term of art, referring to a procedure—a die roll or coin toss—that gives each subject an identical probability of being assigned to the treatment group. The practical drawback of simple random assignment is that when N is small, random chance can create a treatment group that is larger or smaller than what the researcher intended. For example, you could flip a coin to assign each of 10 subjects to the treatment condition, but there is only a 24.6% chance of ending up with exactly 5 subjects in treatment and 5 in control. A useful special case of simple random assignment is *complete random assignment*, where exactly m of N units are assigned to the treatment group with equal probability.[6]

The procedure used to conduct complete random assignment can take any of three equivalent forms. Suppose one has N subjects and seeks to assign treatments to m of them. The first method is to select one subject at random, then select another at random from the remaining units, and so forth until you have selected m subjects into the treatment group. A second method is to enumerate all of the possible ways that m subjects may be selected from a list of N subjects, and randomly select one of the possible allocation schemes. A third method is to randomly permute the order of all N subjects and label the first m subjects as the treatment group.[7]

Beware of the fact that *random* is a word that is used loosely in common parlance to refer to procedures that are arbitrary, haphazard, or unplanned. The problem is that arbitrary, haphazard, or unplanned treatments may follow systematic patterns that go unnoticed. Procedures such as alternation are risky because there may be systematic reasons why certain types of subjects might alternate in a sequence, and indeed, some early medical experiments ran into exactly this problem.[8] We use the term *random* in a more exacting sense. The physical or electronic procedure by which randomization is conducted ensures that assignment to the treatment group is statistically independent of all observed or unobserved variables.

---

6 In Chapters 3 and 4, we discuss other frequently used methods of random assignment: clustered random assignment, where groups of subjects are randomly assigned to treatment and control, and block random assignment (also called stratified random assignment), where individuals are first divided into blocks, and then random assignment is performed within each block. Box 2.5 notes that a defining feature of complete (as opposed to clustered or blocked) random assignment is that all possible assignments of N subjects to a treatment group of size m are equally likely.

7 Cox and Reid 2000, p. 20. The term *complete randomization* is a bit awkward, as the word *complete* does not convey the requirement that exactly m units are allocated to treatment, but this terminology has become standard (see Rosenbaum 2002, pp. 25–26).

8 Hróbjartsson, Gøtzsche, and Gluud 1998.

In practical terms, random assignment is best done using statistical software. Here is an easy procedure for implementing complete random assignment. First, determine N, the number of subjects in your experiment, and m, the number of subjects who will be allocated to the treatment group. Second, set a random number "seed" using a statistics package, so that your random numbers may be reproduced by anyone who cares to replicate your work. Third, generate a random number for each subject. Fourth, sort the subjects by their random numbers in ascending order. Finally, classify the first m observations as the treatment group. Example programs using R may be found at http://isps.research.yale.edu/FEDAI.

Generating random numbers is just the first step in implementing random assignment. After the numbers are generated, one must take pains to preserve the integrity of the assignment process. A deficiency of alternation and many other arbitrary procedures is that they allow those administering the allocation to foresee who will be assigned to which experimental group. If a receptionist seeks to get the sickest patients into the experimental treatment group and knows that the pattern of assignments alternates, he can reorder the patients in such a way as to shuttle the sickest subjects into the treatment group.[9] The same concern arises even when a random sequence of numbers is used to assign incoming patients: random allocation may be undone if the receptionist knows the order of assignments ahead of time, because that enables him to position patients so that they will be assigned to a certain experimental group. In order to guard against potential threats to the integrity of random assignment, researchers should build extra procedural safeguards into their experimental designs, such as blinding those administering the experiment to the subjects' assigned experimental groups.

2.6 The Threat of Selection Bias When Random Assignment Is Not Used

Without random assignment, the identification strategy derived from equation (2.14) unravels. The treatment and control groups are no longer random subsets of all units in the sample. Instead, we confront what is known as a *selection problem*: receiving treatment may be systematically related to potential outcomes. For example, absent random assignment, villages determine whether their councils are headed by women. The villages that end up with female council heads may not be a random subset of all villages.

9 For examples of experiments in which random assignment was subverted, see Torgerson and Torgerson 2008.

To see how nonrandom selection jeopardizes the identification strategy of comparing average outcomes in the treatment and control groups, rewrite the expected difference in outcomes from equation (2.13) by subtracting and adding $E[Y_i(0) | D_i = 1]$:

$$\underbrace{E[Y_i(1) | D_i = 1] - E[Y_i(0) | D_i = 0]}_{\text{Expected difference between treated and untreated outcomes}}$$

$$= \underbrace{E[Y_i(1) - Y_i(0) | D_i = 1]}_{\text{ATE among the treated}} + \underbrace{E[Y_i(0) | D_i = 1] - E[Y_i(0) | D_i = 0]}_{\text{Selection bias}}. \quad (2.15)$$

Under random assignment, the selection bias term is zero, and the ATE among the (randomly) treated villages is the same as the ATE among all villages. In the absence of random assignment, equation (2.15) warns that the apparent treatment effect is a mixture of selection bias and the ATE for a subset of villages.

In order to appreciate the implications of equation (2.15), consider the following scenario. Suppose that instead of randomly selecting villages to receive the treatment, our procedure were to let villages decide whether to take the treatment. Refer back to Table 2.1 and imagine that, if left to their own devices, Village 5 and Village 7 always elect a woman due to villagers' pent-up demand for water sanitation, while the remaining villages always elect a man.[10] Self-selection in this case leads to an exaggerated estimate of the ATE because receiving the treatment is associated with lower-than-average values of $Y_i(0)$ and higher-than-average values of $Y_i(1)$. The average outcome in the treatment group is 25, and the average outcome in the control group is 16. The estimated ATE is therefore 9, whereas the actual ATE is 5. Referring to equation (2.15) we see that in this case the ATE among the treated is not equal to the ATE for the entire subject pool, nor is the selection bias term equal to zero. The broader point is that it is risky to compare villages that choose to receive the treatment with villages that choose not to. In this example, self-selection is related to potential outcomes; as a result, the comparison of treated and untreated villages recovers neither the ATE for the sample as a whole nor the ATE among those villages that receive treatment.

The beauty of experimentation is that the randomization procedure generates a schedule of treatment and control assignments that are statistically independent of

10 When taking expectations over hypothetical replications of an experiment, we consider all possible random assignments. In our example of non-random allocation, however, nature makes the assignment. When taking expectations, we must therefore consider the average of all possible natural assignments. Rather than make up an assortment of possible assignments and stipulate the probability that each scenario occurs, we have kept the example as simple as possible and assumed that the villages "always" elect the same type of candidate. In effect, we are taking expectations over just one possible assignment that occurs with probability 1.

potential outcomes. In other words, the assumptions underlying equations (2.9) to (2.13) are justified by reference to the *procedure* of random assignment, not substantive arguments about the comparability of potential outcomes in the treatment and control groups.

The preceding discussion should not be taken to imply that experimentation invokes no substantive assumptions. The unbiasedness of the difference-in-means estimator hinges not only on random assignment but also on two assumptions about potential outcomes, the plausibility of which will vary depending on the application. The next section spells out these important assumptions.

2.7 Two Core Assumptions about Potential Outcomes

To this point, our characterization of potential outcomes has glossed over two important details. In order to ease readers into the framework of potential outcomes, we simply stipulated that each subject has two potential outcomes, $Y_i(1)$ if treated and $Y_i(0)$ if not treated. To be more precise, each potential outcome depends *solely* on whether the subject *itself* receives the treatment. When writing potential outcomes in this way, we are assuming that potential outcomes respond only to the treatment and not some other feature of the experiment, such as the way the experimenter assigns treatments or measures outcomes. Furthermore, potential outcomes are defined over the set of treatments that the subject itself receives, not the treatments assigned to other subjects. In technical parlance, the "solely" assumption is termed *excludability* and the "itself" assumption is termed *non-interference*.

2.7.1 Excludability

When we define two, and only two, potential outcomes based on whether the treatment is administered, we implicitly assume that the only relevant causal agent is receipt of the treatment. Because the point of an experiment is to isolate the causal effect of the treatment, our schedule of potential outcomes excludes from consideration factors other than the treatment. When conducting an experiment, therefore, we must define the treatment and distinguish it from other factors with which it may be correlated. Specifically, we must distinguish between d_i, the treatment, and z_i, a variable that indicates which observations have been allocated to treatment or control. We seek to estimate the effect of d_i, and we assume that the treatment assignment z_i has no effect on outcomes except insofar as it affects the value of d_i.

The term *exclusion restriction* or *excludability* refers to the assumption that z_i can be omitted from the schedule of potential outcomes for $Y_i(1)$ and $Y_i(0)$. Formally, this assumption may be written as follows. Let $Y_i(z, d)$ be the potential outcome when $z_i = z$ and $d_i = d$, for $z \in (0, 1)$ and for $d \in (0, 1)$. For example, if $z_i = 1$ and $d_i = 1$, the subject is assigned to the treatment group and receives the treatment. We can also envision other combinations. For example, if $z_i = 1$ and $d_i = 0$, the subject is assigned to the treatment group but for some reason does not receive the treatment. The exclusion restriction assumption is that $Y_i(1, d) = Y_i(0, d)$. In other words, potential outcomes respond only to the input from d_i; the value of z_i is irrelevant. Unfortunately, this assumption cannot be verified empirically because we never observe both $Y_i(1, d)$ and $Y_i(0, d)$ for the same subject.

The exclusion restriction breaks down when random assignment sets in motion causes of Y_i other than the treatment d_i. Suppose the treatment in our running example were defined as whether or not a woman council head presides over deliberations about village priorities. Our ability to estimate the effect of this treatment would be jeopardized if nongovernmental aid organizations, sensing that newly elected women will prioritize clean water, were to redirect their efforts to promote water sanitation to male-led villages. If outside aid flows to male-led villages, obviating the need for male village council leaders to allocate their budgets to water sanitation, the apparent difference between water sanitation budgets in councils led by women and councils led by men will exaggerate the true effect of the treatment, as defined above.[11] Even if it were the case that women council leaders have no effect on their own villages' budgets, the behavior of the NGOs could generate different average budgets in male- and female-led villages.

Asymmetries in measurement represent another threat to the excludability assumption. Suppose, for example, that in our study of Indian villages, we were to dispatch one group of research assistants to measure budgets in the treatment group and a different group of assistants to measure budgets in the control group. Each group of assistants may apply a different standard when determining what expenditures are to be classified as contributing to water sanitation. Suppose the research assistants in the treatment group were to use a more generous accounting standard—they tend to exaggerate the amount of money that the village allocates to water sanitation. When we compare average budgets in the treatment and control groups, the estimated treatment effect will be a combination of the true effect of female village heads on budgets and accounting procedures that exaggerate the amount of money spent on water sanitation in those villages. Presumably, when we envisioned the experiment and what we might learn from it, we sought to estimate only the first of these two effects. We wanted to know the effect of female leadership on budgets using a consistent standard of accounting.

11 Whether an excludability violation occurs depends on how a treatment effect is defined. If one were to define the effect of electing a woman to include the compensatory behavior of NGOs, this assumption would no longer be violated.

To illustrate the consequences of measurement asymmetry, we may write out a simple model in which outcomes are measured with error. Under this scenario, the usual schedule of potential outcomes expands to reflect the fact that outcomes are influenced not only by d_i, but also by z_i, which determines which set of research assistants measure the outcome. Suppose that among untreated units we observe $Y_i(0)^* = Y_i(0) + e_{i0}$, where e_{i0} is the error that is made when measuring the potential outcome if an observation is assigned to the control group. For treated units, let $Y_i(1)^* = Y_i(1) + e_{i1}$. What happens if we compare average outcomes among treated and untreated units? The expected value of the difference-in-means estimator from equation (2.14) is

$$E\left[\frac{\sum_1^m Y_i}{m} - \frac{\sum_{m+1}^N Y_i}{N-m}\right] = E[Y_i(1)^* | D_i = 1] - E[Y_i(0)^* | D_i = 0]$$

$$= E[Y_i(1) | D_i = 1] + E[e_{i1} | D_i = 1] - E[Y_i(0) | D_i = 0] - E[e_{i0} | D_i = 0]. \quad (2.16)$$

Comparing equation (2.16) to equation (2.14) reveals that the difference-in-means estimator is biased when the measurement errors in the treated and untreated groups have different expected values:

$$E[e_{i1} | D_i = 1] \neq E[e_{i0} | D_i = 0]. \quad (2.17)$$

In this book, when we speak of a "breakdown in symmetry," we have in mind procedures that may distort the expected difference between treatment and control outcomes.

What kinds of experimental procedures bolster the plausibility of the excludability assumption? The broad answer is anything that helps ensure uniform handling of treatment and control groups. One type of procedure is double-blindness—neither the subjects nor the researchers charged with measuring outcomes are aware of which treatments the subjects receive, so that they cannot consciously or unconsciously distort the results. Another procedure is parallelism in the administration of an experiment: the same questionnaires and survey interviewers should be used to assess outcomes in both treatment and control groups, and both groups' outcomes should be gathered at approximately the same time and under similar conditions. If outcomes for the control group are gathered in October, but outcomes in the treatment group are gathered in November, symmetry may be jeopardized.

The exclusion restriction cannot be evaluated unless the researcher has stated precisely what sort of treatment effect the experiment is intended to measure and designed the experiment accordingly. Depending on the researcher's objective, the control group may receive a special type of treatment so that the treatment vs. control comparison isolates a particular aspect of the treatment. A classic example of a research design that attempts to isolate a specific cause is a pharmaceutical trial in which an experimental pill is administered to the treatment group while an identical sugar pill is administered to the control group. The aim of administering a pill to both groups is to isolate the pharmacological effects of the ingredients, holding constant the effect of merely taking some sort of pill. In the village council example, a researcher may wish to distinguish the effects of female leadership of local councils from the effects of merely appointing non-incumbents to the headship. In principle, one could compare districts with randomly assigned women heads to districts with randomly assigned term limits, a policy that has the effect of bringing non-incumbents into leadership roles. This approach to isolating causal mechanisms is revisited again in Chapter 10, where we discuss designs that attempt to differentiate the active ingredients in a multifaceted treatment.

Protecting the theoretical integrity of the treatment vs. control comparison is of paramount importance in experimental design. In the case of the village budget study, the aim is to estimate the budgetary consequences of having a randomly allocated female village head, not the consequences of using a different measurement standard to evaluate outcomes in treatment and control villages. The same argument goes for other aspects of research activity that might be correlated with treatment assignment. For example, if the aim is to measure the effect of female leadership on budgets per se, bias may be introduced if one sends a delegation of researchers to monitor village council deliberations in women-headed villages only. Now the observed treatment effect is a combination of the effect of female leadership and the effect of research observers. Whether one regards the presence of the research delegation as a distortion of measurement or an unintended pathway by which assignment to treatment affects the outcome, the formal structure of the problem remains the same. The expected outcome of the experiment no longer reveals the causal effect we set out to estimate.

The symmetry requirement does not rule out cross-cutting treatments. For example, one could imagine a version of India's reservation policy that randomly assigned some village council seats to women, others to people from lower castes, and still others to women from lower castes. When we discuss factorial designs in Chapter 9, we will stress what can be learned from deploying several treatments in combination with one another. The point of these more complex designs is to learn about combinations of treatments while still preserving symmetry: randomly assigning treatments both alone and in combination with one another allows the researcher to distinguish empirically between having a female village head and having a female village head who is also from a lower caste.

Finally, let's revisit the case in which other actors intervene in response to your treatment assignments. For example, suppose that in anticipation of greater spending on water sanitation, interest groups devote special attention to lobbying village councils headed by women. Or it may go the other way: interest groups focus greater efforts on villages headed by men because they believe that's where they will meet the most resistance from budget makers. Whether interest group interference violated the assumption of excludability depends on how we define the treatment effect. Interest group activity presents no threat to the exclusion restriction if we define the effect of installing a female council head to include all of the indirect repercussions that it could have on interest group activity. If, however, we seek to estimate the specific effect of having female council heads without any interference by interest groups, our experimental design may be inadequate unless we can find a way to prevent interest groups from responding strategically. These kinds of scenarios again underscore the importance of clearly stating the experimental objectives so that researchers and readers can assess the plausibility of the exclusion restriction.

2.7.2 Non-Interference

For ease of presentation, the above discussion only briefly mentioned an assumption that plays an important role in the definition and estimation of causal effects. This assumption is sometimes dubbed the Stable Unit Treatment Value Assumption, or SUTVA, but we refer to it by a more accessible name, non-interference.[12] In the notation used above, expressions such as $Y_i(d)$ are written as though the value of the potential outcome for unit i depends only upon whether or not the unit itself gets the treatment (whether d equals one or zero). A more complete notation would express a more extensive schedule of potential outcomes depending on which treatments are administered to other units. For example, for Village 1 we could write down all of the potential outcomes if only Village 1 is treated, if only Village 2 is treated, if Villages 1 and 2 are treated, and so forth. This schedule of potential outcomes quickly gets out of hand. Suppose we listed all of the potential outcomes if exactly two of the seven villages are treated: there would now be 21 potential outcomes for each village. Clearly, if our study involves just seven villages, we have no hope of saying anything meaningful about this complex array of causal effects unless we make some simplifying assumptions.

The non-interference assumption cuts through this complexity by ignoring the potential outcomes that would arise if subject i were affected by the treatment of other subjects. Formally, we reduce the schedule of potential outcomes $Y_i(\mathbf{d})$, where $\mathbf{d}$ describes all of the treatments administered to all subjects, to a much simpler schedule $Y_i(d)$, where d refers to the treatment administered to subject i.[13] In the context of our example, non-interference implies that the sanitation budget in one village is unaffected by the gender of the council heads in other villages. Non-interference is an assumption common to both experimental and observational studies.

---

12 The term "stable" in SUTVA refers to the stipulation that the potential outcomes for a given village remain stable regardless of which other villages happen to be treated. The technical aspects of this term are discussed in Rubin 1980 and Rubin 1986.

13 Implicit in this formulation of potential outcomes is the assumption that potential outcomes are unaffected by the overall pattern of actual or assigned treatments. In other words, $Y_i(\mathbf{z}, \mathbf{d}) = Y_i(z, d)$.

Is non-interference realistic in this example? It is difficult to say without more detailed information about communication between villages and the degree to which their budget allocations are interdependent. If the collection of villages were dispersed geographically, it might be plausible to assume that the gender of the village head in one village has no consequences for outcomes in other villages. On the other hand, if villages were adjacent, the presence of a woman council head in one village might encourage women in other villages to express their policy demands more forcefully. Proximal villages might also have interdependent budgets; the more one village spends on water sanitation, the less the neighboring village needs to spend in order to maintain its own water quality.

The estimation problems that interference introduces are potentially quite complicated and unpredictable. Untreated villages that are affected by the treatments that nearby villages receive no longer constitute an untreated control group. If women council heads set an example of water sanitation spending that is then copied by neighboring villages headed by men, a comparison between average outcomes in treatment villages and (semi-treated) control villages will tend to understate the average treatment effect as defined in equation (2.3), which is usually understood to refer to the contrast between treated potential outcomes and completely untreated potential outcomes. On the other hand, if female council heads cause neighboring villages headed by men to free ride on water sanitation projects and allocate less of their budget to it, the apparent difference in average budget allocations will exaggerate the average treatment effect. Given the vagaries of estimation in the face of interference, researchers often try to design experiments in ways that minimize interference between units by spreading them out temporally or geographically. Another approach, discussed at length in Chapter 8, is to design experiments in ways that allow the researcher to detect spillover between units. Instead of treating interference as a nuisance, these more complex experimental designs aim to detect evidence of communication or strategic interaction among units.

SUMMARY

This chapter has limited its purview to a class of randomized experiments in which treatments are deployed exactly as assigned and outcomes are observed for all of the assigned subjects. This class of studies is a natural starting point for discussing core assumptions and what they imply for research design. The chapters that follow will introduce further assumptions in order to handle the complications that arise due to noncompliance (Chapters 5 and 6) and attrition (Chapter 7).

We began by defining a causal effect as the difference between two potential outcomes, one in which a subject receives treatment and the other in which the subject does not receive treatment. The causal effect for any given subject is not directly observable. However, experiments provide unbiased estimates of the average treatment effect (ATE) among all subjects when certain assumptions are met. The three assumptions invoked in this chapter are random assignment, excludability, and non-interference.

1. Random assignment: Treatments are allocated such that all units have a known probability between 0 and 1 of being placed into the treatment group. Simple random assignment or complete random assignment implies that treatment assignments are statistically independent of the subjects' potential outcomes.

   This assumption is satisfied when all treatment assignments are determined by the same random procedure, such as the flip of a coin. Because random assignment may be compromised by those allocating treatments or assisting subjects, steps should be taken to minimize the role of discretion.
2. Excludability: Potential outcomes respond solely to receipt of the treatment, not to the random assignment of the treatment or any indirect by-products of random assignment. The treatment must be defined clearly so that one can assess whether subjects are exposed to the intended treatment or something else.

   This assumption is jeopardized when (i) different procedures are used to measure outcomes in the treatment and control groups and (ii) research activities, other treatments, or third-party interventions other than the treatment of interest differentially affect the treatment and control groups.
3. Non-interference: Potential outcomes for observation i reflect only the treatment or control status of observation i and not the treatment or control status of other observations. No matter which subjects the random assignment allocates to treatment or control, a given subject's potential outcomes remain the same.

   This assumption is jeopardized when (i) subjects are aware of the treatments that other subjects receive, (ii) treatments may be transmitted from treated to untreated subjects, or (iii) resources used to treat one set of subjects diminish resources that would otherwise be available to other subjects. See Chapter 10 for a more extensive list of examples.

Random assignment is different from the other two assumptions in that it refers to a procedure and the manner in which researchers carry it out. Excludability and non-interference, on the other hand, are substantive assumptions about the ways in which subjects respond to the allocation of treatments. When assessing excludability and non-interference in the context of a particular experiment, the first step is to carefully consider how the causal effect is defined. Do we seek to study the effect of electing women to village council positions or rather the effect of electing women from a pool of candidates that consists only of women? When defining the treatment effect of installing a female village council head, is the appropriate comparison a village with male leadership, or a male-led village with no neighboring female-led villages? Attending to these subtleties encourages a researcher to design more exacting experimental comparisons and to interpret the results with greater precision.

Attentiveness to these core assumptions also helps guide experimental investigation, urging researchers to explore the empirical consequences of different research designs. A series of experiments in a particular domain may be required before a researcher can gauge whether subjects seem to be affected by the random assignment over and above the treatment (a violation of excludability) or by the treatments administered to other units (interference).

SUGGESTED READINGS

Holland (1986) and Rubin (2008) provide non-technical introductions to potential outcomes notation. Fisher (1935) and Cox (1958) are two classic books on experimental design and analysis; Dean and Voss (1999) and Kuehl (1999) offer more modern treatments. See Rosenbaum and Rubin (1984) on the distinctive statistical properties of randomly assigned treatments.

EXERCISES: CHAPTER 2

1. Potential outcomes notation:
   (a) Explain the notation "$Y_i(0)$."
   (b) Explain the notation "$Y_i(0) \mid D_i = 1$" and contrast it with the notation "$Y_i(0) \mid d_i = 1$."
   (c) Contrast the meaning of "$Y_i(0)$" with the meaning of "$Y_i(0) \mid D_i = 0$."
   (d) Contrast the meaning of "$Y_i(0) \mid D_i = 1$" with the meaning of "$Y_i(0) \mid D_i = 0$."
   (e) Contrast the meaning of "$E[Y_i(0)]$" with the meaning of "$E[Y_i(0) \mid D_i = 1]$."
   (f) Explain why the "selection bias" term in equation (2.15), $E[Y_i(0) \mid D_i = 1] - E[Y_i(0) \mid D_i = 0]$, is zero when D_i is randomly assigned.
2. Use the values depicted in Table 2.1 to illustrate that $E[Y_i(0)] - E[Y_i(1)] = E[Y_i(0) - Y_i(1)]$.
3. Use the values depicted in Table 2.1 to complete the table below.
   (a) Fill in the number of observations in each of the nine cells.
   (b) Indicate the percentage of all subjects that fall into each of the nine cells. (These cells represent what is known as the joint frequency distribution of $Y_i(0)$ and $Y_i(1)$.)
   (c) At the bottom of the table, indicate the proportion of subjects falling into each category of $Y_i(1)$. (These cells represent what is known as the marginal distribution of $Y_i(1)$.)
   (d) At the right of the table, indicate the proportion of subjects falling into each category of $Y_i(0)$ (i.e., the marginal distribution of $Y_i(0)$).
   (e) Use the table to calculate the conditional expectation that $E[Y_i(0) \mid Y_i(1) > 15]$. (Hint: This expression refers to the expected value of $Y_i(0)$ given that $Y_i(1)$ is greater than 15.)
   (f) Use the table to calculate the conditional expectation that $E[Y_i(1) \mid Y_i(0) > 15]$.

	$Y_i(1)$			
$Y_i(0)$	15	20	30	Marginal distribution of $Y_i(0)$
10				
15				
20				
Marginal distribution of $Y_i(1)$				1.0

4. Suppose that the treatment indicator d_i is either 1 (treated) or 0 (untreated). Define the average treatment effect among the treated, or ATT for short, as $\sum_1^N \tau_i d_i / \sum_1^N d_i$. Using the equations in this chapter, prove the following claim: "When treatments are allocated using complete random assignment, the ATT is, in expectation, equal to the ATE. In other words, taking expectations over all possible random assignments, $E[\tau_i | D_i = 1] = E[\tau_i]$, where τ_i is a randomly selected observation's treatment effect.
5. A researcher plans to ask six subjects to donate time to an adult literacy program. Each subject will be asked to donate either 30 or 60 minutes. The researcher is considering three methods for randomizing the treatment. One method is to flip a coin before talking to each person and to ask for a 30-minute donation if the coin comes up heads or a 60-minute donation if it comes up tails. The second method is to write "30" and "60" on three playing cards each, and then shuffle the six cards. The first subject would be assigned the number on the first card, the second subject would be assigned the number on the second card, and so on. A third method is to write each number on three different slips of paper, seal the six slips into envelopes, and shuffle the six envelopes before talking to the first subject. The first subject would be assigned the first envelope, the second subject would be assigned the second envelope, and so on.
   (a) Discuss the strengths and weaknesses of each approach.
   (b) In what ways would your answer to (a) change if the number of subjects were 600 instead of 6?
   (c) What is the expected value of D_i (the assigned number of minutes) if the coin toss method is used? What is the expected value of D_i if the sealed envelope method is used?
6. Many programs strive to help students prepare for college entrance exams, such as the SAT. In an effort to study the effectiveness of these preparatory programs, a researcher draws a random sample of students attending public high school in the United States, and compares the SAT scores of those who took a preparatory class to those who did not. Is this an experiment or an observational study? Why?
7. Suppose that an experiment were performed on the villages in Table 2.1, such that two villages are allocated to the treatment group and the other five villages to the control group. Suppose that an experimenter randomly selects Villages 3 and 7 from the set of seven villages and places them into the treatment group. Table 2.1 shows that these villages have unusually high potential outcomes.
   (a) Define the term *unbiased estimator*.
   (b) Does this allocation procedure produce upwardly biased estimates? Why or why not?
   (c) Suppose that instead of using random assignment, the researcher placed Villages 3 and 7 into the treatment group because the treatment could be administered inexpensively in those villages. Explain why this procedure is prone to bias.
8. Peisakhin and Pinto[14] report the results of an experiment in India designed to test the effectiveness of a policy called the Right to Information Act (RTIA), which allows citizens to inquire about the status of a pending request from government officials. In their study, the researchers hired confederates, slum dwellers who sought to obtain ration cards (which permit the purchase of food at low cost). Applicants for such cards must fill out a

14 Peisakhin and Pinto 2010.

form and have their residence and income verified by a government agent. Slum dwellers widely believe that the only way to obtain a ration card is to pay a bribe. The researchers instructed the confederates to apply for ration cards in one of four ways, specified by the researchers. The control group submitted an application form at a government office; the RTIA group submitted a form and followed it up with an official Right to Information request; the NGO group submitted a letter of support from a local nongovernmental organization (NGO) along with the application form; and finally, a bribe group submitted an application and paid a small fee to a person who is known to facilitate the processing of forms.

	Bribe	RTIA	NGO	Control
Number of confederates in the study	24	23	18	21
Number of confederates who had residence verification	24	23	18	20
Median number of days to residence verification	17	37	37	37
Number of confederates who received a ration card within one year	24	20	3	5

(a) Interpret the apparent effects of the treatments on the proportion of applicants who have their residence verified and the speed with which verification occurred.

(b) Interpret the apparent effects of the treatments on the proportion of applicants who actually received a ration card.

(c) What do these results seem to suggest about the effectiveness of the Right to Information Act as a way of helping slum dwellers obtain ration cards?

9. A researcher wants to know how winning large sums of money in a national lottery affects people's views about the estate tax. The researcher interviews a random sample of adults and compares the attitudes of those who report winning more than $10,000 in the lottery to those who claim to have won little or nothing. The researcher reasons that the lottery chooses winners at random, and therefore the amount that people report having won is random.

   (a) Critically evaluate this assumption. (Hint: are the potential outcomes of those who report winning more than $10,000 identical, in expectation, to those who report winning little or nothing?)

   (b) Suppose the researcher were to restrict the sample to people who had played the lottery at least once during the past year. Is it now safe to assume that the potential outcomes of those who report winning more than $10,000 are identical, in expectation, to those who report winning little or nothing?

10. Suppose researchers seek to assess the effect of receiving a free newspaper subscription on students' interest in politics. A list of student dorm rooms is drawn up and sorted randomly. Dorm rooms in the first half of the randomly sorted list receive a newspaper at their door each morning for two months; dorm rooms in the second half of the list do not receive a paper.

    (a) University researchers are sometimes required to disclose to subjects that they are participating in an experiment. Suppose that prior to the experiment, researchers distributed a letter informing students in the treatment group that they would be receiving a newspaper as part of a study to see if newspapers make students more interested in politics. Explain (in words and using potential outcomes notation) how this disclosure may jeopardize the excludability assumption.

(b) Suppose that students in the treatment group carry their newspapers to the cafeteria where they may be read by others. Explain (in words and using potential outcomes notation) how this may jeopardize the non-interference assumption.

11. Several randomized experiments have assessed the effects of drivers' training classes on the likelihood that a student will be involved in a traffic accident or receive a ticket for a moving violation.[15] A complication arises because students who take drivers' training courses typically obtain their licenses faster than students who do not take a course.[16] (The reason is unknown but may reflect the fact that those who take the training are better prepared for the licensing examination.) If students in the control group on average start driving much later, the proportion of students who have an accident or receive a ticket could well turn out to be higher in the treatment group. Suppose a researcher were to compare the treatment and control group in terms of the number of accidents that occur within three years of obtaining a license.
   (a) Does this measurement approach maintain symmetry between treatment and control groups?
   (b) Would symmetry be maintained if the outcome measure were the number of accidents per mile of driving?
   (c) Suppose researchers were to measure outcomes over a period of three years starting the moment at which students were randomly assigned to be trained or not. Would this measurement strategy maintain symmetry? Are there drawbacks to this approach?

12. A researcher studying 1,000 prison inmates noticed that prisoners who spend at least three hours per day reading are less likely to have violent encounters with prison staff. The researcher therefore recommends that all prisoners be required to spend at least three hours reading each day. Let d_i be 0 when prisoners read less than three hours each day and 1 when prisoners read more than three hours each day. Let $Y_i(0)$ be each prisoner's potential number of violent encounters with prison staff when reading less than three hours per day, and let $Y_i(1)$ be each prisoner's potential number of violent encounters when reading more than three hours per day.
   (a) In this study, nature has assigned a particular realization of d_i to each subject. When assessing this study, why might one be hesitant to assume that $E[Y_i(0) | D_i = 0] = E[Y_i(0) | D_i = 1]$ and $E[Y_i(1) | D_i = 0] = E[Y_i(1) | D_i = 1]$?
   (b) Suppose that researchers were to test this researcher's hypothesis by randomly assigning 10 prisoners to a treatment group. Prisoners in this group are required to go to the prison library and read in specially designated carrels for three hours each day for one week; the other prisoners, who make up the control group, go about their usual routines. Suppose, for the sake of argument, that all prisoners in the treatment group in fact read for three hours each day and that none of the prisoners

15 See Roberts and Kwan 2001.

16 Vernick et al. 1999.

in the control group read at all during the week of the study. Critically evaluate the excludability assumption as it applies to this experiment.

(c) State the assumption of non-interference as it applies to this experiment.

(d) Suppose that the results of this experiment were to indicate that the reading treatment sharply reduces violent confrontations with prison staff. How does the non-interference assumption come into play if the aim is to evaluate the effects of a policy whereby all prisoners are required to read for three hours?

CHAPTER 3

Sampling Distributions, Statistical Inference, and Hypothesis Testing

Rigorous quantification of uncertainty is a hallmark of scientific inquiry. When analyzing experimental data, the aim is not only to generate unbiased estimates of the average treatment effect but also to draw inferences about the uncertainty surrounding these estimates. Among the most attractive features of experimentation is that random allocation of treatments is a reproducible procedure. Reproducibility allows us to assess the sampling distribution, or collection of estimated ATEs that could have come about under different random assignments in order to better understand the uncertainty associated with the experiment we conducted. One objective of this chapter is to explain how experimental design affects the sampling distribution. We consider ways of designing experiments so as to reduce sampling variability, and we call attention to the fact that the sampling distribution may change markedly depending on the procedures used to randomly allocate subjects to treatment and control conditions.

A second objective is to guide the reader through the calculation and interpretation of key statistical results. When analyzing an experiment, you should consider both the estimated ATE and the uncertainty with which it is estimated. Unless you have prior information about the value of the ATE, the experimental estimate is one's best guess of the true treatment effect, but this guess may be close to or far from the true average causal effect. Statisticians commonly assess uncertainty in two ways. One method is to investigate whether the experimental results are sufficiently informative to refute a determined skeptic who insists that there is no treatment effect whatsoever. Another approach is to identify a range of values that probably bracket the true average treatment effect. This chapter introduces a flexible set of statistical techniques that may be used to assess uncertainty across a wide array of different experimental designs.

What you will learn from this chapter:

1. How to quantify the uncertainty surrounding an experimental estimate.
2. Formulas that suggest ways to design more informative experiments.
3. How to refute a determined skeptic who advances the "null hypothesis" that the treatment has no effect.
4. How to generate confidence intervals that have a 95% chance of bracketing the true sample average treatment effect.

3.1 Sampling Distributions

One of the most important topics in experimental design and analysis is sampling variability. When examining the results of a single experiment, we must bear in mind that we have in front of us just one of the many possible datasets that could have been generated via random assignment. The experiment we happened to conduct yields an estimate of the average treatment effect, but had the same observations been randomly assigned in a different way, our estimate might have been quite different. The term *sampling distribution* refers to the collection of estimates that could have been generated by every possible random assignment.[1]

To illustrate the idea of a sampling distribution, let's return to the village council experiment discussed in the previous chapter. In that study, two of the seven

BOX 3.1

Definition: Sampling Distribution of Experimental Estimates

A sampling distribution is the frequency distribution of a statistic obtained from hypothetical replications of a randomized experiment. For example, if one were to conduct the same experiment repeatedly under identical conditions, the collection of estimated average treatment effects from each replication of the experiment forms a sampling distribution. Under the central limit theorem, the sampling distribution of the estimated average treatment effect takes the shape of a normal distribution as the number of observations in treatment and control conditions increases.

1 The distribution of estimates from all possible randomizations is also called the randomization distribution. See Rosenbaum 1984.

villages were randomly assigned to receive the treatment, which in this case is appointment of a woman to the position of village council head. Empirically, we happen to observe one particular realization of that randomization, illustrated in Table 2.2. But prior to our random allocation, there were 21 different ways to place two of the seven villages into the treatment group, and each of these 21 allocations had the same probability of being selected. Using the schedule of potential outcomes listed in Table 2.1, we may generate the hypothetical experimental results that each of the 21 possible randomizations would have produced. In other words, for each possible randomization, we calculate the average budget allocation in the treatment and control groups, and calculate the difference in means. The results are displayed in Table 3.1.

The first thing to note about the results in Table 3.1 is that the average estimated ATE is 5, which is exactly the same as the true ATE in Table 2.1. This is no coincidence. In fact, the exercise underscores a very important feature of randomized experiments: given three core assumptions discussed in Chapter 2 (random assignment, excludability, and non-interference), the average estimated ATE across all

TABLE 3.1

Sampling distribution of estimated ATEs generated when two of the seven villages listed in Table 2.1 are assigned to treatment

	Estimated ATE	Frequency with which an estimate occurs
	−1	2
	0	2
	0.5	1
	1	2
	1.5	2
	2.5	1
	6.5	1
	7.5	3
	8.5	3
	9	1
	9.5	1
	10	1
	16	1
Average	5	
Total		21

possible random assignments is equal to the true ATE. Any single experiment might give a number that is too high or too low, but the expected value of this estimation procedure is the true ATE. One of the great virtues of experiments is that they generate unbiased estimates of the ATE: merely by subtracting the control group mean from the treatment group mean, we obtain an estimator that on average recovers the true ATE.

The next thing to note about the results in Table 3.1 is that the 21 possible experiments generate quite different results. The largest estimated ATE is 16—we had a 1-in-21 chance of obtaining an estimate that was more than three times the size of the true ATE of 5. Two of the 21 randomizations produce an estimated ATE of -1. Had we obtained an estimate of -1, we might have been led to believe that women village council heads tend to reduce the share of budgets directed at water sanitation, even though the opposite is true. The dispersion of estimates around the true ATE reminds us that while experiments are unbiased, they are not necessarily precise. With just seven observations, our experiment generates results that vary markedly from one randomization to the next.

3.2 The Standard Error as a Measure of Uncertainty

In order to describe the precision with which an experiment recovers the ATE, we need a statistic that characterizes the amount of sampling variability. Sampling variability is typically expressed by reference to the *standard error*. The larger the standard error, the more uncertainty surrounds our parameter estimate. The standard error is the standard deviation of the sampling distribution. It is obtained by calculating the squared deviation of each estimate from the average estimate, dividing by the number of possible randomizations, and taking the square root of the result.

Based on the numbers in Table 3.1, we calculate the standard error as follows:

Sum of squared deviations

$$= (-1 - 5)^2 + (-1 - 5)^2 + (0 - 5)^2 + (0 - 5)^2 + (0.5 - 5)^2 + (1 - 5)^2 + (1 - 5)^2 + (1.5 - 5)^2 + (1.5 - 5)^2 + (2.5 - 5)^2 + (6.5 - 5)^2 + (7.5 - 5)^2 + (7.5 - 5)^2 + (7.5 - 5)^2 + (8.5 - 5)^2 + (8.5 - 5)^2 + (8.5 - 5)^2 + (9 - 5)^2 + (9.5 - 5)^2 + (10 - 5)^2 + (16 - 5)^2 = 445.$$

$$\textit{Square root of the average squared deviation} = \sqrt{\frac{1}{21}(445)} = 4.60. \quad (3.1)$$

When a parameter is estimated using an unbiased estimator (such as difference-in-means), a helpful rule of thumb is that approximately 95% of the experimental outcomes fall within an interval that ranges from two standard errors below the true

BOX 3.2

Definition: Standard Deviation

The standard deviation of a variable X is

$$\sqrt{\frac{1}{N}\sum_{1}^{N}(X_i - \overline{X})^2},$$

where $\overline{X}$ denotes the mean of X. Notice that this formula divides by N. When X is a random sample from a larger population containing N^* subjects whose mean is unknown, the estimate of the population standard deviation is

$$\sqrt{\frac{1}{N-1}\sum_{1}^{N}(X_i - \overline{X})^2}.$$

Definition: Standard Error

The standard error is the standard deviation of a sampling distribution. Suppose there are J possible ways of randomly assigning subjects. Let $\hat{\theta}_j$ represent the estimate (denoted by the "hat" mark: ^) we obtain from the jth randomization, and $\overline{\hat{\theta}}$ represent the average estimate for all J. For example, $\hat{\theta}_j$ may represent the difference-in-means estimate from one of the possible random assignments. Over all J possible random assignments, the standard error of $\hat{\theta}$ is

$$\sqrt{\frac{1}{J}\sum_{1}^{J}(\hat{\theta}_j - \overline{\hat{\theta}})^2}.$$

parameter to two standard errors above it. Given a true parameter of 5 and a standard error of 4.60, this interval stretches from −4.20 to 14.20. (In Table 3.1, we see that in fact this rule of thumb works well for this example: 20 of the 21 estimates, or 95%, fall in this range. The approximation on which this rule of thumb rests tends to become more accurate as sample size increases.) At the low end of this interval, we infer that appointing a female council head diminishes budgets for water sanitation, and at the top of this interval we grossly exaggerate the extent to which women leaders allocate more money toward this policy area.

How can we reduce the standard error? In other words, how can we design our experiment so that it produces more precise estimates of the ATE? In order to answer this question, let's inspect a formula that expresses the standard error as a function of the potential outcomes and the experimental design. Before getting to the formula itself, we first define some key terms. A *variance* of an observed or potential outcome

BOX 3.3

Rule of Thumb for Sampling Distributions of the Estimated ATE

For randomized experiments with a given standard error, approximately 95% of all random assignments will generate estimates of the ATE that fall within ±2 standard errors from the true ATE. For example, if the standard error is 10 and the ATE is 50, approximately 95% of the estimates will be between 30 and 70. This rule of thumb works best when N is large.

for a set of N subjects is the average squared deviation of each subject's value from the mean for all N subjects. For example, the variance of $Y_i(1)$ is:

$$\text{Var}(Y_i(1)) \equiv \frac{1}{N}\sum_1^N\left(Y_i(1) - \frac{\sum_1^N Y_i(1)}{N}\right)^2. \tag{3.2}$$

The higher the variance, the greater the dispersion around the mean. The smallest possible variance is zero, which implies that the variable is a constant. Note that the variance is the square of the standard deviation. For more details on this formula, see Box 3.2.

To obtain the covariance between two variables, such as $Y_i(1)$ and $Y_i(0)$, subtract the mean from each variable, and then calculate the average cross-product of the result:

$$\text{Cov}(Y_i(0), Y_i(1)) \equiv \frac{1}{N}\sum_1^N\left(Y_i(0) - \frac{\sum_1^N Y_i(0)}{N}\right)\left(Y_i(1) - \frac{\sum_1^N Y_i(1)}{N}\right). \tag{3.3}$$

The covariance is a measure of association between two variables. A negative covariance implies that lower values of one variable tend to coincide with higher values of the other variable. A positive covariance means that higher values of one variable tend to coincide with higher values of the other variable.

Applying these formulas to the schedule of potential outcomes listed in Table 2.1, we find the variance of $Y_i(0)$ to be 14.29, and the variance of $Y_i(1)$ to be 42.86. The covariance of $Y_i(0)$ and $Y_i(1)$ is 7.14. In order to obtain the standard error associated with the experimental estimate of the average treatment effect, the remaining step is to decide how many of the N observations are to be treated. We will call m the number of treated units. In our example, $m = 2$ and $N = 7$. In general, we require that $0 < m < N$, because if m were zero or equal to N, we would have just one experimental group rather than two. For the same reason, we also require $N > 1$.

With all of the ingredients in place, we are now ready to write the formula for the standard error of the estimated ATE. The equation places a "hat" mark over the estimand in order to indicate that we are talking about an estimate of the ATE, not the true ATE:

BOX 3.4

Properties of Variances and Covariances

(a) $\text{Cov}(Y_i(0), Y_i(0)) = \text{Var}(Y_i(0)) \geq 0$

(b) $\text{Cov}(Y_i(0), Y_i(1)) = \text{Cov}(Y_i(0), Y_i(0) + \tau_i)$
$= \text{Var}(Y_i(0)) + \text{Cov}(Y_i(0), \tau_i)$

(c) $\text{Cov}(aY_i(0), bY_i(1)) = ab\text{Cov}(Y_i(0), Y_i(1))$

Covariances are bounded by the restriction that:

$$\text{Cov}(Y_i(0), Y_i(1)) \leq \sqrt{\text{Var}(Y_i(0))\text{Var}(Y_i(1))} \leq \frac{\text{Var}(Y_i(0)) + \text{Var}(Y_i(1))}{2}.$$

The correlation between two variables $Y_i(0)$ and $Y_i(1)$ is

$$\frac{\text{Cov}(Y_i(0), Y_i(1))}{\sqrt{\text{Var}(Y_i(0))\text{Var}(Y_i(1))}}.$$

$$SE(\widehat{\text{ATE}}) = \sqrt{\frac{1}{N-1}\left\{\frac{m\text{Var}(Y_i(0))}{N-m} + \frac{(N-m)\text{Var}(Y_i(1))}{m} + 2\text{Cov}(Y_i(0), Y_i(1))\right\}}. \quad (3.4)$$

This formula tells us which factors reduce the size of the standard error.[2] The formula contains five inputs: N, m, $\text{Var}(Y_i(0))$, $\text{Var}(Y_i(1))$, and $\text{Cov}(Y_i(0), Y_i(1))$. Changing each input one by one, while holding the other inputs constant, we can examine how the standard error changes. Here is a summary of the formula's implications for experimental design:

1. The larger the N, the smaller the standard error. Holding constant the other inputs (including m, the size of the treatment group) and increasing N means increasing the control group. As the control group grows, the first and third terms inside the braces of equation (3.4) are diminished by an expanding $N - 1$ denominator. If the control group were infinite in size, the only source of uncertainty would come from the treatment group. Sometimes adding subjects to the control group involves little or no additional cost. For experiments in which treating subjects costs resources but leaving them untreated does not (e.g., sending mail to those in the treatment group while sending nothing to those in the control), bring as many additional subjects as possible into the control group. A similar point holds

2 Equation (3.4) describes the true standard error, $SE(\widehat{\text{ATE}})$, which is not to be confused with the estimated standard error, $\widehat{SE}(\widehat{\text{ATE}})$, calculated based on a particular experiment.

for the treatment group: holding the control group's size constant, increasing m, the size of the treatment group, diminishes the second and third terms in equation (3.4). Of course, where possible, increase the size of both the control group and the treatment group, as this reduces all three terms in equation (3.4).

2. The smaller the variance of $Y_i(0)$ or $Y_i(1)$, the smaller the standard error. To maximize precision, conduct experiments on observations that are as similar as possible in terms of their potential outcomes. This principle has three design implications. First, it encourages researchers to measure outcomes as accurately as possible, as this dampens variability. Second, as discussed later in this chapter, blocking may be used to improve precision by grouping observations with similar potential outcomes.[3] Third, as explained in Chapter 4, another way to reduce variance in outcomes is to measure outcomes in advance of the experimental intervention, sometimes known as a *pre-test*, in addition to measuring outcomes after the intervention via a *post-test*. Instead of defining the experimental outcome to be the score on the post-test, define the experimental outcome to be the change from pre-test to post-test. Change scores usually have less variance than post-test scores.
3. Assuming that $Y_i(0)$ and $Y_i(1)$ do vary, the smaller the covariance between $Y_i(0)$ and $Y_i(1)$, the smaller the standard error.[4] A particularly favorable case occurs when the potential outcomes have negative covariance: that is, where high values of $Y_i(0)$ tend to coincide with low values of $Y_i(1)$. In order to see how this pattern might occur, write $Y_i(1) = Y_i(0) + \tau_i$. Substituting for $Y_i(1)$ allows us to express $\text{Cov}(Y_i(0),Y_i(1))$ as $\text{Var}(Y_i(0)) + \text{Cov}(Y_i(0),\tau_i)$. (See Box 3.4.) So, when $Y_i(0)$ and τ_i are negatively related (e.g., students with low baseline scores are most helped by the treatment), the covariance between $Y_i(0)$ and $Y_i(1)$ may be close to zero or even negative.
4. A subtle implication of the formula is that when the variances of $Y_i(0)$ and $Y_i(1)$ are similar, it is advisable to assign approximately half of the observations to the treatment group, such that $m \approx N/2$. When the potential outcomes have different variances, invest additional observations to the experimental condition with

3 The researcher could restrict attention to subjects with similar background attributes in order to reduce variance in $Y_i(0)$ and $Y_i(1)$, but this approach has the drawback of limiting the generalizations that might be drawn from this narrow set of subjects. In order to overcome this limitation, a researcher may conduct the experiment within several different blocks, each of which contains subjects with similar background attributes.

4 Here is the underlying intuition for why positive covariance leads to larger standard errors. If high values of $Y_i(1)$ tend to coincide with high values of $Y_i(0)$, then selecting a subject with a high potential outcome into the treatment group leaves one fewer subject with high potential outcomes for the control group. Positive covariance between $Y_i(0)$ and $Y_i(1)$ therefore means that results are sensitive to the placement of subjects into treatment or control. On the other hand, if high values of $Y_i(1)$ tend to coincide with low values of $Y_i(0)$, then selecting a subject with high potential outcomes into the treatment group leaves one fewer subject with low potential outcomes for the control group. In this case, there is less sampling variability because the control group is "compensated" for the fact that the treatment group received a high value of $Y_i(1)$. See section 3.6.1.

greater variance. For example, if $\text{Var}(Y_i(1)) > \text{Var}(Y_i(0))$, put a greater share of the observations into the treatment group. In practice, however, researchers seldom know in advance which group is likely to have more variance and therefore place equal numbers of subjects in each condition.

In order to see the formula at work, we fill in the values obtained from Table 2.1 and obtain exactly the same number we calculated from Table 3.1:

$$SE(\widehat{\text{ATE}}) = \sqrt{\frac{1}{6}\left\{\frac{(2)(14.29)}{5} + \frac{(5)(42.86)}{2} + (2)(7.14)\right\}} = 4.60. \qquad (3.5)$$

When we know the full schedule of potential outcomes, equation (3.4) tells us the standard deviation of estimated ATEs from all possible random assignments.

Equation (3.4) can be used to demonstrate that our design of $m = 2$ is less than optimal. Increasing the number of treated units so that $m = 3$ lowers the standard error to 3.7. Raising m to 4 lowers the standard error even further, to 3.3. The reason it is better to put more observations into the treatment group than the control group is that in this example the treated potential outcomes have more variance than the untreated potential outcomes: $\text{Var}(Y_i(1)) > \text{Var}(Y_i(0))$. Raising m to 5 goes too far in that direction and leads to a slight deterioration in precision.

To summarize, standard errors are measures of uncertainty; they indicate the extent to which estimates will vary across all possible random assignments. One consideration when designing experiments is to keep standard errors small. Two types of inputs determine the standard error: the schedule of potential outcomes and the number of observations assigned to treatment and control groups. The schedule of potential outcomes is unobserved, but experimenters sometimes have opportunities to design experiments in ways that limit the mischief that highly variable potential outcomes may cause. Variance in $Y_i(0)$ and $Y_i(1)$ may be reduced by measuring outcomes in a more precise manner or by blocking observations into homogeneous subsets. Standard errors are also reduced by "Robin Hood treatments" that raise outcomes among those with low $Y_i(0)$ while lowering outcomes among those with high $Y_i(0)$. (For example, a regimented physical fitness program in a military academy, where cadets have little spare time, might raise the level of fitness among the least fit while lowering it among top-level athletes.) If the cost per subject is similar in both experimental groups, assign similar numbers to treatment and control, tilting the balance in favor of the group that is expected to have more variable outcomes.

3.3 Estimating Sampling Variability

The previous section illustrated the concept of sampling variability by showing how standard errors can be calculated from a known schedule of potential outcomes. This exercise is important because it suggests ways in which problems of statistical uncertainty can be addressed through experimental design. Now that we have an appreciation for what a standard error is, we take up the question that arises whenever one estimates an ATE for a particular set of subjects.[5] Suppose the researcher wants to know the standard error in order to calibrate the uncertainty associated with this estimate. The researcher has neither the complete schedule of potential outcomes nor results from all of the hypothetical random assignments that could have allocated this set of observations to treatment and control groups. Instead, this researcher has results from a single randomization.

Assessing uncertainty is an estimation problem. The true standard error is unknown. We seek to estimate this unknown quantity using data from a single experiment. What hints can a single experiment provide about how other experimental assignments might have come out? Returning to equation (3.4), we see that the experiment provides information about four of the five inputs that generate standard errors. The number of observations allocated to treatment and control is known. The variance of outcomes in the untreated potential outcomes $Y_i(0)$ can be estimated using the observed outcomes in the control group. Because the control group is assigned at random, the variance that we observe in the control group is an unbiased estimate of the variance in $Y_i(0)$. The same approach can be used to estimate the variance of $Y_i(1)$ based on the observed variance in the assigned treatment group.

The one element in this equation that cannot be estimated empirically is the covariance between $Y_i(0)$ and $Y_i(1)$, as we never observe both potential outcomes for the same subject. The standard approach is to use a conservative estimation formula that is at least as large as equation (3.4) regardless of the covariance between $Y_i(0)$ and $Y_i(1)$.[6] The conservative formula assumes that the treatment effect is the same for all subjects, which implies that the correlation between $Y_i(0)$ and $Y_i(1)$ is 1.0.[7]

5 The average treatment effect is the ATE for the "finite population" of subjects in one's experiment. If the experimental subjects are seen as a sample from a larger population, one may distinguish between the ATE in the sample and the ATE in the population. In Chapter 11, we take up the question of how to use the sample at hand to draw inferences about the ATE in the population.

6 To be more precise, the conservative estimation approach tends to overestimate the true sampling variance, which is the *square* of the standard error depicted in equation (3.4). One note of caution: although the conservative estimation formula overestimates the true sampling variance *on average*, estimation is subject to sampling variability. A given estimate of the sampling variance using the conservative formula may still be smaller than the true sampling variance.

7 See Freedman, Pisani, and Purves 1998, pp. A32–A34. The conservative formula will give unbiased estimates of the sampling variance under either of two scenarios. The first arises when the treatment effect τ_i is the same for all subjects. The second arises when subjects are sampled at random from a large population prior to random assignment, and the objective is to estimate the population average treatment effect. When subjects are sampled randomly from a large population, the selection of one subject for the treatment group has no material effect on the pool of available subjects that can be selected into the control group, which renders the covariance between $Y_i(0)$ and $Y_i(1)$ irrelevant.

The formula for estimating the standard error of the average treatment effect is

$$\widehat{SE} = \sqrt{\frac{\widehat{\text{Var}}(Y_i(0))}{N-m} + \frac{\widehat{\text{Var}}(Y_i(1))}{m}}, \tag{3.6}$$

where the variances are estimated using the m observations of potential outcomes from the treatment group:

$$\widehat{\text{Var}}(Y_i(1)) = \frac{1}{m-1}\sum_1^m\left(Y_i(1) \mid d_i = 1 - \frac{\sum_1^m Y_i(1) \mid d_i=1}{m}\right)^2 \tag{3.7}$$

and the $N - m$ observations of potential outcomes from the control group:

$$\widehat{\text{Var}}(Y_i(0)) = \frac{1}{N-m-1}\sum_{m+1}^N\left(Y_i(0) \mid d_i = 0 - \frac{\sum_{m+1}^N Y_i(0) \mid d_i=0}{N-m}\right)^2. \tag{3.8}$$

Note that when calculating the sample variances in equations (3.7) and (3.8), we divide by one less than the number of observations to take into account the fact that one observation is expended when we calculate the sample mean.[8] In order to avoid dividing by zero when estimating the standard error, we must have at least two subjects in each experimental group.

How closely do the empirical estimates[9] of the standard errors match the true standard errors? Using the conservative formula, the average standard error in our example is 4.65, which is not far from the true standard error of 4.60. Although the estimates vary from one random allocation to the next, on average, the estimator depicted in equation (3.6) does a reasonably good job of estimating the true level of sampling variability.

3.4 Hypothesis Testing

The previous section illustrated the challenges that arise when estimating the sampling uncertainty surrounding a single experiment's estimate of the ATE. Accurate estimation of the standard error requires accurate guesses about the variances and covariance of potential outcomes. If we do not have much data about the variances or if our simplifying assumption about the unobservable covariance is mistaken, our estimated standard errors may be inaccurate.

8 If we know the mean and the outcomes for all but one of the subjects, we can deduce the outcome for the remaining subject. In statistical parlance, calculating the mean expends one *degree of freedom*.

9 The 21 estimated standard errors are {1.581, 1.871, 1.871, 1.871, 1.871, 1.871, 2.236, 2.784, 2.958, 3.122, 3.122, 5.244, 5.339, 7.599, 7.665, 7.730, 7.730, 7.730, 7.730, 7.826, 7.826}.

Much easier is the task of testing the *sharp null hypothesis* that the treatment effect is zero for all observations. What makes it easy is that if the null hypothesis is true, $Y_i(0) = Y_i(1)$. Under this special case, we observe *both* potential outcomes for every observation. We may therefore take the observed outcomes in our dataset and simulate all possible randomizations as though we were working from a complete schedule of potential outcomes. These simulated randomizations provide an exact sampling distribution of the estimated average treatment effect under the sharp null hypothesis. By looking at this distribution of hypothetical outcomes, we can calculate the probability of obtaining an estimated ATE at least as large as the one we obtained from our actual experiment if the treatment effect were in fact zero for every subject.

For example, the randomization depicted in Table 2.2 generated an estimate of the ATE of 6.5. How likely are we to obtain an estimate as large as or larger than 6.5 if the true effect were zero for all observations? The probability, or p-value, of interest in this case addresses a *one-tailed hypothesis*, namely that female village council heads increase budget allocations to water sanitation. If we sought to evaluate the *two-tailed hypothesis*—whether female village council heads either increase or decrease the budget allocation for water sanitation—we would calculate the p-value of obtaining a number that is greater than or equal to 6.5 or less than or equal to -6.5.

Based on the observed outcomes in Table 2.2, we may calculate the 21 possible estimates of the ATE that could have been generated if the null hypothesis were true: $\{-7.5, -7.5, -7.5, -4.0, -4.0, -4.0, -4.0, -4.0, -0.5, -0.5, -0.5, -0.5, -0.5, -0.5, 3.0, 3.0, 6.5, 6.5, 6.5, 10.0, 10.0\}$. Five of the estimates are as large as 6.5. So when evaluating the one-tailed hypothesis that female village heads *increase* water sanitation budgets, we would conclude that the probability of obtaining an estimate as large as 6.5 if the null hypothesis were true is $5/21 = 24\%$. A two-tailed hypothesis test would count all instances in which the estimates are at least as great as 6.5 *in absolute value*. Eight of the estimates qualify, so the two-tailed p-value is $8/21 = 38\%$.

In theory, this type of calculation could be performed on experiments of any size, but in practice the number of possible random assignments becomes astronomical as N increases. For example, an experiment where $N = 50$ and half the observa-

BOX 3.5

Definition: Sharp Null Hypothesis of No Effect

The treatment effect is zero for all subjects. Formally, $Y_i(1) = Y_i(0)$ for all i.

Definition: Null Hypothesis of No Average Effect

The average treatment effect is zero. Formally, $\mu_{Y(1)} = \mu_{Y(0)}$.

tions are assigned to the treatment group potentially generates more than 126 trillion randomizations:

$$\frac{50!}{25!25!} = 126{,}410{,}606{,}437{,}752. \tag{3.9}$$

When the number of possible randomizations is large, one can closely approximate the sampling distribution by sampling at random from the set of all possible random assignments. Whether one uses all possible randomizations or a large sample of them, the calculation of p-values based on an inventory of possible randomizations is called *randomization inference.*

This approach to hypothesis testing has two attractive properties. First, the procedures used to calculate p-values may be applied to a very broad class of hypotheses and applications. The method is not confined to large samples or normally distributed outcomes. Any sample size will do, and the method can be applied to all sorts of outcomes, including counts, durations, or ranks. Second, the method is exact in the sense that the set of all possible random assignments fully describes the sampling distribution under the null hypothesis. By contrast, the hypothesis testing methods discussed in introductory statistics courses rely on an approximation in order to derive the shape of the sampling distribution. (See Box 3.7.) Although exact and approximate methods will tend to give very similar answers when samples are large, we use randomization inference throughout the book so that a single statistical approach may be applied to a broad array of applications without approximations or additional assumptions.

Since the 1930s, it has been conventional to dub p-values that are below 0.05 as *statistically significant* on the grounds that, under the null hypothesis, the researcher had less than a 1-in-20 chance of obtaining the result due to chance. The p-values in the village council experiment were 0.24 and 0.38, which fail to meet this standard. By the 0.05 standard of statistical significance, the estimate of 6.5 does not provide a convincing refutation of the null hypothesis of no effect. The 0.05 standard is a matter of convention, not statistical theory, but it is so deeply entrenched that researchers should be prepared to indicate whether their experimental results are significant at the 0.05 level.

Anticipating this concern, researchers often attempt to forecast the probability that their experiment, when conducted, will lead to the rejection of the null hypothesis at a given significance level, such as 0.05. This probability is termed the *statistical power* of the experiment. To say that an experimental design has 80% power, for instance, means that 80% of all possible random assignments will produce observed results that will lead to the rejection of the null hypothesis in the presence of the posited treatment effect. Forecasting the power of an experiment requires some guesswork. The appendix to this chapter illustrates how assumptions are used when calculating the power of a proposed experiment.

Be careful not to confuse statistical significance with substantive significance. A parameter estimate that falls short of the 0.05 threshold might nevertheless be

BOX 3.6

Definition: One-tailed and Two-tailed Hypothesis Tests

A null hypothesis that specifies that the treatment effect is zero can be rejected by test statistics that are either very large or very small. This is called a two-tailed test. For example, an intervention that is believed to *change* outcomes (either positively or negatively) would be evaluated using a two-tailed test. A null hypothesis that specifies that the effect is zero or less can be rejected by a large positive test statistic. (Similarly, a null hypothesis that specifies that an effect is zero or more is rejected by a large negative test statistic.) This is called a one-tailed test. Therapeutic interventions are advanced in anticipation of finding positive effects, in which case the null hypothesis is that the effect is zero or less, and an appropriate test is one-tailed.

Definition: *p*-value

For a two-tailed test, the *p*-value is the probability of obtaining a test statistic at least as large in absolute value as the observed test statistic, given that the null hypothesis is true. For example, suppose the null hypothesis is that the treatment has no effect, and the test statistic is the estimated ATE. If the estimated ATE is 5 and the *p*-value is 0.20, there is a 20% chance of obtaining an estimate as large as 5 (or as small as -5) simply by chance. If the alternative hypothesis is a positive (or negative) effect, a one-tailed test is used.

Randomization Inference

The sampling distribution of the test statistic under the null hypothesis is computed by simulating all possible random assignments. When the number of random assignments is too large to simulate, the sampling distribution may be approximated by a large random sample of possible assignments. *p*-values are calculated by comparing the observed test statistic to the distribution of test statistics under the null hypothesis.

important and interesting. If female council heads in fact caused a 6.5 percentage-point increase in budgetary allocations to water sanitation, the health consequences of increasing the role of women in local government throughout the Indian countryside could be profound. Although statistical uncertainty remains, the data have taught us something potentially useful. If this were the first experiment of its kind and we had no prior knowledge of the treatment effect, the estimate of 6.5 would still be our best guess of the true ATE, despite the fact that we cannot rule out the

BOX 3.7

Comparing Randomization Inference to T-Tests

Approximate methods for testing the sharp null hypothesis of no effect assume that the sampling distribution of the difference-in-means estimator has a particular shape. For example, the t-test, which should be familiar to those who have taken an introductory statistics course, assumes that the sampling distribution of the estimated difference-in-means follows a t-distribution, which is similar to a normal distribution. The t-test gives accurate *p*-values when outcomes in each of the experimental groups are distributed normally. When outcomes are distributed non-normally, approximate methods become increasingly accurate as the number of subjects in each of the experimental conditions grows.

To illustrate the difference between a t-test and randomization inference, we generated the hypothetical dataset shown below. The treatment and control groups each contain 10 subjects. The treatment is an encouragement to make a charitable donation, and the outcome is the amount of money contributed. This outcome is skewed to the right due to a few large donations. The treatment group average is 80, while the control group average is 10. The sampling distribution under the sharp null hypothesis is bimodal (due to the influence of a few large donations), but conventional methods assume that the sampling distribution is bell-shaped. Repeating the random assignment 100,000 times under the sharp null hypothesis of no effect, we find that the observed average treatment effect of 70 has a one-tailed *p*-value of 0.032. A t-test assuming equal variances puts this *p*-value at 0.082; a t-test allowing for unequal variances declares the *p*-value to be 0.091. The t-test is inaccurate in this case because the outcomes are skewed and the number of observations is small. Readers are encouraged to tinker with this example using the simulation code at http://isps.research.yale.edu/FEDAI to see how the shape of the sampling distribution changes as more subjects are added to the dataset.

Treatment	Donation
1	500
1	100
1	100
1	50
1	25
1	25
1	0
1	0
1	0
1	0
0	25
0	20
0	15
0	15
0	10
0	5
0	5
0	5
0	0
0	0

hypothesis that the ATE is zero. The right way to think about an experimental result that is substantively significant but statistically insignificant is that it warrants further investigation. As we conduct further experiments, our uncertainty will gradually diminish, and we will be able to make a clearer determination about the true value of the ATE. (See Chapter 11 for a discussion of how sampling variability diminishes when pooling a series of repeated experiments.)

Conversely, don't be overly impressed with statistically significant results without reflecting on their substantive significance. All else equal, the standard error declines in proportion to the square root of N. Imagine the number of observations were to grow from 7 to 70,000 villages. This 10,000-fold increase in N would imply a 100-fold decrease in standard error, from 4.6 to 0.046 percentage-points. If this enormous experiment were to indicate that female village council heads increased sanitation budgets by an average of 0.1 percentage points, the finding would be statistically significant at the 0.05 level but substantively trivial.

One final caution about testing the null hypothesis that $Y_i(0) = Y_i(1)$: this equality represents the sharp null hypothesis of *no treatment effect for any observation*, which should not be confused with the null hypothesis of *no average effect*. Suppose the treatment effect were 5 for half of the observations and -5 for the other half. In this scenario, the average effect would be zero, but the sharp null hypothesis would be false. Both null hypotheses have the testable implication that the average outcome in the treatment group will be similar to the average outcome in the control group, but the sharp null hypothesis has the further testable implication that the two groups will have similarly shaped distributions.

In order to conduct a more exacting evaluation of the sharp null hypothesis, we may consider statistics other than the estimated ATE. Because the sharp null hypothesis provides us with the complete schedule of potential outcomes, we can simulate the sampling distribution of any test statistic. For example, we could compare the variances in the treatment and control groups. If treatment effects vary and are uncorrelated with or positively correlated with $Y_i(0)$, the observed variance of $Y_i(1)$ will tend to be larger than the variance of $Y_i(0)$. The sampling distribution under the sharp null hypothesis indicates the probability of observing a difference in variances as large as what we in fact observe in our sample. We simply calculate all possible differences and compute the p-value of the difference we obtained from our experiment. We return to these and other diagnostic tests of heterogeneous treatment effects in Chapter 9.

3.5 Confidence Intervals

When faced with a decision, policy makers may turn to an experiment for guidance about the average effect of an intervention. They typically want to know how big the average treatment effect is, not whether the effect is statistically distinguishable from zero. Testing the null hypothesis is beside the point; their principal objective is to use the results to form a guess about the value of the ATE.

Interval estimation is a statistical procedure that uses the data to generate a probability statement about the range of values within which a parameter is located. For example, suppose a researcher conducts an experiment in an attempt to learn the value of the ATE. Using a formula that will be described momentarily, the researcher generates a 95% confidence interval that ranges from (2, 10). This interval has a 0.95 probability of bracketing the true ATE. In other words, if we imagine a series of hypothetical replications of this experiment under identical conditions, 95 out of 100 random assignments will generate intervals that bracket the true ATE. When interpreting confidence intervals, remember that the location of the interval varies from one experiment to the next, while the true ATE remains constant.

By convention, social scientists construct 95% confidence intervals, but forming a 95% confidence interval based on the results of a single experiment requires some guesswork. Recall that a single experiment reveals the $Y_i(1)$ outcomes for the treated subjects and $Y_i(0)$ outcomes for the untreated subjects. We do not observe the full schedule of potential outcomes. Without the full schedule of potential outcomes, we cannot simulate the sampling distribution of the estimated ATE. The best we can do is approximate the full schedule of potential outcomes by making an educated guess about the unobserved $Y_i(0)$ outcomes for the treated subjects and $Y_i(1)$ outcomes for the untreated subjects.

The most straightforward method for filling in missing potential outcomes is to assume that the treatment effect τ_i is the same for all subjects. For subjects in the control condition, missing $Y_i(1)$ values are imputed by adding the estimated ATE to the observed values of $Y_i(0)$. Similarly, for subjects in the treatment condition, missing $Y_i(0)$ values are imputed by subtracting the estimated ATE from the observed values of $Y_i(1)$. This approach yields a complete schedule of potential outcomes, which we may then use to simulate all possible random allocations. In order to form a 95% confidence interval, we list the estimated ATE from each random allocation in ascending order. The estimate at the 2.5th percentile marks the bottom of the interval, and the estimate at the 97.5th percentile marks the top of the interval.[10]

When the treatment and control groups contain an equal number of subjects ($m = N - m$), this method of forming confidence intervals will tend to be conservative, in the sense that the interval typically will be wider than the interval we

10 A more complex and computationally intensive approach is to "invert" the hypothesis test (Rosenbaum 2002). This method involves hypothesizing an ATE called ATE*, subtracting it from the observed outcomes in the treatment group to approximate $Y_i(0)$, and testing the null hypothesis that the average (adjusted) $Y_i(0)$ in the treatment group is equal to the average $Y_i(0)$ in the control group. Progressively larger values of ATE* are tried until one locates a value that has a p-value of 0.025 and another that has a p-value of 0.975; these values mark the range of the 95% confidence interval. This method tends to be more accurate than the simpler method we describe, but both tend to produce similar results, especially in large samples.

would obtain if we actually observed $Y_i(0)$ and $Y_i(1)$ for all subjects.[11] When the number of subjects in treatment and control differ, this method may produce confidence intervals that are too wide or too narrow, depending on whether $\text{Var}(Y_i(0))$ is larger or smaller than $\text{Var}(Y_i(1))$. The method is conservative when the control group is larger and $\text{Var}(Y_i(0)) > \text{Var}(Y_i(1))$ or when the treatment group is larger and $\text{Var}(Y_i(1)) > \text{Var}(Y_i(0))$. When working with treatment and control groups of markedly different sizes, one should use this method for estimating confidence intervals with caution if outcomes in the smaller group are substantially more variable. A rule of thumb is that caution is required if the smaller group is less than half the size of the larger group, and the smaller group's standard deviation is at least twice as large as the larger group's standard deviation. One reason to assign equal numbers of subjects to treatment and control is that this type of "balanced" experimental design facilitates interval estimation by eliminating these complications.

To illustrate interval estimation, we analyze data from Clingingsmith, Khwaja, and Kremer's study of Pakistani Muslims who participated in a lottery to obtain a visa for the pilgrimage to Mecca.[12] By comparing lottery winners to lottery losers, the authors are able to estimate the effects of the pilgrimage[13] on the social, religious, and political views of the participants. Here, we consider the effect of winning the visa lottery on attitudes toward people from other countries. Winners and losers were asked to rate the Saudi, Indonesian, Turkish, African, European, and Chinese people on a five-point scale ranging from very negative (-2) to very positive ($+2$). Adding the responses to all six items creates an index ranging from -12 to $+12$.

The distribution of responses in the treatment group ($N = 510$) and control group ($N = 448$) is presented in Table 3.2. The average in the treatment group is 2.34, as compared to 1.87 in the control group. The estimated ATE is therefore 0.47. The estimated standard deviations in the treatment and control groups are 2.63 and 2.41, respectively. In this study, the treatment and control conditions contain similar numbers of subjects, and the smaller of the two groups does not have more vari-

11 In small samples, the distance between the estimated ATE and the top (or bottom) of the estimated confidence interval should be widened by a factor of $\sqrt{(N-1)/(N-2)}$ to adjust for the fact that the ATE used to fill in the full schedule of potential outcomes is itself estimated from the data. For example, suppose a study of ten subjects were to produce an estimated interval ranging from 8 to 12. This correction would widen the interval by a factor of 1.061, resulting in an adjusted interval that ranges from 7.88 to 12.12. See Samii and Aronow 2011. As N increases, this correction becomes negligible.

12 Clingingsmith, Khwaja, and Kremer 2009.

13 Our description sidesteps the complications that arise due to the fact that 14% of the lottery losers nonetheless made the pilgrimage and 1% of the lottery winners failed to make the pilgrimage. Anticipating the discussion of noncompliance in Chapters 5 and 6, we estimate the intent-to-treat effect: the effect of winning the lottery. In order to simplify the presentation, we sidestep the problem of clustering (multiple individuals applying for the lottery together) by randomly selecting one person from each cluster. For ease of presentation, we also ignore the slight differences in treatment probabilities across blocks defined by the size and location of the parties applying for a visa. Controlling for blocks has negligible effects on the results.

TABLE 3.2

Pakistani Muslims' ratings of peoples from foreign countries by success in the visa lottery

Ratings of people from other countries	Distribution of responses	
	Control (%)	Treatment (%)
−12	0.00	0.20
−9	0.22	0.00
−8	0.00	0.20
−6	0.45	0.20
−5	0.00	0.20
−4	0.45	0.59
−3	0.00	0.20
−2	1.12	0.98
−1	1.56	2.75
0	27.23	18.63
1	18.30	13.14
2	24.33	25.29
3	8.48	10.98
4	5.80	9.61
5	3.35	3.92
6	3.79	7.25
7	2.23	2.55
8	0.89	1.37
9	0.22	0.78
10	0.45	0.00
11	0.67	0.20
12	0.45	0.98
Total	100	100
N	(448)	(510)

Source: Clingingsmith, Khwaja, and Kremer 2009.

able outcomes, so our method for computing confidence intervals is expected to be reasonably accurate. In order to estimate our 95% interval, we must form a complete schedule of potential outcomes. We add 0.47 to the observed $Y_i(0)$ outcomes in the control group in order to approximate the control group's unobserved $Y_i(1)$ values; we subtract 0.47 from the treatment group's observed $Y_i(1)$ outcomes in order to approximate the treatment group's unobserved $Y_i(0)$ values. Simulating 100,000 random allocations using this schedule of potential outcomes and sorting the estimated ATEs in ascending order, we find that the 2,500th estimate is 0.16 and the 97,501st estimate is 0.79, so the 95% interval is [0.16, 0.79].

The statistical interpretation of this interval is as follows: over hypothetical replications of this experiment, intervals created in this manner have a 95% chance of bracketing the true ATE. So without other information about the true ATE, we conclude that there is a 95% probability that the interval from 0.16 to 0.79 includes the true ATE.

Substantively, we infer that winning the visa lottery led to an increase in positive feelings toward people from foreign countries. Unfortunately, given the data at hand, there is no easy way to translate the estimated ATE of 0.47 into other metrics that have more tangible meaning in terms of societal outcomes or individual behavior. For example, we do not know how positive responses on this scale translate into cooperative diplomatic relations, increased international trade, or friendly behavior toward visitors from these countries. And because the outcome measure is specific to this study, we are unable to compare the 0.47 effect of this intervention to the effect of other interventions. This gap in our understanding sets the stage for further research. Now that this visa lottery has revealed a causal effect, the next steps are to conduct further visa lottery studies using other outcome measures and to measure the effects of other types of interventions using this survey metric.

When presented with experimental results that are not scaled in relation to interpretable outcome metrics or the effects of other interventions, researchers often fall back on the calculation of *standardized effect size*. This approach compares the estimated effect to the naturally occurring degree of variation in outcomes by dividing the estimated ATE by the standard deviation in the control group.[14] Using this formula, the apparent ATE in this study moves people by about one-fifth of a standard deviation. Again, we confront a problem of interpretation. Is a 0.2 movement in standard deviation big or small? Researchers sometimes invoke rules of thumb: effects of less than 0.3 are considered small, between 0.3 and 0.8 are considered medium, and above 0.8 are considered large.[15] One should be cautious about applying these standards for three reasons. First, the standard deviation is a sample-specific statistic; if one's experimental subjects happen to be Pakistanis who share a similar view of foreigners prior to the intervention, the standard deviation will be small, and the standardized effect will seem large. Second, the standard deviation tends to increase when outcomes are measured with error, as is often the case with survey measures of attitudes. Third, even small standardized effects can be substantively important if they alter a hard-to-move dependent variable. The standard deviation of men's height

14 This standardized statistic is known as Glass's Δ, from Glass 1976.

15 Cohen 1988.

is about 2.8 inches, but a dietary supplement that causes a half-inch increase in height would be heralded as remarkable. In much the same way, an intervention that produces a change in attitudes toward foreigners is noteworthy given the difficulty of changing attitudes in this domain.

3.6 Sampling Distributions for Experiments That Use Block or Cluster Random Assignment

The concepts and estimation techniques presented in previous sections may be adapted to experiments in which subjects are assigned randomly but in ways that depart from simple or complete random assignment. In this section, we discuss two such classes of experimental designs: block randomization and cluster randomization.

3.6.1 Block Random Assignment

Block random assignment refers to a procedure whereby subjects are partitioned into subgroups (called blocks or strata), and complete random assignment occurs within each block. For example, suppose we have 20 subjects in our experiment, 10 men and 10 women. Suppose our experimental design calls for 10 subjects to be placed into the treatment condition. If we were to use complete random assignment, chances are that we would end up with unequal numbers of men and women in the treatment group. Block randomization, on the other hand, ensures equal numbers of men and women will be assigned to each experimental condition. First, we partition the subject pool into men and women. From the pool of male subjects, we randomly assign five into the treatment group; from the pool of female subjects, we randomly assign five into the treatment group. In effect, block randomization creates a series of miniature experiments, one per block.

Block randomized designs are used to address a variety of practical and statistical concerns. Sometimes program requirements dictate how many subjects of each type to place in the treatment group. Imagine, for example, that a summer reading program aimed at elementary school students seeks to evaluate its impact on school performance and retention during the following academic year. The school is able to admit only a small fraction of those who apply, and school administrators worry that if too many children with low levels of preparedness are admitted to the program, teachers will find it difficult to manage their classes effectively. These administrators insist that 60% of the children admitted to the program pass an initial test of basic skills. The way to address this concern is by blocking on initial test scores and allocating students within each block so that 60% of the students who are randomly admitted to the program have passed the basic skills test. For example, suppose the school

BOX 3.8

Number of Possible Assignments under Complete Random Assignment and Blocked Random Assignment

Let $0 < m < N$. Of N observations, m are placed into the treatment group, and $N - m$ are placed into the control group. The number of possible randomizations under complete random assignment is

$$\frac{N!}{m!(N - m)!}.$$

For example, the number of randomizations under complete random assignment when $N = 20$ and $m = 10$ is

$$\frac{20!}{10!10!} = 184{,}756.$$

In order to calculate the number of possible random assignments under a blocked design with B blocks, calculate the number of random allocations in each block: $r_1, r_2, \ldots, r_B$. The total number of random assignments is $r_1 \times r_2, \times \ldots \times r_B$. For example, when we randomly assign half of the 10 men to treatment and half of the 10 women to treatment, there are

$$\left(\frac{10!}{5!5!}\right)\left(\frac{10!}{5!5!}\right) = 63{,}504$$

possible random allocations.

can admit 50 of the 100 applicants. Forty of the applicants failed the initial test, and 60 passed. The researcher could create two blocks: one block of students who passed the basic skills test and another block of students who failed. Each block is sorted in random order, and the researcher selects the first 20 students in the block containing those who failed the basic skills test and the first 30 students in the block containing students who passed the test. This procedure ensures that the 60% requirement is satisfied. This design approach comes in handy when resource constraints prevent researchers from treating more than a certain number of subjects from certain regions or when concerns about fairness dictate that treatments be apportioned equally across demographic groups.

Block randomization also addresses two important statistical concerns. First, blocking helps reduce sampling variability. Sometimes the researcher is able to partition the subjects into blocks such that the subjects in each block have similar potential outcomes. For example, students who fail the basic skills test presumably share similar potential outcomes; the same goes for students who pass the basic skills test.

By randomizing within each block, the researcher eliminates the possibility of rogue randomizations that, by chance, place all of the students who fail the basic skills test into the treatment group. Under simple or complete random assignment, these outlandish assignments rarely occur; under block random assignment, they are ruled out entirely. Second, blocking ensures that certain subgroups are available for separate analysis. When analyzing a study involving 10 men and 10 women, a researcher might be interested in comparing the ATE among men to the ATE among women. But what if complete random assignment puts 9 of the 10 men into the treatment group and 9 of the 10 women in the control group? In that case, the treatment effects among men and among women would both be estimated very imprecisely. Blocked randomization guarantees that a specified proportion of a subgroup will be assigned to treatment.

In order to illustrate how blocking works, let's consider a stylized example inspired by Olken's study of corruption in Indonesia.[16] The subjects in this experiment are public works projects, and the treatment is heightened financial oversight by government officials. Outcomes are measured in terms of the amount of money that is unaccounted for (and presumably stolen) when the books are closed on the project. For purposes of illustration, we present in Table 3.3 the complete schedule of potential outcomes for 14 projects, 8 of which are in Region A and 6 in Region B. Because of resource constraints, each region has the capacity to audit only two of its projects. In our example, the ATE is −3 in Region A and −5 in Region B. For both regions combined, the ATE is $(-3)(8/14) + (-5)(6/14) = -3.9$. In general, the relationship between the overall ATE and the ATE within each block j is:

$$\text{ATE} = \sum_{j=1}^{J} \frac{N_j}{N} \text{ATE}_j, \tag{3.10}$$

where J is the number of blocks, the blocks are indexed by j, and the weight N_j/N denotes the share of all subjects who belong to block j.

Before studying the statistical precision of the blocked design, it is useful to consider as a point of comparison the precision of a design that uses complete random assignment. Suppose treatments had been assigned to any 4 of the 14 projects through complete random assignment. Equation (3.4) indicates that the true standard error would have been:

$$SE(\widehat{\text{ATE}}) = \sqrt{\frac{1}{14-1}\left\{\frac{(4)(40.41)}{10} + \frac{(10)(32.49)}{4} + (2)(31.03)\right\}} = 3.50. \tag{3.11}$$

In order to calculate the standard error from a blocked design, we must first calculate the standard error within each block. In our stylized example, the projects within each region have similar potential outcomes. As the variances of $Y_i(0)$ and $Y_i(1)$ are

16 Olken 2007.

TABLE 3.3

Schedule of potential outcomes for public works projects when audited (Y(1)) and not audited (Y(0))

		All subjects		Block A subjects		Block B subjects	
Village	Block	Y(0)	Y(1)	Y(0)	Y(1)	Y(0)	Y(1)
1	A	0	0	0	0		
2	A	1	0	1	0		
3	A	2	1	2	1		
4	A	4	2	4	2		
5	A	4	0	4	0		
6	A	6	0	6	0		
7	A	6	2	6	2		
8	A	9	3	9	3		
9	B	14	12			14	12
10	B	15	9			15	9
11	B	16	8			16	8
12	B	16	15			16	15
13	B	17	5			17	5
14	B	18	17			18	17
	Mean	9.14	5.29	4.00	1.00	16.0	11.0
	Variance	40.41	32.49	7.75	1.25	1.67	17.0
	Cov(Y(0), Y(1))	31.03		2.13		1.00	

much lower within each region than they are when the two regions are combined, the standard error drops markedly when we analyze each region separately. For Region A, the standard error is 1.23; for Region B, it is 2.71. The remaining task is to use these block-specific standard errors to assess the uncertainty of the estimated ATE for subjects in both regions combined. The formula turns out to be straightforward and easily extends to any number of blocks.[17] For two blocks, the standard error is:

$$SE(\widehat{\text{ATE}}) = \sqrt{(SE_1)^2\left(\frac{N_1}{N}\right)^2 + (SE_2)^2\left(\frac{N_2}{N}\right)^2}, \quad (3.12)$$

17 The formula for any number of blocks is

$$SE(\widehat{\text{ATE}}) = \sqrt{\sum_1^J \left(\frac{N_j}{N}\right)^2 SE^2(\text{ATE}_j)}.$$

This formula follows from the general rule about the variance of a sum of independent random variables: $\text{Var}(\alpha A + \beta B) = \alpha^2\text{Var}(A) + \beta^2\text{Var}(B)$.

where SE_j refers to the standard error of the estimated ATE in block j, and N_j refers to the number of observations in block j. Filling in the numbers from our example gives a standard error of:

$$SE(\widehat{\text{ATE}}) = \sqrt{(1.23)^2\left(\frac{8}{14}\right)^2 + (2.71)^2\left(\frac{6}{14}\right)^2} = 1.36. \quad (3.13)$$

The example illustrates the potential benefits of blocking. By making a small design change, we greatly improve the precision with which we estimate the ATE. The standard error plummets from 3.50 to 1.36. The stark difference in sampling distributions is illustrated in Figure 3.1. Under complete random assignment, $141/1001 = 14.1\%$ of the estimated ATEs are greater than zero, which means there is a 14.1% chance that the experiment will indicate that audits are ineffective or actually exacerbate pilfering even though (as we know from Table 3.3) the treatment is effective. Under blocked assignment, just $1/420 = 0.2\%$ of the estimated ATEs are greater than zero.

Let's now consider how one would go about analyzing a block randomized experiment, such as the one reported in Table 3.4, which shows the observed outcomes from a single experiment based on the schedule of potential outcomes from Table 3.3. Estimating the overall ATE is straightforward: first estimate the ATE within

FIGURE 3.1

Sampling distribution under complete randomization (above); sampling distribution under blocked randomization (below)

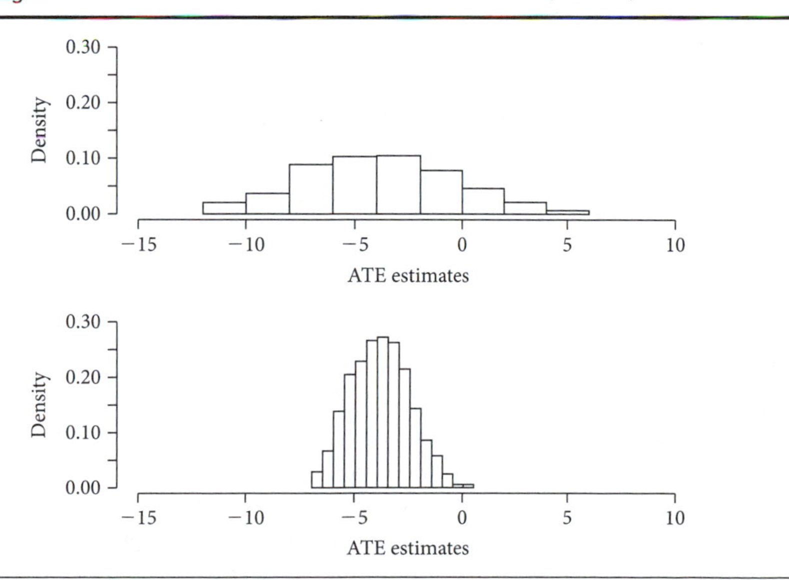

TABLE 3.4
Observed outcomes in a block randomized experiment

All subjects

Block	Y(0)	Y(1)
A	0	?
A	1	?
A	?	1
A	4	?
A	4	?
A	6	?
A	6	?
A	?	3
B	14	?
B	?	9
B	16	?
B	16	?
B	17	?
B	?	17

each block (denoted $\widehat{ATE}_j$ for each block j) and then compute a weighted average based on the ratio of N_j to N:

$$\widehat{ATE} = (\widehat{ATE}_1)\left(\frac{N_1}{N}\right) + (\widehat{ATE}_2)\left(\frac{N_2}{N}\right) = (-1.5)\left(\frac{8}{14}\right) + (-2.75)\left(\frac{6}{14}\right)$$
$$= -2.04. \qquad (3.14)$$

Notice that the weights (N_1/N) and (N_2/N) sum to 1.0.

It should be stressed that this estimator is *not* equivalent in general to a comparison of average outcomes among all subjects in treatment and control. When the probability of being assigned to the treatment group varies by block, as it does in this example, comparing means for all subjects generates a biased estimate of the ATE. In Table 3.4, the probability of being assigned to the treatment group is higher in Region B than in Region A, and Region B tends to have higher average potential outcomes than Region A. Even if there were no treatment effect, if we pool together the two regions, potential outcomes will tend to be higher in the treatment group than in the control group. Unless the probability of assignment to the treatment group is identical for every block, pooling observations across blocks will produce biased estimates of the overall ATE.

In order to estimate the standard error of the $\widehat{\text{ATE}}$, estimate the standard error of each $\widehat{\text{ATE}}_j$ using equation (3.6), and insert the block-level estimates of the standard errors into equation (3.12) above:

$$\widehat{SE}(\widehat{\text{ATE}}) = \sqrt{(1.43)^2\left(\frac{8}{14}\right)^2 + (4.05)^2\left(\frac{6}{14}\right)^2} = 1.92. \tag{3.15}$$

Hypothesis tests and confidence intervals extend directly to blocked designs. When simulating the sampling distribution, remember that randomization takes place within each block. We begin by testing the sharp null hypothesis of no treatment effect. In this example, there are just 420 possible random assignments, 46 of which are less than or equal to the observed estimate of −2.04, implying a one-tailed p-value of 0.11. Confidence intervals may be approximated by simulating the full schedule of potential outcomes within each block by adding the $\widehat{\text{ATE}}j$ to the control group's outcomes and subtracting the $\widehat{\text{ATE}}j$ from the treatment group's outcomes. The 95% interval when treatment effects are assumed to be constant extends from −4.82 to 0.62. One might wonder about the accuracy of this method of estimating the 95% confidence intervals, given that we know from the potential outcomes in Table 3.3 that the constant effects assumption is false. It turns out that applying this estimation procedure to each of the 420 possible random assignments produces an estimated confidence interval that brackets the true ATE in 94.8% of the random allocations. In this example, the estimated 95% confidence intervals do in fact cover the ATE with probability of approximately 0.95.

3.6.1.1 MATCHED PAIR DESIGN

Notice that in order to calculate $\widehat{SE}(\widehat{\text{ATE}}_j)$ for each block, every block must contain at least two observations in treatment and two observations in control. This requirement will not be met by *matched pair designs*, experiments in which every block contains just two subjects, one of which is assigned to treatment. Fortunately, analysis of matched pair experiments is straightforward. The ATE is estimated by subtracting the control outcome from the treatment outcome in each block and averaging over all blocks. (Because the probability of assignment to the treatment group is 0.5 for all blocks, one obtains identical estimates by subtracting the average outcome among the $N/2$ subjects in the control condition from the average outcome among the $N/2$ subjects in the treatment condition). These block-level differences are then used to estimate the standard error:

$$\widehat{SE}(\widehat{\text{ATE}}) = \sqrt{\frac{1}{\frac{N}{2}\left(\frac{N}{2}-1\right)}\sum_{j=1}^{J}(\widehat{\text{ATE}}_j - \widehat{\text{ATE}})^2}$$

$$= \sqrt{\frac{1}{J(J-1)}\sum_{j=1}^{J}(\widehat{\text{ATE}}_j - \widehat{\text{ATE}})^2}. \tag{3.16}$$

The analysis of matched pairs is readily integrated into experiments that contain blocks of varying size. For example, suppose that, as presented in Table 3.5, assignment in the public works audit experiment had taken place within two matched pairs (Blocks B and C) and one larger block containing 10 subjects (Block A). In each matched pair, one subject is assigned to treatment.

The matched pairs generate an $\widehat{\text{ATE}}_j$ of −3 and 0, respectively. The larger block generates an $\widehat{\text{ATE}}_j$ of −2.6. The estimate of the overall ATE is:

$$\widehat{\text{ATE}} = \left(\frac{2}{14}\right)(-3) + \left(\frac{2}{14}\right)(0) + \left(\frac{10}{14}\right)(-2.6) = -2.3. \tag{3.17}$$

In order to calculate the standard error, we first calculate the standard error for the matched pair part of the experiment using equation (3.16):

$$\widehat{SE}(\widehat{\text{ATE}}) = \sqrt{\frac{1}{\frac{4}{2}\left(\frac{4}{2}-1\right)}((-3-(-1.5))^2 + (0-(-1.5))^2)} = 1.5, \tag{3.18}$$

TABLE 3.5

Observed outcomes in a block randomized experiment with blocks of varying size

All subjects		
Block	**Y(0)**	**Y(1)**
A	0	?
A	1	?
A	?	1
A	4	?
A	4	?
A	6	?
B	6	?
B	?	3
A	14	?
A	?	9
A	16	?
A	16	?
C	17	?
C	?	17

and then combine the standard error for the matched pair part of the design with the standard errors from the other block using equation (3.12):

$$\widehat{SE}(\widehat{ATE}) = \sqrt{(1.5)^2\left(\frac{4}{14}\right)^2 + (4.64)^2\left(\frac{10}{14}\right)^2} = 3.3. \tag{3.19}$$

3.6.1.2 SUMMARY OF THE ADVANTAGES AND DISADVANTAGES OF BLOCKING

Blocking provides a way to address two classes of design objectives. The first arises when practical or ethical imperatives dictate the number of treatments that may be assigned to certain groups of subjects. Blocking allows the researcher to build a randomized experiment around these constraints. The second set of objectives has to do with statistical precision. When blocks are formed by grouping subjects with similar potential outcomes, substantial improvements in precision are possible, particularly in small samples. Chapter 4 discusses how one might form blocks based on background information available at the time of randomization.

Does blocking have a downside? Although it is possible to construct perverse examples in which blocking harms precision, blocking rarely has negative consequences in practice.[18] Even blocking subjects into random subsets (which does nothing to group subjects according to their potential outcomes) is an experimental design that is no worse, in terms of the precision with which the ATE is calculated, than complete random assignment. In practice, the biggest downside associated with blocking is the risk that the data will be misanalyzed. The ATE must be estimated block by block when the probability of treatment assignment varies by block, and care must be taken to conduct hypothesis tests in a manner that follows the exact procedure by which the randomization was carried out.

Sometimes block randomization is not feasible. It is not uncommon for field experiments to be conducted under severe time constraints, and background information necessary to form blocks may be unavailable at the time of random allocation. Similarly, the implementation of randomization within blocks may exceed the technical capacity of those implementing random allocation in field settings. Failure to block randomize should not be regarded as a serious design flaw, especially in studies where the treatment and control groups contain more than 100 subjects.[19] As we will see in the next chapter, if you do not use blocking, you can do almost as well in terms of precision through statistical adjustments.

---

18 Imai 2008.

19 Rosenberger and Lachin 2002.

3.6.2 Cluster Random Assignment

To this point, we have considered only experiments in which each of the subjects is *individually* assigned to the treatment or control group. To visualize the difference between individual- and cluster-level assignment, imagine an experiment on the persuasive influence of TV commercials. Suppose a credit card company has purchased records for 200,000 people in each of five cities. A researcher proposes to randomly assign commercials to two of the five cities and assess purchasing behavior among the 1,000,000 people in the database. The N of individuals is 1,000,000, but there are only five clusters. If simple random assignment were used to allocate individuals to treatment or control, the sampling distribution would be very narrow, because chance differences between experimental groups would tend to balance out due to the very large number of subjects in each group. But under cluster assignment, all subjects in the same cluster are placed as a group into either the treatment or control condition; in effect, cluster assignment rules out all of the possible allocations in which individuals in the same cluster are assigned to different experimental conditions. Because all of the inhabitants of a city are assigned as a cluster to treatment or control, the attributes they share are placed into one of the experimental groups. If these cities' potential outcomes differ appreciably, random assignment of just five clusters will produce unbiased but imprecise estimates.

For practical reasons, cluster assignment is sometimes unavoidable. In the case of TV broadcasts, it is currently impossible to administer the treatment to specific individuals; instead, one purchases advertising that targets geographic clusters. The same goes for treatments that are administered by canvassers or election monitors, who may be able to travel only to a few general locations.

Sampling distributions may change markedly when clusters, rather than individuals, are assigned to experimental groups. For the moment, let's consider the simplest scenario: each of the clusters contains the same number of subjects. When clusters are the same size, the usual difference-in-means estimator remains unbiased. Cluster assignment, however, will generate more sampling variability than complete random assignment if subjects within the same cluster tend to share similar potential outcomes.

In order to illustrate the consequences of within-cluster similarity, Table 3.6 presents a schedule of potential outcomes for 12 classrooms located in schools that have varying levels of academic achievement. The average treatment effect in this example is 4 points on a standardized test. Each school contains three classrooms. Using a cluster design, the experimenter randomly assigns two of the four schools to the treatment group. The total number of possible random assignments is:

$$\frac{4!}{2!2!} = 6. \tag{3.20}$$

Differences-in-means based on cluster assignment yield unbiased estimates of the treatment effect: the average estimate is 4. To obtain the standard error, we calculate

TABLE 3.6

Hypothetical schedule of potential outcomes for 12 classrooms in four schools—high sampling variability

		Classroom-level potential outcomes		Cluster-level mean potential outcomes	
School	Classroom	$Y_i(0)$	$Y_i(1)$	$Y_j(0)$	$Y_j(1)$
A	A-1	0	4		
A	A-2	1	5	1	5
A	A-3	2	6		
B	B-1	2	6		
B	B-2	3	7	3	7
B	B-3	4	8		
C	C-1	3	7		
C	C-2	4	8	4	8
C	C-3	5	9		
D	D-1	7	11		
D	D-2	8	12	8	12
D	D-3	9	13		

the standard deviation across all possible random assignments and find it to be 2.9. Suppose, however, that our experiment had instead allocated the 12 classrooms using complete random assignment, half to treatment and half to control. The standard error would have been substantially smaller. The number of possible randomizations is

$$\frac{12!}{6!6!} = 924, \tag{3.21}$$

and across them the standard error is 1.6. Clustering, in other words, led to an 81% increase in the standard error, which is roughly the same increase that would have occurred if our sample size had declined from 12 to 4.

Why were the consequences of clustering so severe? The schools in the preceding example were quite different in terms of their classrooms' average potential outcomes; meanwhile, the classrooms within each school were relatively similar. Let's see what happens when we rearrange the example so that schools are more similar in terms of average classroom performance. Table 3.7 shows another example with four schools that have three classrooms apiece. The classrooms are the same as in Table 3.6, but they are situated in different schools. Across the six possible randomizations of a cluster assignment, the average estimated treatment effect is 4 with a standard error of 0.57, which is actually lower than under complete random assignment.

TABLE 3.7

Hypothetical schedule of potential outcomes for 12 classrooms in four schools—low sampling variability

		Classroom-level potential outcomes		Cluster-level mean potential outcomes	
School	Classroom	$Y_i(0)$	$Y_i(1)$	$Y_j(0)$	$Y_j(1)$
A	A-1	0	4		
A	A-2	3	7	4	8
A	A-3	9	13		
B	B-1	2	6		
B	B-2	3	7	4	8
B	B-3	7	11		
C	C-1	1	5		
C	C-2	4	8	3.3	7.3
C	C-3	5	9		
D	D-1	4	8		
D	D-2	8	12	4.7	8.7
D	D-3	2	6		

The bottom line is that the penalty associated with clustering depends on the variability of the cluster-level means. This point is driven home by the formula for the true standard error assuming a clustered design with equally sized clusters. Equation (3.22) expresses this standard error in a manner that parallels equation (3.4):

$$SE(\widehat{\text{ATE}}) = \sqrt{\frac{1}{k-1}\left\{\frac{m\text{Var}(\overline{Y}_j(0))}{N-m} + \frac{(N-m)\text{Var}(\overline{Y}_j(1))}{m} + 2\text{Cov}(\overline{Y}_j(0),\overline{Y}_j(1))\right\}}, \quad (3.22)$$

where $\overline{Y}_j(0)$ is the average untreated potential outcome in the jth cluster, $\overline{Y}_j(1)$ is the average treated potential outcome in the jth cluster, and k is the number of clusters. If one could somehow create clusters that had nearly identical average potential outcomes, $\text{Var}(\overline{Y}_j(0))$ and $\text{Var}(\overline{Y}_j(1))$ would be small, and the ATE would be estimated with great precision. Unfortunately, researchers are usually stuck with highly variable cluster means and therefore large standard errors. Cluster means tend to vary because clusters tend to be formed based on geography, institutions, age groups, and so forth, and subjects within the same cluster tend to have similar potential outcomes. Comparing equations (3.4) and (3.22) reveals why cluster randomized designs often have lower power than completely randomized designs. Unlike equation (3.4), which divides the elements inside the braces by $N - 1$, equation (3.22) divides by $k - 1$.

In terms of data analysis, the techniques for testing hypotheses and forming confidence intervals must be adjusted when analyzing clustered experiments. Adapting hypothesis testing to clustered experiments is straightforward. Under the sharp null hypothesis, treated and untreated potential values are identical, so the researcher observes the full schedule of potential outcomes. In order to test the sharp null hypothesis, simulate all possible assignments of *clusters* to treatment. The sampling distribution of estimated ATEs under the sharp null hypothesis will usually be much wider under clustered assignment than under complete random assignment. Confidence intervals may be obtained by creating a complete schedule of potential outcomes under the assumption of constant treatment effects. Again, simulate all possible random assignments of clusters in order to obtain the sampling distribution.[20] To obtain standard errors that tend to err on the conservative side, use the formula

$$\widehat{SE}(\widehat{\text{ATE}}) = \sqrt{\frac{N\widehat{\text{Var}}(\overline{Y}_j(0))}{k(N-m)} + \frac{N\widehat{\text{Var}}(\overline{Y}_j(1))}{km}}, \tag{3.23}$$

where k is the number of clusters. When attempting to improve the precision of a cluster randomized design, researchers should look for ways to increase the number of clusters; increasing the number of subjects per cluster typically has little effect on the standard error because it does little to reduce the variance of the cluster-level potential outcomes.

To this point, we have considered the relatively simple case in which clusters contain the same number of subjects. Analysis becomes more complicated when clusters differ in size. If cluster size covaries with potential outcomes, the usual difference-in-means estimator is biased.[21] This bias disappears as the number of clusters increases. Because the full set of potential outcomes is unobserved, the threat of bias is something that can only be assessed indirectly. A straightforward diagnostic approach is to examine whether the estimated ATE appears to change when we restrict attention to large or small clusters.[22] If one suspects bias, an alternative estimator focuses on the difference in total outcomes:

$$\widehat{\text{ATE}} = \frac{k_C+k_T}{N}\left(\frac{\sum Y_i(1)\,|\,d_i=1}{k_T} - \frac{\sum Y_i(0)\,|\,d_i=0}{k_C}\right), \tag{3.24}$$

20 Because estimation of the ATE consumes a degree of freedom, one applies a correction when forming the upper (and lower) bound of the confidence interval. The width of the interval expands by $\sqrt{(k-1)/(k-2)}$, where k is the number of clusters.

21 Middleton and Aronow 2011. The intuition behind this bias is that random assignment of unequal size clusters allows m to vary from one random allocation to the next, which creates bias when cluster size is related to potential outcomes. See exercise 12.

22 Recall that the difference-in-means estimator is unbiased when cluster size is constant, so partitioning the data by cluster size produces approximately unbiased estimates. The drawback to this approach, however, is that subsets of a dataset may lack power to detect a relationship between cluster size and potential outcomes.

where k_C is the number of clusters assigned to control and k_T is the number of clusters assigned to treatment. Unlike the difference-in-means estimator, the estimator in equation (3.24) divides by quantites (k_T, k_C, and N) that are unaffected by which units are assigned to treatment. The estimator in equation (3.24) is equivalent to difference-in-means when clusters are all the same size and is unbiased for any number of clusters or arrangement of cluster sizes; its primary drawback is that it produces estimates with more sampling variability than the conventional difference-in-means estimator because it ignores the number of individual subjects in each cluster.[23] If there are a few clusters with large numbers of subjects and distinctive potential outcomes, estimates may vary markedly depending on the experimental condition to which they are assigned.

In order to illustrate these statistical techniques, let's examine a cluster randomized voter mobilization experiment conducted in Kansas City prior to its 2003 municipal elections.[24] A group called ACORN targeted 28 low-income precincts in an effort to mobilize voters on behalf of a ballot measure designed to fund municipal bus service. ACORN wanted to work within selected precincts in order to make it easier to train and supervise its canvassers. Of the 28 precincts in the sample, 14 were randomly allocated to the treatment group, and ACORN made repeated attempts to canvass and call voters on its target list in those precincts. The 28 precincts contain a total of 9,712 voters, and the number of targeted voters per precinct ranges from 31 to 655; the marked difference in cluster size leaves open the possibility for bias if precincts with large numbers of potential ACORN sympathizers have different potential outcomes from precincts with relatively few.

We begin by estimating the effect using a difference-in-means approach. Turnout was 33.5% in the treatment group ($N = 4{,}933$) and 29.1% in the control group ($N = 4{,}779$), implying an ATE of 4.4 percentage points. In order to test the sharp null hypothesis, we generate a random subset of 100,000 of the 40,116,600 possible random assignments of the 28 precincts and find that 5.0% of the randomizations yield estimates of the ATE that are greater than or equal to 4.4. Thus, the estimated one-tailed p-value is 0.05. We estimate a 95% confidence interval around the estimate of 4.4 by using the constant treatment effect assumption to fill in the schedule of potential outcomes: (−0.5, 9.3). At the low end of this interval, the campaign failed to increase turnout; at the high end, it had a profound effect on turnout. This is the price one pays for a clustered design. The interval is approximately 2.7 times wider than it would have been had researchers randomly assigned individual voters rather than precincts. Fortunately, as we will see in the next chapter, this confidence interval narrows considerably when we control for the voter turnout rates of these subjects in previous elections.

---

23 See Middleton and Aronow 2011. The standard error for the difference-in-totals estimator tends to be large because the variance of a total (as opposed to an average) tends to be large.

24 Arceneaux 2005.

Next, we repeat the analysis using a difference-in-totals approach. Applying equation (3.24) to these data yields an estimated ATE of:

$$\frac{14+14}{9712}\left(\frac{1654}{14}-\frac{1392}{14}\right)=0.054, \tag{3.25}$$

or 5.4 percentage points, which is slightly larger than the estimate obtained using a difference-in-means. This estimate, however, is subject to greater sampling variability. Over a random subset of 100,000 of the 40,116,600 possible random assignments under the sharp null hypothesis, 20% of the estimates are greater than 5.4 percentage points, implying a p-value of 0.20. Assuming a constant treatment effect of 5.4 percentage points, we obtain a 95% confidence interval that ranges from −6 to 17. The loss of precision associated with the difference-in-totals estimator conveys a useful design lesson: when conducting a cluster randomized experiment, try to work with clusters of similar size. If that is not possible, block on cluster size so that the conventional difference-in-means estimator may be used within blocks without risk of bias.

SUMMARY

Sampling variability is a constant concern for the experimental researcher. During the design phase, decisions about how to select subjects, allocate them to experimental conditions, and measure outcomes may have important repercussions for the precision of the experimental estimates. This chapter has considered some of the basic principles of experimental design, and later chapters will build on this framework as we grapple with the complications that arise due to noncompliance and attrition.

The standard error is a measure of statistical uncertainty. A close examination of standard error formulas provides useful design insights. Among design ideas discussed in this chapter are the potential advantages of assigning similar numbers of subjects to treatment and control conditions and of eliminating extraneous sources of variability in outcomes. Comparing the standard errors of different unbiased estimators is another source of insight. Blocked random assignment can improve precision when the blocking variables strongly predict outcomes; cluster assignment may diminish precision when units with similar potential outcomes wind up in the same cluster.

Once the data are collected, sampling variability represents one of the foremost concerns in the interpretation of results. Randomized experiments, in principle, generate empirical results that have two interpretations: either the treatment exerted a causal effect, or the result occurred due to sampling variability. Randomization inference is a valuable method for assessing the sampling variability of estimates. Under the sharp null hypothesis, the observed data reveal all of the potential outcomes, providing the researcher with access to the full sampling distribution. By simulating results from all possible randomizations, the researcher can examine the sampling distribution of any test statistic. Of special interest is the sampling distribution of the estimated average treatment effect and other test statistics that might indicate heterogeneous treatment effects. The latter topic is discussed in detail in Chapter 9.

The use of simulation to conduct hypothesis tests is a flexible approach that may be tailored to fit the details of any random allocation. For example, researchers may divide subjects into blocks and randomly allocate subjects within each block to treatment or control. Or researchers may combine subjects into clusters and allocate all of the members of a cluster to treatment or control. These subtleties of design sometimes profoundly affect the resulting sampling distribution, but the same basic hypothesis testing tools apply. Under the sharp null hypothesis of no treatment effect for any subject, the full schedule of potential outcomes is observed, allowing the researcher to simulate the sampling distribution by following the exact procedures used to assign subjects to experimental conditions.

The estimation of confidence intervals is more challenging because it relies on assumptions about each subject's treatment effect. The method described here invokes the assumption that treatment effects are constant, and that the ATE is equal to the estimated ATE. This assumption works well for experimental designs in which the same number of subjects are in treatment and control conditions. In most cases, the estimated 95% confidence intervals will be conservative in the sense that they have more than a 95% chance of bracketing the true ATE. When applied to designs where one experimental group is much larger than the other, the estimated confidence intervals may be too wide or too narrow depending on whether outcomes in the two groups have different variances. In ambiguous cases, researchers may wish to compare their estimated confidence intervals to the intervals generated by other estimation procedures.

SUGGESTED READINGS

Freedman, Pisani, and Purves (2007) offers a good discussion of the differences between observational and experimental research; see its appendix for a derivation of equation (3.4). See Imai (2008) on the sampling distribution of matched pair experiments. Murray (1998) and Boruch (2005) discuss cluster randomized experiments.

EXERCISES: CHAPTER 3

See http://isps.research.yale.edu/FEDAI for datasets and programs for conducting randomization inference.

1. Important concepts:
   (a) What is a standard error? What is the difference between a standard error and a standard deviation?

(b) How is randomization inference used to test the sharp null hypothesis of no effect for any subject?

(c) What is a 95% confidence interval?

(d) How does complete random assignment differ from block random assignment and cluster random assignment?

(e) Experiments that assign the same number of subjects to the treatment group and control group are said to have a "balanced design." What are some desirable statistical properties of balanced designs?

2. Rewrite equation (3.4) substituting for $Y_i(1)$ using the equation $Y_i(1) = Y_i(0) + \tau_i$. Assume that $N = 2m$, and interpret the implications of the resulting formula for experimental design.

3. Using the equation $Y_i(1) = Y_i(0) + \tau_i$, show that when we assume that treatment effects are the same for all subjects, $\text{Var}(Y_i(0)) = \text{Var}(Y_i(1))$ and the correlation between $Y_i(0)$ and $Y_i(1)$ is 1.0.

4. Consider the schedule of outcomes in the table below. If treatment A is administered, the potential outcome is $Y_i(A)$, and if treatment B is administered, the potential outcome is $Y_i(B)$. If no treatment is administered, the potential outcome is $Y_i(0)$. The treatment effects are defined as $Y_i(A) - Y_i(0)$ or $Y_i(B) - Y_i(0)$.

Subject	$Y_i(0)$	$Y_i(A)$	$Y_i(B)$
Miriam	1	2	3
Benjamin	2	3	3
Helen	3	4	3
Eva	4	5	3
Billie	5	6	3

Suppose a researcher plans to assign two observations to the control group and the remaining three observations to just one of the two treatment conditions. The researcher is unsure which treatment to use.

(a) Applying equation (3.4), determine which treatment, A or B, will generate a sampling distribution with a smaller standard error.

(b) What does the result in part (a) imply about the feasibility of studying interventions that attempt to close an existing "achievement gap"?

5. Using Table 2.1, imagine that your experiment allocates one village to treatment.

(a) Calculate the estimated difference-in-means for all seven possible randomizations.

(b) Show that the average of these estimates is the true ATE.

(c) Show that the standard deviation of the seven estimates is identical to the standard error implied by equation (3.4).

(d) Referring to equation (3.4), explain why this experimental design has more sampling variability than the design in which two villages out of seven are assigned to treatment.

(e) Explain why, in this example, a design in which one of seven observations is assigned to treatment has less sampling variability than a design in which six villages out of seven are assigned to treatment.

6. The Clingingsmith, Khwaja, and Kremer study discussed in section 3.5 may be used to test the sharp null hypothesis that winning the visa lottery for the pilgrimage to Mecca had no effect on the views of Pakistani Muslims toward people from other countries. Assume that the Pakistani authorities assigned visas using complete random assignment. Conduct 10,000 simulated random assignments under the sharp null hypothesis. How many of the simulated random assignments generate an estimated ATE that is at least as large as the actual estimate of the ATE? What is the implied one-tailed p-value? How many of the simulated random assignments generate an estimated ATE that is at least as large *in absolute value* as the actual estimate of the ATE? What is the implied two-tailed p-value?
7. A diet and exercise program advertises that it causes everyone who is currently starting a diet to lose at least seven pounds more than they otherwise would have during the first two weeks. Use randomization inference (the procedure described in section 3.4) to test the hypothesis that $\tau_i = 7$ for all i. The treatment group's weight losses after two weeks are {2, 11, 14, 0, 3} and the control group's weight losses are {1, 0, 0, 4, 3}. In order to test the hypothesis $\tau_i = 7$ for all i using the randomization inference methods discussed in this chapter, subtract 7 from each outcome in the treatment group so that the exercise turns into the more familiar test of the sharp null hypothesis that $\tau_i = 0$ for all i. When describing your results, remember to state the null hypothesis clearly, and explain why you chose to use a one-sided or two-sided test.
8. Naturally occurring experiments sometimes involve what is, in effect, block random assignment. For example, Titiunik studies the effect of lotteries that determine whether state senators in Texas and Arkansas serve two-year or four-year terms in the aftermath of decennial redistricting.[25] These lotteries are conducted within each state, and so there are effectively two distinct experiments on the effects of term length. An interesting outcome variable is the number of bills (legislative proposals) that each senator introduces during a legislative session. The table below lists the number of bills introduced by senators in both states during 2003.
   (a) For each state, estimate the effect of having a two-year term on the number of bills introduced.
   (b) For each state, estimate the standard error of the estimated ATE.
   (c) Use equation (3.10) to estimate the overall ATE for both states combined.
   (d) Explain why, in this study, simply pooling the data for the two states and comparing the average number of bills introduced by two-year senators to the average number of bills introduced by four-year senators leads to biased estimates of the overall ATE.
   (e) Insert the estimated standard errors into equation (3.12) to estimate the standard error for the overall ATE.
   (f) Use randomization inference to test the sharp null hypothesis that the treatment effect is zero for senators in both states.

25 Titiunk 2010.

Texas		Arkansas	
Term Length: 0 = four-year term; 1 = two-year term	# of bills introduced	Term Length: 0 = four-year term; 1 = two-year term	# of bills introduced
0	18	0	11
0	29	0	15
0	41	0	23
0	53	0	24
0	60	0	25
0	67	0	26
0	75	0	28
0	79	0	31
0	79	0	33
0	88	0	34
0	93	0	35
0	101	0	35
0	103	0	36
0	106	0	38
0	107	0	52
0	131	0	59
1	29	1	9
1	37	1	10
1	42	1	14
1	45	1	15
1	45	1	15
1	54	1	17
1	54	1	18
1	58	1	19
1	61	1	19
1	64	1	20
1	69	1	21
1	73	1	23
1	75	1	23
1	92	1	24
1	104	1	28
		1	30
		1	32
		1	34

9. Camerer reports the results of an experiment in which he tests whether large, early bets placed at horse tracks affect the betting behavior of other bettors.[26] Selecting pairs of long-shot horses running in the same race whose betting odds were approximately the same when betting opened, he placed two $500 bets on one of the two horses approximately 15 minutes before the start of the race. Because odds are determined based on the proportion of total bets placed on each horse, this intervention causes the betting odds for the treatment horse to decline and the betting odds of the control horse to rise. Because Camerer's bets were placed early, when the total betting pool was small, his bets caused marked changes in the odds presented to other bettors. (A few minutes before each race started, Camerer cancelled his bets.) While the experimental bets were still "live," were other bettors attracted to the treatment horse (because other bettors seemed to believe in the horse) or repelled by it (because the diminished odds meant a lower return for each wager)? Seventeen pairs of horses in this study are listed below. The outcome measure is the number of dollars that were placed on each horse (not counting Camerer's own wagers on the treatment horses) during the test period, which begins 16 minutes before each race (roughly 2 minutes before Camerer began placing his bets) and ends 5 minutes before each race (roughly 2 minutes before Camerer withdrew his bets).
   (a) One interesting feature of this study is that each pair of horses ran in the same race. Does this design feature violate the non-interference assumption, or can potential outcomes be defined so that the non-interference assumption is satisfied?
   (b) A researcher interested in conducting a randomization check might assess whether, as expected, treatment and control horses attract similarly sized bets prior to the experimental intervention. Use randomization inference to test the sharp null hypothesis that the bets had no effect prior to being placed.
   (c) Calculate the average increase in bets during the experimental period for treatment horses and control horses. Compare treatment and control means, and interpret the estimated ATE.
   (d) Show that the estimated ATE is the same when you subtract the control group outcome from the treatment group outcome for each pair, and calculate the average difference for the 17 pairs.
   (e) Use randomization inference to test the sharp null hypothesis of no treatment effect for any subject. When setting up the test, remember to construct the simulation to account for the fact that random assignment takes place within each pair. Interpret the results of your hypothesis test and explain why a two-tailed test is appropriate in this application.
10. Suppose that 800 students were randomly assigned to classrooms of 25 students apiece, and these classrooms were then randomly assigned as clusters to treatment and control. Assume that non-interference holds. Use equations (3.4) and (3.22) to explain why this clustered design has the same standard error as complete random assignment of individual students to treatment and control.

---

26 Camerer 1998. This example draws on the second of Camerer's studies and restricts the sample to cases in which a treatment horse is compared to a single control horse.

	Total bets on treatment horse 16 minutes before race	Total bets on treatment horse 5 minutes before race (excluding experimental bets)	Treatment change	Total bets on control horse 16 minutes before race	Total bets on control horse 5 minutes before race	Control change	Difference in changes
Pair 1	$533	$1,503	$970	$587	$2,617	$2,030	–$1,060
Pair 2	$376	$1,186	$810	$345	$1,106	$761	$49
Pair 3	$576	$1,366	$790	$653	$2,413	$1,760	–$970
Pair 4	$1,135	$1,666	$531	$1,296	$2,260	$964	–$433
Pair 5	$158	$367	$209	$201	$574	$373	–$164
Pair 6	$282	$542	$260	$269	$489	$220	$40
Pair 7	$909	$1,597	$688	$775	$1,825	$1,050	–$362
Pair 8	$566	$933	$367	$629	$1,178	$549	–$182
Pair 9	$0	$555	$555	$0	$355	$355	$200
Pair 10	$330	$786	$456	$233	$842	$609	–$153
Pair 11	$74	$959	$885	$130	$256	$126	$759
Pair 12	$138	$319	$181	$179	$356	$177	$4
Pair 13	$347	$812	$465	$382	$604	$222	$243
Pair 14	$169	$329	$160	$165	$355	$190	–$30
Pair 15	$41	$297	$256	$33	$75	$42	$214
Pair 16	$37	$71	$34	$33	$121	$88	–$54
Pair 17	$261	$485	$224	$282	$480	$198	$26

11. Use the data in Table 3.3 to simulate cluster randomized assignment.
    (a) Suppose that clusters are formed by grouping observations {1,2}, {3,4}, {5,6}, . . . , {13,14}. Use equation (3.22) to calculate the standard error assuming half of the clusters are randomly assigned to the treatment.
    (b) Suppose that clusters are instead formed by grouping observations {1,14}, {2,13}, {3,12}, . . . , {7,8}. Use equation (3.22) to calculate the standard error assuming half of the clusters are randomly assigned to the treatment.
    (c) Why do the two methods of forming clusters lead to different standard errors? What are the implications for the design of cluster randomized experiments?
12. Below is a schedule of potential outcomes for six classrooms, which are located in three schools. Using a cluster randomized design, researchers will assign one of the three schools (and all the classrooms it contains) to the treatment group.
    (a) What is the average treatment effect among the six classrooms?
    (b) There are three possible randomizations. Is the difference-in-means estimator unbiased?
    (c) In general, cluster random assignment generates biased results when (i) clusters vary in size, (ii) potential outcomes vary by cluster, and (iii) the number of clusters is too small to ensure that m of N units are placed into the treatment condition in each randomization. Show what happens in this example when School A and School B are combined for purposes of random assignment, so that there is a 0.5 probability that either School C is placed in treatment or Schools A and B are placed in treatment. Does this design yield unbiased estimates? What are the implications of this exercise for the design of cluster randomized experiments?

School	Classroom	$Y_i(0)$	$Y_i(1)$
A	A-1	0	0
B	B-1	0	1
B	B-2	0	1
C	C-1	0	2
C	C-2	0	2
C	C-3	0	2

APPENDIX

A3.1 Power

Before running an experiment, the researcher may be interested in calculating the experiment's *power*. In the context of experimental research, power typically refers to the probability that the researcher will be able to reject the null hypothesis of no treatment effect. However, power analysis requires some guesswork: the researcher must furnish values of unknown parameters, such as the size of the true ATE.

To illustrate a power analysis, consider a completely randomized experiment where $N/2$ of N units are selected into a binary treatment. The researcher must now make assumptions about the distributions of outcomes for treatment and for control units. In this example, the researcher assumes that the control group has a normally distributed outcome with mean μ_c, the treatment group has a normally distributed outcome with mean μ_t, and both group's outcomes have a standard deviation σ. The researcher must also choose α, the desired level of statistical significance (typically 0.05).

Under this scenario, there exists a simple asymptotic approximation for the power of the experiment (assuming that the significance test is two-tailed):

$$\beta = \Phi\left(\frac{|\mu_t - \mu_c|\sqrt{N}}{2\sigma} - \Phi^{-1}\left(1 - \frac{\alpha}{2}\right)\right), \qquad \text{(A3.1)}$$

where β is the statistical power of the experiment, $\Phi(\cdot)$ is the normal cumulative distribution function (CDF), and $\Phi^{-1}(\cdot)$ is the inverse of the normal CDF. These functions are easily implemented using R or even a spreadsheet; Microsoft Excel's syntax for $\Phi(\cdot)$ is normsdist(), and its syntax for $\Phi^{-1}(\cdot)$ is normsinv(). For example, if $N = 500$, $\mu_c = 60$, $\mu_t = 65$, $\sigma = 20$ and $\alpha = 0.05$, the power of the experiment, β, would be approximately 0.80. This calculation implies that the proposed experiment has an 80% probability of generating results that lead to the rejection of the null in the presence of a true treatment effect.

Inspection of the formula reveals that power rises as sample size N goes up. One answer to a lack of power is to expand the sample. Power also increases with effect size $|\mu_t - \mu_c|$; strengthening the treatment is another way to remedy insufficient power. Power also increases as σ diminishes, a fact that encourages researchers to reduce noise, obtain prognostic covariates (see Chapter 4), and minimize heterogeneity among subjects.

When assessing experimental designs for which analytical power analysis is not feasible, the researcher may use computer simulation to estimate power. The researcher posits a schedule of potential outcomes for all experimental units. Then, M simulated experiments are generated where treatment is allocated to these units, and p-values are generated for each simulated experiment. The estimated power of the experiment is then the proportion of all M simulated treatment allocations such that the p-value in the experiment is less than α.

CHAPTER 4

Using Covariates in Experimental Design and Analysis

One attractive feature of randomized experiments is that they generate unbiased estimates of the average treatment effect regardless of whether the researcher accounts for other causes of the outcome. The problem of *omitted variables* that bedevils nonexperimental research is addressed by random assignment. Randomly assigned treatments are statistically independent of all observed and unobserved variables. Any correlation between the treatment and other factors that affect outcomes arises purely by chance, due to the way in which observations happened to be allocated to treatment and control groups. Because the experimental researcher controls the assignment process, the sampling distribution of estimated ATEs can be studied in a rigorous manner, and precise inferences may be drawn when evaluating the sharp null hypothesis of no treatment effect. Experimentation makes causal inference possible even when the researcher has a very limited understanding of why some observations have different potential outcomes than others.

Although *covariates*, or supplementary variables that predict outcomes, are not required for unbiased estimation, they may be put to good use when available. This chapter covers several useful roles that covariates play in experimental design and analysis. We begin by discussing three ways in which covariates may be used in the analysis of experimental data. One technique is to use covariates to rescale the dependent variable so that potential outcomes have less variance, which in turn improves the precision with which treatment effects may be estimated. A second technique, regression analysis, uses covariates to eliminate observed differences between treatment and control groups and to reduce the variability in outcomes. The net effect is usually an improvement in the precision with which the treatment effect is estimated. Regression may also be used to check for data-handling errors that potentially undermine random assignment of observations to treatment and control groups. A third use of covariates is block randomized experimental design, which we introduced in Chapter 3. Based on intuitions about which covariates are likely to predict outcomes, the researcher forms a set of relatively homogenous groups, or blocks, each with

What you will learn from this chapter:

1. How to use covariates to obtain more precise estimates of the average treatment effect.
2. How covariates may be used to check the integrity of the random assignment procedure.
3. How block randomized experiments can improve precision, especially in small samples.

different expected outcomes. Random assignment to treatment and control is conducted separately within each block. The final part of the chapter weighs the pros and cons of incorporating covariates into experimental research. Because covariates may be used in many ways, they place additional discretion in the hands of the analyst, who may consciously or unconsciously make decisions that bias the results in a particular direction. Covariates are also a common source of error in the analysis of experimental data, either because researchers attach a causal interpretation to the covariates' apparent "effect" on the outcome or because researchers fail to account for block randomized assignment when analyzing experimental results. Appropriate use of covariates is best worked out during the design phase of an experiment, before the results become known and while the complications that arise from blocking are still fresh in mind.

4.1 Using Covariates to Rescale Outcomes

Perhaps the most intuitive way to begin a discussion of covariates is to consider the advantages of gathering what amounts to a pre-test prior to the launch of an experiment. For example, this pre-test might be an assessment of economic conditions prior to the implementation of an anti-poverty intervention, or of past voter participation in a study designed to assess whether an intervention increases voter turnout. The pre-test need not be a "test" per se but rather any observed variable (X_i) or set of variables ($\mathbf{X}_i$) that is thought to predict potential outcomes.

One of the key assumptions associated with the pre-test and covariates more generally is that they are unaffected by treatment assignment. We need not use potential outcomes notation when describing covariates; it is assumed that the schedule of covariates remains fixed regardless of whether an observation is assigned to treatment or control. The assumption that X_i is fixed casts suspicion on so-called *post-treatment* covariates. Post-treatment covariates are measured after the intervention and are therefore potentially affected by the experimental assignment. Granted, characteristics such as sex or race may be so stable that we may safely assume that they are unaffected by the intervention even if they are measured afterwards. But this is a substantive assumption that will need to be evaluated on a case-by-case basis. For purposes of exposition, we will focus solely on *pre-treatment* covariates. The values of X_i are assumed to be fixed constants that are observed prior to the random assignment of subjects to treatment and control.

If we allocate subjects randomly with equal probability to treatment and control groups, the expected value of X_i in the treatment group will be the same as the expected value of X_i in the control group.

$$E[X_i] = E[X_i | D_i = 1] = E[X_i | D_i = 0], \tag{4.1}$$

where D_i refers to whether a hypothetical assignment causes a subject to receive the treatment. This equality has a number of useful implications. It is the basis for difference-in-differences estimation, which we discuss next, and randomization checks, which we discuss in section 4.3.

What if we were to redefine the outcome measure as the change from pre-test to post-test? Suppose that Y_i represents the post-test, but instead of comparing the average value of Y_i for $d_i = 1$ to the average value of Y_i for $d_i = 0$ we were to compare the average value of $(Y_i - X_i)$ for $d_i = 1$ to the average value of $(Y_i - X_i)$ for $d_i = 0$. The analysis of change scores is called the *difference-in-differences* estimator. Does the difference-in-differences estimator produce unbiased results when treatments are randomly assigned? In order to answer this question, let's consider the expected value of the difference-in-differences estimator in light of the equality in equation (4.1):

$$\begin{aligned} E(\widehat{ATE}) &= E[Y_i - X_i | D_i = 1] - E[Y_i - X_i | D_i = 0] \\ &= E[Y_i | D_i = 1] - E[X_i | D_i = 1] - E[Y_i | D_i = 0] + E[X_i | D_i = 0] \\ &= E[Y_i(1)] - E[Y_i(0)]. \end{aligned} \tag{4.2}$$

The last line is the same expectation as the conventional difference-in-means estimator discussed in Chapter 2. So estimating the ATE using change scores provides another way to obtain unbiased estimates from experimental data.

The two estimators, difference-in-means and difference-in-differences, both generate unbiased estimates. In any finite sample, the two approaches may generate different estimates due to chance. In order to figure out which estimate is more trustworthy, we must figure out which estimation approach is less prone to sampling variability.

In the last chapter, we presented equation (3.4), the formula for the standard error of the difference-in-means estimator. For ease of exposition, let's assume that an experiment places half of the observations into the treatment group such that $m = N/2$. Under this experimental design, equation (3.4) simplifies to:

$$SE(\widehat{\text{ATE}}) = \sqrt{\frac{\text{Var}(Y_i(0)) + \text{Var}(Y_i(1)) + 2\text{Cov}(Y_i(0), Y_i(1))}{N-1}}. \quad (4.3)$$

We can adapt this formula to evaluate the relative merits of the difference-in-differences estimator, simply replacing Y_i with $(Y_i - X_i)$. We replace $\text{Var}(Y_i(0))$ with $\text{Var}(Y_i(0) - X_i) = \text{Var}(Y_i(0)) + \text{Var}(X_i) - 2\text{Cov}(Y_i(0), X_i)$ and $\text{Var}(Y_i(1))$ with $\text{Var}(Y_i(1) - X_i) = \text{Var}(Y_i(1)) + \text{Var}(X_i) - 2\text{Cov}(Y_i(1), X_i)$ to obtain:

$$SE(\widehat{\text{ATE}'})$$
$$= \sqrt{\frac{\text{Var}(Y_i(0)-X_i)+\text{Var}(Y_i(1)-X_i)+2\text{Cov}(Y_i(0)-X_i,\, Y_i(1)-X_i)}{N-1}}$$
$$= \sqrt{\frac{\text{Var}(Y_i(0))+\text{Var}(Y_i(1))+2\text{Cov}(Y_i(0),\, Y_i(1))+4(\text{Var}(X_i)-\text{Cov}(Y_i(0),X_i)-\text{Cov}(Y_i(1),X_i))}{N-1}}. \quad (4.4)$$

Comparing equations (4.3) and (4.4), we see that the difference-in-differences estimator has a smaller sampling variance than the difference-in-means estimator when:

$$\text{Cov}(Y_i(0), X_i) + \text{Cov}(Y_i(1), X_i) > \text{Var}(X_i) \quad (4.5)$$

or

$$\frac{\text{Cov}(Y_i(0), X_i)}{\text{Var}(X_i)} + \frac{\text{Cov}(Y_i(1), X_i)}{\text{Var}(X_i)} > 1. \quad (4.6)$$

In other words, if we were to regress $Y_i(1)$ on X_i and regress $Y_i(0)$ on X_i, the sum of the slope coefficients would have to exceed 1 in order for the rescaling approach to produce more efficient estimates of the ATE.[1] We cannot literally perform these regressions because we lack the full schedule of potential outcomes for all observations, but we can approximate them based on the Y_i that we do observe. The bottom line is that when a covariate strongly predicts potential outcomes, difference scores can produce substantial gains in precision.

In practical terms, this result implies that researchers should take advantage of opportunities to gather background information that may be useful in predicting potential outcomes. Conducting a pre-test or other background survey is not necessary for unbiased estimation, but it can pay dividends in terms of the precision with which the ATE is estimated. Sometimes researchers deliberately focus their interventions on subjects for whom background information is readily available.

1 The connection between equation (4.6) and regression is that the regression coefficient one obtains when regressing Y on X is $\text{Cov}(Y, X)/\text{Var}(X)$.

Researchers may confront trade-offs when considering whether to gather pre-test information. Budget constraints may limit how much data a researcher can collect on subjects prior to the implementation of a treatment. For example, in many cases, researchers must conduct household surveys in order to obtain data on covariates, in which case budget constraints may force a choice between gathering additional subjects and gathering more extensive background information for each subject.

A more serious concern arises when the administration of a pre-test changes the way that participants respond to the treatment and to subsequent attempts to measure outcomes. Suppose the outcome measure is the opinions that subjects express on a survey at the end of an experiment. The administration of the pre-test may suggest to participants that there is a socially desirable way to respond to outcome measures thereafter. Although pre-tests may improve precision, precision takes a backseat to more important concerns, such as the threats that pre-testing may present to core assumptions like excludability. Bias may be introduced if the pre-test provokes different reactions in the treatment and control groups. For example, the control group may shrug off a pre-test because no intervention follows; the treatment group on the other hand may interpret the pre-test as a signal that certain outcomes are expected to change in the wake of the treatment. If this expectation affects outcomes, the experiment measures a combination of the causal effect of the treatment and the causal effect of taking the pre-test. If pre-testing jeopardizes a study's ability to estimate the ATE of interest, researchers should look to other prognostic covariates that can be obtained less obtrusively.

In order to illustrate the potential statistical value of covariates, we present a stylized example inspired by Muralidharan and Sundararaman's study of the educational effects of providing teachers in India with financial bonuses according to their students' academic success.[2] For our purposes, imagine 40 rural primary schools, 20 of which are assigned to the treatment condition: bonus pay for teachers. Academic achievement in each school is assessed by means of a pre-test and a post-test; the pre-test is administered prior to the start of the school year, and the post-test is administered at the end of the school year. For simplicity, assume that the same set of students takes both tests. Finally, in order to illustrate the sampling distributions of different estimators, suppose we know the full schedule of potential outcomes associated with each school.

The schedule of potential outcomes for the 40 schools is presented in Table 4.1. In keeping with the Muralidharan and Sundararaman experiment, our data incorporate a treatment effect of approximately one-quarter of a standard deviation: the treatment effect is 4.0, while the standard deviation in $Y_i(0)$ is 15.5. Our simulation also presupposes a high degree of persistence in scores over the course of a year,

2 Muralidharan and Sundararaman 2011.

TABLE 4.1

Schedule of potential outcomes, pretest scores, and group assignments for simulated teacher incentives experiment

Observation	$Y_i(1)$	$Y_i(0)$	d_i	X_i	X_weak_i
1	5	5	0	6	25
2	15	5	1	8	12
3	12	6	1	5	25
4	19	9	0	13	27
5	17	10	0	9	10
6	18	11	0	15	24
7	24	12	0	16	21
8	11	13	0	17	25
9	16	14	0	19	35
10	25	19	1	23	28
11	18	20	1	28	41
12	21	20	0	28	38
13	17	20	0	9	30
14	24	21	1	16	20
15	27	24	1	23	24
16	26	25	0	15	26
17	30	27	1	23	22
18	37	27	0	33	34
19	43	30	1	42	37
20	39	32	0	31	21
21	36	32	0	29	40
22	27	32	0	28	34
23	33	32	1	35	36
24	37	35	1	28	37
25	48	35	0	41	48
26	39	37	1	37	46
27	42	38	1	32	25
28	37	38	1	37	21
29	53	41	0	36	19
30	50	42	1	44	44
31	51	43	1	48	50
32	43	44	1	43	48
33	55	45	1	55	46
34	49	47	0	53	47
35	48	48	0	51	47
36	52	51	1	43	39
37	59	52	0	57	50
38	52	52	0	51	46
39	55	57	1	49	54
40	63	62	1	55	42

with the pre-test (X_i) predicting 87% of the variance in the $Y_i(0)$ scores.[3] (Later, we consider the consequences of administering a less reliable version of the pre-test that predicts only 50% of the variance in $Y_i(0)$.) The pre-test is also scaled in roughly the same way as the outcome measures: $Y_i(0)$, $Y_i(1)$, and X_i all have standard deviations between 15 and 16.

In order to examine the sampling distribution of the difference-in-means estimator, we simulated 100,000 experiments in which 20 of the 40 observations were assigned to the treatment group. The average estimate of the ATE across the 100,000 experiments was 4.0, with a standard deviation of 4.768. This empirically derived standard error is very close to the 4.773 that we obtain by applying the exact standard error formula in equation (3.4). The top panel of Figure 4.1 shows that the sampling distribution is bell-shaped with a mean of 4, but it is so widely dispersed that 20.4% of all simulated experiments produce negative ATE estimates.

FIGURE 4.1

Sampling distribution of two estimators: difference-in-means and difference-in-differences

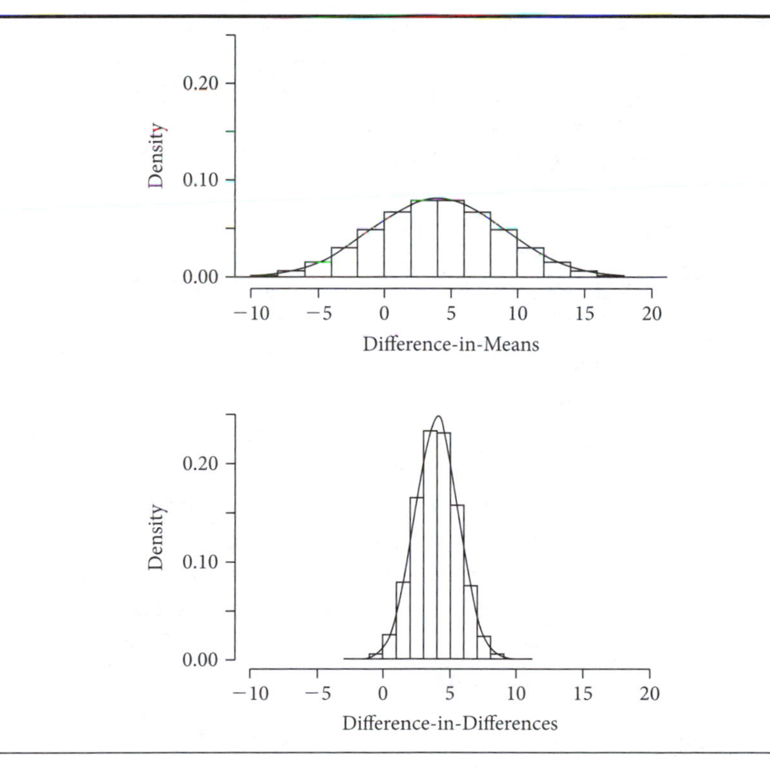

3 This number is the R^2 from a regression of $Y_i(0)$ on X_i using the data in Table 4.1.

By contrast, rescaling the outcome to reflect the change from pre-test to post-test—the difference-in-differences estimator—produces estimates of the ATE that are both unbiased and precise. Based on 100,000 simulated experiments, the bottom panel of Figure 4.1 shows the average estimate to be 4.0 with a standard deviation of 1.53. Just 0.4% of the simulated experiments generate negative ATE estimates. In order to put this improvement into perspective, consider the increase in sample size required to bring about a $1 - (1.53/4.77) = 68\%$ reduction in standard error. One would need to expand the sample by a factor of $1/(1 - 0.68)^2 = 9.8$, which implies expanding the experiment from 40 schools to 392.

Not all experimenters have the luxury of working with highly predictive pre-tests. In the actual Muralidharan and Sundararaman study, the pre-test predicts about half of the variance in the control group's outcomes. A pre-test's predictive accuracy may be diminished by several factors: (1) the pre-test may contain measurement error; (2) the trait that one seeks to measure may change between the administration of the pre-test and the measurement of outcomes, and, in particular, subjects may change at different rates; or (3) the outcomes may be measured with error. Here, we illustrate the first of these possibilities by considering a less reliable version of the pre-test that predicts 50% of the variance in $Y_i(0)$, labeled X_weak_i in Table 4.1. Although this noisy version of the pre-test is less than ideal, it nevertheless improves precision.[4] Rescaling the outcome by this pre-test generates a standard error of 3.48. When compared to the simple difference-in-means estimator, the $1 - (3.48/4.77) = 27\%$ reduction in standard error is considerable. Even the noisy pre-test offers gains in precision equivalent to an 88% increase in sample size.

4.2 Adjusting for Covariates Using Regression

Regression is a flexible tool for analyzing experimental data. Regression allows the researcher to adjust for many covariates simultaneously. For example, if researchers were to administer a pre-test and also gather data on the socioeconomic characteristics of each school, regression could estimate the ATE controlling for both of these background variables. Regression may also be used to estimate the effects of several different treatments, or treatments that are assigned in different dosages (e.g., 100, 200, or 300 hours of instruction), a topic that we revisit in Chapters 9 and 11.

In order to see the connection between regression and the potential outcomes framework, let's rewrite the potential outcome equation (2.2) as a linear regression equation, ignoring covariates for the moment:

4 The noisy pre-test easily satisfies the requirement of equation (4.6). A regression of $Y_i(1)$ on the pre-test gives a coefficient of 0.87, and a regression of $Y_i(0)$ on the pre-test gives a coefficient of 0.95. The sum of the coefficients is therefore greater than 1.0.

$$Y_i = Y_i(0)(1 - d_i) + Y_i(1)d_i = Y_i(0) + (Y_i(1) - Y_i(0))d_i = \underbrace{\mu_{Y(0)}}_{\text{Intercept}}$$
$$+ \underbrace{[\mu_{Y(1)} - \mu_{Y(0)}]d_i}_{\text{ATE}} + \underbrace{Y_i(0) - \mu_{Y(0)} + [(Y_i(1) - \mu_{Y(1)}) - (Y_i(0) - \mu_{Y(0)})]d_i}_{\text{Disturbance}}$$
$$= a + bd_i + u_i. \quad (4.7)$$

Here, the intercept a is equal to $\mu_{Y(0)}$, or the average value of the untreated potential outcomes for all N subjects. The slope b is equal to $\mu_{Y(1)} - \mu_{Y(0)}$ and represents the shift in average potential outcomes, or the ATE. The independent variable is the treatment, d_i. The unobserved part of the regression model is the disturbance term $u_i = Y_i(0) - \mu_{Y(0)} + [(Y_i(1) - \mu_{Y(1)}) - (Y_i(0) - \mu_{Y(0)})]d_i$. The disturbance term comprises the idiosyncratic variation in untreated responses $Y_i(0) - \mu_{Y(0)}$, plus the idiosyncratic variation in treatment effects $[(Y_i(1) - \mu_{Y(1)}) - (Y_i(0) - \mu_{Y(0)})]d_i$. The disturbance term is actually much simpler than it appears. When an observation is assigned to the control group, $d_i = 0$, and the disturbance term equals $Y_i(0) - \mu_{Y(0)}$; when an observation is assigned to the treatment group, $d_i = 1$, and the disturbance term equals $Y_i(1) - \mu_{Y(1)}$.

When ordinary least squares (OLS) regression is used to estimate the parameters of a regression model like equation (4.7), the estimates of the intercept ($\hat{a}$) and slope ($\hat{b}$) are chosen so as to minimize the sum of squared residuals, where a residual is defined as $Y_i - (\hat{a} + \hat{b}d_i)$. A regression of Y_i on d_i produces exactly the same estimate of the ATE as a simple difference-in-means.[5]

In the previous section, we rescaled the dependent variable so that the outcome was defined as the change from pre-test to post-test. In order to see how rescaling looks when expressed as a regression equation, we rewrite the dependent variable as $Y_i^* = Y_i - X_i$:

$$Y_i^* = Y_i - X_i = Y_i(0)(1 - d_i) + Y_i(1)d_i - X_i$$
$$= a + bd_i + u_i - X_i = a + bd_i + u_i^*. \quad (4.8)$$

The last equality reflects the fact that the new disturbance term is $u_i^* = u_i - X_i$.

5 Regression may also be used to estimate the standard error of the treatment effect. When the treatment and control groups contain the same number of subjects, the regression standard error is identical to the estimator in equation (3.6). When the treatment and control groups are of different sizes, regression with "robust" standard errors approximates the estimator in equation (3.6). See Samii and Aronow 2012. When regression is applied to cluster randomized experiments, the standard error may be estimated using "robust cluster" standard errors, which allow disturbances within clusters to be related but assume that disturbances from different clusters are independent. This method of estimating standard errors may be unreliable when the data contain fewer than 20 clusters but becomes increasingly accurate as the number of clusters increases. See Angrist and Pischke 2009, section 8.2.3.

An alternative to rescaling the outcome is to expand the regression model by including one or more covariates as right-hand-side predictors, an approach called *regression adjustment* or *covariate adjustment*. What was before a regression of Y_i on d_i now becomes a regression of Y_i on d_i and X_i. The regression model may be written as follows:

$$Y_i = Y_i(0)(1 - d_i) + Y_i(1)d_i = a + bd_i + cX_i + (u_i - cX_i), \quad (4.9)$$

where the disturbance is the last term in parentheses. This regression adjustment model closely resembles the rescaled regression model presented in equation (4.8), except that now we have included a parameter c in the model whereas equation (4.8) assumed that $c = 1$. If the regression estimate of c is close to 1, regression adjustment and rescaled regression will generate very similar estimates of b.

It should be stressed that the regression estimate $\hat{c}$ has no causal interpretation. The values of Y_i are unaffected by X_i; the middle part of equation (4.9) underscores this point by reiterating that Y_i is solely a function of potential outcomes and treatment status. X_i plays no role in the causal process—using an accounting identity, we added it to the regression model and subtracted it from the disturbance term. The only reason to include X_i as a regressor is that doing so subtracts cX_i from the disturbance. When X_i predicts outcomes, this subtraction reduces the amount of unexplained variation in Y_i, which in turn reduces the standard error of $\hat{b}$. In other words, the decision to include X_i in the regression model is based on whether it is thought to *predict* outcomes; we remain agnostic about whether X_i *affects* outcomes.

What is the advantage of using covariates as regressors? The answer is flexibility. We can include several covariates as right-hand-side variables and let OLS determine their respective scaling parameters. In principle, this approach reduces disturbance variability more effectively than rescaled regression, enabling us to estimate b more precisely.[6]

Are there drawbacks to using regression to control for covariates? In small samples ($N < 20$), controlling for covariates can lead to bias,[7] and rescaling the outcome is probably a safer strategy than regression adjustment. In larger samples ($N \geq 20$), the inclusion of a single covariate typically leads to negligible bias. To avoid bias, regression models that include multiple covariates should have at least 20 more observations than covariates. Whether the inclusion of covariates leads to a reduction in standard error depends on the application. Prognostic covariates may reduce

6 In the schooling example, covariate adjustment using regression produces no gains in precision compared to rescaled regression because the value of c is close to 1.0. Simulations show the standard error of $\hat{b}$ is 1.53 with the rescaled outcomes estimator, as opposed to 1.54 for the regression adjusting for the pre-test as a covariate.

7 Freedman 2008. This bias diminishes rapidly as sample size increases; see Green and Aronow 2011. In samples of any size, regression adjustment does not produce bias when the treatment and control groups are of equal size and treatment effects are constant. Thus, for balanced designs, regression adjustment is unbiased under the sharp null hypothesis of no effect for any subject.

disturbance variability and improve the precision with which treatment effects are estimated. When applying regression to large samples, researchers should include any pre-treatment covariate that previous research (including observational research), pilot testing, or theoretical intuition suggests will predict outcomes. Including covariates that are weakly related to outcomes does little or nothing to reduce sampling variability.[8] However, so long as the researcher does not use the experimental outcomes (the observed values of Y_i) when deciding which covariates to include, the statistical drawbacks of adjusting for covariates are fairly minor, particularly when the experiment involves a large number of subjects.

More serious concerns arise when researchers consider experimental outcomes when deciding which covariates to include in their regression models. For example, after outcome data become available, a researcher might explore a variety of regression models that use different combinations of covariates to predict Y_i. This type of analysis introduces an element of discretion in terms of what results are reported. Perhaps unconsciously, the researcher may settle on a regression model that makes the estimated ATE look impressive or interesting, a decision rule that jeopardizes the unbiasedness of the estimator.

In order to dampen the role of discretion, researchers are encouraged to plan their covariate adjustment procedures in advance of the experiment. By following a predetermined plan, researchers remove the temptation to allow observed outcomes to guide their modeling choices. If the plan specifies a list of what are believed to be prognostic covariates, the covariate-adjusted estimate of the ATE should be regarded as the best guess of the true ATE, because it was expected to be more precise than the difference-in-means estimate. Regardless of whether the analysis is planned in advance or not, researchers should present difference-in-means estimates alongside covariate-adjusted estimates so that readers can judge whether the inclusion of covariates is consequential.

4.3 Covariate Imbalance and the Detection of Administrative Errors

The random assignment of any finite number of observations inevitably produces some degree of *imbalance*, or correlation between the assigned treatment and one or more covariates. By controlling for covariates, whether through rescaling the dependent variable or through regression adjustment, the researcher reestablishes balance.

8 It is possible to construct examples in finite samples where the inclusion of covariates helps or hurts precision, but Lin (2010) shows asymptotically (i.e., as sample size increases) that regression adjustment produces more precise estimates than a simple difference-in-means when similar numbers of subjects are in treatment and control or when treatment effects are uncorrelated with the covariate. When $N \neq 2m$ and treatment effects are correlated with the covariate, regression adjustment may lead to improved precision if the covariate strongly predicts outcomes.

BOX 4.1

Definition: Indicator or "Dummy" Variables

A categorical variable can be recoded into a series of indicator variables, also known as dummy variables. For example, consider a variable defining three experimental blocks {A, B, C}. The three blocks become three variables: Variable 1 is coded 1 if the block variable equals A and is coded 0 otherwise. Variable 2 is coded 1 if the block variable equals B and is coded 0 otherwise. Variable 3 is coded 1 if the block variable equals C and is coded 0 otherwise. When using indicator variables as covariates in a regression, include all but one of the indicator variables. In this example, any two of the three dummy variables may be used.

For example, if one estimates a regression like equation (4.9), the regression will eliminate any linear relationship between the assigned treatment and the pre-test. If one wanted to eliminate both linear and nonlinear relationships between the treatment and the covariate while allowing for possibly nonlinear relationships between the covariate and the outcome, one could either control for a polynomial function of the pre-test (such as X_i, X_i^2, and X_i^3) or code the pre-test as a series of indicator variables in order to examine treatment effects among those who share the same pre-test score. (See Box 4.1.) The broader point is that imbalance on measured indicators is a tractable problem that can be handled by methods such as regression analysis.

Imbalance is a cause for concern not because it presents an insurmountable estimation problem, but because it may call into question the soundness of the random assignment procedure or the administrative handling of the data. Suppose one's randomization of the schooling experiment were to allocate to the treatment group all 20 of the highest-performing schools as gauged by the pre-test. To be sure, such a randomization is conceivable as one of the 137,846,528,820 possible randomizations. On the other hand, it could be a symptom of some sort of clerical error. Maybe someone forgot to randomize the list of schools and left it sorted according to pretest scores? Or maybe there was a miscommunication about which schools were randomly assigned to the treatment?

When concerns like these arise, the first thing to do is retrace the steps that were taken when the observations were randomly assigned. Does the dataset contain the random numbers on which the random assignment was based? If so, do the treatment assignments appear to follow the randomly specified schedule? Is there a computer program that reproduces the random numbers and the process by which they were used to guide the allocation of subjects? These kinds of internal audits may solve the mystery by reference to the randomization procedure itself or other administrative problems that caused some observations to be excluded, double-counted, or miscoded.

Because administrative errors are possible whenever one generates primary data, the nominal standard errors associated with experiments potentially understate the true degree of uncertainty surrounding an experimental result. For that reason, experimenters should carefully investigate the soundness of the randomization procedure, especially if the randomization was conducted by a third party, such as a government agency. When reporting experimental results, researchers are expected to give a step-by-step account of the random assignment process and to describe in detail any instances in which subjects are lost or excluded after allocation (see Chapter 13).

One way to bolster confidence in the integrity of the randomization procedure and the administrative care with which the data were handled is to provide a statistical description of the balance between treatment and control groups on a range of available covariates. Often, one of the first tables in a research article reports the distribution of pre-test scores and other covariates for each of the experimental groups. This table enables the reader to assess two things. First, do prognostic variables, such as pre-test scores, differ appreciably between treatment and control groups? If so, this pattern will help the reader understand why the estimated ATE varies depending on whether covariates are included. Second, for the entire set of covariates, are the imbalances larger than one would expect from chance alone? This question can be answered by means of a statistical test and accompanying simulation. For binary treatments, the statistical test involves a regression of the assigned treatment on all of the covariates and calculation of the F-statistic.[9] In order to calculate the *p*-value of the F-statistic, one uses randomization inference.[10] As explained in section 3.4, randomization inference involves simulating the random assignment procedure many times, calculating the test statistic for each sample. When all of the possible randomizations have been simulated,[11] the collection of test statistics represents the exact

9 Alternatively, one may use logistic regression and calculate the log-likelihood statistic. For experiments with many treatments, the same regression is performed using multinomial logistic regression, and again one calculates a log-likelihood statistic.

10 The F-statistic is used to evaluate the goodness of fit of a regression model. In this case, the statistic is used to test the null hypothesis that the covariates predict the treatment assignments no better than would be expected by chance. Under the null hypothesis, the only predictor of the outcome is the intercept. Under the alternative hypothesis, the q covariates do in fact systematically predict treatment assignments. Call SSR_1 the sum of squared residuals obtained when regressing treatment assignments on an intercept. Call SSR_2 the sum of squared residuals obtained when regressing treatment assignments on an intercept and the q covariates. The F-statistic is

$$\frac{SSR_1 - SSR_2}{q} \Big/ \frac{SSR_2}{N - q - 1}.$$

11 When the number of possible randomizations becomes too large to compute in a timely fashion, one can simulate a large number of random assignments (e.g., 100,000) and treat these as a random sample from the set of possible assignments.

sampling distribution under the null hypothesis that no covariates have any effect on the assigned treatment, which is implied by random assignment. One obtains the p-value of the actual experimental sample's test statistic by finding its location in the sampling distribution of all test statistics. For example, if 43% of the test statistics are at least as large as the observed test statistic, the p-value is 0.43. This procedure works well regardless of the number of covariates, whereas more conventional significance tests break down when the number of covariates is large.[12] Another advantage of this procedure is that it may be used for any randomized design, whether simple, blocked, or clustered.

Suppose this testing procedure reveals that imbalance is greater than one would expect by chance—the p-value is 0.01. This result should prompt a thorough review of the random assignment procedure and any possible data-handling mistakes. If these procedures are checked and found to be satisfactory, proceed on the assumption that the imbalance is due to random chance. One-in-a-hundred outcomes do occur, and if you conduct enough experiments, you can expect to encounter this kind of random anomaly. The imbalance test should be reported, and estimation results should be presented with and without covariate adjustment.

Does random imbalance on measured covariates mean that the experimental estimates are biased? Bias is a property of an estimation procedure, not a specific estimate. The procedure of randomly assigning observations to treatment and control generates unbiased estimates of the ATE, even though some estimates are too high or too low and even though some samples display covariate imbalance. Ex ante (that is, before we conduct our random assignment), we expect our experimental procedure to render, on average, the ATE.

After conducting the randomization and noticing imbalance on one or more prognostic covariates (but before observing outcomes), this expectation may change. For example, suppose we notice that our random assignment has produced a treatment group whose average pre-test scores are higher than the average pre-test scores in the control group. Ex post, we might now expect our estimated ATE to be a higher-than-average draw from the sampling distribution. The teacher incentives study illustrates how these expectations play out numerically. The difference-in-means provides an unbiased estimate of the true ATE, which in this case is 4.0. However, if we restrict our attention to only those randomizations in which the average pre-test scores of the treatment group exceeded the average pre-test scores of the control group, the difference-in-means estimator generates an average estimate of 7.7. Conditional on observing an imbalance on a prognostic variable that favors the treatment group, $E(\widehat{\text{ATE}}) > \text{ATE}$.

Surprisingly, it is possible to recover the ATE even from this pathological subset of experiments. Despite the fact that these randomizations seem to favor the treatment

12 Hansen and Bowers 2008.

group, when we control for pre-test scores using regression, these imbalanced experiments nevertheless yield unbiased estimates: the average estimated ATE is 4.0. The bottom line is that if a covariate is imbalanced due to random assignment, controlling for this imbalance produces unbiased estimates.[13] In practice, if you find evidence of imbalance in a continuous covariate, you may wish to control for this variable using a flexible functional form. For example, if you were to find imbalance on a pre-test score, you might want to control for the pre-test score as well as squared values of the pre-test score.

Sometimes researchers worry that observed imbalance is symptomatic of a broader problem of imbalance among other unobserved factors that affect outcomes. So long as imbalance is solely due to random chance (as opposed to administrative error) and so long as we control for the covariate that is imbalanced, there is no reason to expect imbalance on other covariates or on unmeasured causes of the outcome variable. For example, imagine that poorer villages tend to have lower pre-test scores. Suppose the treatment group has higher pre-test scores on average. By controlling for pre-test scores, we render irrelevant the correlation between wealth and test scores; in effect, we are comparing treatment and control outcomes within schools that have the same pre-test score. When random imbalance turns up in a covariate that is thought to predict outcomes, the solution is to control for that covariate.

4.4 Blocked Randomization and Covariate Adjustment

Random assignment sometimes puts researchers in a quandary. Chance imbalances cause the estimated treatment effect to move around depending on whether and how covariates are controlled. Sometimes the introduction of covariates leads to a decisive improvement in precision, but other times the estimated standard errors are similar, and the choice between the difference-in-means estimator and estimators that use covariate adjustment is less clear cut. The situation becomes especially murky in the absence of a planning document that, prior to the collection of experimental outcomes, lays out the ways in which the researcher intends to analyze the data. The persuasiveness of experimental evidence diminishes when results depend on the analyst's post hoc choices.

Blocked random assignment is a way to prevent this conundrum. Instead of randomly assigning all observations at once, the researcher divides the observations into

13 Regression is subject to the small-sample bias described in section 4.2, but this bias is negligible in this example because the N is fairly large. Note that the benefits of controlling for imbalance do not necessarily hold when imbalance is caused by systematic factors, such as attrition that disproportionately affects one experimental group. See Chapter 7.

more homogeneous blocks and randomly allocates observations within each block. Blocking ensures that all variables used to create strata will be balanced and tends to improve the precision with which the ATE is estimated. For this reason, experimenters tend to follow the dictum "Block what you can, and randomize what you cannot."

With ample preparation time, a researcher can block on several variables simultaneously. For example, a researcher could divide schools according to religion and average levels of poverty. Within each category, observations may be randomly allocated to treatment and control, so that each religion/poverty category allocates $100 * (m/N)\%$ of its observations to the treatment group. If the number of observations is small, the dataset may not contain more than one observation with the same covariate profile, in which case the researcher may block on the most prognostic covariate or an index that summarizes the covariates into a single prognostic score.

Unfortunately, deciding which variables to use in order to construct blocks involves some guesswork. Blocking produces the greatest gains in precision when the variables used to form blocks strongly predict outcomes, but outcomes obviously cannot be observed until the experiment is actually conducted. Researchers therefore look to other studies, including observational studies, for clues about which covariates are the strongest predictors. For example, when planning an experiment that aims to increase elementary students' test scores, one could look at prior research on how these test scores are predicted by region, ethnicity, and scores in prior years. Exercise 4.6 illustrates how one might analyze data for hints about how to form blocks.

Blocking on prognostic covariates helps improve the precision with which treatment effects are estimated. In Chapter 3, we compared the sampling distribution under block random assignment and complete random assignment. We found that when prognostic variables are used to form blocks, block randomization provides

BOX 4.2

Software Routines That Automate Blocking

Many software packages allow researchers to block observations prior to random assignment. One of these packages, blockTools (Moore 2010), is freely available as a package using the R software environment. The blockTools package uses a matching algorithm to automatically group units with similar covariate profiles. It can perform pair-matching or construct blocks of any size, so long as all blocks contain the same number of observations. blockTools is also capable of implementing more complex randomization designs.

An annotated example of blocking in R may be found at http://isps.research.yale.edu/FEDAI. See exercise 4.6.

more precise estimates of the ATE. What happens when we compare block randomization and complete randomization but use regression to control for the variables used to form blocks? For simplicity, suppose subjects in all blocks have the same probability of being assigned the treatment. Blocking eliminates correlation between the assigned treatment and the variables used to form blocks. Returning to the teachers' incentives example in section 4.1, if 75% of the schools in both treatment and control groups are located in predominantly Hindu villages, the correlation between religion and assignment is exactly zero. Because the correlation is zero, regression produces the same estimated ATE regardless of whether we control for the variables used to form blocks.[14] In other words, we obtain the same estimated ATE whether we regress outcomes on d_i or on both d_i and the covariate(s) used to form the blocks. On the other hand, if complete randomization is used, the covariates may be correlated with the treatment; simply by chance, the proportion of predominantly Hindu villages in the treatment group may be higher or lower than 75%. Regression will eliminate this correlation, but the price of doing so is known as the *collinearity penalty*.[15] The larger the R^2 when d_i is regressed on the covariates used to form blocks, the more sampling variability there will be when regression is used to estimate the ATE.

The collinearity penalty tends to decline as the number of subjects increases. In large samples, it is rare to see an appreciable correlation between the assigned treatment and the covariates that could be used to form blocks. For example, suppose we were to conduct an experiment on 75 predominantly Hindu villages and 25 predominantly non-Hindu villages. If half of the 100 villages are randomly selected for treatment, there is a 95% chance that our treatment group will contain between 66% and 84% predominantly Hindu villages. The average R^2 when d_i is regressed on an indicator for village ethnicity is 0.01. Now suppose that the sample were much larger, 750 predominantly Hindu villages and 250 non-Hindu villages. If half of the 1,000 villages are assigned to treatment, there is a 95% chance that our treatment group will contain between 72% and 78% predominantly Hindu villages. The average R^2 when d_i is regressed on an indicator for village ethnicity is now just 0.001. In large samples, the sampling distribution under blocking is very similar to the sampling distribution under regression adjustment. Blocking merely guards against rare instances in which random chance causes covariates to be correlated with d_i.

Does blocking have a downside? The worst-case scenario occurs when blocks are formed using a covariate that fails to predict experimental outcomes. Fortunately,

14 In small samples, it may be impossible to form blocks such that the covariates are completely orthogonal to treatment assignment. In such cases, including the covariates used to form blocks may alter the estimated ATE, although the difference between the estimated ATE with and without covariate adjustment tends to be negligible.

15 The term collinearity refers to the R^2 when d_i is regressed on all of the covariates. If this R^2 were 1.0, regression would not be able to estimate the ATE because no variance would remain in d_i after removing the covariance that d_i shares with the covariates.

even in this case, the estimates remain unbiased and have no more sampling variability than the estimates one would have obtained from a complete randomization.[16] The bottom line is that researchers should block whenever they have reason to believe that a covariate predicts potential outcomes. If a second or third covariate helps predict outcomes, create more refined blocks.

In order to illustrate the potential statistical advantages of blocking, we return to the schools example. In section 4.1, we saw that under complete random assignment, the difference-in-means estimator generated unbiased estimates with a standard error of 4.77. Now, we consider the sampling distribution when schools are blocked according to pre-test scores using the matched pair design described in Chapter 3, section 3.6.1.1. Blocks are created by sorting the 40 schools by pre-test scores in ascending order and forming pairs. The first two schools form the first pair; schools 3 and 4 form the second pair; and so forth. Within each pair, a coin flip determines which of the two schools is placed into the treatment group. This procedure ensures that pre-test scores bear no more than a trivial correlation with treatment assignment.

As noted in Chapter 3, estimation of the ATE involves computing a weighted average of the $\widehat{ATE}_j$ within each block. When the probability of assignment to the treatment condition is the same for every block, as in our matched pair example, regression gives identical results and makes estimation convenient. In order to estimate the ATE and its standard error, apply regression to the following model:

$$Y_i = a + bd_i + c_1X_{1i} + c_2X_{2i} + \ldots + c_{19}X_{19i} + u_i, \tag{4.10}$$

where the 19 indicator variables X_{1i} through X_{19i} mark each of the blocking pairs. For example, X_{1i} is scored 1 when the observations belong to the first two schools on the sorted list, and 0 otherwise. The indicator variable marking the twentieth pair of observations is omitted from the model because the intercept for this omitted pair is represented by a, the constant term.

The statistical advantages of blocking depend on the prognostic quality of the covariates used to generate blocks. In this example, blocking results in a substantial improvement in precision because the pre-test strongly predicts outcomes. When the difference-in-means estimator is applied to block randomized data, the estimated ATE is 4.0 with a standard error of 1.35. If the covariate is a weak predictor of outcomes, blocking will do less to improve precision. For example, if we were to repeat the blocking exercise using the noisy pre-test as the basis for forming blocks, the improvement in precision is much less dramatic: the standard error is 3.18.

Figure 4.2 compares the sampling distributions of the difference-in-means estimator across three experimental designs: complete random assignment, blocked

16 Bruhn and McKenzie 2009. Another potential downside of blocking is that it makes it easier for those implementing an experiment to anticipate and subvert the treatment assignment; see Hewitt and Torgerson 2006.

FIGURE 4.2

Comparison of sampling distributions based on completely randomized and block randomized designs

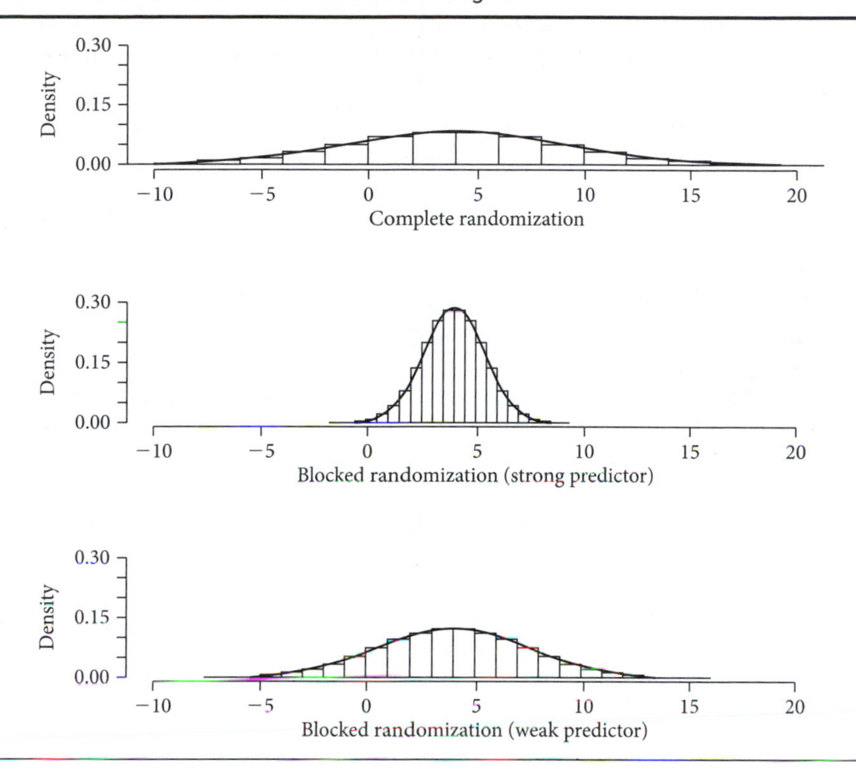

random assignment using the strongly prognostic pre-test, and blocked random assignment using the noisy pre-test. The graph illustrates the principal advantage of blocking. By excluding many of the randomizations that produce outlandish results, blocking narrows the sampling distribution. The advantages of blocking diminish when less prognostic covariates are used to form blocks. Although the noisy pre-test measure predicts 50% of the variance in $Y_i(0)$, many observations that have the same noisy pre-test score turn out to have very different potential outcomes. If an even weaker pre-test were administered such that the scores were nothing but random noise, blocked randomization would generate as much sampling variability as complete randomization.

In terms of precision, the advantages of blocking are less dramatic when the alternative is covariate adjustment using regression. Like blocking, covariate adjustment controls for differences in pre-test scores. Blocking's advantage over controlling for covariates using regression derives from the fact that blocking eliminates randomly occurring correlation between the treatment and the covariate used to form blocks. The smaller the sample, the bigger blocking's advantage will be. For experiments with more than 100 observations in both the treatment and control groups, there are seldom appreciable differences in terms of precision between block randomized designs and complete randomized designs that are analyzed using regression with controls for covariates.[17]

In our running example, we have just 20 observations in each experimental group, and blocking's advantages are noticeable. If, instead of blocking, we had conducted a complete randomization but controlled for each pair of sorted pre-test scores using regression, the estimated treatment effect would average 4.00 with a standard error of 2.05. Due to the collinearity penalty described above, covariate adjustment is not as precise as blocking, which generates estimates whose standard error is 1.35. Nevertheless, covariate adjustment is a dramatic improvement over the difference-in-means estimator, which has a standard error of 4.77.

The practical implications of this discussion may be summarized as follows:

1. Blocking is a useful design tool for reducing sampling variability when the researcher has access to covariates that are thought to predict outcomes. When difference-in-means estimation is used, blocked randomization tends to lead to substantially more precise estimates than complete randomization. Blocking's advantage over regression-based covariate adjustment tends to be modest in samples with more than 100 observations. In small samples, especially when $N < 20$, blocking has two advantages over covariate adjustment: blocking improves precision, and the difference-in-means estimator remains unbiased, whereas regression may be biased.
2. Blocking has the advantage of making transparent the researcher's prior beliefs about which covariates are most likely to predict outcomes. The manner in which blocks are formed provides a clear template for analyzing the experiment's results.
3. By eliminating the correlation between the blocked covariates and treatment assignment, blocking sidesteps the problems of interpretation that arise when adjusted and unadjusted estimates differ.
4. When using randomization inference to test the sharp null hypothesis on block-randomized data, the simulated randomizations must occur within blocks. When simulating randomizations, always follow the procedure that was originally implemented.

17 Rosenberger and Lachin 2002. Some calculations using the binomial distribution illustrate how covariate imbalance and the resulting collinearity penalty tend to diminish with sample size. Recalling our previous example of 10 male and 10 female subjects, half of whom are randomly assigned to treatment, there is a 0.503 probability that random assignment creates a treatment group with 8 or fewer men or 12 or more men. If our sample size increases to 100 men and 100 women, there is just a 0.006 probability that the treatment group will contain 80 or fewer men or 120 or more men.

5. Precision gains from blocking are limited when one's blocks do not differ appreciably in terms of their average potential outcomes. When the covariates used to form blocks have no predictive value whatsoever, the sampling distribution of the estimated ATE under blocked random assignment is approximately the same as under complete random assignment.

BOX 4.3

Randomization-based Inference versus Large-sample Approximations

Chapters 3 and 4 present a randomization-based framework for testing hypotheses and constructing confidence intervals. The three attractive features of this approach are that (1) it grows directly out of the random allocation procedure and is easily tailored to designs that are blocked or clustered, (2) it may be adapted to a wide array of estimation problems, and (3) when calculating critical values, it does not rely on assumptions about the shape of the sampling distribution.

Notwithstanding the advantages of randomization inference, relatively few social scientists currently use randomization-based methods, preferring instead to test hypotheses and construct confidence intervals using a normal approximation based on the standard errors reported by regression. Under simple or complete randomization, the numbers reported by regression software are frequently quite close to what one would obtain using randomization-based methods. When samples are large or when the disturbances in a regression equation are drawn from a normal distribution, regression tends to generate hypothesis tests and confidence intervals that match what one obtains using randomization inference. When neither condition holds, regression's assumptions about the shape of the sampling distribution may produce misleading results. Under these same conditions, one also finds a close correspondence between randomization inference and regression-based confidence intervals when weighted least squares is applied to a block randomized design. Under cluster assignment, regression generates similar hypothesis tests and confidence intervals to randomization inference so long as the number of clusters is large and one instructs regression software to calculate "robust cluster" standard errors, which assume that disturbances are independent across clusters but possibly correlated within clusters. In sum, regression offers a convenient way to test hypotheses and construct confidence intervals under the assumption of constant treatment effects. The assumptions that regression makes about the shape of the sampling distribution may break down, however, especially when the number of observations (or clusters) is small.

4.5 Analysis of Block Randomized Experiments with Treatment Probabilities That Vary by Block

When blocking the teacher incentives experiment, we implemented a blocking scheme in which one member of each pair of schools was assigned to the treatment group. This design implies that the probability of being assigned to the treatment group was identical from pair to pair. The same would be true if we had created four blocks of five schools each, randomly assigning three members of each block to the treatment group. When treatment probability is identical across blocks, estimation is straightforward: compare the average outcome in the treatment group to the average outcome in the control group.

As stressed in Chapter 3, estimation becomes more complicated when the probability of being assigned to the treatment varies from block to block. For example, suppose we were to divide the 40 schools in the teacher incentives experiment into two blocks based on their pre-test scores. The 24 schools with the lowest pre-test scores form one block, and the remaining 16 schools form the other block. Suppose we were to randomly assign 16 of the low-scoring schools to the treatment and the remaining 8 to control. Among high-scoring schools, we randomly allocate 4 schools to the treatment and the remaining 12 to control. In the first block, the probability of treatment is $2/3$, whereas in the second block this probability is $1/4$. If we simply compare the treatment and control groups without controlling for block, our estimator will be biased. The 20 schools in the treatment group have an expected outcome of 29.2. The 20 schools in the control group have an expected outcome of 35.3. Despite the fact that the true ATE is 4.0, the estimated ATE produced by a difference-in-means comparison is, on average, −6.1! Pooling observations across blocks leads to severe bias in this example because treatment assignment is correlated with test scores. By design, schools with higher scores are less likely to receive the treatment.

Random assignment of subjects occurred within blocks, and the appropriate way to estimate the ATE is to compare subjects within each block, not across blocks. The ATE among the 24 schools in the low-achieving block is 3.625. The ATE among the 16 schools in the high-achieving block is 4.5625. In order to obtain the overall ATE, we calculate a weighted average of the two ATEs, with weights equal to each block's share of the overall sample. When estimating the sample average treatment effect using actual data, one uses the estimator presented in equation (3.10). Suppose the estimated ATEs in each block were equal to 3.625 and 4.5625:

$$\left(\frac{N_{Low}}{N}\right)\widehat{ATE}_{Low} + \left(\frac{N_{High}}{N}\right)\widehat{ATE}_{High} = \left(\frac{24}{40}\right)3.625 + \left(\frac{16}{40}\right)4.5625 = 4. \quad (4.11)$$

The estimator in equation (3.10) can be implemented using regression, applying weights to each observation. Each treated observation is weighted by the inverse of the proportion of subjects in its block who were assigned to the treatment condition, and each control subject is weighted by the inverse of the proportion of subjects in its block who were assigned to the control condition. In order to express these weights compactly, define $p_{ij} \equiv m_j/N_j$ as the probability that subject i in block j is assigned to treatment, and weight each observation by:

$$w_{ij} \equiv \left(\frac{1}{p_{ij}}\right)d_i + \left(\frac{1}{1-p_{ij}}\right)(1 - d_i) = \frac{d_i}{p_{ij}} + \frac{1-d_i}{1-p_{ij}}. \tag{4.12}$$

Most regression software makes it easy for users to impose weights such as w_{ij}. See exercise 9 for a detailed example.

One reason to use regression in order to implement the estimator in equation (3.10) is that one can easily augment the model to control for additional covariates. For example, a researcher may block schools based on information that is readily available at the time of random assignment and then control for covariates that become available afterward. Returning to the stylized example presented in Table 4.1, suppose a researcher had blocked the schools according to whether the student body is predominantly Hindu, Muslim, or Sikh. Random assignment occurs with different probabilities within each block. Later, the researcher obtains pre-test scores for each school. Table 4.2 reports each school's random assignment. Within the predominantly Hindu block, the probability of treatment assignment is 0.6, and so the 10 control group schools are weighted by $1/(1 - 0.6) = 2.5$, and the 15 treatment group schools are weighted by $1/0.6 = 1.67$. Within the predominantly Muslim block, the probability of treatment assignment is 0.5, and so the 5 control group schools are weighted by $1/(1 - 0.5) = 2$, and the 5 treatment group schools are also weighted by $1/0.5 = 2$. Finally, within the predominantly Sikh block, the probability of treatment assignment is 0.4, and so the 3 control group schools are weighted by $1/(1 - 0.4) = 1.67$, and the 2 treatment group schools are weighted by $1/0.4 = 2.5$. Notice that (1) within each block, the weighted N in the treatment group equals the weighted N in the control group (so that the weighted probability of treatment is 0.5), and (2) the sum of the weights in the treatment group is N, and the sum of the weights in the control group is N.

Our regression model for these weighted observations is:

$$Y_i = a + bd_i + cX_i + u_i, \tag{4.13}$$

where d_i is the treatment indicator and X_i is the noisy pre-test (this variable is labeled X_weak_i in Table 4.2). Using the observed outcomes in Table 4.2, weighted least squares regression produces an estimated ATE of 3.925. This relationship can be visualized by creating a scatterplot, shown in Figure 4.3: the X-axis represents the residuals from a weighted regression of d_i on X_i, and the Y-axis represents the residuals from a weighted regression of Y_i on X_i. This plot depicts the weights by varying the size of the hollow circles used to represent each subject. The slope of the regression line is 3.925. Assuming a constant effect of 3.925 in order to form a full schedule

TABLE 4.2

Block random assignment of teacher incentives experiment

Observation	Y_i	d_i	X_weak_i	Block	Probability	Weight*
1	5	0	25	1: Predominantly Hindu	0.6	2.5
2	15	1	12	1: Predominantly Hindu	0.6	1.667
3	6	0	25	1: Predominantly Hindu	0.6	2.5
4	19	1	27	1: Predominantly Hindu	0.6	1.667
5	10	0	10	1: Predominantly Hindu	0.6	2.5
6	11	0	24	3: Predominantly Sikh	0.4	1.667
7	24	1	21	1: Predominantly Hindu	0.6	1.667
8	11	1	25	1: Predominantly Hindu	0.6	1.667
9	16	1	35	1: Predominantly Hindu	0.6	1.667
10	25	1	28	1: Predominantly Hindu	0.6	1.667
11	18	1	41	1: Predominantly Hindu	0.6	1.667
12	20	0	38	3: Predominantly Sikh	0.4	1.667
13	20	0	30	1: Predominantly Hindu	0.6	2.5
14	24	1	20	1: Predominantly Hindu	0.6	1.667
15	24	0	24	1: Predominantly Hindu	0.6	2.5
16	26	1	26	1: Predominantly Hindu	0.6	1.667
17	30	1	22	1: Predominantly Hindu	0.6	1.667
18	27	0	34	1: Predominantly Hindu	0.6	2.5
19	43	1	37	3: Predominantly Sikh	0.4	2.5
20	32	0	21	3: Predominantly Sikh	0.4	1.667
21	32	0	40	1: Predominantly Hindu	0.6	2.5
22	27	1	34	1: Predominantly Hindu	0.6	1.667
23	32	0	36	1: Predominantly Hindu	0.6	2.5
24	37	1	37	2: Predominantly Muslim	0.5	2
25	35	0	48	2: Predominantly Muslim	0.5	2
26	39	1	46	3: Predominantly Sikh	0.4	2.5
27	42	1	25	1: Predominantly Hindu	0.6	1.667
28	37	1	21	1: Predominantly Hindu	0.6	1.667
29	41	0	19	1: Predominantly Hindu	0.6	2.5
30	42	0	44	2: Predominantly Muslim	0.5	2
31	51	1	50	2: Predominantly Muslim	0.5	2
32	44	0	48	2: Predominantly Muslim	0.5	2
33	45	0	46	1: Predominantly Hindu	0.6	2.5
34	49	1	47	1: Predominantly Hindu	0.6	1.667
35	48	1	47	1: Predominantly Hindu	0.6	1.667
36	52	1	39	2: Predominantly Muslim	0.5	2
37	52	0	50	2: Predominantly Muslim	0.5	2
38	52	1	46	2: Predominantly Muslim	0.5	2
39	55	1	54	2: Predominantly Muslim	0.5	2
40	62	0	42	2: Predominantly Muslim	0.5	2

* Weight will vary with the randomization. $Pr(D = 1 \mid Block = 1) = 15/25 = 0.6$; $Pr(D = 1 \mid Block = 2) = 5/10 = 0.5$; $Pr(D = 1 \mid Block = 3) = 2/5 = 0.4$.

FIGURE 4.3

Scatterplot of residuals from weighted regressions of d_i on X_weak_i and Y_i on X_weak_i. Estimated ATE = 3.925; plotted circles are sized to reflect each observation's weight.

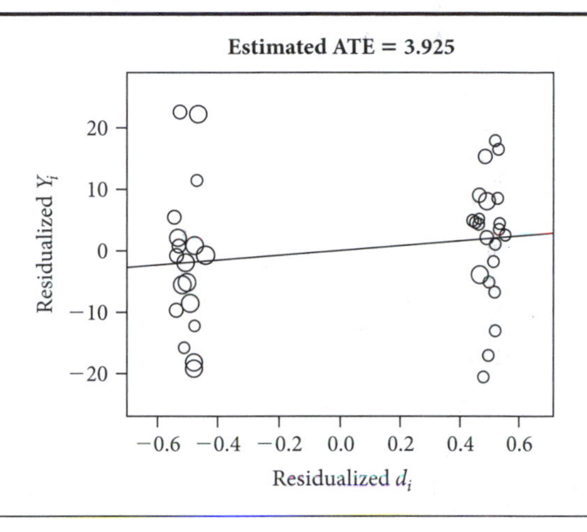

of potential outcomes and simulating 100,000 random assignments, we obtain a 95% confidence interval for the $\widehat{ATE}$ that ranges from −2.75 to 10.67. When we generate the full schedule of potential outcomes assuming the sharp null hypothesis of no treatment effect, we find that 13.3% of the simulated random assignments are larger than 3.925, implying a p-value of 0.13.

This weighting method raises a subtle point about the use of regression to estimate the ATE when experiments employ blocking. A common estimation approach is to use unweighted regression: researchers often regress outcomes on the treatment, controlling for a series of dummy variables that indicate each block. This unweighted regression can be shown to be equivalent to a weighted regression in which each block's estimated ATE is weighted by $(N_j/N)P_j(1 - P_j)$, where N_j/N refers to the share of the total sample that is in block j, and where P_j refers to the proportion of subjects in block j that are assigned to the treatment group.[18] When the probability of assignment to the treatment group is such that $P_j(1 - P_j)$ is constant across blocks, regression and the weighted average estimator in equation (3.10) give identical results. When this condition does not hold, the two estimators may give different results. In our example, the $P_j(1 - P_j)$ for the three blocks are similar (6/25 for blocks 1 and 3, and 1/4 for block 2), so, on average, regression produces an estimate of 3.96, which is close to the ATE. Regression with dummy variables makes sense

18 Each of the weights is normalized by dividing by the sum of all weights; the sum of the normalized weights is therefore 1. See Humphreys 2009 for derivation and illustration of this point.

BOX 4.4

Displaying Covariate-Adjusted Estimates of the ATE under Simple or Complete Random Assignment

In order to depict regression estimates of the ATE when the regression model includes covariates, perform the following steps:

1. Regress d_i on *all* covariates and compute the residual for each observation. Call these residuals e_{di}.
2. Regress Y_i on the covariates (do not include d_i) and compute the residual for each observation. Call these residuals e_{Yi}.
3. Plot the values of e_{di} and e_{Yi}, placing e_{di} on the X-axis and e_{Yi} on the Y-axis.

Verify that you have followed the steps correctly by regressing e_{Yi} and e_{di}; this regression should generate the same estimated ATE as a regression of Y_i on d_i and the covariates.

Displaying Covariate-Adjusted Estimates of the ATE under Block Random Assignment

Follow the same three steps, but in steps (1) and (2), perform a weighted regression using the weights in equation (4.12). In step (3), plot each observation using a hollow circle that is sized to reflect each observation's weight. See Figure 4.3 and the online appendix at http://isps.research.yale.edu/FEDAI for annotated examples.

when you are prepared to assume that treatment effects are the same for all subjects; in that case, you want to weight each block by the precision with which the treatment effect is estimated.

Regardless of whether one controls for blocks using weighted regression or regression with indicators for blocks, the key principle is to compare treatment and control subjects within, not between, blocks. The same principle applies to the presentation of statistics describing covariate balance when the probability of treatment assignment varies by block; presenting treatment and control averages only makes sense *within* each block. In other words, when presenting statistical descriptions of the treatment and control groups, subdivide the table so that each block's results are presented separately.[19] This principle also applies to figures that illustrate the distribution of

19 If the number of blocks is so numerous that block-by-block tables become cumbersome, weight each observation by w_{ij} as defined in equation (4.12) and, for the treatment and control groups, present the weighted means for each covariate.

BOX 4.5

Restricted Random Assignment

Sometimes the time constraints under which field experiments are conducted make blocking impossible. When using complete random assignment, a researcher may quickly approximate blocking by conducting a test of balance and redoing the randomization if the balance criterion is not satisfied (e.g., one obtains a *p*-value below 0.05 when using the F-statistic from a regression to test the null hypothesis that none of the covariates predict treatment assignment). This "catch-and-release" rule results in randomly assigned groups that have similar covariate profiles.

When using restricted randomization, remember to make two adjustments when analyzing the results of your experiment. First, if your treatment and control groups are different sizes, your randomization procedure may cause the probability of being assigned to the treatment group to vary from one subject to the next. For example, subjects with very unusual covariate profiles may be less likely to be assigned to the smaller of the two experimental groups, because their presence in that assigned group would cause a potential random assignment to fail the balance test. In order to obtain unbiased estimates of the ATE, you must weight the observations by w_i, the inverse of the probability of assignment to the observed experimental group. These weights are estimated by simulating a large number of random assignments and tabulating the fraction of random assignments that assign each subject to treatment or control. Second, when conducting randomization inference, you must throw out any random allocations that would have failed your balance test. It is important to document this rejection rule in detail so that randomization tests can exactly reproduce the set of admissible randomizations. See exercise 4.5 for an example of how to estimate the ATE and conduct hypothesis tests under restricted randomization.

outcomes in the treatment and control groups. If the number of blocks is small, these figures should depict relationships for each block separately. If you have a large number of blocks, weight the observations using equation (4.12), and depict more heavily weighted observations with larger symbols.

SUMMARY

By making judicious use of covariates, experimental researchers can improve the precision with which average treatment effects are estimated. Sometimes the improvement is dramatic, equivalent to a marked increase in the number of observations.

The challenge is to use covariates in a principled way, because any gain in precision can be offset by procedures that potentially introduce bias. Covariates should be chosen from the set of variables that exist prior to the random intervention. Ideally, the particular covariates that the researcher wishes to control should be specified in advance of seeing the experimental outcomes; otherwise, the selective use of covariates raises concerns about whether the researcher is trying to guide the estimates toward a particular result. Block randomized designs are particularly appealing in this regard. Although blocking often produces a relatively small gain in precision over after-the-fact regression adjustment of completely randomized experiments, blocking has the virtue of embodying ex ante expectations about prognostic variables and a specific model for estimating treatment effects. If the probability of treatment assignment varies by block, be sure to take this aspect of the design into account when analyzing the data; comparing unweighted outcomes in treatment and control groups may produce severely biased estimates if treatment probabilities are correlated with potential outcomes.

Covariates are of special value to researchers who must conduct experiments under severe budget constraints. Squeezing every drop of precision out of an experiment requires a bit of extra effort and ingenuity. Researchers should be on the lookout for target populations for which extensive background information is readily available. The Swedish government, for example, makes available to researchers extraordinarily detailed information about its citizens' school records, employment history, and health records. Researchers conducting surveys or social network mapping may have rich data on individuals who could later be the subjects of an experimental intervention. Firms that distribute credit or discount cards sometimes have extensive information about the purchasing behavior of their customers. Experiments that are built on these databases start with a wealth of prognostic covariates.[20]

Covariates also assist in the detection of administrative errors. Mistakes are inevitable in the collection and processing of primary data, and one virtue of random assignment is that it creates well-defined statistical patterns that can be detected with the aid of covariates. No method of error detection is foolproof, but when covariates are strongly related to treatment assignment, researchers should review the procedures used to allocate observations to experimental groups.

One danger in mixing covariates into the analysis of experimental data is the temptation to interpret covariates in causal terms. It is not uncommon to encounter research articles in which authors offer a causal interpretation for all of the regression coefficients they present, including the coefficients associated with pre-test scores, blocking indicators, and other factors that are not subject to random assignment.

20 These covariates may also prove valuable in the detection of heterogeneous treatment effects, a topic to which we return in Chapter 9.

Attaching a causal story to these coefficients is at odds with the experimental approach, which uses random assignment to build up a stock of defensible causal claims. The coefficients associated with covariates are uninterpretable without strong substantive assumptions. Predominantly Hindu schools might perform better on average than Muslim schools due to Hindu childrearing practices, or it could just be that religion is a marker for a host of unmeasured attributes that are correlated with educational outcomes. The set of covariates included in an experimental analysis need not be a complete list of factors that affect outcomes; the fact that some factors are left out or are poorly measured is not a source of bias when the aim is to measure the average treatment effect of the random intervention. Omitted variables and mismeasurement, however, can lead to severe bias if the aim is to draw causal inferences about the effects of covariates. Causal interpretation of the covariates encounters all of the threats to inference associated with analysis of observational data.

SUGGESTED READINGS

For an accessible introduction to the mechanics of regression, see Freedman, Pisani, and Purves (2007) and, for more technical detail, Angrist and Pischke (2009). The use of regression to adjust for covariates is critiqued in Freedman (2008), but Schochet (2010) and Green and Aronow (2011) suggest that regression tends to produce approximately unbiased estimates unless samples are very small. Randomization inference for checking balance is discussed in Hansen and Bowers (2008). Lock (2011) discusses the properties and drawbacks of restricted randomization. Humphreys (2009) discusses the use of unweighted regression to analyze block randomized experiments with assignment probabilities that vary by block. Wooldridge (2002) discusses methods for correcting regression standard errors to account for clustering.

EXERCISES: CHAPTER 4

See http://isps.research.yale.edu/FEDAI for datasets and programs for conducting randomization inference.

1. Important concepts:
   (a) Define "covariate." Explain why covariates are (at least in principle) measured prior to the random allocation of subjects to treatment and control.
   (b) Define "disturbance term."
   (c) In equation (4.2), we demonstrated that rescaling the outcome by subtracting a pre-test leads to unbiased estimates of the ATE. Suppose that instead of subtracting the pre-test X_i, we subtracted a rescaled pretest cX_i, where c is some positive constant. Show that this procedure produces unbiased estimates of the ATE.
   (d) Show that the parameter b in equation (4.7) is identical to the ATE.
2. A researcher working with Israeli elementary school students sought to improve students' ability to solve logic puzzles.[21] Students in the treatment and control group initially took

21 Dan Gendelman conducted this study in 2004 and shared it with us via personal communication.

a computer-administered test, and the number of correctly solved puzzles was recorded. A few days later, students assigned to the control group were then given 30 minutes to improve their puzzle-solving skills by playing on a computer. During the same allotment of time, students in the treatment group listened to an instructor describe some rules of thumb to keep in mind when solving logic puzzles. All subjects then took a computer-administered post-test, and the number of correctly solved puzzles was recorded. The table below shows the results for each subject.

Subject	d_i	Pre-test	Post-test	Improvement
1	1	10	10	0
2	1	9	11	2
3	1	5	6	1
4	1	3	6	3
5	1	3	6	3
6	1	6	7	1
7	1	6	7	1
8	1	5	6	1
9	1	6	7	1
10	0	9	9	0
11	0	6	7	1
12	0	11	10	−1
13	0	4	5	1
14	0	3	3	0
15	0	10	10	0
16	0	7	8	1
17	0	7	7	0
18	0	8	10	2

(a) As a randomization check, use randomization inference to test the null hypothesis that the pre-test scores are unaffected by treatment assignment.

(b) Use difference-in-means estimation to estimate the effect of the treatment on the post-test score. Form a 95% confidence interval.

(c) Use difference-in-differences estimation to estimate the effect of the treatment on the post-test score. Form a 95% confidence interval, and compare it to the interval in part (b).

3. The table below illustrates the problems that may arise when researchers exercise discretion over what results to report to readers. Suppose the true ATE associated with a given treatment were 1.0. The table reports the estimated ATE from nine experiments, each of which involves approximately 200 subjects. Each study produces two estimates, one based on a difference-in-means and another using regression to control for covariates. In principle, both estimators generate unbiased estimates, and covariate adjustment has a slight edge in terms of precision. Suppose the researchers conducting each study use the following decision rule: "Estimate the ATE using both estimators and report whichever estimate is larger." Under this reporting policy, are the reported estimates unbiased? Why or why not?

Study	No covariates	With covariates
1	5	4
2	3	3
3	2	2
4	6	5
5	1	1
6	0	0
7	−3	−1
8	−5	−4
9	0	−1
Average	1.00	1.00
Standard Deviation	3.54	2.83

4. Table 4.1 contains a column of treatment assignments, d_i, that reflects a complete random assignment of 20 schools to treatment and 20 schools to control.
   (a) Use equation (2.2) to generate observed outcomes based on these assigned treatments. Regress Y_i on d_i and interpret the slope and intercept. Is the estimated slope the same as the estimated ATE based on a difference-in-means?
   (b) Regress treated and untreated outcomes on X_i to see whether the condition in equation (4.6) appears to hold. What do you infer about the advisability of rescaling the dependent variable so that the outcome is a change score (i.e., $Y_i - X_i$)?
   (c) Regress Y_i on d_i and X_i. Interpret the regression coefficients, contrasting these results with those obtained from a regression of Y_i on d_i alone.
   (d) With the estimates obtained in part (a), use randomization inference (as described in Chapter 3) to evaluate the sharp null hypothesis of no effect for any school. To obtain the sampling distribution under the sharp null hypothesis, simulate 100,000 random assignments, and for each simulated sample, estimate the ATE using a regression of Y_i on d_i. Interpret the results.
   (e) With the estimates obtained in part (c), use randomization inference to evaluate the sharp null hypothesis of no effect for any school. To obtain the sampling distribution under the sharp null hypothesis, simulate 100,000 random assignments, and for each simulated sample, estimate the ATE using a regression of Y_i on d_i and X_i. Interpret the results.
   (f) Use the estimated ATE in part (a) to construct a full schedule of potential outcomes for all schools, assuming that every school has the same treatment effect. Using this simulated schedule of potential outcomes, construct a 95% confidence interval for the sample average treatment effect in the following way. First, randomly assign each subject to treatment or control, and estimate the ATE by a regression of Y_i on d_i. Repeat this procedure until you have 100,000 estimates of the ATE. Order the estimates from smallest to largest. The 2,500th estimate marks the 2.5th percentile, and the 97,501st estimate marks the 97.5th percentile. Interpret the results.
   (g) Use the estimated ATE in part (c) to construct a full schedule of potential outcomes for all schools, assuming that every school has the same treatment effect. Using this

simulated schedule of potential outcomes, simulate the 95% confidence interval for the sample average treatment effect estimated by a regression of Y_i on d_i and X_i. Interpret the results. Is this confidence interval narrower than one you generated in response to question (f)?

5. Randomizations are said to be "restricted" when the set of all possible random allocations is narrowed to exclude allocations that have inadequate covariate balance. Suppose, for example, that the assignment of treatments (d_i) in Table 4.1 was conducted subject to the restriction that a regression of d_i on X_i (the pre-test) generates an F-statistic whose reported p-value is greater than 0.05. In other words, had the researcher found that the assigned d_i were significantly predicted by X_i, the random allocation would have been conducted again, until the d_i met this criterion.
   (a) Conduct a series of random assignments in order to calculate the weighting variable w_i; for units in the treatment group, this weight is defined as the inverse of the probability of being assigned to treatment, and for units in the control group, this weight is defined as the inverse of the probability of being assigned to control. See Table 4.2 for an example. Does w_i appear to vary within the treatment group or within the control group?
   (b) Use randomization inference to test the sharp null hypothesis that d_i has no effect on Y_i by regressing Y_i on d_i and comparing the estimate to the sampling distribution under the null hypothesis. Make sure that your sampling distribution includes only random allocations that satisfy the restriction mentioned above. If the probability of treatment varies from one subject to the next, estimate the ATE by weighting each observation by w_i. Calculate the p-value and interpret the results.
   (c) Use randomization inference to test the sharp null hypothesis that d_i has no effect on Y_i by regressing Y_i on d_i and X_i (weighting the observations by w_i, if necessary) and comparing the estimate to the sampling distribution under the null hypothesis. Calculate the p-value, and interpret the results.
   (d) Compare the sampling distributions under the null hypothesis in parts (b) and (c) to the sampling distributions obtained in exercises 4(d) and 4(e), which assumed that the randomization was unrestricted.

6. One way to practice your experimental design skills is to undertake a mock randomization of an existing nonexperimental dataset. In this exercise, the existing dataset is treated as though it were a baseline data collection effort that an experimental researcher gathered in preparation for a random intervention. The actual data in question come from a panel study of Russian villagers.[22] Villagers from randomly selected rural areas of Russia were interviewed in 1995 and re-interviewed in 1996 and 1997. Our attention focuses on the 462 respondents who were interviewed in all three waves and provided answers to questions about their income, church membership, and evaluation of national conditions (i.e., how well are things going in Russia?). Imagine that an experimental intervention occurred after the 1996 survey and that national evaluations in the 1997 survey were the experimental outcome of interest. The dataset provided at http://isps.research.yale.edu/FEDAI contains the following pre-treatment covariates that may be used for blocking:

22 O'Brien and Patsiorkovski 1999.

sex, church membership, social class, and evaluations of national conditions in 1995 and 1996. As you design your experiment, imagine that "post-intervention" evaluations of national conditions in 1997 were unknown.

(a) One way to develop a sense of which variables are likely to predict post-intervention evaluations of national conditions in 1997 is to regress evaluations of national conditions in 1996 on sex, church membership, social class, and evaluations in 1995. Which of these variables seem to most strongly predict evaluations of national conditions in 1996? What is the R^2 from this regression?

(b) Suppose you were to design a block random assignment in order to predict evaluations in 1997. Use the R package blockTools (for example code, see http://isps.research.yale.edu/FEDAI) to perform a block random assignment, blocking on sex, church membership, social class, and evaluations in 1996. Decide for yourself how many subjects to include in each block. Compare the treatment and control groups to verify that blocking produced groups that have the same profile of sex, church membership, social class, and evaluations in 1996.

(c) Suppose you wanted to assess how well your blocking design performed in terms of increasing the precision with which treatment effects are estimated. Of course, there was no actual treatment in this case, but imagine that shortly after the survey in 1996, a treatment were administered at random to half of the respondents. (Here is an instance in which the sharp null hypothesis of no effect is known to be true!) The outcome from this imaginary experiment is evaluations of national conditions in 1997. Compare the sampling distribution of the estimated treatment effect (which should be centered on zero) under complete random assignment to the sampling distribution of the estimated treatment effect under block random assignment.

(d) Calculate the sampling distribution of the estimated treatment effect under complete random assignment using regression to control for the variables that would have otherwise been used to form blocks. Compare the resulting distribution to the sampling distribution of the estimated treatment effect under block random assignment. Does blocking produce an appreciable gain in precision over what is achieved by covariate adjustment?

7. Researchers may be concerned about using block randomization when they are unsure whether the variable used to form the blocks actually predicts the outcome. Consider the case in which blocks are formed randomly—in other words, the variable used to form the blocks has no prognostic value whatsoever. Below is a schedule of potential outcomes for four observations.

(a) Suppose you were to use complete random assignment such that $m = 2$ units are assigned to treatment. What is the sampling variance of the difference-in-means estimator across all six possible random assignments?

(b) Suppose you were to form blocks by randomly pairing the observations. Within each pair, you randomly allocate one subject to treatment and the other to control so that $m = 2$ units are assigned to treatment. There are three possible blocking schemes; for each blocking scheme, there are four possible random assignments. What is the sampling variance of the difference-in-means estimator across all twelve possible random assignments?

(c) From this example, what do you infer about the risks of blocking on a non-prognostic covariate?

Subject	Y(0)	Y(1)
A	1	2
B	0	3
C	2	2
D	5	5

8. Sometimes researchers randomly assign subjects from lists that are later discovered to have duplicate entries. Suppose, for example, that a fundraising experiment randomly assigns 500 of 1,000 names to a treatment that consists of an invitation to contribute to a charitable cause. However, it is later discovered that 600 names appear once and 200 names appear twice. Before the invitations are mailed, duplicate invitations are discarded, so that no one receives more than one invitation.
   (a) What is the probability of assignment to the treatment group among those whose names appeared once in the original list? What is the probability of assignment to the treatment group among those whose names appeared twice in the original list?
   (b) Of the 800 unique names in the original list, how many would you expect to be assigned to treatment and control?
   (c) What estimation procedure should one use in order to obtain unbiased estimates of the ATE?
9. Gerber and Green conducted a mobilization experiment in which calls from a large commercial phone bank urged voters in Iowa and Michigan to vote in the November 2002 election.[23] The randomization was conducted within four blocks: uncompetitive congressional districts in Iowa, competitive congressional districts in Iowa, uncompetitive congressional districts in Michigan, and competitive congressional districts in Michigan. Table 4.3 presents results only for one-voter households in order to sidestep the complications of cluster assignment.
   (a) Within each of the four blocks, what was the apparent effect of being called by a phone bank on voter turnout?
   (b) When all of the subjects in this experiment are combined (see the rightmost column of the table), turnout seems substantially higher in the treatment group than the control group. Explain why this comparison gives a biased estimate of the ATE.
   (c) Using the weighted estimator described in Chapter 3, show the calculations used to provide an unbiased estimate of the overall ATE.
   (d) When analyzing block randomized experiments, researchers frequently use regression to estimate the ATE by regressing the outcome on the treatment and indicator variables for each of the blocks (omitting one indicator variable if the regression includes an intercept.) This regression estimator places extra weight on blocks that allocate approximately half of the subjects to the treatment condition (i.e., $P_j = 0.5$)

23 Gerber and Green 2005.

TABLE 4.3

Data from the Iowa and Michigan voter mobilization experiment

	BLOCK #1: Uncompetitive Iowa congressional districts		BLOCK #2: Competitive Iowa congressional districts		BLOCK #3: Uncompetitive Michigan congressional districts		BLOCK #4: Competitive Michigan congressional districts		Total	
	Control	Treatment	Control	Treatment	Control	Treatment	Control	Treatment	Control	Treatment
% Voting	48.34	49.30	56.57	55.79	43.62	42.26	48.24	49.07	46.54	48.97
N	39,618	6,935	136,074	7,007	581,919	7,548	154,385	7,229	911,996	28,719
Total N	46,553		143,081		589,467		161,614		940,715	
Share of total N in each block (W_j)	0.049487		0.1520981		0.626616		0.171799			
Estimated ATE	0.00964		−0.007829		−0.01362		0.008271			
Estimated SE	0.006505		0.0060725		0.005745		0.006013			
Weighted average of estimated ATEs with weights equal to W_j									−0.007828	
Standard error (square root of weighted average of estimated variances with weights equal to $W_j^2 / \sum_j^J W_j^2$)									0.00387071	
Proportion of subjects assigned to treatment (P_j)	0.14897		0.0489723		0.012805		0.04473			
Weights used by OLS: $W_j P_j (1 - P_j) / \sum (W_j P_j (1 - P_j))$	0.219216		0.2475177		0.276768		0.256498			
Weighted average of estimated ATEs with weights equal to $W_j P_j (1 - P_j) / \sum (W_j P_j (1 - P_j))$									−0.001472823	
OLS estimate, controlling for blocks (standard error in parentheses)									−0.0014728 (.0030273)	

because these blocks tend to estimate the within-block ATE with less sampling variability. Compare the four OLS weights to the weights w_j used in part (c).

(e) Regression provides an easy way to calculate the weighted estimate of the ATE in part (c) above. For each treatment subject *i*, compute the proportion of subjects in the same block who were assigned to the treatment group. For control subjects, compute the proportion of subjects in the same block who were assigned to the control group. Call this variable q_i. Regress outcomes on treatment, weighting each observation by $1/q_i$, and show that this type of weighted regression produces the same estimate as weighting the estimated ATEs for each block.

10. The 2003 Kansas City voter mobilization experiment described in Chapter 3 is a cluster randomized design in which 28 precincts comprising 9,712 voters were randomly assigned to treatment and control.[24] The study contains a wealth of covariates: the registrar recorded whether each voter participated in elections dating back to 1996. The dataset may be obtained at http://isps.research.yale.edu/FEDAI.

(a) Assess the balance of the treatment and control groups by looking at whether past turnout predicts treatment assignment. Regress treatment assignment on the entire set of past votes, and calculate the sum of squared residuals. Use randomization inference to test the null hypothesis that none of the past turnout variables predict treatment assignment. Remember that to simulate the distribution of the F-statistic, you must generate a large number of random *cluster assignments* and calculate the F-statistic for each simulated assignment. Judging from the *p*-value of this test, what does the F-statistic seem to suggest about whether subjects in the treatment and control groups have comparable background characteristics?

(b) Regress turnout in 2003 (after the treatment was administered) on the experimental assignment and the full set of covariates. Interpret the estimated ATE. Use randomization inference to test the sharp null hypothesis that experimental assignment had no effect on any subject's decision to vote.

(c) When analyzing cluster randomized experiments with clusters of varying size, one concern is that difference-in-means estimation is prone to bias. This concern also applies to regression. In order to sidestep this problem, researchers may choose to use the difference-in-totals estimator in equation (3.24) to estimate the ATE. Estimate the ATE using this estimator.

(d) Use randomization inference to test the sharp null hypothesis that treatment assignment had no effect, using the difference-in-totals estimator.

(e) The difference-in-totals estimator can generate imprecise estimates, but its precision can be improved by incorporating information about covariates. Create a new outcome variable that is the difference between a subject's turnout (1 = vote, 0 = abstain) and the average rate of turnout in all past elections. Using this "differenced" outcome variable, estimate the ATE using the difference-in-totals estimator, and test the sharp null hypothesis of no effect.

24 Arceneaux 2005.

CHAPTER 5

One-Sided Noncompliance

In previous chapters, we discussed the design, analysis, and interpretation of experiments that were presumed to be implemented as planned. Sometimes researchers have the luxury of working with such pristine experiments, but quite often problems of implementation arise: treatments are not administered to the assigned experimental groups, or outcomes are not measured for all subjects. Implementation problems may have profound consequences for the analysis and interpretation of experimental results. The next few chapters discuss the most common types of implementation problems. Our aim is to help the reader understand the implications of each type of problem and to suggest experimental designs that may prevent implementation failures or contain their ill effects.

In Chapters 5 and 6, we discuss the complications that arise when some subjects do not receive the treatment to which they were assigned. In Chapter 5, we consider the case of *one-sided noncompliance* or *failure-to-treat*, which occurs when some of the subjects assigned to the treatment group do not actually receive the treatment. In Chapter 6, we consider the more complex case of *two-sided noncompliance*, which occurs when some subjects in the assigned treatment group go untreated and some subjects in the assigned control group receive the treatment.

When conducting experiments in natural settings, researchers are sometimes unable to administer the treatment to all subjects assigned to the treatment group. Failure-to-treat may occur for a variety of reasons. Logistical snags are common. Treatments planned for a certain area may be scrubbed because of miscommunication, manpower shortages, or transportation problems. Sometimes subjects targeted for treatment prove difficult to reach. For example, voter mobilization experiments that send canvassers to speak with registered voters about an upcoming election often find that a large fraction of subjects are not home when canvassers arrive. In other experiments subjects may refuse the treatment. An *encouragement design* invites subjects in the assigned treatment group to participate in a program offered by the researchers. Only some of those invited may elect to participate. School choice

What you will learn from this chapter:

1. Expanded potential outcomes notation that draws the distinction between assigned treatments and actual treatments.
2. New terms: intent-to-treat effect, Compliers and Never-Takers, and the Complier average causal effect.
3. A modeling approach that establishes whether causal estimands may be recovered given an experimental design.
4. Estimation approaches that gauge the average causal effects among Compliers, those subjects who would be treated if and only if assigned to the treatment group.
5. Research designs that use placebos (i.e., intentionally ineffective treatments) in order to improve the precision with which causal effects are estimated.
6. Core assumptions necessary to recover causal effects from experiments that encounter one-sided noncompliance.

experiments, for example, provide lottery winners with vouchers that can be used to defray tuition at private schools, but only some of the lottery winners actually use the vouchers. Failure-to-treat, in short, means that the assigned treatment group is no longer the same as the group that is actually treated.

In the vocabulary of experiments, the term *compliance* is used to describe whether the actual treatment coincides with the assigned treatment. Under full compliance, all subjects assigned to the treatment group receive the treatment, and no subject assigned to the control group receives the treatment. Noncompliance occurs when subjects who were assigned to receive the treatment go untreated or when subjects assigned to the control group are treated inadvertently. In everyday language, the terms *compliance* and *noncompliance* have connotations of agreeableness on the one hand and disobedience on the other. The experimental terminology has no normative connotations. As the face-to-face canvassing example shows, noncompliance may occur without any refusal on the part of the subjects.

In this chapter we focus on *one-sided* noncompliance, which occurs when there is noncompliance for only one of the assigned groups. We restrict our focus to the case most relevant for social science applications, which occurs when no subject assigned to the control group is treated but some of those assigned to the treatment group go untreated. A common experimental design that displays one-sided noncompliance is a study in which the treatment is available only through the experiment. There is no opportunity for the control group to be treated without the cooperation of the experimenters.

To appreciate the challenges posed by one-sided noncompliance, suppose that you are interested in assessing the effect of a face-to-face canvassing effort on voter turnout, which is measured for each subject using public records of who voted. Imagine 2,000 subjects are randomly divided into two equal groups (1,000 treatment and 1,000 control), and canvassers are sent out to contact the subjects in the treatment group. In the familiar case of full compliance, 100% of those assigned to the treatment group receive the treatment. Recall from Chapter 2, which considered experiments with full compliance, that an unbiased estimate of the average causal effect of the treatment can be calculated by subtracting the average outcome in the control group from the average outcome in the treatment group. However, in face-to-face canvassing experiments, not everyone will be found at their listed address when the canvassers visit. Typically, only 25% of those assigned to the treatment group can be reached by the canvassers; the remaining 75% of the treatment group goes untreated. Applying these rates to our hypothetical canvassing example, we observe the average outcomes for three groups of subjects: the 250 subjects assigned to the treatment group who are actually treated, the 750 subjects assigned to the treatment group who remain untreated, and the 1,000 (untreated) subjects in the control group. Now suppose you were the principal investigator of this study. How should the data from these three groups be analyzed? Which causal effects can be estimated using these data?

Let's consider two possible approaches. One option is to ignore the issue of noncompliance and compare the average outcome for the entire treatment group (all 1,000 observations) with the average outcome of the 1,000 subject control group. This comparison is what we examine when analyzing an experiment with full compliance: we subtract the control group average from the treatment group average. But recall that only 25% of the treatment group was treated. If the apparent difference in average group outcomes is interpreted as an average treatment effect for the *entire* group of subjects, this method implicitly assumes that the average treatment effect is zero for the untreated portion of the treatment group. To assume that the untreated members of the treatment group have an ATE of zero is to assert that their average outcomes would be no different if they had been treated, which seems implausible.

As an alternative to comparing the average outcomes based on which group a subject is assigned to, how about comparing the average outcome among subjects in the treatment group who receive the treatment (250 subjects) to the average outcome among the control group (1,000 subjects)?[1] Despite the intuitive appeal of this approach, there is a serious problem with comparing the *actually* treated to

1 Yet another option, which has the intuitive appeal of not wasting data, is to group together all the untreated subjects (the 1,000 control group subjects and the 750 untreated subjects from the treatment group) to form a large group of the untreated. These 1,750 untreated subjects could then be compared to the 250 treated subjects. The latter approach calls to mind an observational study in which people who receive a treatment are compared to those who do not.

the untreated. The subjects who are actually treated are a non-random subset of the original treatment group. This point cannot be emphasized enough. Unlike groups formed by random assignment, groups formed *after* random assignment will not in general have the same expected potential outcomes. Comparing these groups opens the door to biased inference.

Applying this general point to the case of a canvassing experiment, there are many reasons why those treated by canvassers may be different from those who go untreated. Some subjects assigned to the treatment group may no longer live at the address listed on the official list of registered voters; these subjects will not be successfully reached by the canvassers, and their turnout rates will be very low. Subjects who are successfully canvassed exclude all people who have moved away, but the control group will include some movers because no one attempted to contact the control group and took note of who moved. Comparison of the voting rate for the subset of treated subjects (all non-movers) with the voting rate of the entire control group (movers and non-movers) exaggerates the apparent effect of canvassing. Even if there were no treatment effect, this comparison would make canvassing look effective. Given the many ways in which a researcher may misanalyze this type of experiment, it pays to acquire a detailed understanding of noncompliance and appropriate methods for addressing it.

This chapter provides a broad introduction to the analysis of experiments that encounter one-sided noncompliance. We begin by describing the problem in formal terms, introducing the idea that the subject pool consists of two groups: subjects who will be treated if and only if they are assigned to the treatment group and subjects who will never be treated regardless of which group they are assigned to. The upshot of this formal discussion is that experiments with one-sided noncompliance tell us only about the average treatment effect among subjects who take the treatment if assigned to be treated. This limitation forces researchers to interpret their results more cautiously. Failure-to-treat also has important implications for experimental design, particularly when researchers have reason to anticipate high rates of noncompliance.

5.1 New Definitions and Assumptions

In Chapter 2, we introduced the potential outcomes framework for defining a causal effect. Let's briefly review this basic setup. Whether subject i receives the treatment is denoted d_i, where $d_i = 1$ indicates that subject i is treated and $d_i = 0$ indicates that subject i is not treated. For each subject, we define potential outcomes, the set of outcomes that might occur. The outcome measure of greatest interest to the researcher is typically labeled Y_i. The potential outcome $Y_i(d)$ refers to the outcome that subject i would exhibit when $d_i = d$. Specifically, $Y_i(1)$ is the potential outcome when subject i is treated, and $Y_i(0)$ is the potential outcome when subject i is not treated. When writing potential outcomes as $Y_i(d)$ rather than $Y_i(\mathbf{d})$ we are invoking the non-interference assumption. Potential outcomes are written solely in terms of the treatment that subject i receives; treatments received by other subjects are assumed to be irrelevant. Assuming non-interference holds, the causal effect of the treatment on subject i is $Y_i(1) - Y_i(0)$. Recall that the problem of estimating causal effects stems from the fact that we never observe both $Y_i(1)$ and $Y_i(0)$; we observe each subject either in a treated or untreated state.

In order to build noncompliance into this modeling framework, we expand the notation to account for the possibility that assigned treatment may not always coincide with actual treatment. The set of variables that may be influenced by treatment assignment now includes not only the outcome variable (Y_i), but also whether an individual is treated (d_i).

In order to distinguish assigned treatment from actual treatment, let the experimental assignment of subject i be denoted z_i. When $z_i = 1$, the subject is assigned to the treatment group, and when $z_i = 0$, the subject is assigned to the control group. In previous chapters, $d_i = z_i$: all subjects assigned to the treatment group are treated, and no subjects assigned to the control group are treated. We now relax this constraint and consider actual treatment to be a potential outcome. Let $d_i(z)$ represent whether subject i is actually treated when treatment assignment is z. For compactness, we write $d_i(z = 1)$ as $d_i(1)$ and $d_i(z = 0)$ as $d_i(0)$. For example, if a subject would be treated if assigned to the treatment group, $d_i(1) = 1$. If a subject would receive no treatment if assigned to the treatment group, $d_i(1) = 0$.

For each subject, a pair of potential outcomes, $d_i(1)$ and $d_i(0)$, indicates whether the subject will receive the treatment if assigned to the treatment or control group. For the case of one-sided noncompliance, $d_i(0)$ is set to 0 for all i: subjects are never treated when they are assigned to the control group. $d_i(1)$, however, can be either 0 or 1.

BOX 5.1

Potential Outcomes

When referring to potential outcomes, researchers usually have in mind the way that the "dependent variable" of an experimental study responds to treatment. The term "potential outcome" is actually broader in its meaning. Potential outcomes arise when any variable responds to assigned or actual treatment. For example, whether a subject actually receives treatment when assigned to the treatment group is a potential outcome.

BOX 5.2

Actual Treatment and Assigned Treatment

Bear in mind the distinction between the assigned treatment group and the group of subjects who actually receive the treatment. The potential outcome $d_i(z)$ indicates whether subject i is actually treated when treatment assignment is z. There are four possible configurations:

- $d_i(1) = 0$ means that a subject assigned to the treatment group is untreated.
- $d_i(0) = 0$ means that a subject assigned to the control group is untreated.
- $d_i(1) = 1$ means that a subject assigned to the treatment group is treated.
- $d_i(0) = 1$ means that a subject assigned to the control group is treated. This last potential outcome is ruled out in this chapter, which assumes one-sided noncompliance, but will be revisited in Chapter 6.

Subjects in experiments with one-sided noncompliance can be divided into two groups. Subjects are called *Compliers* if their potential outcomes meet two conditions: they receive the treatment if assigned to the treatment group ($d_i(1) = 1$), and they do not receive the treatment if assigned to the control group ($d_i(0) = 0$). These subjects "comply" with their treatment assignment inasmuch as they receive the treatment when assigned to the treatment group and do not receive the treatment when assigned to the control group. In contrast, those for whom $d_i(1) = 0$ and $d_i(0) = 0$ are called *Never-Takers*. They never take the treatment regardless of whether they are assigned to the treatment group or control group. Because $d_i(0) = 0$ for all subjects, when discussing one-sided noncompliance, researchers sometimes refer to Never-Takers using the shorthand $d_i(1) = 0$ and refer to Compliers using the shorthand $d_i(1) = 1$. For example, the expression $\text{ATE} | (d_i(1) = 1)$ would be read "average treatment effect among Compliers."

When classifying subjects as either Compliers or Never-Takers, three points should be kept in mind. First, this terminology has nothing whatsoever to do with outcomes, Y_i. The terms *Complier* and *Never-Taker* refer only to whether subjects would take the treatment if assigned to the treatment group.

Second, the definition of d_i given above presupposes an abstract treatment that is either administered or not. When conducting an actual experiment, a researcher must define the criteria that will be used to classify each subject as treated or untreated. In some cases, the definition of treatment is unambiguous. In a study designed to test the efficacy of a vaccine, subjects are either injected with the vaccine or not. In other

BOX 5.3

Types of Subjects in an Experiment with One-Sided Noncompliance

Subjects may be classified into two groups according to their potential outcomes $d_i(z)$. This notation refers to the potential outcome for d_i when the input is an assigned treatment z, which is either 0 or 1. Angrist, Imbens, and Rubin (1996) suggested the following terminology:

Subjects for whom $d_i(1) = 1$ and $d_i(0) = 0$ are called *Compliers*. These subjects receive the treatment when assigned to the treatment group and do not receive the treatment when assigned to the control group. Some authors define Compliers as subjects for whom $d_i(1) > d_i(0)$. The two definitions are equivalent.

Subjects for whom $d_i(1) = 0$ and $d_i(0) = 0$ are called *Never-Takers*. They are untreated regardless of whether they are assigned to treatment or control.

cases, formulating a definition is less straightforward. For example, what if the treatment is a yearlong class, but some subjects only attend for a few months? Later in this chapter, we will see that how the researcher defines the treatment will have important consequences for estimation and interpretation. The definition of treatment affects the definition of a Complier.

Finally, the classification of subjects as either Compliers or Never-Takers reflects not only the subjects' background attributes but also the experimental context and experimental design. For example, if a face-to-face canvassing effort is conducted only on weekends, a subject who is home every weeknight but gone every weekend will be a Never-Taker. If the experimenter instead instructs canvassers to work weeknights but not weekends, this same subject would now be a Complier. Each subject's potential outcomes $d_i(z)$ depend on how the treatment is administered. For this reason, it is important for researchers to clearly describe their procedures and the context in which the experiment occurred. Otherwise, it will be difficult for readers to get a sense of who the Compliers are when interpreting the results.

5.2 Defining Causal Effects for the Case of One-Sided Noncompliance

The new notation allows us to restate two core assumptions, non-interference and excludability, in ways that allow for noncompliance.

5.2.1 The Non-Interference Assumption for Experiments That Encounter Noncompliance

This assumption consists of two parts. Part A stipulates that whether a subject is treated depends only on the subject's own treatment group assignment. Other subjects' assignments are assumed to have no bearing on whether one receives the treatment. In order to depict this assumption formally, define $\boldsymbol{z}$ as a list of treatment assignments for each of the N subjects. The treatment assignment of subject i is one of the N elements of this list. Imagine altering the treatment assignments of some or all of the other subjects while keeping the treatment assignment of subject i the same. Call any such altered list of assignments $\boldsymbol{z}'$. Part A of the non-interference assumption states that:

$$d_i(z) = d_i(z') \text{ if } z = z'_i, \quad (5.1)$$

where the notation $z_i = z_i'$ means that subject i keeps the same treatment assignment even when the assignments of other subjects change.

Part B says that the potential outcomes are affected by (1) the subject's own assignment and (2) the treatment that the subject receives as a consequence of that assignment. Other subjects' assignments and treatments are assumed to have no bearing on one's outcomes.

$$Y_i(\boldsymbol{z}, \boldsymbol{d}) = Y_i(\boldsymbol{z}', \boldsymbol{d}') \text{ if } \boldsymbol{z}_i = \boldsymbol{z}_i' \text{ and } \boldsymbol{d}_i = \boldsymbol{d}_i'. \quad (5.2)$$

The plausibility of the non-interference assumption depends on the specifics of a given experiment. As discussed in detail in Chapter 8, non-interference may be violated in many experimental contexts. The rule of thumb when assessing the validity of this assumption is to reflect on whether potential outcomes vary depending on how subjects are allocated to experimental groups or how treatments are actually administered. Imagine a canvassing experiment in which two of the subjects live on high peaks, and suppose that if one high-elevation subject is assigned to the treatment group, canvassers will be too tired to walk to the other high-elevation subject's house. Regardless of whether the first high-elevation subject is treated, Part A of the non-interference assumption is violated because one subject's assignment affects whether the other subject receives the treatment. Part B would be violated if one subject's potential outcomes are affected by other subjects' assigned or actual treatments. Non-interference would be violated if, as a consequence of being canvassed, treated subjects tell their neighbors about the upcoming election, changing these subjects' propensity to vote. Assuming non-interference greatly simplifies the schedule of potential outcomes, allowing us to write them solely in terms of the treatment that subject i is assigned or receives.

We next distinguish the causal effect of treatment assignment from the causal effect of receiving the treatment. The causal effect of assignment to the treatment group is called the *intent-to-treat* effect because it reflects the intended assignments, not the actual treatments. The intent-to-treat effect can be defined for any set of potential outcome that may be affected by treatment assignment, such as $d_i(z)$, $Y_i(d(z))$, or $Y_i(z, d(z))$.

The intent-to-treat effect of z_i on d_i for each subject is defined as:

$$\text{ITT}_{i,D} \equiv d_i(1) - d_i(0). \tag{5.3}$$

When we calculate the average $\text{ITT}_{i,D}$ across all subjects, we obtain ITT_D, which is the proportion of subjects who are treated in the event that they are assigned to the treatment group, minus the proportion who would have been treated even if they had instead been assigned to the control group:

$$ITT_D \equiv E[ITT_{i,D}] = E[d_i(1)] - E[d_i(0)]. \tag{5.4}$$

Since we are assuming one-sided noncompliance, $E[d_i(0)] = 0$, and the expression for the ITT_D simplifies to $E[d_i(1)]$.

The intent-to-treat effect of z_i on Y_i for each subject is defined as:

$$\text{ITT}_{i,Y} \equiv Y_i(z = 1) - Y_i(z = 0). \tag{5.5}$$

Another way to write the $ITT_{i,Y}$ is:

$$\text{ITT}_{i,Y} \equiv Y_i(z = 1, d(1)) - Y_i(z = 0, d(0)). \tag{5.6}$$

The latter way of expressing the $\text{ITT}_{i,Y}$ makes explicit the fact that outcomes may respond to treatment assignment or to the treatments that result from treatment assignment.

The average $ITT_{i,Y}$ is the change in expected potential outcomes that occurs when subjects move from an assigned control group ($z = 0$) to an assigned treatment group ($z = 1$):

$$\text{ITT}_Y \equiv \frac{1}{N}\sum_{i=1}^{N}(Y_i(z = 1, d(1)) - Y_i(z = 0, d(0)))$$
$$= E[Y_i(z = 1, d(1))] - E[Y_i(z = 0, d(0))] = E[\text{ITT}_{i,Y}]. \tag{5.7}$$

For experiments with 100% compliance, treatment assignment is the same as treatment status, and so the ITT_Y is the same as the average treatment effect (ATE). Since the average intent-to-treat effect on Y_i is by far the more important of the two intent-to-treat effects, we will drop the subscript and simply call it by the shorthand ITT. The ITT is a measure of the average effect of experimental assignment on outcomes, regardless of the fraction of the treatment group that is actually treated. It is commonly used to describe the effectiveness of a program when the main concern is the extent to which the program changed outcomes. If all you care about is whether the *program* "made a difference," noncompliance is in some sense irrelevant. Regardless of whether the program treated a large or small proportion of its intended targets, did it change the average outcome?

Often, however, researchers seek to estimate the average treatment effect, not the average effect of assignment to treatment. In other words, they want to know the average effect of d_i on Y_i, not the average effect of z_i on Y_i. The expanded notation for writing potential outcomes allows $Y_i(z, d(z))$ to respond to two inputs: the assigned treatment (z) and the treatment that results from assignment ($d_i(z)$). To isolate the effect of the treatment from the effect of assignment, we extend the excludability assumption that was first introduced in Chapter 2.

5.2.2 The Excludability Assumption for One-Sided Noncompliance

The excludability assumption stipulates that potential outcomes respond to treatments, not treatment assignments. If we assume non-interference, the excludability assumption may be stated with reference to the treatment that subject i is assigned and receives. For all subjects, values of z, and values of d,

$$Y_i(z, d) = Y_i(d). \tag{5.8}$$

Untreated subjects have the same potential outcomes regardless of their assignments: $Y_i(z = 0, d = 0) = Y_i(z = 1, d = 0)$. The same goes for subjects who receive the treatment: $Y_i(z = 0, d = 1) = Y_i(z = 1, d = 1)$. Under the excludability assumption, only d matters; we therefore write the potential outcomes according to whether the subject received the treatment and disregard the assigned treatment.

This assumption is called excludability or the exclusion restriction because it maintains that experimental assignment (z) may be "excluded" as a cause of Y_i since it has no influence on potential outcomes except insofar as it affects treatment (d). For example, if the exclusion restriction holds, Never-Takers' Y_i are the same regardless of whether they are assigned to the treatment or control group. For Never-Takers, treatment assignment always results in $d = 0$, and the assignment (z) has no opportunity to affect outcomes.

The plausibility of the exclusion restriction must be assessed based on a close inspection of each experiment. The researcher's definition of what constitutes the treatment often points to possible violations of the exclusion restriction. Consider an experiment in which only subjects in the treatment group get a letter inviting them to participate in a new program, and d_i is whether the subject actually participates in the program. The exclusion restriction states that the letter has no effect on potential outcomes other than through the subject's participation in the program. Among those who participate in the program, it is assumed that $Y_i(z = 0, d = 1) = Y_i(z = 1, d = 1)$. Among those who do not participate, it is assumed that $Y_i(z = 0, d = 0) = Y_i(z = 1, d = 0)$. Whether these assumptions are credible depends on one's intuitions about the recruitment letter—does it seem plausible that the letter affects outcomes apart from affecting whether subjects participate in the program? The experiment, as implemented, cannot speak to this issue because everyone in the treatment group received the letter. If this possible violation of the exclusion restriction is taken to be a serious challenge to the claim that participation per se affects results, it may be necessary to conduct a new experiment in which recruitment is encouraged in some other way. For example, a recruitment letter could be sent to both the treatment and control groups, but with the treatment group encouraged to participate sooner.

Exclusion restrictions sometimes provoke controversy when researchers implement multifaceted interventions but interpret the experiment as though it reveals the effect of one component of the intervention. To see how this interpretation may lead to a violation of the exclusion restriction, let's again consider an experiment in which eligible voters are encouraged to participate in an upcoming election. Canvassers are sent to the homes of those in the treatment group. If the targeted voter answers the door, the canvasser gives a brief speech about the importance of voting and hands the subject a leaflet that indicates where and when to vote. If no one answers, the leaflet is slipped under the door. If the researcher interprets the study as an investigation of the effects of the verbal message delivered by the canvassers, then the subjects who answer the door are considered $d_i(1) = 1$. The exclusion restriction implies that $Y_i(z = 0, d = 0) = Y_i(z = 1, d = 0)$ and that $Y_i(z = 0, d = 1) = Y_i(z = 1, d = 1)$; the subject is assumed to be affected by the speech but not by the leaflet (or any other aspect of treatment assignment). Notice that this complication arises even when *all* subjects are treated when assigned to the treatment group. The important implication of this example is that when the experiment involves compound treatments, the researcher must make a choice. Either the definition of the treatment must be adjusted to include the entire package of interventions (leafleting, sometimes in conjunction with a speech), or the researcher must be prepared to stipulate the effect of certain components, such as leafleting. A design implication is that if researchers seek to estimate the distinct effects of each component, they may need to perform a more complex experiment in which different treatment groups receive different interventions. One experimental condition might provide canvassing and leaflets, while another experimental condition might provide only canvassing.

5.3 Average Treatment Effects, Intent-to-Treat Effects, and Complier Average Causal Effects

Noncompliance limits what researchers can learn from an experiment. Ordinarily, researchers would aim to estimate the average treatment effect:

$$\text{ATE} \equiv \frac{1}{N}\sum_{i=1}^{N}(Y_i(1) - Y_i(0)) = E[Y_i(d = 1) - Y_i(d = 0)]. \quad (5.9)$$

But, as we will see momentarily, experiments that encounter noncompliance do not generate information required to identify the ATE. A more realistic goal is the estimation of the Complier Average Causal Effect (CACE), which is defined as:

$$\text{CACE} \equiv \frac{\sum_{i=1}^{N}(Y_i(1) - Y_i(0))d_i(1)}{\sum_{i=1}^{N} d_i(1)} = \underbrace{E[(Y_i(d=1) - Y_i(d=0))}_{\text{average treatment effect}}\,|\,\underbrace{d_i(1) = 1]}_{\text{among Compliers}}. \quad (5.10)$$

The CACE is the average treatment effect for a subset of the subjects, the Compliers.[2]

To solidify our understanding of the new definitions, Table 5.1 provides an example of nine subjects along with their potential outcomes $Y_i(z, d(z))$ and $d_i(z)$. The schedule of potential outcomes depicted in Table 5.1 is designed to satisfy both the non-interference assumption and the exclusion restriction. In other words, $Y_i(d = 1)$ and $Y_i(d = 0)$ are solely a function of whether the subject receives the treatment. Using Table 5.1, we can calculate the ITT, the ITT_D, the ATE, and the CACE. Notice that Subjects 1, 3, 4, 5, 7, and 8 are Compliers, while Subjects 2, 6, and 9 are Never-Takers. For all subjects, the value of Y_i observed when a subject is assigned to the control group is $Y_i(d(0) = 0)$. Whether $Y_i(d(1) = 1)$ or $Y_i(d(1) = 0)$ is observed when a subject is assigned to the treatment group depends on whether the subject is a Complier or a Never-Taker.

First, consider the average treatment effect (ATE) for this set of subjects. The ATE is calculated by taking the average of $(Y_i(d = 1) - Y_i(d = 0))$. For the nine subjects in Table 5.1, the ATE is equal to: $(2 + 6 + 4 + 2 + 4 + 8 + 3 + 3 + 4)/9 = 4$.

Next, what is the ITT? This quantity is the average difference in $Y_i(d(1)) - Y_i(d(0))$ when subjects are assigned to the treatment group versus the control group. To calculate the ITT, calculate the difference in Y_i when a subject is assigned to the treatment versus control group subject by subject, sum these differences, and divide by the number of subjects. For the first observation, $Y_i(d(1))$ is 6 because when Subject 1 is assigned to the treatment group, the treatment is actually administered. $Y_i(d = 1)$ is therefore the relevant column. When assigned to the control group, $Y_i(d(0)) = 4$, and so the difference when assigned to treatment versus control for this observation is 2. Subject 2 is a Never-Taker. For this subject $d(1) = 0$, which means that if a treatment is assigned, it will not be administered. In that case, $Y_i(d(1)) = Y_i(d = 0) = 2$. For this subject, $Y_i(d(0)) = 2$ as well, so the effect of changing the treatment assignment from control to treatment is zero. Applying this reasoning to the nine subjects shows that the ITT $= (2 + 0 + 4 + 2 + 4 + 0 + 3 + 3 + 0)/9 = 2$.

2 Angrist, Imbens, and Rubin (1996) refer to the Complier average causal effect as the local average treatment effect, or LATE.

TABLE 5.1

Hypothetical schedule of potential outcomes assuming one-sided noncompliance

Observation	$Y_i(d=0)$	$Y_i(d=1)$	$d_i(z=0)$	$d_i(z=1)$	Type
1	4	6	0	1	Complier
2	2	8	0	0	Never-Taker
3	1	5	0	1	Complier
4	5	7	0	1	Complier
5	6	10	0	1	Complier
6	2	10	0	0	Never-Taker
7	6	9	0	1	Complier
8	2	5	0	1	Complier
9	5	9	0	0	Never-Taker

Notice that the calculation of the ITT is unaffected by the treatment effects among subjects who are Never-Takers. The ITT compares outcomes when subjects are assigned to treatment rather than control. Regardless of their assigned condition, Never-Takers are never treated and always exhibit their untreated potential outcomes.

Finally, what is the CACE, the average treatment effect for the subset of subjects who are Compliers? For the six Compliers, the ATE is $(2 + 4 + 2 + 4 + 3 + 3)/6 = 3$.

5.4 Identification of the CACE

In practice we do not have access to the complete schedule of potential outcomes for the subjects. Instead, we must make the most of what we observe after conducting an experiment with one-sided noncompliance. This section calls attention to an important theoretical relationship between the ITT, ITT_D, and CACE, which in turn sheds light on how one might estimate these quantities using experimental data.

Theorem 5.1 shows that the CACE can be calculated by forming the ratio of two intent-to-treat parameters. Returning to Table 5.1, we can apply the theorem to this particular set of potential outcomes. The $\text{ITT} = 2$, and the $\text{ITT}_\text{D} = 2/3$. According to the CACE theorem, $\text{CACE} = \text{ITT}/\text{ITT}_\text{D} = 3$. This number matches our earlier line-by-line calculation of the average treatment effect among Compliers.

Figure 5.1 illustrates the geometry behind the CACE theorem. Figure 5.1 also provides some important intuitions about the properties of the CACE and the challenges of estimating it using an experiment that encounters noncompliance. The

THEOREM 5.1

Identification of the Complier Average Causal Effect under One-sided Noncompliance

Using the definitions of ITT_D, ITT, and CACE in equations (5.4), (5.7), and (5.10), which assume non-interference:

$$ITT_D = E[d_i(z = 1) - d_i(z = 0)]. \tag{5.11}$$

$$\begin{aligned} ITT &= E[Y_i(z = 1)] - E[Y_i(z = 0)] \\ &= E[Y_i(z = 1, d(1))] - E[Y_i(z = 0, d(0))]. \end{aligned} \tag{5.12}$$

$$CACE = E[(Y_i(d = 1) - Y_i(d = 0)) \mid d_i(1) = 1]. \tag{5.13}$$

Assuming the excludability of treatment assignments and $ITT_D > 0$, the Complier average causal effect $= ITT/ITT_D$.

Proof: Expected potential outcomes among those in the treatment group can be written as a weighted average of treated potential outcomes among Compliers and untreated potential outcomes among Never-Takers.

$$\begin{aligned} E[Y_i(z = 1, d(1))] = {} & E[Y_i(z = 1, d = 1) \mid d_i(1) = 1]ITT_D \\ & + E[Y_i(z = 1, d = 0) \mid d_i(1) = 0](1 - ITT_D). \end{aligned} \tag{5.14}$$

Similarly, expected outcomes in the control group can be written as a weighted average of untreated potential outcomes among Compliers and Never-Takers.

$$\begin{aligned} E[Y_i(z = 0, d(0))] = {} & E[Y_i(z = 0, d = 0) \mid d_i(1) = 1]ITT_D \\ & + E[Y_i(z = 0, d = 0) \mid d_i(1) = 0](1 - ITT_D). \end{aligned} \tag{5.15}$$

By substitution, the ITT, too, may be expressed as a weighted average of the ITT among Compliers and the ITT among Never-Takers.

$$\begin{aligned} ITT = {} & E[\{Y_i(z = 1, d = 1) - Y_i(z = 0, d = 0)\} \mid d_i(1) = 1]ITT_D \\ & + E[\{Y_i(z = 1, d = 0) - Y_i(z = 0, d = 0)\} \mid d_i(1) = 0](1 - ITT_D). \end{aligned} \tag{5.16}$$

The exclusion restriction implies that the second part of this expression is zero, because treatment assignment has no effect on the untreated potential outcomes of Never-Takers. The assumption that $ITT_D > 0$ implies that there is at least one Complier. Applying the exclusion restriction to Compliers and dividing the ITT by the ITT_D gives:

$$\frac{ITT}{ITT_D} = E[\{Y_i(d = 1) - Y_i(d = 0)\} \mid d_i(1) = 1] = CACE. \tag{5.17}$$

FIGURE 5.1

Estimating the Complier Average Causal Effect from an experiment with one-sided noncompliance.

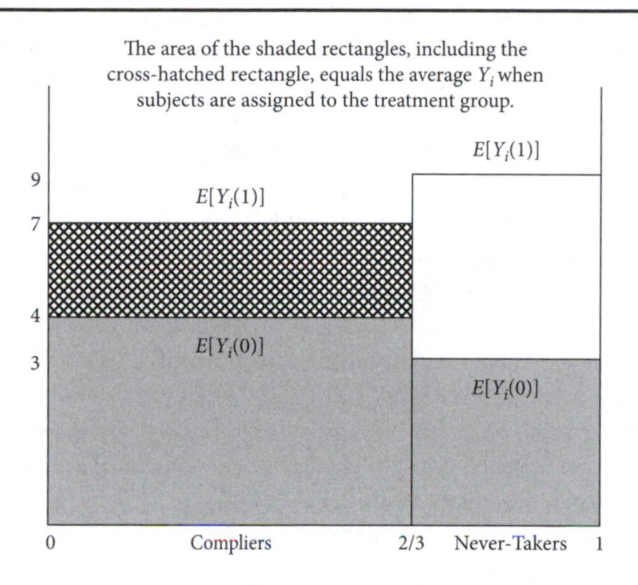

Panel A: Treatment group

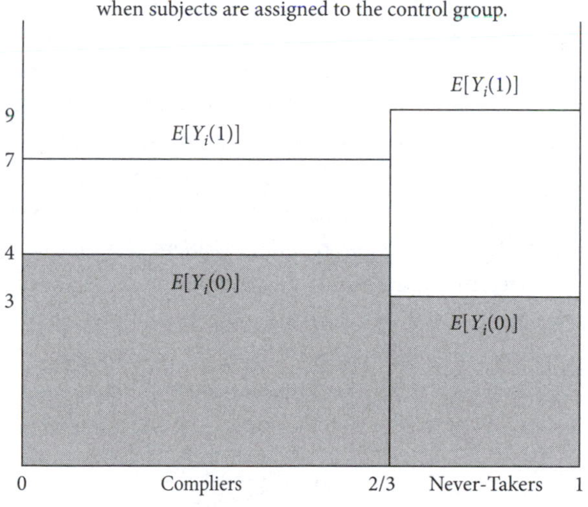

Panel B: Control group

Note: ITT = (7 − 4) * (2/3) = 2, the area of the cross-hatched section. $E[d_i(1)] = \text{ITT}_D = 2/3$. CACE = 2/(2/3) = 3, the area of the cross-hatched section divided by the width of the Compliers column, which is the height of the cross-hatched section.

figure depicts a subject pool comprised of Compliers and Never-Takers. The X-axis indicates the proportion of each type in the subject pool. The Y-axis reflects the expected potential outcome. Panel A depicts the subjects' outcomes when they are assigned to the treatment condition, and Panel B shows the subjects' outcomes when they are assigned to the control condition. The height of each column represents the average potential outcome for each type of subject, and column widths represent the proportion of each type of subject in the subject pool.

The diagram illustrates the ITT, ITT_D, and CACE. First, the total shaded area in Panel A (including the cross-hatched region) represents the average outcome for the subject pool if all subjects were assigned to the treatment group. The shaded area in Panel B represents the average outcome if all subjects were assigned to the control group. The difference between the shaded areas in Panels A and B equals the average effect of being assigned to treatment, the ITT. In other words, the geometric analogue to the change in outcomes produced by assignment to the treatment condition is the difference between the total areas of the shaded rectangles for the treatment group and the control group. The ITT is depicted as the cross-hatched region in Panel A. This cross-hatched area $(3 \cdot (2/3) = 2)$ represents the change in average outcomes that occurs when Compliers are treated.

Second, the width of each Complier rectangle indicates the ITT_D. The ITT_D is the proportion of subjects who are Compliers.

Finally, the CACE is obtained by subtracting the shaded area in the control group from the shaded area in the treatment group (the ITT) and dividing the difference by the proportion of Compliers in the sample pool (the ITT_D). Algebraically, this ratio reflects the equality that $CACE = ITT/ITT_D = 2/(2/3) = 3$. Intuitively, this operation is equivalent to redrawing the figure so as to exclude Never-Takers. If Compliers ranged from 0 to 1 on the X-axis, the shaded area in the treatment group would be 7, the shaded area in the control group would be 4, and the difference would be 3.

Figure 5.1 also indicates why the ATE cannot be estimated from an experiment that fails to treat some subjects in the treatment group. The empty box in Panel A represents the average treatment effect among Never-Takers. If Never-Takers were (somehow) exposed to the treatment, the empty region and the cross-hatched region would together sum to the ATE for the entire sample. But the empty box never materializes because our experiment fails to treat Never-Takers. The same point can be illustrated algebraically. The average treatment effect for the entire subject pool may be viewed as a weighted average of the treatment effect among Compliers and Never-Takers:

$$\begin{aligned} ATE &= E[Y_i(d=1) - Y_i(d=0) \mid d_i(1) = 1]ITT_D \\ &\quad + E[Y_i(d=1) - Y_i(d=0) \mid d_i(1) = 0](1 - ITT_D) \\ &= (3)\left(\frac{2}{3}\right) + (6)\left(\frac{1}{3}\right) = 4. \end{aligned} \tag{5.18}$$

The first term is the ITT (the area of the cross-hatched region), which is estimated by comparing average outcomes in the treatment and control groups. The second term, on the other hand, cannot be estimated. The experiment provides no information about the average treatment effect among Never-Takers.

In the next section we estimate the Complier average causal effect using experimental data. Theorem 5.1 says that the CACE is equal to the ITT divided by the ITT_D. We do not observe the quantities, but experiments enable us to construct unbiased estimates of the ITT and ITT_D, and the ratio of these estimates provides a consistent estimate of the CACE. Before proceeding to estimation and examples, a few points should be emphasized.

1. An experiment with one-sided noncompliance enables the researcher to estimate the average effect of assignment to treatment (the ITT) and the ATE for a subset of the subject pool. This subset consists of Compliers, those subjects who would be treated if assigned to the treatment group.
2. Whether a subject is a Complier or a Never-Taker is in part a function of the experimental design and the context in which the experiment is conducted. An experimenter can alter the share of Compliers by providing incentives, making the treatment more appealing, or working harder to contact subjects assigned to the treatment group.
3. This chapter has yet to discuss random assignment. When deriving the relationship between the ITT and the CACE, we made frequent reference to z, which denotes assignment to the treatment or control group, but we made no assumptions about *how* subjects are assigned. Independence of Z_i and potential outcomes (a property which follows from random assignment of subjects to groups) allows for unbiased *estimation* of ITT (Chapter 2), but the theorem linking ITT and CACE holds no matter how Z_i is assigned.
4. Increasing the treatment rate (increasing ITT_D) does not necessarily lower the Complier average causal effect. Confusion sometimes arises because the ITT_D is in the denominator of equation (5.17). However, as should be clear from Figure 5.1, increasing the share of Compliers may also change the numerator (the ITT, represented by the area of the cross-hatched rectangle), depending on how these extra Compliers respond to the treatment. Raising the treatment rate may cause the CACE to increase or decrease.
5. The overall ATE is a weighted average of the ATE for each subject type, with the weights equal to the proportion of each type within the subject pool. The ATE for Compliers may be larger than, smaller than, or equal to the ATE for Never-Takers. When making generalizations based on the estimated CACE, bear in mind that the average effect among Compliers may be a poor guide to the average effect among Never-Takers. Programs that work well among Compliers may

FIGURE 5.2

Illustration of an exclusion restriction violation.

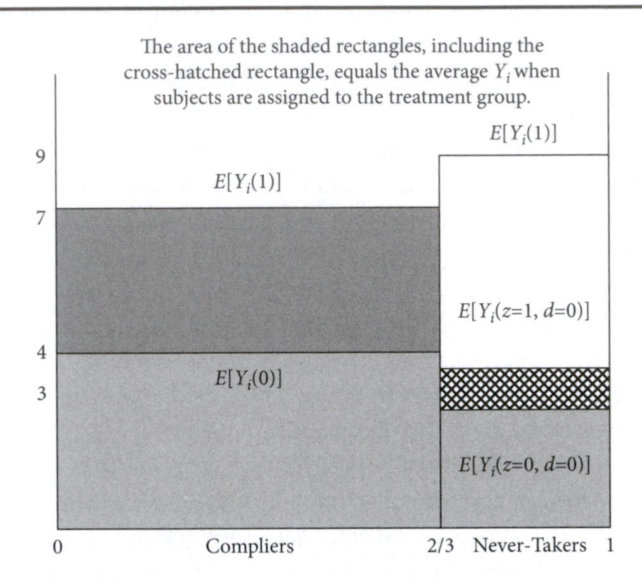

Panel A: Treatment group

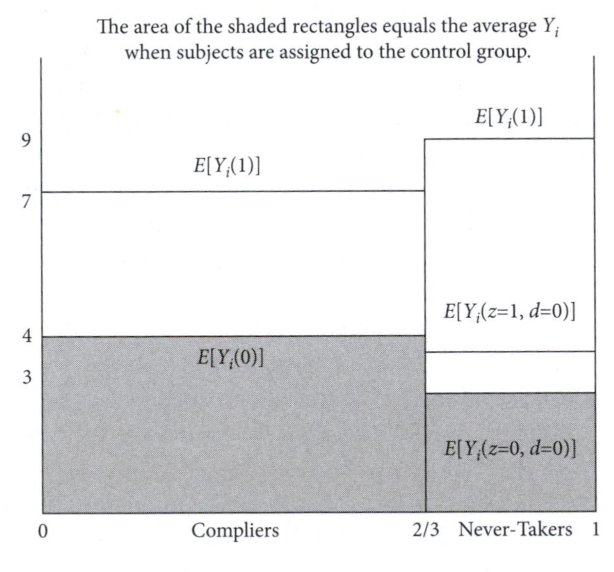

Panel B: Control group

Note: The exclusion restriction is violated here because Never-Takers' potential outcomes are affected by treatment assignment, as depicted by the cross-hatched region over the Never-Takers column in Panel A.

produce disappointing results when applied to society as a whole, which is comprised of both Compliers and Never-Takers.

6. The exclusion restriction plays a critical role in the CACE theorem. This assumption states that treatment assignment has no effect beyond the treatment that is actually received. One implication is that treatment assignment should have no effect on Never-Takers. Figure 5.2 illustrates the consequences of violating the exclusion restriction. In this scenario, the average untreated potential outcome among Never-Takers increases when they are assigned to the treatment group. The estimation strategy based on the CACE formula no longer works. In the example depicted in Figure 5.2, violation of the exclusion restriction raises the average outcome in the treatment group, but the ITT_D remains the same. As a result, the ratio $\text{ITT}/\text{ITT}_\text{D}$ overestimates the true CACE. Exclusion restriction violations may take various forms and may lead to positive or negative biases.
7. When the ITT_D is close to zero, even a slight violation of the exclusion restriction may severely bias the estimation of the CACE. Consider, again, Figure 5.2. Suppose, holding all else equal, we narrow the column of Compliers (reducing the ITT_D) and widen the column of Never-Takers. Our estimate of the CACE becomes increasingly biased when Never-Takers are affected by treatment assignment. This point can also be seen algebraically. The ratio of the ITT to ITT_D under a violation of the exclusion restriction is:

$$\frac{ITT}{ITT_\text{D}} = \text{CACE} + \left(\frac{1 - \text{ITT}_\text{D}}{\text{ITT}_\text{D}}\right) E[Y_i(z = 1, d = 0) - Y_i(z = 0, d = 0)]. \tag{5.19}$$

All else being equal, as the ITT_D decreases, a violation of excludability leads to more severe bias.

5.5 Estimation

In this section, we show how experiments with one-sided noncompliance can be used to estimate the Complier average causal effect. Random assignment now plays a crucial role. We expand the random assignment assumption from Chapter 2 to accommodate noncompliance:

$$Z_i \perp\!\!\!\perp Y_i(z, d(z)) \text{ and } Z_i \perp\!\!\!\perp D_i(z). \tag{5.20}$$

In other words, assignment is independent of potential outcomes. Even if you knew the schedule of potential outcomes prior to a hypothetical random assignment, you could not predict who would be assigned to the treatment group.

To illustrate estimation in the presence of one-sided noncompliance, we present results from a voter mobilization experiment conducted in New Haven.[3] During the weeks leading up to the 1998 general elections, more than 30,000 registered voters were randomly assigned to treatment and control groups. The sample employed in this example is restricted to 7,090 individuals living in one-voter households. (We exclude households with more than one voter in order to avoid complications arising from clustered random assignment.) These individuals were assigned either to a control group, which received no contacts from the campaign, or a treatment group, which was visited by canvassers who stressed the importance of voting. After the election, researchers consulted public records to determine which registered voters cast ballots in this election. Table 5.2 shows the results of the experiment.

In Chapter 2 we estimated the average effect of being assigned to the treatment group versus the control group by calculating the difference in means. Using the same approach, we obtain unbiased estimates of the intent-to-treat effect:

$$\widehat{ITT} = 41.38 - 37.54 = 3.84 . \quad (5.21)$$

Those assigned to the treatment group were 3.84 percentage points more likely to vote. The estimated ITT may be a useful thing to know. If you are conducting an evaluation of a program, you can use the ITT to assess the program's output in relation to its cost. If you wish to project how many votes would be produced by a similar intervention, using a similar level of canvassing intensity, conducted under similar circumstances, you can obtain an estimate by multiplying the estimated ITT by the size of the group targeted for canvassing.

Sometimes researchers are less interested in evaluating the overall effectiveness of programs, and focus instead on estimating average treatment effects among Compliers. The first step in estimating the CACE is to write down a series of equations that makes explicit the posited relationship between observed quantities and unknown parameters in our application. In the New Haven study, three observed

TABLE 5.2

Voter turnout by experimental group, New Haven voter mobilization experiment

	Treatment group	Control group
Turnout rate among those contacted by canvassers	54.43 (395)	
Turnout rate among those not contacted by canvassers	36.48 (1,050)	37.54 (5,645)
Overall turnout rate	41.38 (1,445)	37.54 (5,645)

Note: Entries are percent voting, with number of observations in parentheses. Sample restricted to households containing a single registered voter.

3 Gerber and Green 2000.

quantities are the voter turnout rate in the treatment group, the voter turnout rate in the control group, and the rate at which subjects in the assigned treatment group were actually treated. These three quantities enable us to form unbiased estimates of ITT and ITT_D. However, it is not the case that the ratio of two unbiased estimators is an unbiased estimator for the ratio of the two estimands.[4] The ratio of $\widehat{\text{ITT}}$ and $\widehat{\text{ITT}}_\text{D}$ does provide a *consistent* estimator of the CACE: as the sample size grows, the estimated ratio converges to the true value of the CACE.

In the New Haven experiment, what proportion of the subject pool was Compliers? As Table 5.2 indicates, canvassers routinely found it difficult to speak with subjects in the treatment group. In order to estimate the ITT_D, we compare the average value of d_i in the treatment and control groups:

$$\widehat{\text{ITT}}_\text{D} = \frac{395}{1445} - \frac{0}{5{,}645} = 0.273. \quad (5.22)$$

The experiment was able to treat 27.3% of the treatment group. No one in the control group was treated. We therefore estimate that Compliers make up 27.3% of the subject pool.

Using the CACE theorem, an estimate of the CACE is formed by the ratio of these two estimates:

$$\widehat{\text{CACE}} = \frac{\widehat{\text{ITT}}}{\widehat{\text{ITT}}_\text{D}} = 3.84/.273 = 14.1. \quad (5.23)$$

The estimated average treatment effect of the canvassing treatment among Compliers is a 14.1 percentage point increase in the probability of voting.

---

4 For example, suppose coin A is flipped and comes up heads with probability 0.5. Coin B is flipped independently and also comes up heads with probability 0.5. Score each coin's outcome as 2 if heads and 1 if tails. The expectation of each coin is 1.5. What is the expectation of the ratio of A/B? The answer is not 1 but rather:

$$\frac{\left(\frac{1}{1}+\frac{1}{2}+\frac{2}{1}+\frac{2}{2}\right)}{4} = 1.125.$$

The general formula for the expectation of a ratio is:

$$E\left(\frac{A}{B}\right) = \frac{E(A) - \text{Cov}\left(\frac{A}{B}, B\right)}{E(B)},$$

assuming that $B > 0$. In this example, this formula gives:

$$E\left(\frac{A}{B}\right) = \frac{1.5 - \left(-\frac{0.75}{4}\right)}{1.5} = 1.125.$$

The same principle explains why $\frac{\widehat{\text{ITT}}}{\widehat{\text{ITT}}_\text{D}}$ gives a biased estimate of the CACE. This bias tends to be negligible in large samples because the covariance between $\frac{\widehat{\text{ITT}}}{\widehat{\text{ITT}}_\text{D}}$ and $\widehat{\text{ITT}}_\text{D}$ diminishes as the sampling variability of $\widehat{\text{ITT}}$ and $\widehat{\text{ITT}}_\text{D}$ diminishes.

5.6 Avoiding Common Mistakes

We are now in a position to return to some of the issues raised at the beginning of the chapter. As Table 5.2 shows, an experiment with one-sided noncompliance provides data on average outcomes for three groups of subjects: the control group, the treated subjects in the treatment group, and the untreated subjects in the treatment group. We now know how to use these results to estimate the average treatment effect among Compliers: compare those *assigned* to the treatment and control groups and divide this ITT estimate by the estimated proportion of Compliers. What was wrong with the alternatives described at the start of the chapter?

What if the analyst ignores the failure-to-treat and compares the treatment and control groups? This comparison estimates the ITT, not the ATE. The ITT can be informative, as it summarizes the net impact of an intended intervention, but quite often researchers want to learn about the effectiveness of the treatment that was administered, not the treatment that was assigned.

What about comparing the control group to those who actually receive the treatment? This comparison, in expectation, produces the following:

$$\begin{aligned} &E[Y_i(d=1)\,|\,(D_i(1)=1] - E[Y_i(z=0)] \\ &\quad = E[Y_i(d=1)\,|\,D_i(1)=1] - E[Y_i(d=0)\,|\,D_i(1)=1] \\ &\quad + E[Y_i(d=0)\,|\,D_i(1)=1] - E[Y_i(z=0)] \\ &\quad = \text{CACE} + E[Y_i(d=0)\,|\,D_i(1)=1] - E[Y_i(z=0)]. \end{aligned} \tag{5.24}$$

Since

$$E[Y_i(z=0)] = E[Y_i(z=0)\,|\,D_i(1)=1]\,\text{ITT}_\text{D} + E[Y_i(z=0)\,|\,D_i(1)=0](1-\text{ITT}_\text{D}),$$

substitution gives

$$\text{CACE} + E[Y_i(d=0)\,|\,D_i(1)=1] - \{E[Y_i(z=0)\,|\,D_i(1)=1]\,\text{ITT}_\text{D} + E[Y_i(z=0)\,|\,D_i(1)=0](1-\text{ITT}_\text{D})\}. \tag{5.25}$$

This expression may be simplified further by noting that

$$E[Y_i(z=0)\,|\,D_i(1)=1] = E[Y_i(d=0)\,|\,D_i(1)=1]$$

and that

$$E[Y_i(z=0)\,|\,D_i(1)=0] = E[Y_i(d=0)\,|\,D_i(1)=0],$$

giving

$$\text{CACE} + \{E[Y_i(d=0)\,|\,D_i(1)=1] - E[Y_i(d=0)\,|\,D_i(1)=0]\}(1-\text{ITT}_\text{D}). \tag{5.26}$$

Estimating the CACE by comparing the treated with the control group may lead to severe bias. This estimator will, in expectation, reveal the CACE plus an extra term. This bias term is the product of the share of Never-Takers in the subject pool times the difference between the average untreated potential outcomes among Compliers and Never-Takers. Bias disappears when $ITT_D = 1$ (the sample contains only Compliers) or when the untreated Never-Takers are a perfect proxy for the untreated Compliers. The formula illustrates the problem with comparing the actually treated with the entire control group. When one conducts an experiment with one-sided noncompliance, the control group contains both Compliers and Never-Takers. The dubious rationale behind calculating the treatment effect by comparing the treated to the control group is that the control group might provide an estimate of what the outcome would have been for the actually treated had they not been treated. This approach presupposes that Never-Takers and Compliers have the same average untreated potential outcomes. This assumption has no connection to the experimental design and should be viewed with skepticism.

Returning to the example of canvassing, it seems implausible to assume that subjects who can be successfully reached by canvassers have the same average untreated potential outcomes as those who cannot be reached. Reflecting on how Compliers and Never-Takers might be different suggests some intuitions about why voting rates among these types of subjects may differ. Never-Takers may be out of town. They may tend to work long hours and rarely be at home. They may have moved away. They may prefer not to answer the door when a stranger knocks. Any or all of these factors might alter the probability that a subject votes. And in fact, we see that while 37.5% of the control group voted, only 36.5% of those who were in the treatment group but not canvassed voted. A comparison of the treated with the control group would have exaggerated the effectiveness of canvassing: $54.4 - 37.5 = 16.9$ percentage points, as opposed to the estimate of 14.1 percentage points obtained in equation (5.23). This estimate would have been even larger if Never-Takers in the treatment group had been lumped in with the control group, a mistake that researchers occasionally make when analyzing experiments with one-sided noncompliance.

When experiments encounter high rates of noncompliance, a question is sometimes raised about whether the results are "biased" because the subjects who are treated are not representative of the pool of subjects. Although this use of the term *bias* is somewhat loose, the query does highlight an important issue: the feasible estimand for an experiment with noncompliance is the CACE, not the ATE. The methods described here produce consistent estimates of the CACE, the average treatment effect among that particular subset of the experiment's subject pool that moves from "untreated" to "treated" when assigned to the treatment condition. The CACE may be quite different from the ATE in many applications. Experiments that use encouragement designs estimate the average treatment effect among those who voluntarily enlist for a program; those who do not enlist may have different average treatment effects—indeed, the subjects may have elected not to participate because they sensed that the program would do little for them.

Given the difficulty of drawing inferences about Never-Takers' ATE, researchers sometimes adduce indirect evidence. One tactic is to demonstrate that among subjects in the treatment group, those who receive the treatment (Compliers) have the same background attributes as those who go untreated (Never-Takers). This evidence is not sufficient to show that the CACE is similar to the average treatment effect. The relevant comparison is the treatment effect, not the covariates or the level of $Y_i(d = 0)$. Figure 5.1 could be adjusted so that all groups have the same average $Y_i(d = 0)$ values but different average values of $Y_i(d = 1) - Y_i(d = 0)$.

A more instructive approach is to change the experimental design to raise the proportion of Compliers. As Figure 5.1 illustrates, if the untreated share of the population approaches zero (the Never-Takers column narrows), the treatment effect for this type would have to be very different from the other subjects in order to produce enough "area" for there to be a large difference between the CACE and ATE. Although raising the share of the treatment group that is successfully treated typically reduces the difference between the CACE and ATE, in theory anything could happen. In a pathological case, the marginal treated individual may have a more atypical treatment effect than the average of those "easily" treated, in which case the gap between ATE and CACE will grow as treatments are applied to the first few extra Compliers. As shown in the exercises, the CACE and ATE gap can be investigated empirically by observing treatment effects when the effort expended to reach subjects is varied randomly.[5] Investigations of this kind are instructive, but unless compliance rates approach 100%, there remains some uncertainty about whether the experiment provides an unbiased estimate of the ATE.

The implications of measuring the CACE rather than the ATE depend on the research objectives and whether treatment effects vary across individuals. Sometimes the treatment effect among those who are treated is exactly what the researcher is interested in, and so failure-to-treat is a feature of the experiment, not a bug. For example, if a campaign is interested in the returns to a particular type of canvassing sweep through a neighborhood, the campaign wants to know the response of the people whom the effort will likely reach, not the hypothetical responses of people who do not open the door to canvassers or who have moved away.

Confusion about the estimand is sometimes apparent when critics of field experiments cite noncompliance as a problem that is overcome by laboratory experiments, which ordinarily treat everyone assigned to the treatment group. This comparison is misleading. Although the inability to treat everyone in the assigned treatment group

5 This approach parallels the strategy of investigating the effects of survey non-response by making extra efforts to interview non-respondents and assessing differences between the lower and higher response rate samples (Flemming and Parker 1998).

is a conspicuous problem in many field experiments, it is a common, if disguised, problem in laboratory experiments as well. The typical laboratory experiment randomly allocates subjects after compliance is assured and therefore achieves a 100% treatment rate. This research design does not "solve" the problem of measuring the average treatment effect for a pool of subjects (ATE) as opposed to the average treatment effect among those who comply if assigned to treatment (CACE). The estimand for a laboratory experiment is the ATE for the particular group of people who show up for the experiment, pass the screening requirements, sign the consent form, and so forth. This group of subjects' ATE may be different from the ATE among the "Never-Takers" who never got to the point where they were assigned to a treatment or control group. Although compliance is 100% among those who are assigned to experimental groups, failure-to-treat occurs at the subject recruitment and retention stages. Lab experiments, like field experiments that encounter noncompliance, estimate Complier average causal effects.

5.7 Evaluating the Assumptions Required to Identify the CACE

The equivalence of the ITT/ITT_D ratio and the CACE depends on *substantive assumptions*. By substantive assumptions, we mean conjectures about how subjects are affected by their treatment status and the treatment status of others in the context of the experiment. The plausibility of these assumptions and the degree to which they may be violated (and consequences of the violation) need to be assessed based on the particular features of the experiment: how the treatments are delivered, the context in which subjects are treated, the subjects' attributes, the procedures used to measure outcomes, and other details of the research. Let's consider the plausibility of key substantive assumptions in the canvassing experiment.

5.7.1 Non-Interference Assumption

This assumption states that the potential outcomes for each individual are unaffected by the assignment or treatment of any other individual. This assumption will be violated if subject i's potential outcomes change depending on whether her neighbor j is canvassed. There are several ways this might happen. For example, j might tell i about the canvassing effort or her newfound enthusiasm for participating in the upcoming election. Suppose that as a result, subject i becomes more inclined to vote. To see how contagious enthusiasm might produce a downward bias in this application, suppose that all of those who are untreated have their untreated outcome raised by a fixed amount. Since the control group contains more untreated people than the treatment group, contagious enthusiasm will tend to reduce the apparent difference between the turnout rates of the treatment and control groups ($\widehat{ITT}$), and, since the $\widehat{ITT}_D$ estimate is unaffected, reduce the $\widehat{CACE}$. In other words, the average effect of the experimental intervention on Compliers will be underestimated, and increased voting among the control group will not be credited to the intervention. Potential violations of the non-interference assumption often involve contagion, displacement, or communication. If the treatment spreads from the treated to the untreated, the distinction between the treatment and control group may be blurred, dampening the apparent effect of the treatment. Biases due to interference may be positive or negative depending on the ways in which potential outcomes are affected by second-hand treatment.

Experiments can be designed to reduce their vulnerability to interference. When everyone in a particular place is canvassed except for a sliver of the population, it is quite likely that some of the canvassing effects will be transmitted to those not canvassed. In this case, interference may be reduced by keeping the density of treatment to a low level. Another approach is to measure the outcome soon after the treatment is administered, before it has an opportunity to spread.

5.7.2 Exclusion Restriction

The exclusion restriction states that a subject's treatment assignment does not matter once we account for whether a subject is actually treated. Violations of the exclusion restriction could lead us to systematically over- or underestimate the CACE of the canvassing treatment. First consider an example of how a violation might lead one to underestimate the CACE. Suppose a zealous political group notices that one's experimental canvassing effort has skipped certain houses. These houses, which were assigned to the control group, become the focus on this political group's compensatory mobilization efforts. If the political group's efforts are as successful as the experimental mobilization campaign's, the control group will vote at the same rate as the treatment group. The estimated ITT (and therefore the estimated CACE) will be zero even if the experimental canvassing campaign were highly effective.

Now consider an example of an exclusion restriction violation that leads to an overestimate of the CACE. Suppose that the attempt to treat someone assigned to the treatment group has an effect on the subject apart from the effect of the treatment itself. For instance, if the canvasser leaves a persuasive handwritten note at all the houses where the target subject is not at home, Never-Takers assigned to the treatment group will have a different potential outcome in the untreated state than Never-Takers who are assigned to the control group, since only those assigned to the treatment group get the handwritten note. This violation of the exclusion restriction may exaggerate the apparent effect of canvassing. Suppose the handwritten notes increase turnout: the $\widehat{ITT}$ increases, the $\widehat{ITT}_D$ remains unchanged, and therefore the

$\widehat{CACE}$ increases. In effect, some people received a treatment but were not classified as treated. The entire boost in the treatment group's outcomes is credited to canvassing, which exaggerates its effectiveness.

Every experiment involves *some* threat to the exclusion restriction because the treatment as defined by the researcher inevitably departs in some way from the treatment that is actually administered. Canvassers deliver the treatment—a short speech encouraging people to vote—but inevitably they also say and do other things as well. They walk through the neighborhoods of those assigned to the treatment group. They carry clipboards. They ring doorbells. They dodge aggressive pets. Substantive assumptions inevitably come into play when a researcher invokes the exclusion restriction and, in effect, stipulates that countless factors that coincide with the assigned treatment have negligible effects on potential outcomes.

5.8 Statistical Inference

In order to obtain parameter estimates, it is convenient to use regression. Two kinds of regression models are presented in this section. The first is ordinary least squares regression, which is used to estimate the ITT and ITT_D. The second is two-stage least squares regression, which is used to estimate the CACE.

To estimate the ITT, regress the outcome variable (Y_i) on the assigned treatment (z_i). The outcome variable in the New Haven voter mobilization experiment is VOTED, which is scored 1 if the subject voted and 0 otherwise. The assigned treatment is called ASSIGNED, which is scored 1 if the subject was assigned to receive a visit from a canvasser and 0 otherwise. Box 5.4 presents the results of this reduced form regression.[6] Because this regression is equivalent to a difference-in-means estimator, the results are identical to what we obtain when we subtract the voting rate in the control group from the voting rate in the treatment group.

The estimated coefficient of 0.03845 implies that *assignment* to the canvassing condition generated a 3.845 percentage point increase in turnout rates. In order to obtain a confidence interval around the estimated ITT, we use the method described in Chapter 3. We simulate the full schedule of potential outcomes by assuming a constant effect of ASSIGNED_i and conduct 100,000 random assignments. The 95% confidence interval extends from 1.04 to 6.66 percentage points. Because the rate of voting in the control group is 37.54%, assignment to treatment appears to have generated an increase in turnout of between 2.8% and 17.7%.

---

6 The term *reduced form regression* refers to a regression of outcomes on assigned treatments and covariates. This type of analysis contrasts with *structural estimation*, which attempts to gauge parameters such as the CACE.

BOX 5.4

Estimation of the ITT Using Data from the 1998 New Haven Voter Mobilization Experiment

```
itt_fit <- lm (VOTED ~ ASSIGNED)
coeftest (itt_fit,vcovHC (itt_fit))
```

OLS Regression with Robust Standard Errors

	Estimate	Std. Error	t value	Pr(>|t|)
(Intercept)	0.375376	0.006446	58.2344	< 2.2e-16
ASSIGNED	0.038464	0.014479	2.6565	0.007914

In this analysis, voter turnout (VOTED) is regressed on treatment assignment. Turnout is scored 1 if the subject votes and 0 otherwise. Treatment assignment (ASSIGNED) is scored 1 if the subject is assigned to the treatment group and 0 otherwise.

Source: Gerber and Green 2000.

In order to test the sharp null hypothesis that the ITT is zero for all subjects—a null hypothesis that also implies that the CACE is zero for all subjects—we use randomization inference. The procedure is identical to the one described in Chapter 3: for every subject, we assume $Y_i(1) = Y_i(0)$. We simulate a large number of simple random assignments and calculate the number of hypothetical experiments that generate an estimated ITT at least as large as what we in fact observe. Here, we use a one-sided test because the canvassing experiment was conducted with the expectation that mobilization would produce a positive effect on turnout. The results show that just 416 of the 100,000 estimates are as large as .038464 (the observed coefficient), which means that the p-value is 0.004. This p-value leads us to reject the sharp null hypothesis of no effect for any subject.

The proportion of Compliers is estimated by calculating the ITT_D. Regress actual treatment (d_i) on assigned treatment (z_i). In the context of this experiment, regress TREATED, which is scored 1 if canvassers spoke to the subject and 0 otherwise, on ASSIGNED. Box 5.5 presents the results of this so-called "first stage" regression. The results again reproduce the calculations based on the percentages from Table 5.2. The intercept of zero in this equation indicates that no one in the control group was contacted, in keeping with the definition of one-sided noncompliance. The coefficient 0.273 indicates that assignment to the treatment group caused 27.3% of the targeted subjects to be treated. In other words, the estimated share of Compliers in

BOX 5.5

Estimation of the ITT_D Using Data from the 1998 New Haven Voter Mobilization Experiment

```
itt_d_fit <- lm (TREATED ~ ASSIGNED)
coeftest (itt_d_fit,vcovHC (itt_d_fit))
```

OLS Regression with Robust Standard Errors

	Estimate	Std. Error	t value	Pr(>|t|)
(Intercept)	5.8120e-15	2.3349e-16	24.892	< 2.2e-16
ASSIGNED	2.7336e-01	1.1733e-02	23.299	< 2.2e-16

In this analysis, actual treatment (TREATED) is regressed on treatment assignment. Treatment is scored 1 if the subject spoke with a canvasser and 0 otherwise. Treatment assignment (ASSIGNED) is scored 1 if the subject is assigned to the treatment group and 0 otherwise.

Source: Gerber and Green 2000.

the subject pool is 27.3%. The 95% confidence interval suggests that this proportion ranges from 25.0% to 29.6%.

A convenient way to estimate the CACE is to estimate a two-stage least squares (2SLS) regression. The regression model involves two equations. The first is a model of the outcome variable:

$$\text{VOTED}_i = \beta_0 + \beta_1 \text{TREATED}_i + u_i. \tag{5.27}$$

The key parameter of interest is β_1, the CACE. Notice that this model of outcomes invokes the exclusion restriction; ASSIGNED_i does not appear in equation (5.27) because it is assumed to have no effect on outcomes over and above the effects of actual treatment. The second model accounts for the possibility that TREATED_i is "endogenous," or related to unobserved factors (u_i) that affect outcomes. In the second equation, ASSIGNED_i is an "instrumental variable," a statistical term that refers to a variable that predicts TREATED_i and is assumed to be independent of u_i. When ASSIGNED_i is randomly assigned and the exclusion restriction holds, this assumption is satisfied by design.

$$\text{TREATED}_i = \alpha_0 + \alpha_1 \text{ASSIGNED}_i + e_i. \tag{5.28}$$

In this application (and in most experiments), 2SLS regression is equivalent to instrumental variables regression. The two estimators are the same whenever the number of variables excluded from the outcome equation is equal to the number of endogenous independent variables in the outcome equation. Applying 2SLS to the New Haven data generates an estimated coefficient of 0.1407, which confirms the estimate of 14.1 percentage points that we obtained when we calculated the CACE by hand in equation (5.23).

In order to form a confidence interval, we need to create a full schedule of potential outcomes, which is problematic because potential outcomes now include both $Y_i(d(z))$ and $d_i(z)$. Ideally, we would like to be able to stipulate a CACE for the Compliers and some other average treatment effect for the Never-Takers; unfortunately, for subjects in the control condition, we have no way of distinguishing these two types of subjects. We therefore fall back on conventional methods for estimating confidence intervals, which assume a normal sampling distribution. Multiplying the estimated standard error (0.052434) by 1.96 and adding and subtracting this product from the estimated CACE (0.140711) gives a 95% confidence interval that ranges from 3.8 to 24.3 percentage points. Although the hypothesis test performed above indicates that canvassing increased turnout, the estimated CACE is subject to considerable sampling uncertainty. Our best guess is that canvassing increased turnout rates among Compliers by 14.1 percentage points, but the true CACE could plausibly be as low as 3.8 percentage points or as high as 24.3 percentage points.

BOX 5.6

Estimation of the CACE Using Data from the 1998 New Haven Voter Mobilization Experiment

```
cace_fit <- ivreg (VOTED ~ TREATED,~ASSIGNED)
coeftest (cace_fit, vcovHC (cace_fit))

Instrumental Variables Regression with Robust Standard Errors
---------------------------------------------------------------------------
                 Estimate      Std. Error       t value        Pr(>|t|)
---------------------------------------------------------------------------
(Intercept)      0.375376        0.006446       58.2344          <2e-16
TREATED          0.140711        0.052434        2.6836          0.0073
---------------------------------------------------------------------------
```

In this instrumental variables regression model, voter turnout is regressed on actual treatment (TREATED) using treatment assignment (ASSIGNED) as an instrument. The results suggest that contact by canvassers increased turnout among Compliers by 14.1 percentage points.

Source: Gerber and Green 2000.

5.9 Designing Experiments in Anticipation of Noncompliance

Noncompliance not only prevents researchers from estimating the ATE, it also makes estimation of the CACE challenging. Although 2SLS provides a consistent estimator for the CACE, the precision of this estimator deteriorates as the rate of noncompliance increases. When an experiment is large and the CACE is close to zero, the standard error of the 2SLS estimator is approximately equal to the standard error of the ITT estimator divided by the proportion of Compliers:

$$SE(\widehat{CACE}) \approx \frac{SE(\widehat{ITT})}{ITT_D}. \quad (5.29)$$

This relationship is illustrated by the estimated standard errors from the New Haven experiment. Regression estimates the standard error for the $\widehat{ITT}$ to be 0.0145, and the estimated standard error for the $\widehat{CACE}$ is 0.0524. The ratio of the two estimated standard errors is approximately 0.273, the estimated ITT_D.[7]

The formula for the standard error of the CACE has several important implications. The confidence interval around the estimated CACE is approximately equal to the ITT confidence interval scaled by the ITT_D. The two intervals coincide when there is no noncompliance ($ITT_D = 1$). All else being equal, as the proportion of Compliers shrinks, the standard error of the estimated CACE increases. If the ITT_D were to fall from 1.00 to 0.10, the standard error would increase by a factor of 10. In order to compensate for a tenfold increase in the standard error, one needs an experimental sample that is roughly 100 times as large. This unfortunate fact accounts for the enormous size of experiments that are designed to offset high rates of noncompliance.

Another way to mitigate statistical uncertainty is to employ a placebo design. This type of experiment proceeds in two steps. First, subjects are recruited to receive a treatment. Second, given compliance, subjects are randomly allocated to two groups. The treatment group receives the treatment in the usual way. The placebo group receives a "non-treatment" that is assumed to have no effect on the outcome of interest. For example, using a placebo design, Nickerson conducted a canvassing experiment in which subjects who answered the door were randomly allocated to a treatment group (which encouraged voting) or a placebo group (which encouraged recycling).[8] The CACE can be estimated by comparing the outcomes for those given the canvassing treatment and those given the "non-treatment." The function of the

---

7 Note that these estimated standard errors are the so-called robust standard errors generated by most statistical packages. The estimator for the robust standard error of a binary treatment is approximately equal to the estimator in equation (3.6).

8 Nickerson 2008.

placebo arm is to isolate a random sample of Compliers whose untreated potential outcomes can be measured.

It is instructive to explore the statistical logic underlying the placebo design. Recall that one of the challenges in estimating the Complier average causal effect is that the control group is a mixture of Compliers and Never-Takers. In effect, the placebo design screens out the Never-Takers. Compliers in the treated state can then be compared directly to Compliers in the untreated state, which eliminates the noise generated by the presence of Never-Takers in both the treatment and control groups. By isolating the Compliers, the placebo design moves us from an experiment with noncompliance to the familiar situation where there are two assigned groups and "full compliance." From a practical standpoint, the downside to this design is that not all the Compliers receive an actual treatment; half of those who might be treated are given a placebo, which means that the resources expended to contact them are wasted.[9]

Regardless of whether one uses a conventional design or a placebo design, the estimand is the same. Both designs recover the CACE. Choosing between the two designs sometimes comes down to figuring out which one provides more precise estimates of the CACE for a given budget. This tradeoff can be evaluated mathematically. Consider the case where a certain percentage of subjects (ITT_D) are Compliers. In a conventional design, N subjects are randomly assigned in equal numbers to the treatment group or control group. Thus, the treatment group consists of m subjects, of whom approximately ($m \cdot \text{ITT}_D$) are actually treated. If the variance of the potential outcomes in the subject pool is approximated by σ^2 for both the treated and untreated state, the variance of the 2SLS estimator is approximately:

$$\text{Var}\left(\frac{\widehat{\text{ITT}}}{\widehat{\text{ITT}}_D}\right) \approx \frac{2\sigma^2}{m \cdot (\text{ITT}_D)^2}. \tag{5.30}$$

Using a placebo design, the researcher divides the number of treated subjects ($m \cdot \text{ITT}_D$) equally between treatment and placebo groups. The CACE is estimated by comparing the average outcome among those who receive the treatment with the average outcome among those who receive the placebo. The variance of this estimator of the CACE is approximately:

$$\text{Var}\{\hat{E}[(Y_i | D_i(1) = 1, D_i = 1) - (Y_i | D_i(1) = 1, D_i = 0)]\} \approx \frac{4\sigma^2}{m \cdot (\text{ITT}_D)}. \tag{5.31}$$

Comparing these two equations reveals that the conventional design is preferable when $\text{ITT}_D > 1/2$. In other words, under a fixed budget, the conventional design provides estimates with less sampling variability when Compliers make up at least half of the

9 However, the placebo design has a further practical advantage. Two sets of researchers working on unrelated topics may share costs. One researcher's treatment serves as the other researcher's placebo, on the assumption that both sets of potential outcomes are unaffected by placebos.

sample. Bear in mind that the typical canvassing study encounters compliance rates of well below 50%. In Nickerson's canvassing study, the compliance rate was just 19%.[10]

When implementing a placebo design, care must be taken to avoid two sources of bias: the placebo message may affect outcomes, and the "placebo Compliers" may not be the same as the "treatment Compliers." These problems can usually be avoided through careful experimental design. For example, if the placebo message used in a canvassing experiment were a discussion of some issue or cause of local concern, this intervention could encourage heightened political awareness and participation in the upcoming election. Although it is hard to imagine an intervention that would uncover Complier status and have absolutely no possible effect on the subject, it is easy to avoid interventions that are asking for trouble. Second, the placebo and treatment messages must produce an equivalent set of Compliers. Technically, a sufficient condition for unbiased estimation is that both $Y_i(d(z))$ and $D_i(z)$ are independent of whether the subject is assigned the treatment or placebo message. In a canvassing experiment, independence would be violated if in high-turnout neighborhoods

BOX 5.7

Combining Conventional and Placebo Designs

This chapter has discussed two designs for estimating the CACE. The conventional design compares the average outcome in the treatment and control groups (the ITT) and divides by the estimated proportion of Compliers (the ITT_D). The placebo design compares the average outcome among those who receive the treatment to the average outcome among those who receive the placebo. Is there some way to combine the two approaches to do better than either strategy alone?

Yes. Consider a third experimental design. Subjects are randomly assigned to three groups: control, placebo, and treatment. This design permits the researcher to estimate the CACE in two different ways. Both estimators produce consistent estimates of the CACE, and combining them produces a more precise estimate of the CACE. This design is especially useful when control observations are "free" in the sense that no additional cost is incurred in gathering control group outcomes. An additional advantage to the three-group approach is that it measures the effect of the placebo on outcomes, which can be used to verify that the placebo has no effect. For further details about the three-group design, see Gerber, Green, Kaplan, and Kern (2010).

10 Nickerson 2008.

canvassers prefer to mobilize voters rather than encourage recycling, and are aware of a subject's treatment assignment as they approach the subject's door. A dispirited canvasser approaches the door of a placebo subject, knocks lightly, and leaves; had the subject been assigned to the voter mobilization treatment, the canvasser would have knocked vigorously, pushed the doorbell, and tossed a pebble at an upper-story window. This asymmetry can be minimized by design, such as blinding those administering the treatment to subjects' experimental condition until after compliance status is determined. This problem can also be detected empirically: the researcher should confirm that the rate of successful message delivery is the same for the placebo and the treatment group and that background attributes are similarly predictive of compliance in both groups.

5.10 Estimating Treatment Effects When Some Subjects Receive "Partial Treatment"

At the start of this chapter, we stressed the importance of clearly defining "treatment" so that subjects could be classified as treated or not. However, in many experiments it is possible for subjects to be partially treated. If the treatment is viewing a five-part TV series, some subjects will decline to participate, others will complete the full program, and some will watch only some of the episodes. If the treatment is exposure to a canvasser's voter mobilization message, subjects might slam the door after hearing only part of the intended message. How does partial treatment alter our analysis and interpretation of experiments with noncompliance?

If subjects are assigned to either treatment or control groups (no subjects are assigned to a partial treatment group), it will not be possible to empirically distinguish the average causal effect of partial treatment from the average causal effect of full treatment. Imagine, for example, we were to divide the subject population into three groups: Compliers, Partial Compliers, and Never-Takers. We can estimate the distribution of types in the subject pool by looking at the results of our attempt to treat the assigned treatment group. The problem, however, is that the control group's outcomes are a weighted average of the three types' untreated potential outcomes. With just two randomly assigned groups, we cannot isolate the contribution of each of the three types. As discussed in the exercises, to identify the effect of partial treatment we need an augmented experimental design that varies whether subjects are encouraged to receive partial or full treatment.

Suppose, however, that our experimental design merely assigns subjects to treatment and control conditions. Some of the subjects assigned to the treatment condition receive full treatment, and others partial treatment. One approach is to define $d = 1$ as full treatment but to err on the side of underestimating the CACE. To implement this approach, the researcher simply considers all partially treated subjects as fully treated. Ordinarily, one would assume that the effect of partial treatment is smaller than the effect of full treatment. The estimated CACE moves toward zero under this definition of treatment: the ITT is unaffected by the definition of treatment, but the ITT_D increases when the definition of compliance becomes more permissive.

An approach similar in spirit is to bound the CACE by alternative classifications for the partially treated. The lower bound is obtained by classifying the partially treated as treated, and the upper bound is obtained by classifying the partially treated as untreated. The resulting estimates bound the true CACE for full treatment if we assume that the average treatment effect for the partially treated lies between zero and the CACE. Notice that classifying the partially treated as untreated leads to a violation of the exclusion restriction when partial treatment affects outcomes. In this case, partial treatment affects the ITT, but because the partially treated are classified as untreated, partial treatment will not affect the estimate of ITT_D. If we assume that the average treatment effect of partial treatment is the same sign as the CACE, the absolute size of the treatment effect will be exaggerated by classifying the partially treated as not treated. The magnitude of this bias increases with the size of the treatment effect from partial treatment and the share of subjects partially treated.

A final approach is to scale the treatment according to how much is delivered and assign "partial credit" for partial treatments. A person who attends three-quarters of the program sessions might be classified as $d_i = 0.75$. This scoring method has an ad hoc feel to it because it rests on substantive assumptions about the underlying metric along which treatment is measured. Researchers using this approach should assess whether their conclusions are sensitive to different classification schemes. When interpreting the results, researchers should also bear in mind that the choice of classification schemes introduces additional uncertainty beyond the statistical uncertainty reflected in confidence intervals.

SUMMARY

This chapter draws the important distinction between a subject's assigned experimental condition and the treatment that this subject actually receives. The practical importance of this distinction depends on your research objectives. If you seek to evaluate a program that sometimes fails to deliver the treatment, noncompliance is a feature of the program, and an experiment that randomly assigns subjects to the program may be adequate for your purposes. The intent-to-treat effect is the average effect of assigning subjects to the intervention, whether or not they actually receive it. Estimation of the ITT uses the same set of tools described in previous chapters.

Extracting meaningful estimates becomes more challenging when the aim is to estimate the average causal effect of treatment, rather than assigned treatment.

Noncompliance narrows the scope of what can be estimated and reduces the precision of those estimates. The ATE is no longer identified. By invoking an excludability assumption, the researcher can identify the average causal effect among a subgroup, Compliers. This subgroup is defined by its potential outcomes; it consists of subjects who take the treatment if and only if assigned to the treatment group. Whether this subgroup is of interest depends on your research objectives. If, for example, your experiment is designed to test the effects of an intervention that will become available to those who seek it out, it makes sense to focus your attention on Compliers, who follow through when encouraged to seek out the experimental treatment. If, on the other hand, your aim is to test the effects of an intervention that will be mandatory for everyone, your ability to draw generalizations will be limited by the fact that your experiment provides no information about the ATE among Never-Takers.

An experiment's capacity to render useful estimates of the CACE depends on its design. Low rates of compliance usually mean disappointingly large standard errors. More importantly, when compliance rates are low, relatively minor violations of the excludability assumption can lead to substantial bias. Unless offset by large sample sizes or addressed through the use of placebo designs, the uncertainty introduced by noncompliance may be debilitating. Partial compliance is also challenging because a design that assigns subjects to treatment or control groups does not allow the researcher to identify the effects of different gradations of treatment without additional assumptions. In anticipation of partial compliance, researchers may use more nuanced experimental designs in which treatment groups receive encouragements of varying strength.

When conducting experiments in settings where noncompliance is a potential concern, researchers should consider some of the following design recommendations:

1. Conduct a small pilot study in order to see whether noncompliance problems arise and, if so, whether these problems can be overcome by adjusting the treatment or the manner in which it is delivered. If compliance rates are expected to remain low, consider the feasibility of a placebo design.
2. Define the treatment and the criteria used to classify subjects as treated or untreated (or partially treated). Ensure that systematic procedures are in place to measure actual receipt of treatment. Unless you ascertain the proportion of the assigned treatment group that actually receives the treatment, you will not be able to calculate the ITT_D or the CACE.
3. Avoid targeting an excessive number of subjects for treatment. When collaborating with an organization that is enthusiastic about the intervention, researchers are sometimes urged to allocate more subjects to the treatment group than the organization can realistically treat. This arrangement sometimes leads to failure-to-treat not because subjects failed to cooperate, but because the organization could not deliver the assigned treatments.

4. In situations where pilot testing is impossible and the diplomacy of collaboration demands an unrealistically large treatment group, randomly allocate subjects to be treated in different phases. Insist that subjects be treated in the order of assignment, so that the treatment begins in (randomly chosen) Precinct A before proceeding to B, C, D, and so forth. So long as those administering the treatment respect the randomly assigned order and stop treatment when they run out of time or resources (as opposed to stopping because they preferred not to treat the next precinct on the list, a choice that might reveal something about that precinct's potential outcomes), this approach partitions the treatment group into two random subsets: the top of the list, where treatment was attempted, and the bottom of the list, which is the same, in expectation, as the control group.[11]

Each of these approaches requires careful management, supervision, and often negotiation. Timely adjustments in design can make an experiment vastly more informative.

When planning an experiment and accompanying statistical analysis, address the issue of noncompliance systematically, starting with a well-reasoned model. Begin by defining compliance. Next, write out the expected outcomes in each of the randomly assigned groups, making explicit the latent types of subjects that comprise each group. This exercise helps clarify which subjects' average treatment effects may be estimated and how the core assumptions of non-interference, excludability, and independence come into play. This approach also helps guard against common mistakes, such as lumping untreated subjects in the treatment group together with the control group.

In terms of estimation, many of the principles from previous chapters apply. Covariates aid in assessing balance and may improve the precision with which average treatment effects are estimated. Randomization inference may be used to test whether the ITT, and by extension, the CACE, are distinguishable from zero. This chapter has also introduced some new analytic methods, including instrumental variables regression. This technique enables researchers to easily estimate the CACE, with or without controlling for covariates. The range of applications of this technique is quite broad and is the subject of further elaboration in Chapter 6.

SUGGESTED READINGS

Angrist, Imbens, and Rubin (1996) provide a derivation of the CACE theorem, and Angrist and Pischke (2009) offer an accessible technical discussion of instrumental variables regression and the complications posed by weak instruments. See Gerber et al. (2010) on efficient estimation of causal effects for designs that include treatment, control, and placebo groups.

---

11 This design hinges on the assumption that the stopping rule is unrelated to subjects' potential outcomes, regardless of where they appear on the list. See Nickerson 2005 for a discussion of research designs to address noncompliance that occurs due to logistical snafus rather than factors that might be related to subjects' potential outcomes.

EXERCISES: CHAPTER 5

1. Using the data in Table 5.2:
   (a) Estimate the following quantities: $E[d_i(1)]$, $E[Y_i(0) \mid d_i(1) = 0]$, $E[Y_i(0) \mid d_i(1) = 1]$, and $E[Y_i(1) \mid d_i(1) = 1]$.
   (b) Using these estimates and assuming that $E[Y_i(1) \mid d_i(1) = 0] = 0.5$, construct a figure that follows the format of Figure 5.1. Show the apparent proportion of Compliers, the ITT, and the CACE.
2. Make up a hypothetical schedule of potential outcomes for three Compliers and three Never-Takers in which the ATE is positive but the CACE is negative. Suppose that an experiment were conducted on your pool of subjects. In what ways would the estimated CACE be informative or misleading?
3. Explain whether each of the following statements is true or false for the case of one-sided noncompliance, assuming that an experiment satisfies non-interference and excludability.
   (a) If the ITT is negative, the CACE must be negative.
   (b) The smaller the ITT_D, the larger the CACE.
   (c) One cannot identify the CACE if no one in the experiment receives the treatment.
4. Explain whether each of the following equalities follows as a consequence of the excludability assumption.
   (a) $E[Y_i(z = 1)] = E[Y_i(z = 1) \mid d_i(1) = 1]$
   (b) $E[Y_i(z = 0, d = 0) \mid d_i(1) = 0] = E[Y_i(z = 1, d = 0) \mid d_i(1) = 0]$
   (c) $E[Y_i(z = 1, d(1))] = E[Y_i(z = 1), d(0))]$
   (d) $E[Y_i(z = 0, d(0))] = E[Y_i(z = 1), d(0))]$
5. Critically evaluate the following statement: "If you are conducting an experiment that encounters one-sided noncompliance, you will never know which of your subjects are Compliers and which of your subjects are Never-Takers."
6. Suppose that a researcher hires a group of canvassers to contact a set of 1,000 voters randomly assigned to a treatment group. When the canvassing effort concludes, the canvassers report that they successfully contacted 500 voters in the treatment group, but the truth is that they only contacted 250. When voter turnout rates are tabulated for the treatment and control groups, it turns out that 400 of the 1,000 subjects in the treatment group voted, as compared to 700 of the 2,000 subjects in the control group (none of whom were contacted).
   (a) If you believed that 500 subjects were actually contacted, what would your estimate of the CACE be?
   (b) Suppose you learned that only 250 subjects were actually treated. What would your estimate of the CACE be?
   (c) Do the canvassers' exaggerated reports make their efforts seem more or less effective? When formulating your answer, you may define effectiveness in terms of either the ITT or the CACE.
7. Make up a schedule of potential outcomes that would generate Figure 5.2, which illustrates the consequences of an exclusion restriction violation. Hint: you will need to allow for potential outcomes that respond to both d and z.
8. Cotterill et al. report the results of an experiment conducted in an area of the United Kingdom where only half of the local residents recycle their trash.[12] Canvassers visited

12 Cotterill et al. 2009.

homes and encouraged residents to recycle. Outcomes were measured by whether the home put out a recycling bin on at least one occasion during the following three weeks. We restrict our attention here to homes that did not recycle trash during a pre-experimental period of observation. When implementing the intervention, researchers encountered one-sided noncompliance: 1,015 of the 1,849 homes assigned to the treatment group were successfully canvassed; none of the 1,430 homes assigned to the control group were canvassed. These researchers found that 591 homes in the treatment group recycled, as opposed to 377 in the control group. The researchers also observed that 429 of 1,015 homes that were successfully canvassed recycled, as opposed to 539 of the 2,264 homes that were not canvassed.

(a) Estimate the ITT, and interpret the results.
(b) Estimate the ITT_D, and interpret the results.
(c) Using the equations in Theorem 5.1 as a guide, write down a model of the expected recycling rate among those assigned to the control group. Do the same for the expected recycling rate among those assigned to the treatment group. Show that under the assumptions of Theorem 5.1, the CACE can be identified based on the design of this experiment.
(d) Estimate the CACE, and interpret the results.
(e) Explain why comparing the recycling rates of the treated and untreated subjects tends to produce misleading estimates of the CACE and ATE.

9. One way to detect heterogeneous treatment effects across subgroups is to employ a design that randomly manipulates the level of compliance. One such study was conducted in Michigan in 2002.[13] Subjects were randomly allocated to three experimental groups. The first treatment group was targeted for a phone call that encouraged subjects to vote in the upcoming November election. The second treatment group was targeted for the same call using the same script on the same day, but more attempts were made to reach subjects. No attempts were made to contact the control group. The table below shows the contact rates and voting rates for each of the three assigned groups.

	Control	Treatment group #1 (minimal effort)	Treatment group #2 (maximal effort)
Percent reached by callers	0	29.97	47.31
Percent voting	55.89	55.91	56.53
N	317,182	7,500	7,500

(a) Define two types of Compliers: those who respond when called with minimal (or maximal) effort and those who respond only when called with maximal effort. Write down a model expressing the expected voting rate among those assigned to the control group as a weighted average of potential outcomes among Minimal Compliers, Maximal Compliers, and Never-Takers. Do the same for the expected rate of voting among those assigned to each of the treatment groups.
(b) Show that the CACE for each of the treatments can be identified based on the design of this experiment.

13 Gerber and Green 2005.

(c) Estimate the share of the subject pool that Maximal Compliers comprise. Estimate the share of the subject pool that Minimal Compliers comprise.
(d) Estimate the average treatment effect among each type of Complier, and interpret the results.

10. Guan and Green report the results of a canvassing experiment conducted in Beijing on the eve of a local election.[14] Students on the campus of Peking University were randomly assigned to treatment or control groups. Canvassers attempted to contact students in their dorm rooms and encourage them to vote. No contact with the control group was attempted. Of the 2,688 students assigned to the treatment group, 2,380 were contacted. A total of 2,152 students in the treatment group voted; of the 1,334 students assigned to the control group, 892 voted. One aspect of this experiment threatens to violate the exclusion restriction. At every dorm room they visited, even those where no one answered, canvassers left a leaflet encouraging students to vote.
    (a) Using the dataset at http://isps.research.yale.edu/FEDAI, estimate the ITT.
    (b) Use randomization inference to test the sharp null hypothesis that the ITT is zero for all observations, taking into account the fact that random assignment was clustered by dorm room. Interpret your results.
    (c) Assume that the leaflet had no effect on turnout. Estimate the CACE.
    (d) Assume that the leaflet raised the probability of voting by one percentage point among both Compliers and Never-Takers. In other words, suppose that the treatment group's turnout rate would have been one percentage point lower had the leaflets not been distributed. Write down a model of the expected turnout rates in the treatment and control groups, incorporating the average effect of the leaflet.
    (e) Given this assumption, estimate the CACE of canvassing.
    (f) Suppose, instead, that the leaflet had no effect on Compliers (who heard the canvasser's speech and ignored the leaflet) but raised turnout among Never-Takers by 3 percentage points. Given this assumption, estimate the CACE of canvassing.

11. Nickerson describes a voter mobilization experiment in which subjects were randomly assigned to one of three conditions: a baseline group (no contact was attempted), a treatment group (canvassers attempted to deliver an encouragement to vote), and a placebo group (canvassers attempted to deliver an encouragement to recycle).[15] Based on the results presented below, calculate the following:
    (a) Estimate the proportion of Compliers based on subjects' responses to the treatment. Estimate the proportion of Compliers based on subjects' responses to the placebo. Assuming that the individuals are assigned randomly to the treatment and placebo groups, are these rates of compliance consistent with the null hypothesis that both groups have the same proportion of Compliers?
    (b) Do the data suggest that Never-Takers in the treatment and placebo groups have the same rate of turnout? Is this comparison informative?
    (c) Estimate the CACE of receiving the placebo. Is this estimate consistent with the substantive assumption that the placebo has no effect on turnout?

14 Guan and Green 2006.
15 Nickerson 2005, 2008.

(d) Estimate the CACE of receiving the treatment using two different methods. First, use the conventional method of dividing the ITT by the ITT_D. Second, compare turnout rates among Compliers in both the treatment and placebo groups. Interpret the results.

Treatment assignment	Treated?	N	Turnout
Baseline	No	2572	31.22%
Treatment	Yes	486	39.09%
	No	2086	32.74%
Placebo	Yes	470	29.79%
	No	2109	32.15%

12. Imagine a math tutoring program that involves two daylong sessions with instructors. Given the likely possibility that some of the students who are randomly assigned to the program will attend only the first of the two sessions, suppose that administrators are primarily interested in finding out whether two sessions improve performance on end-of-year tests but are also interested in assessing the effectiveness of the first session alone.
    (a) Propose an experimental design that addresses the possibility that some students will only attend the first session.
    (b) Show that your experimental design is capable of identifying the causal effects of the full program and the abbreviated program.
    (c) Suppose that one-sided noncompliance occurred in the following way: everyone in the treatment group received the treatment, and some of the control group was inadvertently treated. Show that by modifying the CACE theorem (for example, replacing Never-Takers with Always-Takers) one can still identify the CACE in this case.

CHAPTER 6

Two-Sided Noncompliance

In Chapter 5 we explained how to analyze experiments that fail to treat all of the subjects assigned to the treatment group. This type of noncompliance is labeled "one-sided" because assigned and actual treatment differ in one direction only: some subjects assigned to the treatment condition do not receive treatment, but all subjects assigned to the control condition go untreated. We described how to estimate average treatment effects for the subgroup of subjects called Compliers, who are treated if assigned to the treatment condition but not if assigned to the control condition. Never-Takers, by contrast, are never treated, regardless of whether they are assigned to treatment or control conditions. Because the treatment assignment affects only the Compliers, average outcomes in the treatment and control groups differ systematically when Compliers respond to the treatment. Dividing the difference in observed treatment and control group outcomes by the proportion of subjects who are Compliers allows us to estimate the average effect of treatment on this subgroup, the Complier average causal effect (CACE).

Our discussion of one-sided noncompliance emphasized several important points. First, experiments with one-sided noncompliance allow us to estimate the average causal effect among Compliers, not the average causal effect among the entire subject pool. This type of experiment cannot estimate the average treatment effect among Never-Takers because we never observe them in their treated state. Second, the proper way to estimate the CACE is to compare subjects *assigned* to the treatment and control conditions; one should not compare subjects who actually receive the treatment to subjects who go untreated. Groups formed by random assignment have comparable potential outcomes; groups formed after random assignment often do not. As we explained at length in Chapter 5, a comparison of treated and untreated subjects conflates the effect of the treatment with pre-existing differences between these groups that have nothing to do with the treatment effect. One of the main reasons to conduct an experiment is to avoid comparing groups that are formed by some unknown selection process.

What you will learn from this chapter:

1. New terms: Always-Takers, Defiers, and the Monotonicity Assumption.
2. A modeling approach that establishes whether causal estimands may be recovered given two-sided noncompliance.
3. An estimation approach that, under two-sided noncompliance and a set of core assumptions, gauges the average causal effect among Compliers, those subjects who would be treated if and only if assigned to the treatment group.
4. The idea of tracing the "downstream" consequences of an experimental intervention and the assumptions that come into play when this type of analysis is undertaken.

This chapter extends our discussion of noncompliance to a more general case, *two-sided noncompliance*, which occurs when some subjects assigned to the control group receive treatment, and some subjects assigned to the treatment group go untreated. This type of noncompliance is quite common. Many field experiments use an "encouragement" design: subjects are invited to participate in a program or engage in an activity that is available to them even without the researcher's invitation. Subjects assigned to the treatment group receive an encouragement to participate, and those who actually participate in the program are classified as treated. Subjects assigned to the control group receive no encouragement (or may be urged to engage in some other activity that diverts them from the treatment); nevertheless, some subjects in the control group may participate in the program and are classified as treated. For example, experiments designed to gauge the effects of private versus public schools award randomly selected students with vouchers that can be used to pay tuition at private schools. Students who win the vouchers comprise the treatment group, and treatment is defined as attending a private school. Some subjects who win a voucher nevertheless remain in public school, and some subjects who do not win a voucher attend private school anyway.

Two-sided noncompliance is also common in experiments involving naturally occurring randomizations. Lotteries determine the allocation of prize money, judges in criminal trials, college roommates, and military conscription. Here the researcher is a bystander with no control over access to the treatment. Many researchers, for example, exploit the naturally occurring experiment afforded by the Vietnam draft lottery to study the effect of military service on veterans' wages and health. Men eligible for military service were randomly assigned a low or a high draft number, and those with a low number were very likely to be drafted. Two-sided noncompliance occurs because some men with high numbers did in fact serve in the military, while some men with low numbers did not.

Experiments that encounter one-sided noncompliance permit the researcher to calculate three average outcomes: the average outcome among subjects assigned to the control group, the average outcome among subjects assigned to the treatment group who go untreated, and the average outcome among subjects assigned to the treatment group who receive the treatment. In the more general case of two-sided noncompliance, where some control group subjects may be treated, we have four groups: (i) subjects assigned to the control group but treated, (ii) subjects assigned to the control group and untreated, (iii) subjects assigned to the treatment group but untreated, and (iv) subjects assigned to the treatment group and treated. If the objective is to estimate an average treatment effect, how should the researcher use the available information? Initial intuition might suggest that we estimate the treatment effect by comparing the treated and the untreated subjects. By this point, however, you have probably grown skeptical of this suggestion. As you may have noticed, cautioning the reader to resist the temptation to make such comparisons is a recurrent theme of this book. As we will show again in this chapter, a comparison of the treated and the untreated will conflate the treatment effect with pre-existing differences between treated and untreated subjects. In order to estimate causal effects, the experimental researcher begins by comparing groups formed by random assignment.

In order to lay the groundwork for the analysis of experiments with two-sided noncompliance, this chapter introduces some new terms and a new assumption. The new terminology is meant to describe all possible ways in which subjects might respond to treatment assignment: some subjects may take the treatment regardless of their assignment, while other subjects take the treatment if and only if they are assigned to the control group. We then revisit the CACE theorem from Chapter 5 and show the conditions under which it applies to experiments with two-sided noncompliance. An empirical example illustrates estimation and interpretation of the CACE. The analytic framework laid out in this section is applicable to a broad array of social science applications, such as estimating the "downstream effects" of random interventions. The concluding section discusses a pair of empirical examples and suggests experimental designs that address challenges to estimation and inference.

6.1 Two-Sided Noncompliance: New Definitions and Assumptions

In Chapter 5, we defined two types of subjects based on their potential outcomes $d_i(z)$, where z refers to the assigned treatment (1 if assigned to treatment and 0 if assigned to control) and d_i indicates whether this subject is actually treated (1 if treated and 0 if untreated). Never-Takers are subjects who never receive the treatment regardless of their treatment assignment. Their configuration of potential outcomes is $d_i(1) = 0$

and $d_i(0) = 0$. Compliers are subjects who are treated if and only if assigned to the treatment group: $d_i(1) = 1$ and $d_i(0) = 0$.

Under two-sided noncompliance, four patterns of potential outcomes are possible. In addition to Compliers and Never-Takers, we add two new groups, *Always-Takers* and *Defiers*. Always-Takers do as their label implies: they take the treatment regardless of their treatment assignment. Their potential outcomes are $d_i(1) = 1$ and $d_i(0) = 1$. Defiers' treatment status is always the *opposite* of their assignment. Their potential outcomes are $d_i(1) = 0$ and $d_i(0) = 1$. Notice that treatment assignment has no effect on whether Always-Takers or Never-Takers are treated. Defiers and Compliers, on the other hand, respond to treatment assignment, but in opposite ways.

When discussing two-sided noncompliance, researchers sometimes refer to Defiers using the shorthand $d_i(1) < d_i(0)$ and refer to Compliers using the shorthand $d_i(1) > d_i(0)$. For example, the expression $\text{ATE} \mid (d_i(1) > d_i(0))$ would be read "average treatment effect among Compliers." This notation is a bit more elaborate now that we are considering two-sided rather than one-sided noncompliance. In Chapter 5, it was sufficient to refer to Compliers as those with $d_i(1) = 1$; $d_i(0) = 0$ followed automatically from the fact that those in the control group could not receive the treatment. Under two-sided noncompliance, $d_i(1) = 1$ is true of both Compliers and Always-Takers. In this chapter, therefore, we denote Compliers by referring to their distinguishing characteristic, $d_i(1) > d_i(0)$.

We illustrate these definitions using an experiment conducted by Mullainathan, Washington, and Azari on the effects of viewing candidate debates.[1] Campaigns are a central part of democratic representation, and candidate debates are among the most celebrated events in the campaign season, but do these high profile engagements affect viewers' opinions? This study investigates the effect of watching a televised political debate on subjects' attitudes about the candidates. Subjects were contacted by phone prior to a televised debate between candidates for New York City mayor. Those assigned to the treatment group were encouraged to watch the debate, and those assigned to the control group were encouraged to watch a different, nonpolitical TV program airing at the same time as the debate.

The first step in defining compliance is defining what it means to be treated. Here subjects are classified as treated if they report viewing the debate when interviewed a day or two later.[2] Based on this definition, subjects may be partitioned into the four compliance types. Always-Takers will report watching the debate with or without encouragement, while Never-Takers will report not watching the debate regardless of encouragement. Compliers and Defiers change their behavior in response to encouragement. Compliers report watching the debate if and only if they are encouraged to

1 Mullainathan, Washington, and Azari 2010.

2 The researchers also administered a small quiz about the debate's moderator in order to verify that self-reported viewers actually viewed the show.

BOX 6.1

Classification of Subject Types in Experiments with Two-sided Noncompliance

Subjects may be classified according to their potential outcomes $d_i(z)$. Balke and Pearl (1993) and Angrist, Imbens, and Rubin (1996) suggested the following terminology:

Subjects for whom $d_i(1) = 1$ and $d_i(0) = 0$ are called *Compliers*. These subjects receive the treatment when assigned to the treatment group and do not receive the treatment when assigned to the control group. Compliers may also be described as subjects for whom $d_i(1) - d_i(0) = 1$. Another description is $d_i(1) > d_i(0)$.

Subjects for whom $d_i(1) = 0$ and $d_i(0) = 0$ are called *Never-Takers*. They are untreated regardless of whether they are assigned to treatment or control.

Subjects for whom $d_i(1) = 1$ and $d_i(0) = 1$ are called *Always-Takers*. They are treated regardless of whether they are assigned to treatment or control.

Subjects for whom $d_i(1) = 0$ and $d_i(0) = 1$ are called *Defiers*. These subjects receive the treatment when assigned to the control group and do not receive the treatment when assigned to the treatment group. Defiers may also be described as subjects for whom $d_i(1) - d_i(0) = -1$. Another description is $d_i(1) < d_i(0)$.

do so. Defiers do just the opposite: they report watching the debate if and only if they are encouraged to watch the non-political program.

Given the experimental design, a subject's compliance type is a fixed attribute. We can imagine a schedule of potential outcomes that determines which observed outcomes are revealed when a subject is assigned to the treatment or control condition. This fixed response schedule mechanically transforms inputs (treatment assignments) into outputs (TV viewing and opinion change). That said, the classification of subjects into the four compliance types in this particular study would not necessarily carry over to other studies involving the same subjects. When designing experiments, incentives are sometimes used to increase the share of Compliers. Suppose that rather than a telephone encouragement, the researchers had offered the subjects a large monetary reward for watching the debate or the alternative television program. A greater proportion of the subjects would have been Compliers.

Because subjects were randomly assigned to treatment and control groups, we may estimate the shares of Never-Takers, Always-Takers, Compliers, and Defiers in the subject pool. Let's start by introducing some notation that will make the formal presentation more concise. Always-Takers' share of the subject pool is denoted π_{AT}. Formally,

$$\pi_{AT} \equiv \frac{1}{N}\sum_{i=1}^{N} d_i(1)\, d_i(0). \qquad (6.1)$$

This formula adds up the number of Always-Takers and divides by the total number of subjects. The proportion of subjects who are Never-Takers is defined as :

$$\pi_{NT} \equiv \frac{1}{N}\sum_{i=1}^{N} (1 - d_i(1))(1 - d_i(0)). \qquad (6.2)$$

Similarly, for Compliers:

$$\pi_{C} \equiv \frac{1}{N}\sum_{i=1}^{N} d_i(1)(1 - d_i(0)). \qquad (6.3)$$

Since we require that all four proportions add up to 1, the proportion of Defiers is represented as the leftover share:

$$\pi_{D} \equiv \frac{1}{N}\sum_{i=1}^{N} (1 - d_i(1))\, d_i(0) = 1 - \pi_{AT} - \pi_{NT} - \pi_{C}. \qquad (6.4)$$

Let's consider what a given experiment reveals about the distribution of subject types. Under random assignment, the assigned treatment group has the same *expected* shares of Always-Takers, Never-Takers, Compliers, and Defiers as the assigned control group. In the control group, the untreated subjects are either Never-Takers or Compliers. The study of the New York City mayoral debates found that 84% of the control group reported not watching the debate, so $\hat{\pi}_{NT} + \hat{\pi}_{C} = 0.84$. Subjects in the control group who watched the debate are either Always-Takers or Defiers, and $\hat{\pi}_{AT} + \hat{\pi}_{D} = 0.16$. In the treatment group, 37% of the subjects reported watching the debate. These subjects must be either Always-Takers or Compliers, so $\hat{\pi}_{AT} + \hat{\pi}_{C} = 0.37$. The remaining 63% who did not watch the debate when encouraged to do so are either Never-Takers or Defiers; thus, $\hat{\pi}_{NT} + \hat{\pi}_{D} = 0.63$. Unfortunately, these equations do not provide enough information to solve for $\hat{\pi}_{AT}$, $\hat{\pi}_{NT}$, and $\hat{\pi}_{C}$.

This indeterminacy is an inherent feature of experiments with noncompliance. In order to circumvent this problem, researchers commonly invoke an additional assumption: the sample contains no Defiers. The plausibility of this assumption will depend on the application, but in most cases where encouragements are used, it is reasonable to regard Defiers' potential outcomes as outlandish and rare. Although one can dream up a scenario under which Defiers exist in the debate study (see our

BOX 6.2

Definition: Monotonicity Assumption

Let potential outcomes $d_i(z)$ represent the treatment received when the treatment assignment is z. For all i, $d_i(1) \geq d_i(0)$.

The monotonicity assumption rules out Defiers, as their potential outcomes are characterized by $d_i(1) < d_i(0)$.

attempt to do so below), it seems unlikely that more than a handful of subjects would report watching the debates if and only if they were assigned to the control group.

The formal name for the "No Defiers Assumption" is *monotonicity*. The assumption takes its name from the requirement that $d_i(1) \geq d_i(0)$; whenever a subject moves from the control condition to the treatment condition, d_i either remains unchanged or increases.

When we rule out Defiers and set $\pi_D = 1 - \pi_{AT} - \pi_{NT} - \pi_C = 0$, it becomes possible to estimate the proportions of Always-Takers, Never-Takers, and Compliers. Earlier, we noted that $\hat{\pi}_{AT} + \hat{\pi}_D = 0.16$, and this equation now implies that $\hat{\pi}_{AT} = 0.16$. Similarly, $\hat{\pi}_{NT} + \hat{\pi}_D = 0.63$ simplifies to $\hat{\pi}_{NT} = 0.63$. We can solve for the proportion of Compliers using either $\hat{\pi}_{AT} + \hat{\pi}_C = 0.37$ or $\hat{\pi}_{NT} + \hat{\pi}_C = 0.84$. Either formula yields $\hat{\pi}_C = 0.20$.

6.2 ITT, ITT$_D$, and CACE under Two-Sided Noncompliance

In Chapter 5, we presented the CACE theorem for the case of one-sided noncompliance. Theorem 5.1 shows that, when we assume non-interference and excludability, the average treatment effect among compliers (CACE) is equal to ITT/ITT$_D$. We now extend this theorem so that it addresses the case of two-sided noncompliance.

In sum, once Defiers are ruled out, the CACE is obtained using the same formula that was introduced in Chapter 5. Although two-sided noncompliance introduces the possibility that some subjects are Always-Takers, they pose no identification problems. Always-Takers have no effect on the ITT, and the share of Always-Takers is differenced away when we calculate the ITT$_D$. The key distinction between this theorem and the CACE theorem from Chapter 5 is the addition of the monotonicity assumption.

THEOREM 6.1

Identification of the Complier Average Causal Effect under Two-sided Noncompliance

Using the definitions of ITT_D and ITT in equations (5.4) and (5.7), which assume non-interference:

$$ITT_D = E[d_i(z = 1) - d_i(z = 0)]. \tag{6.5}$$

$$\begin{aligned} ITT &= E[Y_i(z = 1)] - E[Y_i(z = 0)] \\ &= E[Y_i(z = 1, d(1))] - E[Y_i(z = 0, d(0))]. \end{aligned} \tag{6.6}$$

The definition of the CACE from equation (5.10) may be adapted to the case of two-sided noncompliance. Compliers are now defined as subjects for which $d_i(1) - d_i(0) = 1$. Using this definition:

$$CACE = E[(Y_i(d = 1) - Y_i(d = 0)) \,|\, d_i(1) - d_i(0) = 1]. \tag{6.7}$$

Assuming the excludability of treatment assignments, monotonicity, and $ITT_D > 0$, the Complier average causal effect $= ITT/ITT_D$.

Proof: Under the exclusion restriction, the ITT may be written in terms of potential outcomes that respond only to the treatment that is actually administered:

$$\begin{aligned} ITT &= E[Y_i(z = 1, d(1))] - E[Y_i(z = 0), d(0)] \\ &= E[Y_i(d(1))] - E[Y_i(d(0))]. \end{aligned} \tag{6.8}$$

The ITT may be expressed as a weighted average of the ITT among Never-Takers, Always-Takers, Compliers, and Defiers using the definitions of π_{NT}, π_{AT}, π_C, and π_D from equations (6.1) to (6.4):

$$\begin{aligned} ITT = {} & E[(Y_i(d(1)) - Y_i(d(0))) \,|\, d_i(1) = d_i(0) = 0]\pi_{NT} \\ & + E[(Y_i(d(1)) - Y_i(d(0))) \,|\, d_i(1) = d_i(0) = 1]\pi_{AT} \\ & + E[(Y_i(d(1)) - Y_i(d(0))) \,|\, d_i(1) - d_i(0) = 1]\pi_C \\ & + E[(Y_i(d(1)) - Y_i(d(0))) \,|\, d_i(1) - d_i(0) = -1]\pi_D. \end{aligned} \tag{6.9}$$

The first two terms are zero because Always-Takers and Never-Takers never change their treatment status when assignments change. The last term drops out due to the monotonicity assumption, which implies that $\pi_D = 0$. Thus, the ITT reduces to:

$$E[(Y_i(d(1)) - Y_i(d(0))) \,|\, d_i(1) - d_i(0) = 1]\pi_C. \tag{6.10}$$

The ITT_D can be similarly decomposed into the contribution made by each of the four subgroups. Following the same steps and dropping terms equal to zero,

$$ITT_D = E[(d_i(1) - d_i(0)) \,|\, d_i(1) - d_i(0) = 1]\pi_C = \pi_C. \tag{6.11}$$

Dividing the ITT by the ITT_D gives:

$$\frac{ITT}{ITT_D} = E[(Y_i(1) - Y_i(0)) \,|\, d_i(1) - d_i(0) = 1] = CACE. \tag{6.12}$$

6.3 A Numerical Illustration of the Role of Monotonicity

In order to appreciate the importance of monotonicity in the context of two-sided noncompliance, consider the hypothetical schedule of potential outcomes depicted in Table 6.1. Building on the example we presented in Table 5.1, this collection of subjects includes not only Compliers and Never-Takers but also Always-Takers and Defiers. Continuing with the example of an experiment to measure the effect of a political debate, suppose that Y_i is a measure of opinion change, d_i indicates whether the subject watched a candidate debate, and z_i indicates whether the subject is encouraged to watch the debate or a non-political show.

In this example, the Complier average treatment effect is the ATE for the first four subjects:

$$\frac{10+10+13+7}{4} = 10. \tag{6.13}$$

The question is whether we can recover the CACE using the formula $\text{ITT}/\text{ITT}_\text{D}$. The ITT_D is the difference between the average $d_i(1)$ and the average $d_i(0)$: $0.5 - 0.3 = 0.2$. The ITT is the average effect of assigned treatment ($Y_i(z = 1) - Y_i(z = 0)$), which is:

$$\frac{10+10+13+7-4-6+0+0+0+0}{10} = 3. \tag{6.14}$$

TABLE 6.1

Schedule of hypothetical potential outcomes for 10 subjects

Observation	$Y_i(d=0)$	$Y_i(d=1)$	$d_i(z=0)$	$d_i(z=1)$	Subject type	$Y_i(z=0)$	$Y_i(z=1)$
1	24	34	0	1	Complier	24	34
2	18	28	0	1	Complier	18	28
3	19	32	0	1	Complier	19	32
4	19	26	0	1	Complier	19	26
5	18	22	1	0	Defier	22	18
6	22	28	1	0	Defier	28	22
7	10	20	1	1	Always-Taker	20	20
8	11	12	0	0	Never-Taker	11	11
9	8	15	0	0	Never-Taker	8	8
10	11	18	0	0	Never-Taker	11	11

Note: The exclusion restriction is assumed, so potential outcomes for Y are solely a function of D.

Unfortunately, an experiment performed on these subjects will not recover the CACE. The formula ITT/ITT_D returns $3/0.2 = 15$. What went wrong?

The presence of two Defiers violates the monotonicity assumption. The ITT for the sample as a whole is a weighted average of the ITT for each of the four types of subjects. Among Compliers, the ITT is 10. Among Defiers, the ITT is −5. For Always-Takers and Never-Takers, the ITT is, of course, always 0. Writing the ITT as a weighted average, we find:

$$(10)\left(\frac{4}{10}\right) + (-5)\left(\frac{2}{10}\right) + (0)\left(\frac{1}{10}\right) + (0)\left(\frac{3}{10}\right) = 3. \quad (6.15)$$

If we were to exclude the two Defiers, the ITT among the remaining eight subjects would instead be:

$$(10)\left(\frac{4}{8}\right) + (0)\left(\frac{1}{8}\right) + (0)\left(\frac{3}{8}\right) = 5. \quad (6.16)$$

Excluding Defiers also changes the ITT_D. Five of the eight subjects are treated if assigned to the treatment group, and only one of the eight is treated if assigned to the control group. Thus, $ITT_D = 0.5$. When we remove Defiers from the sample, the usual formula correctly estimates the CACE: $ITT/ITT_D = 5/0.5 = 10$.

Another way to illustrate the role of monotonicity is to represent average potential outcomes geometrically. Figure 6.1 provides a geometric depiction of average outcomes when a pool of subjects is assigned to the treatment group (Panel A) and control group (Panel B). The X-axis, which extends from 0 to 1, shows the distribution of the four compliance types. The share of compliers, for example, is represented by the horizontal distance between zero and π_C, here 0.4. The Y-axis measures average treated and untreated potential outcomes, $E[Y_i(d = 1)]$ and $E[Y_i(d = 0)]$, which are depicted for each of the four subject types by the height of each rectangle. The vertical distance between $E[Y_i(d = 1)]$ and $E[Y_i(d = 0)]$ shows the average treatment effect for each type of subject.

The average outcome when subjects are assigned to treatment is represented by the area of the shaded (and cross-hatched) regions in Panel A. For example, the area of the rectangle for Compliers is the width (π_C) times the height, or $(E[Y_i(d = 1) \mid d_i(1) > d_i(0)])$, or $(4/10)(30) = 12$. Summing the area for all four groups gives 21.

The average outcome when subjects are assigned to the control condition is represented by the area of the shaded (including the cross-hatched) regions in Panel B. The difference between the shaded and cross-hatched area in Panel A and Panel B is the ITT. For example, the area of the rectangle for Compliers is the width (π_C) times the height $(E[Y_i(d = 0) \mid d_i(1) > d_i(0)])$, or $(4/10)(20) = 8$. Summing the area for all four groups gives 18. Subtracting the shaded area in Panel B from the shaded area in Panel A gives $21 - 18 = 3$, which is the ITT.

FIGURE 6.1

Illustration of estimation of CACE without monotonicity

The area of the shaded rectangles, including cross-hatched rectangles, equals the average Y_i when subjects are assigned to the treatment group

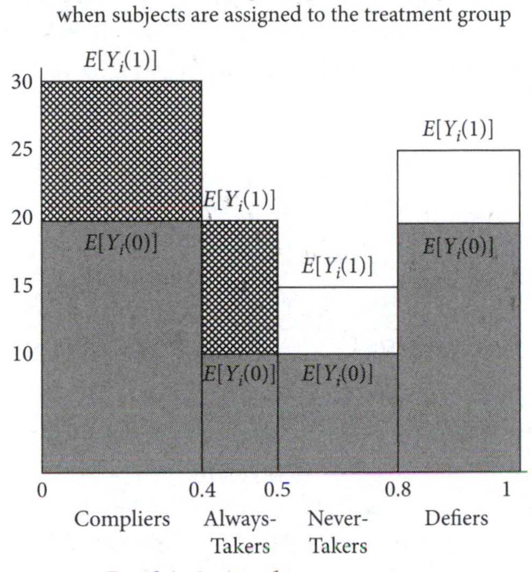

Panel A: Assigned to treatment group

The area of the shaded rectangles, including cross-hatched rectangles, equals the average Y_i when subjects are assigned to the control group

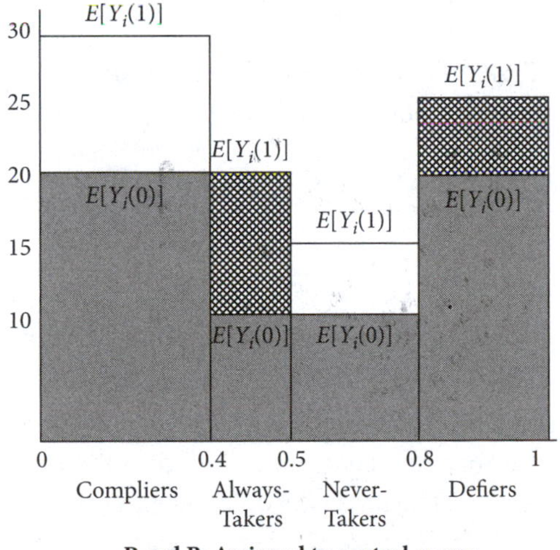

Panel B: Assigned to control group

Note: The upper bar in each column, labeled $E[Y_i(1)]$, indicates the average treated potential outcome for each type of subject. The lower bar, labeled $E[Y_i(0)]$, indicates the average untreated potential outcome for each type of subject.

Another way to visualize the ITT is to compare the areas of the cross-hatched regions, which represent the *change* in average potential outcomes that occurs when a subject is assigned to a particular experimental condition. Notice that there is no cross-hatched region for Never-Takers, because they never receive the treatment. Always-Takers have identical cross-hatched regions in both Panel A and Panel B, so the difference between the cross-hatched areas in the two panels is unaffected by Always-Takers. Defiers have a cross-hatched region only in Panel B, because they are treated only if assigned to the control group. Subtracting the total cross-hatched area in Panel B:

$$\left(\frac{1}{10}\right)(10) + \left(\frac{2}{10}\right)(5) = 2 \qquad (6.17)$$

from the total hatched region in Panel A:

$$\left(\frac{1}{10}\right)(10) + \left(\frac{4}{10}\right)(10) = 5 \qquad (6.18)$$

gives 3, the ITT.

When monotonicity is violated by the presence of Defiers, the ITT combines the increment from Compliers with the decrement from Defiers. In terms of Figure 6.1, the hatched area for Defiers is subtracted from the hatched area for Compliers. The presence of Defiers distorts the numerator of the ratio used to recover the CACE. Is this distortion corrected by the denominator, the ITT_D? Only under special conditions. The difference between the proportion treated when assigned to the treatment group and the proportion treated when assigned to the control group is:

$$\text{ITT}_\text{D} = (\pi_C + \pi_{AT}) - (\pi_D + \pi_{AT}) = \pi_C - \pi_D. \qquad (6.19)$$

If we divide the ITT by the difference in the proportion treated in the treatment and control group, we no longer obtain a treatment effect estimate for any type of subject:

$$\frac{\text{ITT}}{\text{ITT}_\text{D}} = \frac{(\text{ATE}\,|\,\text{Compliers})\pi_C - (\text{ATE}\,|\,\text{Defiers})\pi_D}{\pi_C - \pi_D}. \qquad (6.20)$$

This ratio equals the CACE only when the Defiers and the Compliers have the same average treatment effect or when there are no Defiers ($\pi_D = 0$). The latter scenario is what is assumed under monotonicity. Suppose we did not impose the monotonicity assumption. Under what conditions does the presence of Defiers severely bias the CACE estimator? This formula suggests that two conditions must hold: the ATEs for Defiers and Compliers must be markedly different and the ratio of Defiers to Compliers must be substantial. Rarely do researchers have any information about the relative size of the two ATEs. Intuitions about bias tend to come down to the researcher's sense about the relative share of Defiers. A few Defiers rarely make much mischief—unless the sample contains few Compliers. This point has an important design implication.

One reason to maximize compliance is to dampen the consequences of violating monotonicity. The greater the share of Compliers, the more resilient the CACE estimator will be to the presence of Defiers.

6.4 Estimation of the CACE: An Example

To illustrate estimation of the CACE using experimental data, we return to our running example of New York mayoral debates. Prior to a candidate debate in the New York City mayoral election, the researchers contacted 1,000 subjects by phone who were successfully interviewed a few days later. Approximately half of the subjects were encouraged to watch the upcoming debate (the treatment group), and half were encouraged to watch a non-political show (the control group). After the mayoral candidate debate, respondents were called back. All subjects were asked whether they had watched the debate and whether their views of the candidates had recently changed.

Table 6.2 shows the results of the study. Notice that a substantial proportion of the 495 subjects assigned to the control group (80/495) was treated. Also notice that there was substantial noncompliance among the 505 subjects assigned to the treatment group: despite being assigned to the treatment group, only 185/505 were treated.

Using the CACE theorem for two-sided noncompliance, a consistent estimate of the CACE can be obtained by calculating the ratio of two estimates: the ITT estimate and the ITT_D estimate.

$$\widehat{ITT} = .475 - .418 = .057 \tag{6.21}$$

$$\widehat{ITT_D} = .366 - .162 = .204 \tag{6.22}$$

The difference between average outcomes in the treatment and control group provides an unbiased estimate of the ITT. The estimated ITT suggests that assignment to the treatment caused a 5.7 percentage point increase in the outcome. To estimate the

TABLE 6.2

Percent reporting changed views of one or more candidates since watching the New York City mayoral debate

	Treatment group	Control group
% Reporting change (N treated)	59.5 (185)	50.0 (80)
% Reporting change (N untreated)	40.6 (320)	40.2 (415)
% Reporting change (total N)	47.5 (505)	41.8 (495)

Source: Mullainathan, Washington, and Azari 2010.

ITT_D, we calculate the proportion viewing the debate in the treatment group (36.6%) and subtract the rate viewing in the control group (16.2%). This difference provides an unbiased estimate of the ITT_D. Assuming monotonicity, the $\widehat{\text{ITT}}_D$ indicates the proportion of the subjects who are Compliers, which is estimated to be 20.5%. Evidently, the phone call prior to the debate encouraging the subjects to watch the debate did in fact boost the share of respondents who reported watching the debate.

In order to estimate the CACE, we calculate:

$$\frac{\widehat{\text{ITT}}}{\widehat{\text{ITT}}_D} = \frac{0.057}{0.205} = 0.279. \tag{6.23}$$

Under the assumptions of non-interference, excludability, and monotonicity, this estimator provides consistent estimates of the ATE among Compliers. The estimate suggests that watching the debates raises the rate at which Compliers report opinion change by 28 percentage points. This estimate implies a substantial effect among Compliers. The ITT is much smaller than the CACE because Compliers only account for approximately one-fifth of the sample.

We now show how the same estimates can be obtained using regression. In order to estimate the $\widehat{\text{ITT}}_D$, we regress viewing the debate (a variable called TREATED, scored 1 if the subject reported watching the debate, 0 otherwise) on an indicator variable called ASSIGNED, which is scored 1 if the subject was encouraged to watch the debate (see Box 6.3). This regression estimates that Compliers make up 20.5% of the subject pool. A

BOX 6.3

Regression Estimate of the ITT_D in the New York Debate Study

```
lm(formula = TREATED ~ ASSIGNED)

Residuals:
    Min       1Q   Median       3Q      Max
-0.3663  -0.3663  -0.1616   0.6337   0.8384

Coefficients:
---------------------------------------------------------------------------
                  Estimate      Std. Error       t value        Pr(>|t|)
---------------------------------------------------------------------------
(Intercept)        0.16162         0.01931         8.367       < 2e-16 ***
ASSIGNED           0.20472         0.02718         7.532      1.12e-13 ***
---------------------------------------------------------------------------

Signif. codes:      0 `***' 0.001 `**' 0.01 `*' 0.05 `.' 0.1 ` ' 1

Residual standard error: 0.4297 on 998 degrees of freedom
Multiple R-squared: 0.05379,  Adjusted R-squared: 0.05284
F-statistic: 56.73 on 1 and 998 DF, p-value: 1.116e-13
```

Source: Mullainathan, Washington, and Azari 2010.

test of the strict null hypothesis of no effect suggests a p-value that is less than 0.00001. Because we have assumed monotonicity, we can be quite confident that Compliers make up some positive fraction of the subject pool, which satisfies the requirement that $ITT_D > 0$. Still, Compliers appear to make up a relatively small fraction of all subjects: the 95% confidence interval ranges from 15.1% to 25.8%. Since the $\widehat{ITT}_D$ appears in the denominator of the $\widehat{CACE}$ estimator, this interval suggests that the $\widehat{ITT}$ could have been multiplied by a number ranging from 4 to 7.

The next regression (see Box 6.4) estimates the ITT, the effect of treatment assignment on opinion change. Here we find a coefficient of 0.057, which confirms our hand calculations based on a comparison of treatment and control group means. A rigorous test of the sharp null hypothesis of no treatment effect is conducted by means of randomization inference. After repeating the randomization 100,000 times, we find that 3,911 of the simulated random assignments generate an estimated ITT of at least 0.05707, which implies a one-tailed p-value of 0.039. We therefore reject the sharp null hypothesis that the assigned treatment has no effect on outcomes.

Estimates of the CACE are obtained from a 2SLS regression in which opinion change is regressed on viewing the debate, using encouragement as an instrumental variable. The estimate (0.279) confirms our earlier calculation that the treatment raised the rate of opinion change among Compliers by 27.9 percentage points. Confidence intervals are estimated using the normal approximation introduced in section 5.8. This interval ranges from $0.2787 - (1.96)(0.15299) = -0.02$ to $0.2787 + (1.96)(0.15299) = 0.58$.

BOX 6.4

Regression Estimate of the ITT in the New York Debate Study

```
lm(formula = Y ~ ASSIGNED)

Residuals:
     Min       1Q   Median       3Q      Max
 -0.4753  -0.4753  -0.4182   0.5248   0.5818

Coefficients:
----------------------------------------------------------------------------
                 Estimate      Std. Error      t value      Pr(>|t|)
----------------------------------------------------------------------------
(Intercept)       0.41818        0.02233        18.725       <2e-16 ***
ASSIGNED          0.05707        0.03143         1.816       0.0697 .
----------------------------------------------------------------------------

Signif. codes:     0 '***' 0.001 '**' 0.01 '*' 0.05 '.' 0.1 ' ' 1

Residual standard error: 0.4969 on 998 degrees of freedom
Multiple R-squared: 0.003293, Adjusted R-squared: 0.002294
F-statistic: 3.297 on 1 and 998 DF, p-value: 0.06969
```

Source: Mullainathan, Washington, and Azari 2010.

BOX 6.5

2SLS Regression Estimate of CACE in the New York Debate Study

```
> summary(tsls(Y~TREATED,~ASSIGNED))

2SLS Estimates

Model Formula: Y ~ D

Instruments: ~ASSIGNED
---------------------------------------------------------------------------
                Estimate      Std. Error      t value       Pr(>|t|)
---------------------------------------------------------------------------
(Intercept)      0.3731        0.04346         8.585         0.00000
TREATED          0.2787        0.15299         1.822         0.06876
---------------------------------------------------------------------------

Residual standard error: 0.4952 on 998 degrees of freedom
```

Source: Mullainathan, Washington, and Azari 2010.

In addition to estimating treatment effects, we can use the monotonicity assumption to home in on average potential outcomes for some types of subjects. The estimated average potential outcome of Always-Takers in their treated state is 50.0%; we infer this percentage from the fact that only Always-Takers are treated when assigned to the control group. Subjects who are treated when assigned to the treatment group are a mixture of Always-Takers and Compliers. But since we have estimates of the proportion of Compliers and Always-Takers, and the average outcome when treated for the Always-Takers, we can use the estimates we have gathered to estimate the average outcome for Compliers when treated. The calculation uses the fact that:

$$E[Y_i(z = 1, d = 1) \\ = \frac{E[Y_i(1)|D_i(1) > D_i(0)]\pi_C + E[Y_i(1)|D_i(1) = D_i(0) = 1]\pi_{AT}}{\pi_C + \pi_{AT}}. \quad (6.24)$$

Swapping in estimates of the proportions of Compliers and Always-Takers as well as the average treated potential outcomes of Always-Takers gives:

$$0.595 \approx \frac{\hat{E}[Y_i(1)|D_i(1) - D_i(0) = 1](0.205) + (0.500)(0.162)}{(0.205) + (0.162)}. \quad (6.25)$$

Solving for $\hat{E}[Y_i(1)|D_i(1) - D_i(0) = 1]$ gives 0.670. The results suggest that Compliers have a 67% rate of opinion change when treated. The average rate of opinion change among Compliers when untreated is estimated to be:

$$\hat{E}[Y_i(1)|D_i(1) - D_i(0) = 1] - \widehat{CACE} = 0.670 - 0.279 = 0.391. \quad (6.26)$$

TABLE 6.3

Estimates of average potential outcomes from the New York mayoral debate experiment

	Compliers	Always-Takers	Never-Takers	Defiers
Average $Y_i(1)$	67.0%	50.0%	unknown	unknown
Average $Y_i(0)$	39.1%	unknown	40.6%	unknown
Percentage of subject pool	20.5%	16.2%	63.4%	0%*

*Defiers are 0% by assumption of monotonicity.
Source: Mullainathan, Washington, and Azari 2010.

Suppose that instead of conducting an experiment, the authors had simply conducted a post-election survey, asking respondents if they had watched the debate and if they had recently changed their opinions of the candidates. We can approximate this type of observational study by restricting our attention to subjects in the control group of the experiment. Comparing control group subjects who are treated to control group subjects who are not treated is a comparison of the average outcome among Always-Takers when they are treated (50.0) to the average outcome among a combination of Compliers and Never-Takers when they are untreated (40.2). This method produces an estimated ATE of about 10 percentage points. This example illustrates the hazards of nonexperimental comparisons. It would appear from this comparison that watching the debate does not produce a lot of reported opinion change, but this type of comparison throws together different types of subjects, and the estimate it generates is an uninterpretable jumble of treatment effects and pre-existing differences between types.

Table 6.3, which assembles our estimates of average potential outcomes, shows that the Always-Takers and Compliers have very different average outcomes when treated. It is easy to forget that the CACE is the treatment effect for a particular subgroup, Compliers, and to extrapolate to the entire subject pool. We have no information about the ATE among Always-Takers. For all we know, it may be the same as the ATE among Compliers. However, when average potential outcomes appear as different as they do for Compliers and Always-Takers, one should be especially cautious when extrapolating from the former to the latter.

6.5 Discussion of Assumptions

The conclusion that watching the debate caused substantial reported opinion change rests on some important assumptions, primarily monotonicity, excludability, and independence. We review these assumptions, consider their plausibility, and evaluate the possible effects of their violation. This exercise serves two purposes. First, we learn more about the assumptions when we see them in action. Second, careful examination of assumptions is an important way to assess non-statistical sources of uncertainty. Much of this discussion is nitpicky and designed to illustrate the implications of the assumptions, not the deficiencies of a creative piece of experimental research.

6.5.1 Monotonicity

The monotonicity assumption rules out Defiers, subjects who would watch the debate when encouraged to watch some other show, but would watch some other show when encouraged to watch the debate. Such behavior seems downright perverse, and intuition suggests that the fraction of subjects fitting this description cannot be large.

For the sake of argument, however, suppose that some subjects are initially more attracted to the non-debate program than the debate. Recall that the non-debate show described to the control group aired at the same time as the debate. Suppose that if these subjects are called and encouraged to watch the debate, they do not turn on the television. If they are asked to watch the non-debate show, they decide to turn on the television to check it out. Suppose these subjects become bored and flip from channel to channel, eventually landing on this unexpectedly interesting political debate, which they then watch. Notice that these hypothetical subjects are Defiers: when assigned to the treatment group, they are not treated, but when assigned to the control group, they are treated.

When the subject pool contains Defiers, the ratio of ITT/ITT_D no longer recovers the CACE. Rather, this formula becomes a weighted average of the treatment effects among Compliers and Defiers. This mixture of treatment effects will typically not be the quantity the researcher set out to discover. For example, suppose that the reason you seek to estimate the CACE is to obtain a ballpark estimate of the effect of a public service announcement encouraging people to tune in to the debates. The kinds of people who will be induced by the announcement to watch the debate are likely to be similar to the Compliers in your experiment, because the public service announcement will not increase viewership among Defiers (who, according to the scenario described above, tune in for the "placebo" show and then surf). To approximate the effect of a public service announcement on listeners, you want the CACE.

How big a problem is this? It depends. If the ratio of Defiers to Compliers is small, the distortion to the estimated CACE will be small. And even if the ratio is large, the bias is small when the ATE among Defiers is similar to the ATE among Compliers. Without further investigation, it is difficult to say whether people who tune into debates inadvertently are as strongly influenced by them as people who tune in when encouraged by survey interviewers.

6.5.2 Exclusion Restriction

The exclusion restriction assumption states that the effect of the treatment assignment on potential outcomes occurs due to the treatment itself; the effect is not transmitted through any other channel. Based on this assumption, the observed difference between treatment and control group outcomes is "credited" to the treatment.

This assumption might fail for at least two reasons. First, suppose that those who are encouraged to watch the political debate respond to this request by becoming more attentive to New York City politics so that they will seem more knowledgeable when surveyed again about the election. Paying more attention to politics may cause the subject to acquire information, thereby inducing opinion change. Second, a subtle effect of the encouragement might be to alter reports of opinion change. Subjects encouraged to watch the political debate might have the same viewing experience as those who watch without encouragement; however, when *asked* about their experience, subjects who received prior encouragement might be inclined to answer in a more civic-minded way, reporting more active engagement with the debate and, accordingly, more opinion change.

When the ITT_D is small, even modest violations of the exclusion restriction can have a large effect on the estimated treatment effect. The estimated CACE is the ITT scaled by the ITT_D. In the debate experiment, the observed ITT is divided by approximately 0.2, which is the same as multiplying the ITT by 5. Suppose that subjects who are encouraged to view the debate respond by paying more attention to local politics over the next few days, and, as a result, the treatment group becomes 3 percentage points more likely to report changing their opinion of the candidates. Boosting the average outcome in the treatment group by 3 percentage points raises the estimated CACE by 15 percentage points. Under this scenario, approximately half of the 27.9 percentage point effect estimated above would be due to a violation of the exclusion restriction.

How might one detect violations of the exclusion restriction? There are no easy answers, but more elaborate experimental designs would help address some of the concerns raised above. You could resurvey a random sample of the treatment and control groups just *prior* to the debate to detect shifts in self-reported opinion caused by factors other than exposure to the debate. You might also ask respondents directly about attentiveness to local politics to assess whether subjects encouraged to view the debate differ from subjects assigned to the control group. In order to reduce pressure to report opinion change, one might also introduce an additional experimental condition, such as a group that is encouraged to view the debate on some other pretext, such as evaluating the fairness of the moderators. If the main concern is the obtrusiveness of the encouragement itself, one could alter the design fundamentally, using advertisements directed to different localities or households to induce debate viewership. The challenge, in that case, would be to induce high enough rates of compliance to allow for precise estimation of the CACE.

6.5.3 Random Assignment

In the debate study, subjects were randomly assigned to treatment conditions, but not all of the subjects who were assigned to experimental conditions were successfully re-interviewed after the debate. When outcomes are missing for some subjects, an experiment is said to suffer from *attrition*. As explained in Chapter 7, attrition threatens to undo random assignment. If subjects' potential outcomes predict whether they drop out of the study, results may be biased: the average potential outcomes of those who remain in the treatment and control groups are no longer equivalent. In the context of this study, a concern is that attrition will be more likely to eliminate the politically inattentive in the treatment group, as those encouraged to view the debate may be embarrassed to participate in a post-debate interview if they failed to watch. Chapter 7 reviews a variety of methods for assessing attrition-related bias and estimating average treatment effects in the presence of attrition.

6.5.4 Design Suggestions

The imaginative encouragement design employed by Mullainathan, Washington, and Azari provides an instructive example of the issues that arise when experiments encounter two-sided noncompliance. When critically evaluating a study of this kind, it is important to think about the ways in which assumptions may break down, but no less important to think about ways of augmenting the research design so as to address these concerns empirically. How might one go about increasing the share of Compliers? If one intends to conduct a series of experiments, one should experiment with different ways of encouraging compliance. In the debate study, for example, the researchers randomly assigned some subjects to an experimental condition in which they were encouraged both to watch the debate and to pledge to do so. So long as the number of different encouragements does not grow so large as to be administratively unmanageable or incapable of producing reasonably precise statistical results, diversifying the array of different encouragements may lead to discoveries that facilitate follow-up research. How might one rule out Defiers? Perhaps one could employ a third treatment group that is encouraged to do something other than watch TV or perhaps is not encouraged at all. In order to shore up excludability, perhaps cloak the primary aim of the study (studying the effects of debate viewing) amid questions about unrelated topics, so that subjects in the treatment group will not feel obligated to report a change in their opinions. The thorniest problem may be attrition; one design idea that we propose in the next chapter is a supplementary sample of people who drop out of the study. The larger point is that every experiment raises questions about assumptions, and much of the creativity that goes into research is devoted to finding feasible ways of addressing lingering questions.

6.6 Downstream Experimentation

Experiments that experience two-sided noncompliance are usually presumed to be flawed in some way. Although some studies encounter two-sided noncompliance due to lax experimental procedures that lead to inadvertent treatment of the control group, noncompliance is an expected feature of well-executed experiments that employ encouragement designs. Subjects in the treatment condition are encouraged but not required to take a treatment; subjects in the control group are neither encouraged nor barred access to the treatment. Noncompliance occurs because the random intervention is just one of many "encouragements" that cause people to take the treatment. A bit of reflection reveals that the world abounds with encouragement designs. Random interventions affect an outcome variable, which in turn may cause other outcomes.

The modeling framework discussed above for analyzing encouragement designs applies to the investigation of *downstream experiments*, studies that trace the indirect consequences of a randomly generated causal effect.[3] Examples of downstream experiments may be found in a wide array of social science disciplines. Kling estimates the effects of incarceration on subsequent earnings by taking advantage of the fact that judges' sentencing philosophies vary and therefore incarceration is predicted by the judge to which defendants are randomly assigned.[4] Green and Winik also use randomly assigned judges as a treatment predicting prison sentences of drug offenders and gauge the effects of prison terms on rates of recidivism.[5] Beaman, Chattopadhyay, Duflo, Pande, and Topalova examine the downstream effects of an electoral law that places women into leadership roles in randomly selected Indian villages on villagers' attitudes toward women.[6] Chauchard uses a similar design to study the effects of randomly assigned political representation for lower castes on Indian villagers' attitudes and behaviors toward lower-caste people.[7] Earlier, we alluded to scholarly efforts to use draft lotteries to gauge the effects of military service on earnings,[8] long-term health outcomes,[9] and criminal convictions[10] in the United States. This line of research has been extended to other countries as well. Galiani, Rossi, and Schargrodsky, for example, examine the effect of the draft lottery in Argentina on criminal activity.[11]

3 Green and Gerber 2002.
4 Kling 2006.
5 Green and Winik 2010.
6 Beaman, Chattopadhyay, Duflo, Pande, and Topalova 2009.
7 Chauchard 2010.
8 Angrist 1990.
9 Dobkin and Shabani 2009.
10 Lindo and Stoecker 2010.
11 Galiani, Rossi, and Schargrodsky 2010. See also Davenport 2011 and Erikson and Stoker 2011 on the Vietnam draft lottery's effects on political attitudes and behavior.

Electoral outcomes and political behavior have attracted a good deal of attention as well. Bagues and Esteve-Volart study the effect of randomly generated income windfalls in Spain associated with the Christmas lottery on electoral support for incumbents.[12] Gerber, Green, and Shachar examine the consequences of randomly induced increases in turnout on voters' subsequent chances of voting,[13] and Sondheimer and Green consider the effects of experimentally induced changes in high school graduation rates on subsequent voter turnout.[14]

In order to see the connection between two-sided noncompliance and the investigation of downstream effects, let's consider a simplified example inspired by the Sondheimer and Green study. Imagine an experiment in which subjects are randomly assigned to a treatment (Z_i) that in turn generates an intent-to-treat effect on an outcome (Y_i). For concreteness, let $z = 1$ refer to assignment of elementary school students to classrooms of 15 students, and let $z = 0$ refer to assignment to typical classrooms of 24 students. And suppose that Y_i were a measure of educational attainment, such as whether a student graduates from high school. If the experiment were truly successful in raising graduation rates, one could investigate the downstream effect of randomly increased graduation rates on other outcomes, such as participation in elections. Let's call this downstream outcome ψ_i. Although the notation has been rearranged, the framework is identical to the case of two-sided noncompliance described above. The experimental intervention encourages high school graduation, which in turn affects voting. In order to map the new notation back to our usual notation, call z_i the assignment of elementary school class size, call d_i graduation from high school, and call Y_i voter turnout at some later date.

Tracing effects in this way parallels an encouragement design, and the assumptions spelled out earlier in this chapter apply in the usual way. Non-interference presumes that subjects' potential outcomes are unaffected by the assignment or treatment of other subjects. In this application, turnout is assumed to respond to each subject's own class size and educational attainment, and not anyone else's. Excludability requires that assigned class size influences voting only through high school graduation. This assumption would be violated if class size affected voter turnout for reasons other than high school graduation. For example, class size may affect self-confidence, which in turn may affect turnout in ways that are not transmitted through educational attainment. Monotonicity presupposes that small class sizes never cause someone who would otherwise graduate to drop out of high school. Finally, complete random assignment implies that the assigned treatment is unrelated to potential outcomes. When the experiment is launched, this assumption may be satisfied initially by the use of a random allocation procedure, but subjects may drop out of

12 Bagues and Esteve-Volart 2010.
13 Gerber, Green, and Shachar 2003.
14 Sondheimer and Green 2010.

the study over time. If smaller class size causes students to stay in school, it may also affect whether students' outcomes are measureable years later. Suppose that less self-confident students have lower average potential outcomes when it comes to voter turnout. If less self-confident students are more likely to remain in the study when they are assigned to the treatment group, the estimate of the CACE may be downwardly biased (see Chapter 7). Whether these concerns actually lead to appreciable bias will depend on the details of a particular experiment; the point in reviewing these assumptions is to call attention to the kinds of arguments that a researcher must contend with when proposing and defending downstream analysis.

When these assumptions are satisfied, the usual $\widehat{ITT}/\widehat{ITT}_D$ estimator provides a consistent estimate of the average effect of graduation on turnout among Compliers. In the context of this experiment, Compliers are subjects who graduate if and only if they are assigned to a small classroom. By contrast, Never-Takers are subjects who drop out of high school regardless of assignment, and Always-Takers graduate regardless of assignment. The CACE therefore assesses the effect of high school graduation on turnout for subjects who are induced to graduate by their assigned treatment. The study cannot speak to the question of whether high school graduation increases participation among the relatively high-achieving Always-Takers or the low-achieving Never-Takers.

Let's now consider some simulated data that illustrate the mechanics of downstream analysis and the ways that key assumptions come into play. Table 6.4 presents a schedule of potential outcomes relating assigned treatments, actual treatments, and outcomes. Again, we can think of z_i as assignment to a small classroom, d_i as graduation from high school, and Y_i as voting in a subsequent election. Non-interference is built into the simulation insofar as outcomes for subject i depend only on this subject's own assignment and treatment. The exclusion restriction is also assumed, at least for the moment, because potential outcomes respond only to assignment-induced treatment, not assignments themselves. The column labeled "N" refers to the number of observations in the simulated dataset with a specific configuration of potential outcomes. By changing the values of N, readers can adapt this schedule of potential outcomes to simulate an array of different scenarios.

The Ns listed in Table 6.4 generate the $\widehat{ITT}_D$ and $\widehat{ITT}$ estimates depicted in the first column of Table 6.5. Because Defiers are given weights of zero, this hypothetical illustration imposes the monotonicity assumption. Simulation 1 in Table 6.5 indicates that graduation rates increase from 60% to 85% when subjects are assigned to the treatment condition, implying an $\widehat{ITT}_D$ of 0.25. Voter turnout also increases from 49.5% among those assigned to the control group to 54.5% among those assigned to the treatment group, implying an $\widehat{ITT}$ of 0.05. The estimator $\widehat{ITT}/\widehat{ITT}_D = 0.20$, correctly recovering the ATE among Compliers. Assuming excludability, the interpretation is that among those who graduate if and only if they are placed in small classrooms as youngsters, completing high school raises voter turnout by 20 percentage

TABLE 6.4

Hypothetical schedule of potential outcomes and assigned treatments for illustration of a downstream experiment

$Y_i(d=0)$	$Y_i(d=1)$	$d_i(z=0)$	$d_i(z=1)$	z	N	Subject Type
0	0	1	1	0	100	Always-Takers
0	1	1	1	0	100	Always-Takers
1	0	1	1	0	100	Always-Takers
1	1	1	1	0	300	Always-Takers
0	0	1	1	1	100	Always-Takers
0	1	1	1	1	100	Always-Takers
1	0	1	1	1	100	Always-Takers
1	1	1	1	1	300	Always-Takers
0	0	0	1	0	125	Compliers
0	1	0	1	0	55	Compliers
1	0	0	1	0	5	Compliers
1	1	0	1	0	65	Compliers
0	0	0	1	1	125	Compliers
0	1	0	1	1	55	Compliers
1	0	0	1	1	5	Compliers
1	1	0	1	1	65	Compliers
0	0	1	0	0	0	Defiers
0	1	1	0	0	0	Defiers
1	0	1	0	0	0	Defiers
1	1	1	0	0	0	Defiers
0	0	1	0	1	0	Defiers
0	1	1	0	1	0	Defiers
1	0	1	0	1	0	Defiers
1	1	1	0	1	0	Defiers
0	0	0	0	0	100	Never-Takers
0	1	0	0	0	25	Never-Takers
1	0	0	0	0	0	Never-Takers
1	1	0	0	0	25	Never-Takers
0	0	0	0	1	100	Never-Takers
0	1	0	0	1	25	Never-Takers
1	0	0	0	1	0	Never-Takers
1	1	0	0	1	25	Never-Takers

TABLE 6.5

Estimated ITT_D, ITT, and CACE from simulated downstream experiments

	SIMULATION 1	SIMULATION 2	SIMULATION 3	SIMULATION 4
	Assumes monotonicity and exclusion restriction	Assumes monotonicity and exclusion restriction, adds extra Never-Takers and Always-Takers	Violates monotonicity but assumes exclusion restriction	Assumes monotonicity but violates exclusion restriction
% Graduating from high school in control group	60.0%	68.6%	63.6%	60.0%
% Graduating from high school in treatment group	85.0%	82.9%	77.3%	85.0%
*ITT_D (standard error)	0.250 (0.019)	0.143 (0.014)	0.136 (0.019)	0.250 (0.019)
% Voting in subsequent election in control group	49.5%	52.6%	49.6%	49.5%
% Voting in subsequent election in treatment group	54.5%	55.4%	54.1%	56.5%
*ITT (standard error)	0.050 (0.022)	0.029 (0.017)	0.045 (0.021)	0.070 (0.022)
**CACE (standard error)	0.200 (0.085)	0.200 (0.112)	0.333 (0.147)	0.280 (0.084)
N	2,000	3,500	2,200	2,000
ATE | Always-Takers (N)	0 (1,200)	0 (2,400)	0 (1,200)	0 (1,200)
ATE | Compliers (N)	0.20 (500)	0.20 (500)	0.20 (500)	0.20 (500)
ATE | Defiers (N)	N/A (0)	N/A (0)	0 (200)	N/A (0)
ATE | Never-Takers (N)	0.17 (300)	0.17 (600)	0.17 (300)	0.17 (300)

Note: *Calculated from OLS regression. **Calculated from 2SLS regression.

points. The 95% confidence interval estimated using instrumental variables regression ranges from 0.03 to 0.37.

The second column of Table 6.5 shows how the precision of these estimates diminishes as we add more Never-Takers and Always-Takers to the sample. In contrast to Simulation 1, where Compliers comprise 25% of the 2,000 observations, Simulation 2 reports a scenario where Compliers comprise 14% of 3,500 observations. (The second scenario simply doubles the N in each row of Always-Takers and Never-Takers in Table 6.4.) The $\widehat{\text{ITT}}_{\text{D}}$ drops to 0.143, because the graduation rates now increase from 68.6% to 82.9% among those assigned to control and treatment conditions. The $\widehat{\text{ITT}}$ also drops to 0.029, as it represents a weighted average of all four subject types, and Always-Takers and Never-Takers contribute nothing to the ITT. The estimator $\widehat{\text{ITT}}/\widehat{\text{ITT}}_{\text{D}} = 0.20$, recovering the correct CACE, but now the sampling distribution is much wider: the 95% confidence interval ranges from −0.02 to 0.42. The broader lesson is that in order to extract precise estimates from a downstream experiment, one should focus on treatments that strongly influence the "immediate" outcome d_i or on samples that contain a high proportion of Compliers.[15] Standard errors tend to decline as the $\widehat{\text{ITT}}_{\text{D}}$ rises.

Simulation 3 of Table 6.5 shows what happens when the monotonicity assumption breaks down. To keep the algebra simple, we consider a case in which the ATE among Defiers is 0, and the ratio of Compliers to Defiers is 5:2. Recalling equation (6.20), we can calculate the consequences for the $\widehat{\text{CACE}}$:

$$\frac{\widehat{\text{ITT}}}{\widehat{\text{ITT}}_{\text{D}}} = \frac{(\text{ATE}\,|\,\text{Compliers})\pi_C - (\text{ATE}\,|\,\text{Defiers})\pi_D}{\pi_C - \pi_D}$$

$$= \frac{(0.20)\left(\frac{500}{2200}\right) - (0)\left(\frac{200}{2200}\right)}{\left(\frac{500}{2200}\right) - \left(\frac{200}{2200}\right)} = \frac{1}{3}. \qquad (6.27)$$

The presence of Defiers in this instance reduces the denominator without changing the numerator, producing an upwardly biased $\widehat{\text{CACE}}$. Had the ATE among Defiers been larger than the ATE among Compliers, the bias would have gone in the opposite direction.

The final simulation in Table 6.5 illustrates the consequences of violating the exclusion restriction. The simulation assumes that z_i directly influences outcomes, raising Y_i by 2 percentage points in the assigned treatment group. Substantively, this kind of bias could creep in if the intervention had turnout-promoting effects other

15 When searching for strong $\widehat{\text{ITT}}_{\text{D}}$ effects that could facilitate a downstream analysis, bear in mind that the $\widehat{\text{ITT}}_{\text{D}}$ effect must be "real" and not merely a sampling fluke. If the $\widehat{\text{ITT}}_{\text{D}}$ is large due to a lucky draw from the sampling distribution, the $\widehat{\text{CACE}}$ may also be an unrepresentative draw.

BOX 6.6

Incorrect Application of Regression as a Test of the Exclusion Restriction, Leading to Misleading Results

```
coef(summary(lm(Y ~ D+Z)))
```

	Estimate	Std. Error	t value	Pr(>|t|)
(Intercept)	0.23418367	0.02061000	11.362621	4.898133e-29
D	0.43469388	0.02416360	17.989618	3.538618e-67
Z	-0.05867347	0.02157878	-2.719035	6.603925e-03

than raising high school graduation rates. The simulation illustrates how this type of violation increases the $\widehat{ITT}$ and therefore leads to an overestimate of the CACE. In this case, the share of Compliers is 25%, so the bias to the CACE is multiplied by 4. Many downstream analyses work with an $\widehat{ITT}_D$ of 0.10 or less, which greatly amplifies any violation of the exclusion restriction. Unfortunately, exclusion restrictions are not directly testable. One should *not* assess whether Y_i is directly affected by z_i by regressing Y_i on z_i and d_i. This type of analysis will not generate unbiased estimates of the effect of z_i because d_i is not randomly assigned. What if we had applied this type of misguided regression analysis to the data from Simulation 1, where the exclusion restriction is valid? As Box 6.6 indicates, we would have incorrectly concluded that z_i has a statistically significant negative effect on Y_i. We also would have grossly misestimated the CACE, obtaining an estimate of 0.43, when the true CACE is 0.20.

Before concluding our discussion of downstream investigation, let's consider an empirical example that illustrates one further design principle: the use of multiple treatments facilitates the identification of downstream effects. The initial experiment was conducted by Gerber, Green, and Larimer and involved four different mailings to encourage voter turnout prior to an August 2006 primary election.[16] As described in detail in Chapter 10, the mailings employed varying degrees of "social pressure" to urge compliance with the norm of voting. The first mailing encouraged recipients to do their civic duty; the second mailing, in addition, informed recipients that whether they voted was being monitored by researchers (testing the so-called Hawthorne effect, or the effect of being studied); the third mailing indicated that voter turnout is a matter of public record and disclosed the past turnout of everyone in the household; and the fourth mailing disclosed not only the household's record of voter turnout but also the turnout record of other households on the same block. The researchers found that voter turnout increases as social pressure intensifies. As

16 Gerber, Green, and Larimer 2008.

TABLE 6.6

Estimates of the downstream effect of voting in August 2006 on voting in August 2008 and August 2010, by treatment condition

	Treatment administered prior to the 2006 August election					
	Control	Civic duty	Hawthorne	Self	Neighbor	Entire sample
% Voting in August 2006	31.67	33.63	34.44	36.85	40.41	
ITT_D (standard error)		1.96 (0.35)	2.77 (0.35)	5.18 (0.36)	8.75 (0.36)	
% Voting in August 2008	34.79	34.58	35.15	35.79	36.07	
ITT (standard error)		−0.21 (0.36)	0.36 (0.36)	1.00 (0.36)	1.28 (0.36)	
CACE (standard error)		−10.81 (18.82)	12.98 (12.54)	19.37 (6.68)	14.67 (3.97)	15.69 (3.63)
% Voting in August 2010	46.78	46.91	47.16	47.61	48.05	
ITT (standard error)		0.14 (0.37)	0.39 (0.37)	0.83 (0.37)	1.28 (0.37)	
CACE (standard error)		6.94 (18.45)	14.00 (12.85)	16.06 (6.87)	14.60 (4.09)	14.91 (3.74)
N	165,297	32,992	33,047	33,037	32,983	297,356

Note: Standard errors account for clustered assignment by households. Estimates of the ITT, ITT_D, and the CACE are calculated based on OLS and IV regressions that compare a single treatment group to the control group. The estimate of the CACE for the entire sample is based on an overidentified 2SLS regression of voting in 2008 (or 2010) on voting in 2006; the four treatment assignments serve as instrumental variables predicting voting in 2006.

shown in Table 6.6, turnout rates increase significantly in every treatment condition, especially the last two, which disclose records of voter turnout.[17] The largest effect is 8.75 percentage points, which represents a 28% increase in turnout over the voting rate in the control group.

What are the downstream consequences of this effect? One hypothesis is that the act of voting creates or solidifies voting habits, which then manifest themselves in subsequent elections. For example, one might suppose that many registered voters are unaccustomed to voting in an August election. By inducing August turnout in 2006, the random intervention also increases turnout in subsequent August elections. This hypothesis implies that we should see an intent-to-treat effect of the mailings on turnout in August of 2008. Table 6.6 shows that in fact a significant

17 This table is adapted from Davenport, et al. (2010). We restrict the sample to voters who were still registered to vote at the same address two years later, on the assumption that treatment assignment has no effect on moving or re-registering. Empirically, we find no relationship between treatment assignment and changing address.

intent-to-treat effect appears to exist for two of the treatments, the treatments that showed the most powerful effects on voting in August 2006. When tracking outcomes over time, one concern is that persistent treatment effects are just a sampling fluke: random assignment may have chosen an unusually participatory group of voters for the two seemingly effective treatments, and these subjects vote at elevated rates in election after election. In a smaller sample, this conjecture might have some credence; here, the sample sizes are so enormous that the pattern cannot plausibly be attributed to chance.

The ITT estimates suggest that the effects of the treatment persist. As we prepare to estimate the CACE, let's evaluate whether the application seems to satisfy core assumptions. Non-interference and independence seem to hold in this application. Researchers studying this and other datasets have found that social pressure mailings generate weak spillover effects to nearby neighbors. Independence is satisfied because of random assignment and because sample attrition is apparently unrelated to treatment assignment. Monotonicity is a bit more uncertain. It is hard to know whether the sample contains Defiers, subjects who vote in the 2006 election if and only if they are not sent a mailing. If Defiers exist, it remains unclear whether their ATE (i.e., the average effect of voting in the first election on voting in the second election) is substantially different from the ATE among Compliers.

In this application, the most difficult assumption to satisfy is excludability. The main concern is that voters may remember receiving the mailings two years later.[18] If the mailings themselves or the prospect of receiving future mailings like them precipitate voting, the estimated CACE may incorrectly attribute to habit an effect that is in fact due to social pressure. Without additional data on whether subjects remember the mailings two years later, this hypothesis cannot be ruled out.[19] Thus, we are left with three explanations for why the initial treatments may generate different apparent downstream effects: the estimates are subject to sampling variability; different sets of Compliers to each treatment have different average downstream effects; and violations of excludability bias some or all of the downstream estimates.

Although the exclusion restriction cannot be tested directly, sometimes claims about a particular violation have testable implications. In the case of voting habits, one might reason that memories of the mailings fade over time, but habits endure. When we track these subjects for an additional two years, do we continue to find elevated turnout rates? The bottom rows of Table 6.6 show that treatment effects remain

18 For a discussion of other critiques related to the exclusion restriction as applied to the study of habit formation, see Gerber, Green, and Shachar 2003. For example, turnout may increase because those who vote in the current election receive greater attention from political campaigns in subsequent elections.

19 When faced with this kind of uncertainty, researchers sometimes conduct a *sensitivity test*, varying the fraction of the ITT that is due to a violation of the exclusion restriction and examining the p-value of the resulting 2SLS estimate. A sensitivity test provides an indication of how far one can bend an assumption without changing one's statistical inference. We present an example of this type of sensitivity analysis at http://isps.research.yale.edu/FEDAI.

highly significant four years after the original treatment. The ITT estimates for 2010 appear to be only slightly weaker than those for 2008.

Because this experiment deployed four versions of the treatment, in theory, we could estimate average treatment effects for four somewhat different sets of Compliers. For example, we could compare the Civic Duty group to the control group in order to estimate an ITT_D and an ITT. We could repeat this exercise for the Hawthorne, Self, and Neighbors groups. The results of this analysis are instructive. The least precisely estimated $\widehat{CACE}$ emerges from the comparison of the Civic Duty and control groups. This estimate is noisy (and in fact negative, although statistically indistinguishable from zero), reflecting the weak $\widehat{ITT}_D$ of just 1.96 percentage points. Notice that the standard errors of the $\widehat{CACE}$ diminish as we move to progressively larger values of $\widehat{ITT}_D$. The most precisely estimated $\widehat{CACE}$ suggests that inducing Compliers to vote in 2006 causes a 14.7 percentage point increase in their turnout two years later.

One can look at the four $\widehat{CACE}$ estimates in one of two ways. The first is to consider estimates of different estimands. Each treatment induces a different set of Compliers to vote in 2006, and the downstream effect of this participation need not be the same for different types of people. Another way to look at the four $\widehat{CACE}$ estimates is to regard them as four different estimates of the same parameter. From this perspective, the four groups of Compliers are sufficiently similar that one may reasonably stipulate that all of them should exhibit the same average "habit formation" effect. Arguably, habit formation is a broad hypothesis that presumably applies generally to different types of voters. If the habit formation hypothesis is true, no matter what causes elevated turnout in one election, this uptick in turnout should persist in future elections. From this standpoint, it is a design advantage to track an experiment that increased turnout in 2006 using a variety of treatments.

When the same causal parameter can be estimated in more than one way, the parameter is said to be *overidentified*. The surplus of information is reflected in the two-equation system that we estimate using 2SLS. The second-stage equation is the usual model of the outcome variable:

$$Y_i = \beta_0 + \beta_1 VOTED_i + u_i. \tag{6.28}$$

The key parameter of interest is β_1, the CACE, which is assumed to be the same across all sets of Compliers. In the first-stage equation, $CIVIC_i$, $HAWTHORNE_i$, $SELF_i$, and $NEIGHBORS_i$ are dummy variables that mark each of the four treatment conditions.

$$VOTED_i = \alpha_0 + \alpha_1 CIVIC_i + \alpha_2 HAWTHORNE_i + \alpha_3 SELF_i + \alpha_4 NEIGHBORS_i + e_i. \tag{6.29}$$

These four variables are instrumental variables that are assumed to be independent of u_i. This assumption follows from the exclusion restriction and the fact that simple random assignment was used to allocate subjects to treatments.

The 2SLS regression estimator finds the combination of coefficients in equation (6.29) that produce the best-fitting prediction of $VOTED_i$. These predictions ($\widehat{VOTED}_i$) from the first-stage equation (6.29) are substituted for $VOTED_i$ in the second stage equation (6.28) and regression is used to estimate β_1. The rightmost column of Table 6.6 shows the estimated $\widehat{CACE}$ of 15.7 with a standard error of 3.6. Interestingly, quite similar results emerge when we repeat the exercise using 2010 turnout as the outcome. The estimated $\widehat{CACE}$ is 14.9 with a standard error of 3.7. It should be stressed that observed turnout in 2008 plays no part in the analysis of 2010 outcomes. The similarity between the two sets of results reflects the fact that the ITT estimates for 2008 happen to resemble those for 2010.

Overidentification confers two advantages. One is extra precision: the pooled estimate has less sampling variability than the estimates derived from comparing each treatment group to the control group. The other advantage is the ability to assess goodness-of-fit. The null hypothesis is that outcomes in all of the treatment groups are generated by a single CACE, and sampling accounts for the observed variability in the estimates. The alternative hypothesis is that the CACE varies by treatment. Failure to reject the null hypothesis is consistent with the hypothesis that a similar process of habit formation holds generally across the four groups. As shown in Figure 6.2, the confidence intervals associated with each treatment's CACE overlap as one would expect from sampling variability, and a statistical test of overidentification fails to reject the null for both the 2008 outcome and the 2010 outcome.[20] This finding does not prove that a single CACE is at work, but it does suggest that the observed variation among the four CACE estimates is not greater than one would expect given sampling variability.

The broader design principle is the idea of using several treatments to generate an experimental effect, which in turn provides an opportunity for testing the same downstream hypothesis in multiple ways. Ideally, to detect a general phenomenon such as habit formation while at the same time meeting concerns about the exclusion restriction, researchers should strive to increase turnout in a variety of different ways, some memorable and others not. The same principle applies to any experiment that is suspected of violating the excludability assumption. Assuming that the same treatment effect operates on all those who would comply with an initial intervention, the design solution is to administer alternative interventions in an effort to isolate the factors that produce the apparent effect. The more widely varying the interventions, the less likely it is that an exclusion restriction violation associated with a specific intervention will go undetected.

---

20 The Hansen *J* statistic provides a test of the null hypothesis that a single CACE generated all four CACE estimates. A nonsignificant statistic suggests that the observed variation across CACE estimates is not greater than what one would expect due to sampling variability. Like the methods used to calculate confidence intervals in this chapter, this test relies on assumptions about the distributions of the error term in equations (6.28) and (6.29) in order to approximate the sampling distribution of this test statistic.

FIGURE 6.2

Assessing treatment effect heterogeneity in the Complier average causal effect using multiple treatments as instrumental variables.

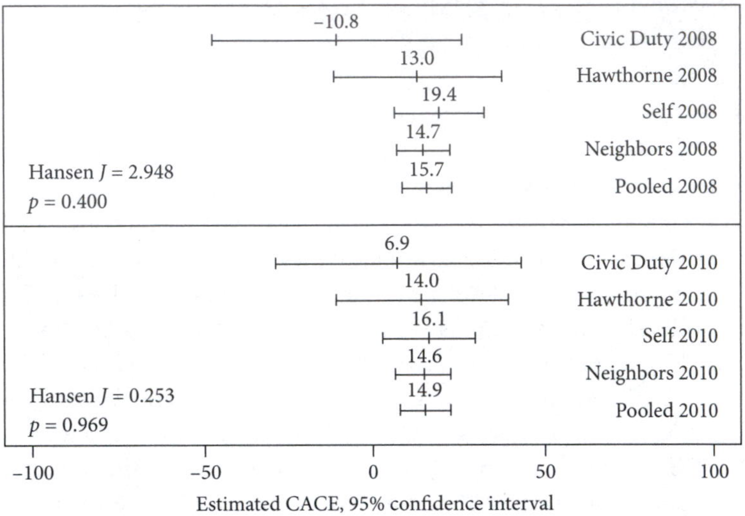

-10.8 Civic Duty 2008
13.0 Hawthorne 2008
19.4 Self 2008
14.7 Neighbors 2008
15.7 Pooled 2008
Hansen $J = 2.948$
$p = 0.400$

6.9 Civic Duty 2010
14.0 Hawthorne 2010
16.1 Self 2010
14.6 Neighbors 2010
14.9 Pooled 2010
Hansen $J = 0.253$
$p = 0.969$

-100 -50 0 50 100

Estimated CACE, 95% confidence interval

Note: Horizontal lines indicate the 95% confidence intervals associated with each estimated CACE. The Hansen *J* statistic provides a test of the null hypothesis that a single CACE generated all four CACE estimates. The fact that the statistic is non-significant suggests that the observed variation across CACE estimates is not greater than one would expect due to sampling variability.

SUMMARY

Two-sided noncompliance occurs when some subjects in the control group are treated and some subjects in the treatment group go untreated. Two-sided noncompliance is common when subjects have access to treatments and discretion about whether to take them. Naturally-occurring randomizations, such as lotteries that change the probability that certain subjects receive a treatment (e.g., private schooling or induction into the military), often give rise to two-sided noncompliance. So do experiments that use encouragement designs to coax randomly selected subjects into taking a treatment, such as watching a TV show.

Two-sided noncompliance involves many of the same conceptual challenges as one-sided noncompliance. The analyst must distinguish between assigned and actual treatment when estimating causal effects, taking care to compare randomly assigned groups, not groups that are formed after random assignment. Comparing average outcomes among those who do and do not receive treatment is a nonexperimental research strategy that is prone to bias.

Noncompliance changes the interpretation of the experimental estimates. Instead of estimating the average treatment effect, the researcher estimates the intent-to-treat effect or the Complier average causal effect. Intent-to-treat effects refer to the effect of assignment to treatment on outcomes, ignoring the rate at which assignment results in actual treatment. The Complier average causal effect refers to the ATE among subjects with a particular configuration of potential outcomes: they are treated if assigned to the treatment group but not if they are assigned to the control group. Which subjects are Compliers remains unknown in the context of two-sided noncompliance. The interpretation of experimental results is constrained in two ways. First, generalizations are limited by the fact that experimental estimates refer to causal effects among Compliers, not the sample as a whole. Second, uncertainty about who the Compliers are means that researchers must be cautious when generalizing to other interventions, subjects, and settings.

The main additional complication posed by two-sided noncompliance is the possibility that the subject pool contains Defiers, or subjects who are treated if assigned to the control group but not if assigned to the treatment group. When analyzing the results of an experiment with two-sided noncompliance, researchers invoke a new assumption, monotonicity, which states that any subject who would not be treated if assigned to the treatment condition would also not be treated if assigned to the control condition.[21] Formally, for all i, $d_i(1) \geq d_i(0)$. Monotonicity rules out Defiers. When the core assumptions of monotonicity, excludability, non-interference, and independence hold, the Complier average causal effect may be estimated empirically from experimental data. The estimator is the ratio of the estimated intent-to-treat effect of random assignment on outcomes divided by the estimated intent-to-treat effect of random assignment on actual treatment. This estimator is consistent, which is to say, it converges to the true CACE as the sample size increases. Estimation is aided when compliance rates are high; all else being equal, as the proportion of Compliers increases, standard errors decline, and the estimator becomes less susceptible to bias due to violations of the excludability assumption.

One important class of applications in which two-sided noncompliance arises is the downstream analysis of random interventions. Here, a randomized intervention produces an experimental effect on an outcome Y_i; the researcher then examines whether this exogenous change in Y_i in turn affects other outcomes of interest, perhaps in a very different substantive domain. The investigation of downstream effects inevitably raises questions about the validity of the excludability assumption. One response is to address these concerns empirically by designing experiments that vary the manner in which Y_i is manipulated.

21 An alternative assumption is that the ATE is the same for Defiers and for Compliers, in which case the presence of Defiers does not introduce bias.

SUGGESTED READINGS

Angrist (2006) provides an accessible discussion of two-sided noncompliance in reference to the Minnesota domestic violence experiment. A more technical discussion may be found in Angrist, Imbens, and Rubin (1996). For more on the monotonicity assumption, see Bhattacharya, Shaikh, and Vytlacil (2008). See Green and Gerber (2002) for an introduction to downstream experimentation.

EXERCISES: CHAPTER 6

1. The following three quantities are similar in appearance but refer to different things. Describe the differences.

$$E[Y_i(d(1)) \mid D_i = 1]$$
$$E[Y_i(d(1)) \mid d_i(1) = 1]$$
$$E[Y_i(d(1)) \mid d_i(1) = d_i(0) = 1]$$

2. The following expression appears in the proof of the CACE theorem. Interpret the meaning of each term in the expression, and explain why the expression as a whole is equal to zero:

$$E[Y_i(d(1)) \mid d_i(1) = d_i(0) = 0] - E[Y_i(d(0)) \mid d_i(1) = d_i(0) = 0].$$

3. Assuming that the excludability and non-interference assumptions hold, are the following statements true or false? Explain your reasoning.
   (a) Among Compliers, the ITT equals the ATE.
   (b) Among Defiers, the ITT equals the ATE.
   (c) Among Always-Takers and Never-Takers, the ITT and ATE are zero.
4. When analyzing experiments with two-sided noncompliance, why is it incorrect to define Compliers as "those who take the treatment if assigned to treatment"?
5. Suppose that a sample contains 30% Always-Takers, 40% Never-Takers, 15% Compliers, and 15% Defiers. What is the ITT_D?
6. Suppose that, in violation of the monotonicity assumption, a sample contains both Compliers and Defiers. Let π_C be the proportion of subjects who are Compliers, and let π_D be the proportion of subjects who are Defiers. Show that the CACE is nevertheless identified if (i) the ATE among Defiers equals the ATE among Compliers and (ii) $\pi_C \neq \pi_D$.
7. In experiments with one-sided noncompliance, the ATE among subjects who receive the treatment (sometimes called the average treatment-on-the-treated effect, or ATT) is the same as the CACE, because only Compliers receive the treatment. Explain why the ATT is not the same as the CACE in the context of two-sided noncompliance.
8. In the Milwaukee domestic violence experiment, researchers working in collaboration with police officers randomly assigned one of three treatments when officers responded to an incident involving domestic violence.[22] Officers were instructed to arrest the perpetrator and hold him overnight, arrest the perpetrator but release him after a brief period,

22 Sherman et al. 1992.

TABLE 6.7

Assigned treatment, actual treatment, and outcomes in the Milwaukee domestic violence experiment

		Assigned treatment		
		Full arrest	Brief arrest	Warning
Actual Treatment	Full arrest	400	13	1
	Brief arrest	1	384	1
	Warning	3	1	396
	Total N	**404**	**398**	**398**
Subsequent outcomes	Calls to hotline to report perpetrator	296	301	261
	Perpetrators later arrested	146	157	151

or issue a warning. The full breakdown of assigned and actual treatments is presented in Table 6.7, along with observed rates of later arrest in the three treatment conditions.

(a) Consider a simplified coding of the assigned and actual treatment, dividing subjects into two categories: arrest or non-arrest. Evaluate the plausibility of the non-interference, excludability, and monotonicity assumptions in this application.

(b) Assume that the core assumptions hold, and calculate the $\widehat{ITT}_D$, $\widehat{ITT}$, and $\widehat{CACE}$ given the simplified treatment categorization. Interpret the results.

(c) Suppose monotonicity were not assumed. What do the results of the simplified treatment suggest about the maximum and minimum values of π_{NT}, π_{AT}, and π_C?

(d) More complexity is introduced when we consider the full array of three treatment assignments and three forms of actual treatment. In the case of two assigned treatments and two actual treatments, we have four types of subjects (Compliers, Defiers, Never-Takers, and Always-Takers). How many types of subjects are there with three treatment assignments and three forms of actual treatment?

(e) How many types are there if you make the following "monotonicity" stipulations: (i) anyone who is fully arrested if assigned to be warned would also be fully arrested if assigned to be briefly arrested or fully arrested; (ii) anyone who is fully arrested if assigned to be briefly arrested would also be fully arrested if assigned to be fully arrested; (iii) anyone who is briefly arrested if assigned to be warned would also be briefly arrested if assigned to be briefly arrested; and (iv) anyone who is warned if assigned to be arrested would also be warned if assigned to be warned.

9. In their study of the effects of conscription on criminal activity in Argentina, Galiani, Rossi, and Schargrodsky use official records of draft lottery numbers, military service, and prosecutions for a cohort of men born between 1958 and 1962.[23] Draft eligibility is

23 Galiani, Rossi, and Schargrodsky 2010.

scored 1 if an individual had a draft lottery number that caused him to be drafted, and 0 otherwise. Draft lottery numbers were selected randomly by drawing balls from an urn. Military service is scored 1 if the individual actually served in the armed services, and 0 otherwise. Subsequent criminal activity is scored 1 if the individual had a judicial record of prosecution for a serious offense. For a sample of 5,000 observations, the authors report an $\widehat{ITT}_D$ of 0.6587 (SE = 0.0012), an $\widehat{ITT}$ of 0.0018 (SE = 0.0006), and a $\widehat{CACE}$ of 0.0026 (SE = 0.0008). The authors note that the $\widehat{CACE}$ implies a 3.75% increase in the probability of criminal prosecution with military service.

(a) Interpret the $\widehat{ITT}_D$, $\widehat{ITT}$, $\widehat{CACE}$, and their standard errors.

(b) The authors note that 4.21% of subjects who were not draft eligible nevertheless served in the armed forces. Based on this information and the results shown above, calculate the proportion of Never-Takers, Always-Takers, and Compliers under the assumption of monotonicity.

(c) Discuss the plausibility of the monotonicity, non-interference, and excludability assumptions in this application. If an assumption strikes you as implausible, indicate whether you think the $\widehat{CACE}$ is biased upward or downward.

10. In her study of election monitoring in Indonesia, Hyde randomly assigned international election observers to monitor certain polling stations.[24] Here, we consider a subset of her experiment where approximately 20% of the villages were assigned to the treatment group. Because of difficult terrain and time constraints, observers monitored 68 of the 409 polling places assigned to treatment. Observers also monitored 21 of the 1,562 stations assigned to the control group. The dependent variable here is the number of ballots that were declared invalid by polling station officials.

    (a) Is monotonicity a plausible assumption in this application?

    (b) Under the assumption of monotonicity, what proportion of subjects (polling locations) would you estimate to be Compliers, Never-Takers, and Always-Takers?

    (c) Explain what the non-interference assumption means in the context of this experiment.

    (d) Download the sample dataset at http://isps.research.yale.edu/FEDAI and estimate the ITT and the CACE. Interpret the results.

    (e) Use randomization inference to test the sharp null hypothesis that there is no intent-to-treat effect for any polling location. Interpret the results. Explain why testing the null hypothesis that the ITT is zero for all subjects serves the same purpose as testing the null hypothesis that the ATE is zero for all Compliers.

11. A large-scale experiment conducted between 2002 and 2005 assessed the effects of Head Start, a preschool enrichment program designed to improve school readiness.[25] The assigned treatment encouraged a nationally representative sample of eligible (low-income) parents to enroll their four-year-olds in Head Start. Of the 1,253 children assigned to the Head Start treatment, 79.8% actually enrolled in Head Start; 855 of the children assigned to the control group (13.9%) nevertheless enrolled in Head Start. One of the outcomes of interest is pre-academic skills, as manifest at the end of the yearlong intervention. The

24 Hyde 2010.

25 Puma et al. 2010. We focus here on one part of the study, the sample of four-year-old subjects.

principal investigators report that scores averaged 365.0 among students assigned to the treatment group and 360.5 among students assigned to the control group, with a two-tailed p-value of .041. Two years later, students completed first grade. Their first grade scores on a test of academic skills averaged 447.7 in the treatment group and 449.0 in the control group, with a two-tailed p-value of .380.

(a) Estimate the CACE for this experiment, using pre-academic skills scores as the outcome.

(b) Estimate the CACE for this experiment, using academic skills in first grade as an outcome.

(c) Estimate the average downstream effect of pre-academic skills on first grade academic skills. Hint: Divide the estimated ITT (from a regression of first grade academic skills on assigned treatment) by the estimated ITT_D (from a regression of pre-academic skills on assigned treatment). Interpret your results. Are the assumptions required to identify this downstream effect plausible in this application? If not, would you expect the apparent downstream effect to be overestimated or underestimated?

CHAPTER 7

Attrition

Recall from Chapter 2 that simple random assignment of subjects to treatment and control groups implies that the average outcome in the treatment group is an unbiased estimator of the average $Y_i(1)$ in the sample pool. Likewise, the average outcome in the control group is an unbiased estimate of the average $Y_i(0)$ in the subject pool. The difference between the treatment and control group average therefore provides an unbiased estimator of the average treatment effect. Implicit in this description, however, is the assumption that the researcher observes outcomes for all of the experimental subjects who are allocated either to treatment or control.

Attrition occurs when outcome data are missing. Certain forms of attrition pose a grave threat to unbiased inference. When attrition is systematically related to potential outcomes, removing observations from the data set means that remaining subjects assigned to the treatment or control group no longer constitute random samples of the original collection of subjects, and therefore a comparison of group averages may no longer be an unbiased estimator of the average treatment effect.

Why might an experiment fail to measure outcomes for some subjects? The sources of attrition vary widely:

- Subjects may refuse to cooperate with researchers. Experiments that measure outcomes using surveys routinely find that some subjects are unwilling to fill out a post-treatment questionnaire.
- Researchers lose track of experimental subjects. Substantial attrition often occurs, for example, when researchers investigate the long-term effects of an intervention using administrative data because subjects change address or name.
- Firms, organizations, or government agencies block researchers' access to outcomes. This problem is particularly common among experiments that focus on sensitive topics, such as corruption or electoral violence.
- The outcome variable may be intrinsically unavailable for some subjects. For instance, an evaluation of a job training program might aim to measure subjects' wages six months later, but this outcome will go unmeasured for subjects without jobs.

What you will learn from this chapter:

1. Terms used to classify different types of attrition.
2. Statistical conditions under which attrition leads to biased estimates of the ATE.
3. Nonparametric methods for analyzing experiments with missing data.
4. Research designs for mitigating the effects of attrition.

- Researchers deliberately discard observations. Perhaps ill-advisedly, laboratory researchers sometimes exclude from their analysis subjects who seem not to understand the instructions or who fail to take the experimental situation seriously.

When attrition occurs, the researcher observes outcomes for certain subjects and missing data for others. The researcher may choose to exclude subjects with missing outcomes when analyzing the experimental results. This approach is risky. If attrition is systematically related to a subject's potential outcomes, analyzing the remaining observations may produce biased estimates of the average treatment effect.

In order to get some sense of how attrition complicates the task of drawing causal inferences from experimental results, consider the example of the RAND Health Insurance Experiment.[1] During the mid-1970s, a large sample of American families in six geographic regions was recruited into an experiment that replaced their current health insurance coverage or public health benefits with a randomly assigned insurance plan. The experimental plans covered 5%, 50%, 75%, or 100% of each family's health costs; in order to insure families in the first three experimental groups against catastrophic financial loss, costs over $1,000 (or, for low income families, over a specified fraction of annual income) were covered by the experimental insurance plan.[2] In order to encourage subjects in each group to participate in the study, generous financial incentives were distributed so that families "could not lose financially by participating."[3] This experiment cost approximately $80 million in 1974 dollars, which, adjusted for inflation, is more than $300 million in 2010 dollars.

The basic findings from this experiment are striking. After three to five years in the study, participants in plans that required them to pay a larger share of their health costs out-of-pocket made significantly fewer physician visits and had significantly lower rates of hospital admissions. Overall, subjects whose insurance plan paid 100% of health costs consumed 46% more in health services than those whose insurance

1 Newhouse 1989.

2 When inflation is taken into account, $1,000 in 1975 dollars translates into roughly $4,300 in 2010 dollars.

3 Newhouse 1989, p. 35.

plan covered only 5%. More important, those receiving free health care were on average no healthier, based on a wide array of different health assessments. Policy analysts interpreted the results to mean that insurance arrangements that require people to pay a sizable share of their health costs "did not have an adverse impact on health outcomes for the average person."[4]

These conclusions, however, are based on statistical analyses that excluded people who dropped out of the study. Newhouse reports that after completing the baseline survey, 8% of those who were assigned to the free care group refused to enroll in the program, as opposed to a 25% refusal rate among those assigned to a plan that covered only 5% of their health costs.[5] After enrollment in the experiment, further attrition occurred over the course of the three- to five-year study period. Newhouse reports that withdrawal from the study was more common in experimental groups where subjects were required to pay for health services.[6] Among 1,294 adult participants assigned to the free plan, 0.4% left the experiment voluntarily; of the 2,664 who were assigned to one of the cost-sharing plans, 6.7% voluntarily left the experiment.[7]

This pattern of attrition raises the concern that subjects assigned to experimental conditions that required them to pay a portion of their health expenses dropped out of the study in order to maintain their existing coverage because they anticipated serious health problems. Given this selection problem, the apparent increase in service utilization in the free plan might in part reflect the fact that people who chose to participate and remain in the co-payment plans anticipated fewer health problems, while those who anticipated serious health problems dropped out of experimental groups that required them to make co-payments. On the other hand, this speculation could be mistaken. Attrition could be benign: although the specter of co-payments undoubtedly caused some subjects to drop out of the study, the unmeasured health outcomes and consumption of medical care among those who dropped out of the study on average may have been no different from those who remained. The problem with missing outcome data is that several interpretations are often plausible, and we often lack a definitive way of figuring out which interpretation is correct.

The attrition-related issues that arise in this study preview the main themes of this chapter. Attrition forces the researcher to make assumptions about the statistical properties of missingness. One key issue is whether subjects with missing outcomes have, on average, the same expected potential outcomes as subjects for whom outcome data are available.

---

4 Gruber 2006, p. 4.

5 Newhouse 1993, p. 18.

6 Newhouse 1993, p. 24.

7 Involuntary withdrawal from the experiment occurred because of military service, institutionalization, Medicare disability, or death. Failure to complete data collection forms also led to expulsion from the study (Newhouse 1993, p. 24).

When missingness is independent of potential outcomes, ignoring missing data and comparing group means will still yield unbiased inferences, although attrition reduces the effective sample size and thus increases standard errors. Because missing outcomes are, by definition, unavailable, one cannot directly assess whether missingness is systematically related to potential outcomes, although indirect evidence may be marshaled. For example, Newhouse reports that in two of the six sites, refusal to participate in the study is positively correlated with rates of hospitalization in the preceding year among those assigned to one of the cost-sharing plans, but the relationship is weak and statistically insignificant.[8] Newhouse also points out that average pretreatment health outcomes and rates of health care utilization are similar across experimental groups among subjects who do not drop out of the study. If one assumes that prior health outcomes are indicative of subsequent unmeasured potential outcomes, one could interpret this evidence to mean that missingness is independent of potential outcomes, and thus the missing data could be ignored without introducing bias.

Another modeling approach supposes that attrition is independent of potential outcomes within subgroups defined by the subjects' background attributes. For example, Newhouse reports that refusal to enroll in the assigned program is correlated with age and education.[9] If attrition were unrelated to potential outcomes once we focus our attention on people with specified ages and levels of education, one could obtain unbiased estimates of average treatment effects by reweighting the data to "fill in" the age/education cells that were depleted by attrition. Reweighting is the first method of addressing attrition we discuss in this chapter.

A second way to address the attrition problem is to guess the missing values of those who left the study. One approach is to explore worst-case scenarios, filling in the most extreme possible values—in effect, assuming that those who disappear from different experimental groups are extremely healthy or extremely ill. A related approach is to "trim" the observations in the free-care group, discarding its healthiest (or, conversely, its least healthy) subjects until its attrition rate is as high as the attrition rate in the cost-sharing groups.

A final method for addressing attrition is to gather more data from missing subjects. In the health insurance experiment, for example, although outcome measures were not gathered for those who refused to participate, health outcomes were eventually gathered for 77% of those who initially enrolled in the study but later withdrew. Because it may be prohibitively expensive to track down a large fraction of missing subjects, particularly when the rate of missingness is high, we discuss the statistical properties of a research design that focuses an intensive measurement effort on a random sample of those with missing outcomes. We show that under certain conditions this random sampling strategy may be a cost-effective way to address the threat of bias.

8 Newhouse 1993, p. 23.
9 Newhouse 1993, p. 18.

This chapter begins with a formal discussion of attrition and its statistical consequences. After explaining the conditions under which attrition leads to bias, we describe a variety of statistical approaches. Some of the most common statistical remedies invoke strong assumptions about the relationship between missingness and potential outcomes.[10] We focus instead on methods that make weaker assumptions but provide more equivocal results, merely placing bounds around the location of the average treatment effect for the entire sample or specific subgroups. Finally, we propose a data collection strategy that attempts to fill in missing values for a randomly chosen subset of missing observations. We conclude by addressing common sources of confusion about attrition and ways to contain the problems it creates. In sum, there is no ideal solution to problems caused by attrition. One can try to contain attrition-related problems through data analysis and supplementary data collection, but the best solution is to design and implement studies that minimize attrition or eliminate it altogether.

7.1 Conditions Under Which Attrition Leads to Bias

In order to facilitate our discussion of attrition and its consequences for estimation, we introduce some new notation. The key idea behind this notation is that "missingness" is itself a potential outcome—whether a subject's outcome is reported may depend on the experimental group to which the subject was assigned. For ease of presentation, this section assumes that the assigned treatment (z_i) is identical to the treatment that each subject actually receives (d_i).

For each subject i we define potential outcomes $Y_i(z)$ for $z \in (0, 1)$. Because we assume in this chapter that treatment assigned is the same as treatment received, $z = 0$ when the subject i is not treated and 1 when the subject is treated. When there is no attrition, we observe $Y_i = Y_i(0)(1 - z_i) + Y_i(1)z_i$. That is, we observe either $Y_i(0)$ or $Y_i(1)$, depending on whether the subject is assigned to the treatment or the control group. Attrition prevents us from measuring Y_i for some subjects. To capture this possibility, we define a new potential outcome, $r_i(z)$, which denotes whether or not the outcome data for subject i is reported when the treatment assignment is z. Let $r_i = 1$ when the outcome is reported, and let $r_i = 0$ when the outcome is missing. Whether a subject's outcomes are reported or missing may depend on the subject's group assignment. Notice that the non-interference assumption has been applied to attrition: the potential outcomes $r_i(1)$ and $r_i(0)$ depend only on the subject's own treatment assignment. Collecting the notation, the observed reporting outcome r_i is determined by each subject's treatment assignment and potential outcomes:

$$r_i = r_i(0)(1 - z_i) + r_i(1)z_i. \tag{7.1}$$

10 See suggested readings.

Attrition occurs when some values of r_i are zero. Because the observed r_i depends on treatment assignment, it is possible for some subjects to have the potential for attrition even if no attrition actually occurs. Following the notation convention we used when discussing treatments and treatment assignments, we use lowercase r_i when describing whether subject i reported an outcome in a past experiment and use uppercase R_i to refer to whether subject i reports an outcome in a hypothetical experiment. The expression r_i refers to a fixed quantity, whereas R_i refers to a random variable.

The observed outcomes Y_i require a bit more explanation. The model of missingness presented here presupposes that there is an underlying $Y_i(0)$ or $Y_i(1)$ whose value will either become known to the researcher or not based on whether r_i is 1 or not. Rather than use the complicated notation $Y_i(z, r(z))$ to refer to potential outcomes, we assume that potential outcomes $Y_i(0)$ or $Y_i(1)$ are unaffected by whether these outcomes are reported. Formally, this simplifying assumption amounts to an exclusion restriction: $Y_i(z) = Y_i(z, r(z) = 1) = Y_i(z, r(z) = 0)$. These potential outcomes are translated into observed outcomes according to the following rule:

$$Y_i = Y_i(0) + [Y_i(1) - Y_i(0)]z_i \text{ if } r_i = 1; \\ Y_i \text{ is missing if } r_i = 0. \quad (7.2)$$

With this notation in place, let's see how attrition may lead to bias. Recall that the average treatment effect (ATE) is the average difference in the potential outcomes $Y_i(1)$ and $Y_i(0)$ for the entire collection of subjects. When the outcomes for the entire treatment and control groups are observed, the treatment group's outcomes are a random sample of $Y_i(1)$ for the pool of subjects, and the control group's outcomes are a random sample of $Y_i(0)$ values. The average outcome in each experimental group provides an unbiased estimate of $\mu_{Y(1)}$ and $\mu_{Y(0)}$. When attrition occurs, the subjects for whom we have recorded outcomes may no longer be representative of the subject pool. As a result, the difference in the average of the observed values of $Y_i(1)$ and $Y_i(0)$ will not, in general, produce an unbiased estimate of the ATE.

For example, the expected potential outcome under treatment may be written as a weighted average of outcomes among those who are observable and those who are missing:

$$E[Y_i(1)] = E[R_i(1)]E[Y_i(1) \mid R_i(1) = 1] + \{1 - E[R_i(1)]\}E[Y_i(1) \mid R_i(1) = 0]. \quad (7.3)$$

Similarly, the expected outcome in the control group is:

$$E[Y_i(0)] = E[R_i(0)]E[Y_i(0) \mid R_i(0) = 1] + \{1 - E[R_i(0)]\}E[Y_i(0) \mid R_i(0) = 0]. \quad (7.4)$$

Thus, the ATE may be expressed as:

$$\begin{aligned} E[Y_i(1) - Y_i(0)] = \; & E[R_i(1)]E[Y_i(1) \mid R_i(1) = 1] + \{1 - E[R_i(1)]\}E[Y_i(1) \mid R_i(1) = 0] - \\ & E[R_i(0)]E[Y_i(0) \mid R_i(0) = 1] - \{1 - E[R_i(0)]\}E[Y_i(0) \mid R_i(0) = 0]. \end{aligned} \quad (7.5)$$

By comparison, the expected difference in potential outcomes when we restrict the average to subjects with non-missing values is:

$$E[Y_i(1) \mid (R_i(1) = 1)] - E[Y_i(0) \mid (R_i(0) = 1)]. \qquad (7.6)$$

Comparing equations (7.5) and (7.6) we see that, depending on the relationship between $R_i(z)$ and $Y_i(z)$, the ATE for all subjects and the ATE for non-missing subjects may be quite different. One special case in which the two quantities are the same arises when the random variables $R_i(z)$ and $Y_i(z)$ are independent. In that case:

$$E[Y_i(1) \mid R_i(1) = 1] = E[Y_i(1) \mid R_i(1) = 0]. \qquad (7.7)$$

and

$$E[Y_i(0) \mid R_i(0) = 1] = E[Y_i(0) \mid R_i(0) = 0]. \qquad (7.8)$$

In this special case, equation (7.5) simplifies to equation (7.6), indicating that when missingness is unrelated to potential outcomes for Y_i, the ATE for the subject pool is identical to the ATE among the non-missing.

In order to get a feel for how bias creeps in when the special conditions of equations (7.7) and (7.8) do not apply, consider the hypothetical schedule of potential outcomes described in Table 7.1. The table shows how potential outcomes $R_i(z)$ and $Y_i(z)$ manifest themselves as observed outcomes. The average treatment effect for all eight subjects may be calculated by comparing the average potential outcomes in the treated ($Y_i(1)$) and untreated state ($Y_i(0)$). Averaging the values in the respective columns yields: $E[Y_i(0)] = 44/8 = 5.5$ and $E[Y_i(1)] = 60/8 = 7.5$. The difference between these quantities is the ATE, which is equal to 2.

When there is attrition, however, we no longer observe Y_i values for all subjects. In Table 7.1, attrition is indicated by the entry "missing" in the last two columns. For Subject 5, neither $Y_i(0)$ nor $Y_i(1)$ is reported. For Subjects 1, 2, and 7,

TABLE 7.1

Hypothetical potential outcomes for eight subjects

Observation	$Y_i(0)$	$Y_i(1)$	$r_i(0)$	$r_i(1)$	$Y_i(0) \mid r_i(0)$	$Y_i(1) \mid r_i(1)$
1	3	8	0	1	Missing	8
2	3	7	0	1	Missing	7
3	8	10	1	1	8	10
4	7	8	1	1	7	8
5	6	6	0	0	Missing	Missing
6	5	8	1	1	5	8
7	6	6	1	0	6	Missing
8	6	7	1	1	6	7

whether the outcome is missing depends on whether the subject is treated. If we compare the average values of the non-missing potential outcomes, we find that $E[Y_i(0) \mid R_i(0) = 1] = 32/5 = 6.4$ and $E[Y_i(1) \mid R_i(1) = 1] = 48/6 = 8$. The difference between the average observed potential outcomes in the treated and untreated state is 1.6. Notice that this quantity is not equal to the true average treatment effect, which is 2. Were we to randomly assign subjects to treatment and control groups and compare average *observed* values of Y_i in each group (ignoring the missing observations), we would on average find a difference of 1.6, not 2. Due to attrition, the difference-in-means estimator is biased.

In this example, attrition causes bias because missingness is correlated with $Y_i(z)$. Inspection of the table reveals a tendency for lower values of $Y_i(z)$ to be missing, and more subjects are missing when assigned to the control group rather than the treatment group. For example, the low values of $Y_i(0)$ for Subjects 1 and 2 are never observed, which inflates the average $Y_i(0)$ outcome among the control group, in turn reducing the observed difference between the treatment and control group averages. This pattern explains why the estimated ATE using only the non-missing data is lower than the true ATE. Another way to trace the bias is to fill in equation (7.5) with the corresponding values from Table 7.1:

$$E[Y_i(1) - Y_i(0)] = \left(\frac{6}{8}\right)\left(\frac{48}{6}\right) + \left(\frac{2}{8}\right)\left(\frac{12}{2}\right) - \left(\frac{5}{8}\right)\left(\frac{32}{5}\right) - \left(\frac{3}{8}\right)\left(\frac{12}{3}\right) = 2. \quad (7.9)$$

Comparing only the observed outcomes in the treatment and control groups is equivalent to comparing only the two shaded values in this equation.

In this example, does the comparison of observed group means reveal the average treatment effect of *any* interesting subgroup? Recall that in Chapters 5 and 6 we worked with subgroups formed by partitioning the subject pool according to their potential compliance with treatment group assignment. We found that sometimes the comparison of outcomes in the treatment and control groups could be used to calculate an estimate of the causal effect of treatment for one subgroup, the "Compliers." This analogy raises the question of whether equation (7.9) is related to the ATE for some subgroup. Equation (7.6) can be rewritten as:

$$\begin{aligned} &E[Y_i(1) \mid (R_i(1) = 1) - Y_i(0) \mid (R_i(0) = 1)] = \\ &\quad \underbrace{E[Y_i(1) \mid (R_i(0) = 1) - Y_i(0) \mid (R_i(0) = 1)]}_{\text{ATE for subjects who report when untreated}} + \\ &\quad \underbrace{E[Y_i(1) \mid (R_i(1) = 1) - Y_i(1) \mid (R_i(0) = 1)]}_{\text{Heterogeneity}}. \quad (7.10) \end{aligned}$$

The left-hand side of the equation is the expected difference in outcomes among those who are non-missing. The right-hand side of the equation contains two terms. The first is the ATE among subjects who are observed when untreated. The second term represents "heterogeneity": the difference in average treated outcomes between those who would be observed if treated ($R_i(1) = 1$) and those observed when untreated ($R_i(0) = 1$). In the example from Table 7.1, the effect of heterogeneity is positive because subjects with low potential outcomes tend to be missing, and the treatment effect is positive. Thus, the treatment causes some observations to be observed that would go missing if assigned to the control group.

Applying formula (7.10) to Table 7.1 shows how the observed difference (1.6) may be partitioned into two components: (1) the ATE for those subjects whose outcomes would be recorded if they were placed in the control group, and (2) heterogeneity.

$$1.6 = \frac{7}{5} + \left(8 - \frac{39}{5}\right) = 1.4 + 0.2. \tag{7.11}$$

The bottom line is rather sobering. Not only will the difference-in-means estimator not recover the ATE for the entire subject pool, it will not recover the ATE for any meaningful subgroup. In sum, an analysis that simply excludes missing observations may generate misleading results. Beware of the fact that most statistical software, by default, excludes observations with missing outcomes when calculating summary statistics or regression estimates.

7.2 Special Forms of Attrition

One relatively innocuous form of attrition occurs when data are *missing independent of potential outcomes*. Stated formally, data are missing independent of potential outcomes (MIPO) if:

$$Y_i(z) \perp\!\!\!\perp R_i(z). \tag{7.12}$$

In words, this independence condition implies that learning whether a subject's outcomes are potentially missing gives you no clues about the values of $Y_i(1)$ or $Y_i(0)$.[11] Occasionally, this condition is satisfied by the research design itself. For example, survey experiments sometimes divide subjects into random subgroups and measure each subgroup's outcomes with a different set of questions.[12] In this case, a random procedure creates missingness, and random missingness implies that $R_i(z)$ is independent of not only potential outcomes but background attributes as well. More often, the claim that data are missing independent of outcomes is rooted not in a

11 This assumption is somewhat different from Little and Rubin's (1987) assumption of "missing completely at random," which refers to independence of missingness and the values of all potential outcomes *and* covariates.

12 Allison and Hauser 1991.

random procedure but rather an assumption about the unknown process by which some observations are recorded while others go missing.

When data are missing independent of potential outcomes, the special conditions of equations (7.7) and (7.8) apply. As we saw above, these conditions imply that the ATE among the observed outcomes equals the ATE for the entire sample:

$$E[Y_i(1) | R_i(1) = 1] - E[Y_i(0) | R_i(0) = 1] = E[Y_i(1)] - E[Y_i(0)]. \tag{7.13}$$

Intuitively, if there were no systematic relationship between missingness and potential outcomes, an average of observed outcomes in the treatment group would in expectation equal the average of the $Y_i(1)$ potential outcomes. A parallel argument applies to the control group. As a thought exercise, suppose you have a collection of five measurements. If you adopted the procedure of randomly discarding two of them, would the expected value of the three remaining observations be equal to the average of the five observations? Yes, and this is why the difference-in-means estimator remains an unbiased estimator of the ATE in the presence of random attrition.

Although missingness independent of potential outcomes cannot be verified directly, one can gather some circumstantial evidence about its plausibility. This indirect approach starts with a model of random missingness and evaluates its empirical adequacy. If missingness were literally brought about by a random procedure that deleted outcome data, we would expect to find no systematic relationship between r_i and the subjects' background attributes or their experimental assignment. Applying the logic of the randomization check described in Chapter 4, we could turn the random missingness hypothesis into a statistical test that we evaluate using randomization inference: a regression of r_i on prognostic covariates and experimental assignment should produce an F-test statistic that is non-significant when compared to the sampling distribution of F statistics. This analysis is instructive but also has its limitations. Random missingness is assumed to hold unless the statistical evidence shows otherwise. Failure to reject the null of random missingness is not the same as proving that missingness is unrelated to potential outcomes. The covariates at one's disposal may not include the systematic sources of missingness that are predictive of potential outcomes. Conversely, rejecting the null hypothesis of random missingness does not necessarily imply that the difference-in-means estimator is biased. In principle, attrition could produce estimates of $E[Y_i(1)]$ and $E[Y_i(0)]$ with identical biases, in which case there would be no net bias when estimating the ATE.

Since data from actual experiments are rarely deleted according to some random procedure, MIPO is usually invoked as an analogy. Researchers confronting attrition use their understanding of how and why missingness occurs in order to assess whether this assumption is a reasonable approximation. In voter mobilization experiments, for example, sometimes a town clerk fails to record who voted in a timely fashion, resulting in missing outcome data for the subjects residing in that town. This idiosyncratic behavior is unlikely to be related to the size of these missing subjects' potential outcomes, although one never knows for sure. As a rule of thumb, the case for MIPO is strengthened when subjects have little discretion over whether their outcomes will be reported. For example, consider an experiment designed to measure the effect of providing extra tutoring in math to second graders at the beginning of the school year. The outcome is measured by performance on a midyear math assessment. Some second graders may be absent from school the day the exam is given, but it is unlikely that many of them are missing school in order to avoid the exam. Rules of thumb provide no guarantee against bias, however. Even this example underscores how strong the MIPO assumption is: it is possible that the frequency of school absences is related to how well the child is doing academically, which could be affected by the treatment. If the treatment causes some low-performing students to attend the test, average treatment scores will decline. Attrition may therefore lead the researcher to underestimate the treatment effect.

An alternative assumption is that attrition is unrelated to potential outcomes conditional on pre-treatment covariates, which are denoted by $\boldsymbol{X}_i$. This assumption is called *missing independent of potential outcomes given X*, or MIPO|X. In the previous example, X_i (one variable from the set of covariates $\boldsymbol{X}_i$) might be "attendance record prior to the treatment intervention." For example, suppose that students with poor attendance are both more likely to be missing on the day of the assessment and more likely to benefit from the intervention. MIPO|X means that if one partitions the experimental sample by prior attendance, missingness is random within each subgroup. Among students whose record of attendance is poor, there is no relationship between missing school on the day of the assessment and the subject's potential outcomes. The same goes for students with a good record of attendance.

Stated formally, data are MIPO|X if:

$$\{Y_i(z) \perp\!\!\!\perp R_i(z)\} \,|\, \boldsymbol{X}_i = \boldsymbol{x} \text{ for all } \boldsymbol{x}. \tag{7.14}$$

Before explaining how the MIPO|X assumption guides data analysis, we first review the mechanics of how data may be reweighted in order to compensate for different rates of missingness across subgroups. Let's start with a simple example involving weights. Suppose we merely want to calculate the average of a collection of observations. We can add all the observations and then divide by the number of observations. Equivalently, we can partition the observations into groups, calculate an average for each of the groups, and then produce an overall average by weighting each group's average by the group's proportion of the total number of observations. For an example using the latter method of computing the overall average, suppose there are 40 observations. We could divide the collection into a group of 30 (men) and a group of 10 (women). The overall average would be calculated as (30/40) times the average among men plus (10/40) times the average among women.

Applying MIPO|X to the example with men and women, suppose that 15 of the men were missing at random. Because these men are missing at random, the average of the remaining 15 observations provides an unbiased estimate of the average among all 30 men. An estimate for the overall average for the subject pool of 30 men and 10 women is $\frac{3}{4}$ (average among the 15 observed men) + $\frac{1}{4}$ (average among the 10 observed women). Since half of the men are missing, the observation for each of the 15 observed men is counted $1/(1 - 1/2) = 2$ times. This weighting adjusts the data to count "30" men in the overall average by counting each of the observed men twice.

Applying this reasoning to the task of estimating average treatment effects when MIPO|X holds suggests how we might produce an estimate of the ATE using the observed values. To estimate the ATE, we need to estimate $E[Y_i(1)]$ and $E[Y_i(0)]$. When MIPO|X holds, $E[Y_i(1)]$ may be written as a weighted average of the treated potential outcomes among those with a particular covariate profile:

$$E[Y_i(1)] = \frac{1}{N}\sum_{i=1}^{N} \frac{Y_i(1)\, r_i(1)}{\pi_i(z = 1, \boldsymbol{x})}, \tag{7.15}$$

where $\pi_i(z = 1, \boldsymbol{x})$ is the share of non-missing subjects among those who are treated and have a covariate profile $\boldsymbol{x}$. Subjects with missing outcomes appear nowhere in this formula, whereas subjects with reported outcomes are weighted by a factor of $1/\pi_i(z = 1, \boldsymbol{x})$. In effect, this weighting operation replaces the subjects with missing values with multiple copies of subjects with non-missing values. This method produces accurate estimates of average treatment effects when the non-missing observations are in fact good substitutes for the observations that are missing outcomes. Because observations are weighted by the inverse probability that a subject with a set of characteristics $\boldsymbol{X}_i = \boldsymbol{x}$ is observed, this weighting scheme is called *inverse probability weighting.*

To illustrate inverse probability weighting, Table 7.2 presents a schedule of potential outcomes, potential missingness, and a covariate X_i for eight hypothetical subjects. To motivate the example, suppose the treatment depicted in Table 7.2 is an educational program being offered at a community center; $x = 1$ if the subject lives near the community center and $x = 0$ if the subject lives far away. After the intervention has ended, all subjects are asked to return to the community center for a follow-up evaluation (perhaps a test on the material presented in the program). Some subjects do not show up. In this example, those who live far away ($x = 0$) are less likely to show up for the follow-up evaluation. In addition, those who live in the neighborhood close to the community center have a different pattern of potential outcomes than those who live far away. To keep the example simple, we have set things up so that those who live far away have noticeably higher potential outcomes than those who live near the community center.

Because we have the complete schedule of potential outcomes in Table 7.2, we can calculate that the average treatment effect is equal to 3.5. Given attrition, are we able to recover this ATE? First, notice that MIPO does not hold. Learning that a subject is missing provides clues about the subject's potential outcomes. The dis-

TABLE 7.2

Complete set of potential outcomes and covariates for eight subjects

Observation	$Y_i(0)$	$Y_i(1)$	$r_i(0)$	$r_i(1)$	$Y_i(0)\|r_i(0)$	$Y_i(1)\|r_i(1)$	X_i
1	3	4	1	1	3	4	1
2	4	7	1	1	4	7	1
3	3	4	1	1	3	4	1
4	4	7	1	1	4	7	1
5	10	14	0	0	Missing	Missing	0
6	12	18	0	0	Missing	Missing	0
7	10	14	1	1	10	14	0
8	12	18	1	1	12	18	0

tributions of potential outcomes for those subjects with missing outcome data are quite different from the typical subject. When MIPO fails, comparison of observed group means may lead to biased estimates. The potential for bias is apparent from the schedule of outcomes shown in Table 7.2. The ATE among those whose outcomes are observed is 3, which is less than the 3.5 ATE for the entire subject pool.

However, MIPO|X holds in this example. Dividing the sample into two groups according to whether the covariate X_i is equal to 0 or 1, we see that within these two groups the potential outcomes of subjects whose data are missing perfectly parallels the potential outcomes of subjects for whom outcomes are observed. Using equation (7.15), we can calculate $E[Y_i(1)]$ and $E[Y_i(0)]$ and the implied ATE. The relevant proportions of non-missingness are $\pi(z = 0, x = 1) = 1$, $\pi(z = 1, x = 1) = 1$, $\pi(z = 1, x = 0) = 0.5$, and $\pi(z = 0, x = 0) = 0.5$. Applying the weighted average formula (7.15):

$$E[Y_i(1)] = \left(\frac{1}{8}\right)\left(\frac{4}{1} + \frac{7}{1} + \frac{4}{1} + \frac{7}{1} + \frac{14}{0.5} + \frac{18}{0.5}\right) = 10.75,$$

$$E[Y_i(0)] = \left(\frac{1}{8}\right)\left(\frac{3}{1} + \frac{4}{1} + \frac{3}{1} + \frac{4}{1} + \frac{10}{0.5} + \frac{12}{0.5}\right) = 7.25,$$

$$E[Y_i(1)] - E[Y_i(0)] = 3.5. \tag{7.16}$$

In sum, when MIPO|X holds, inverse probability weighting recovers the true ATE.[13]

Weighting can have a downside. If the MIPO|X assumption is incorrect, applying inverse probability weights may produce misleading estimates. Inverse probability

13 When applied to the actual data generated by an experiment (as opposed to a schedule of potential outcomes), inverse probability weighting produces a difference-in-means estimator that is biased but consistent. Bias occurs because the weighted number of observations in each experimental group varies according to the random assignment.

weights, by definition, assign the largest weights to subgroups that have the highest rates of attrition. If attrition biases the estimate of the ATE for a subgroup, then this biased estimate may be given greater weight in the overall estimate. In some instances, the inverse probability weighted average may be more biased than an analysis of unweighted data. A lesser but still noteworthy concern is that reweighting schemes such as inverse probability weighting may increase the sampling variability of the estimates, because extra weight is placed on subsamples with a large share of missing observations.

Thus far, our discussion of MIPO|X has focused on showing how weighting schemes play out in hypothetical scenarios in which potential outcomes are known. In practice, the researcher will see only the revealed outcomes and will be forced to make assumptions about attrition. Weighting the observed data to restore the subject pool's covariate proportions fails to address what is typically the greatest concern regarding attrition: the possibility that missingness remains related to potential outcomes even after conditioning on the observables. For example, suppose the rate of attrition turns out to be greater for men than women; to restore the original proportions in the subject pool, the researcher reweights the outcomes so as to place greater weight on men. But this approach fails to address the threat of bias that looms if the men who attrite have different average potential outcomes than the men who do not, or if the women who attrite have different average potential outcomes than the women who do not. Unfortunately, the relationship between $Y_i(z,x)$ and $R_i(z,x)$ is not directly observable. Evaluating MIPO|X, like evaluating MIPO, involves a combination of statistical detective work and theoretical speculation.

To implement inverse probability score weighting in the analysis of actual data, a two-stage procedure is used. In the first stage, one uses logistic regression (see Chapter 11) to estimate $\hat{\pi}(z,\boldsymbol{x})$, the predicted probability of reporting an outcome given treatment assignment and a covariate profile. The second stage is a weighted difference-in-means calculation or a weighted regression. In this stage, one restricts attention to the sample with observed outcomes, weighting each observation by $1/[\hat{\pi}(z,x)]$. Whether the weighted results differ from the unweighted results depends on the relationship between attrition rates in the control and treatment groups. If attrition rates are similar for subjects with different covariate profiles, weights will be relatively constant across observations.

7.3 Redefining the Estimand When Attrition Is Not a Function of Treatment Assignment

Under some conditions a simple comparison of observed group averages may be highly informative. If all subjects are either *never* missing outcomes or *always* missing outcomes, an unweighted comparison of the average treatment and control group outcomes provides an unbiased estimate of the treatment effect for a subset of the subject pool, called *Always-Reporters*: subjects who report their outcomes regardless of group assignment. The average treatment effect for this subgroup will not generally be the same as the overall subject pool ATE. We first describe this new estimand and how it is estimated, and then discuss conditions under which this special pattern of attrition might occur.

A special form of attrition occurs when $r_i(1) = r_i(0)$ for all i. In this case, missingness is unaffected by treatment assignment. Suppose we want to estimate the ATE for the subset of subjects for whom $r_i(0) = r_i(1) = 1$. When subjects are randomly assigned to treatment and control groups, the expected average of the observed outcomes is $E[Y_i(0)|R_i(0) = 1]$ and $E[Y_i(1)|R_i(1) = 1]$, respectively. Because $r_i(1) = r_i(0)$ for all i, the expected difference between observed treatment and control group averages can be rewritten as $E[(Y_i(1) - Y_i(0))|R_i(0) = R_i(1) = 1]$. In the special case where subjects are either Always-Reporters or Never-Reporters, the difference between the treatment and control group averages is an unbiased estimator of the treatment effect for a particular group of subjects, Always-Reporters. An example where $r_i(1) = r_i(0)$ applies may be found in Table 7.2; there the ATE among Always-Reporters is 3.

In practice, researchers frequently encounter situations where attrition is thought to be related to potential outcomes but not to experimental groups to which subjects are assigned. When the outcome variable is an administrative record, it is often reasonable to suppose that subjects' group assignments have no effect on whether their outcomes are missing. In voter turnout experiments, some jurisdictions are slow to update their records after the election. All those who live in the affected geography, whether treated or untreated, are missing outcome data and would be missing regardless of treatment assignment. A simple comparison of treatment and control group averages among the remaining subjects provides the ATE among those who always have outcome data (those subjects from places where administrative records are available). Notice that potential outcomes for subjects in places where records go missing may be unusual, in which case the MIPO assumption may not hold. For example, political participation may be unusually low in places where administrators lose track of voter turnout records. Voters in poorly managed precincts may, on the whole, be less (or more) responsive to voter mobilization efforts. Because potential outcomes may be unusual among the missing, the ATE among those with non-missing outcomes may differ from the overall ATE.

Another situation in which attrition may be unrelated to treatment assignment occurs when there are delays between the intervention and measurement of the outcome. Outcomes may be missing because subjects have moved away. For instance, consider a study of an elementary school reading program. Suppose researchers return a year after the intervention and are able to test only those students who still live in the same state. If moving out of state is independent of a child's group assignment, a comparison of treated and untreated subjects that is restricted to subjects with outcome data (that is, restricted to the Always-Reporters who remain behind) provides an unbiased estimate of the effect of the intervention among Always-Reporters. This point holds even if Always-Reporters have unusual potential outcomes and treatment effects.

Obtaining an unbiased estimate of the ATE among Always-Reporters can be valuable. First, treatment effect estimates are useful in assessing theoretical predictions about interventions. These predictions may be stated in very general terms, in which case the estimated ATE of any group, including Always-Reporters, may help shed light on whether the prediction holds. Second, the treatment effect among those who always have outcome data may be what the researcher seeks when evaluating an intervention. For example, if one of the goals of the reading experiment were to find out what works best for long-term residents of the community, the treatment effect among the non-movers may be of special interest.

7.4 Placing Bounds on the Average Treatment Effect

When attrition seems neither random nor random conditional on X, the researcher is forced to fall back on other ways of extracting inferences from the available outcomes. There are two main strategies. First, the researcher may be able to place bounds on the treatment effect, estimating the largest and smallest ATEs that would obtain if the missing information were filled in with extremely high or low outcomes. This cautious approach has the virtue of imposing few assumptions. A second approach is to fashion a statistical model that reflects the specific sources of missingness in a given application. Because these models often invoke strong assumptions about the functional form linking inputs and outputs, the distribution of unobserved variables, and homogeneity of treatment effects, the latter approach is in tension with the agnostic style of experimental investigation. For that reason, we focus on the first approach.

Extreme value bounds gauge the potential consequences of attrition by examining how the estimated ATE varies depending on how one fills in missing potential outcomes.[14] To see how this works, suppose that the $Y_i(z)$ measured in Table 7.1 can range from 0 to 10. It is assumed that 0 and 10 represent the range of possible values for *all* subject outcomes, missing or not. Suppose we randomly assign the subject pool described by Table 7.1 to treatment and control groups. For example, let subjects 2, 3, 5, and 7 be assigned to treatment and 1, 4, 6, and 8 be assigned to control. Had there been no attrition, the average for the treatment group would

14 Manski 1989.

be $(7 + 10 + 6 + 6)/4 = 29/4 = 7.25$, and the control group average would be $(3 + 7 + 5 + 6)/4 = 21/4 = 5.25$. If all potential outcomes were observed, the estimated treatment effect would be 2. Due to attrition, however, not all the potential outcomes are observed. Instead, we have for the treatment group $(7 + 10 + ? + ?)/4 = ?$ and for the control group $(? + 7 + 5 + 6)/4 = ?$, where question marks denote quantities unknown due to attrition. To find the upper bound on the treatment effect estimate, substitute 10 for the missing values in the treatment group and 0 for the missing value in the control group. The lower bound is formed by filling in the missing values in the treatment group with 0 and the missing value in the control group with 10:

$$\text{Upper bound: } \frac{37}{4} - \frac{18}{4} = \frac{19}{4}. \tag{7.17}$$

$$\text{Lower bound: } \frac{17}{4} - \frac{28}{4} = -\frac{11}{4}. \tag{7.18}$$

First, notice that the upper and lower bounds contain the true ATE of 2. Although the extreme value bounds are in this case sample estimates and therefore subject to sampling variability, extreme value bounds tend to be successful in bracketing the true ATE. Second, the bounds are wide. This is frequently the case with extreme value bounds due to the assumption-free manner in which the bounds are constructed. For experiments with a modest share of missing observations and a narrow range of feasible outcomes, these bounds can be quite useful. As the rate of attrition increases or as the range of possible Y_i values expands, extreme value bounds become less informative. Indeed, extreme value bounds are undefined when Y_i has infinite range. However, bounds can be formed for transformations of Y_i when this is the case; redefine the outcome variable to be 1 if Y_i is greater than a certain value and 0 otherwise. By "coarsening" the data, one obtains extreme value bounds on an average treatment effect that is now defined in terms of a binary outcome.

A somewhat more restrictive approach imposes an assumption about the attrition process that allows the researcher to estimate bounds on the treatment effect by *trimming* the observed data.[15] Attrition is said to be *monotonic* if:

$$r_i(1) \geq r_i(0) \text{ or } r_i(1) \leq r_i(0) \text{ for all } i. \tag{7.19}$$

Suppose, for example, that $r_i(1) \geq r_i(0)$. This requirement states that any subject who would not be missing if assigned to the control group would also not be missing if assigned to the treatment group. In other words, there are no subjects for whom $r_i(1) = 0$ and $r_i(0) = 1$.

Recall that in Chapters 5 and 6 we classified subjects according to their pattern of potential outcomes for receiving the treatment based on their group assignment.

15 Lee 2005.

TABLE 7.3

Latent subgroups defined by their potential observability, assuming monotonicity

$z=0$	$z=1$	Type of subject
$r_i(0) = 1$	$r_i(1) = 1$	Always-Reporters
$r_i(0) = 0$	$r_i(1) = 1$	If-Treated-Reporters
$r_i(0) = 0$	$r_i(1) = 0$	Never-Reporters

Note: No entry is presented for If-Untreated-Reporters, for whom $r_i(0) = 1$ and $r_i(1) = 0$, because this group is ruled out by monotonicity, which in this example implies that $r_i(1) \geq r_i(0)$.

Similarly, we may classify subjects according to their potential missingness given their group assignment. Table 7.3 shows the complete set of possibilities when we assume $r_i(1) \geq r_i(0)$. Although there are four logical combinations of $r_i(0)$ and $r_i(1)$, only three types remain when we assume that missingness is monotonic. By this assumption, no subjects have observed outcomes when untreated but missing outcomes when treated.

Under monotonicity, we can bound the average treatment effect for subjects earlier classified as Always-Reporters. For this type of subject, the ATE is defined as:

$$E[Y_i(1) | R_i(1) = 1, R_i(0) = 1] - E[Y_i(0) | R_i(1) = 1, R_i(0) = 1]. \quad (7.20)$$

Although experimental data do not provide a direct estimate of equation (7.20), they do enable a researcher to estimate certain parts of the equation and form bounds around the rest. For example, it is straightforward to estimate the average untreated potential outcome for Always-Reporters, $E[Y_i(0) | R_i(1) = 1, R_i(0) = 1]$, because when subjects are assigned to the control group, we observe $Y_i(0)$ only for those subjects who are Always-Reporters; by definition, If-Treated-Reporters and Never-Reporters produce only missing values, and under monotonicity we assume there are no If-Untreated-Reporters. The challenge is to estimate the first part of the expression, the average treated potential outcome for Always-Reporters. When subjects are assigned to the treatment group, we observe $Y_i(1)$ for both Always-Reporters and If-Treated-Reporters. We want to extract the average $Y_i(1)$ values of the Always-Reporters from this mixture, but there is no way to do so without additional assumptions. We can, however, estimate the relative shares of Always-Reporters and If-Treated-Reporters in the subject pool. The subject pool share of Always-Reporters may be estimated based on the non-attrition rate in the control group. The greater share of non-missing subjects in the treatment group reflects the fact that the treatment group contains both Always-Reporters and If-Treated-Reporters. Using the estimated ratio of Always-Reporters and If-Treated-Reporters, we can place bounds on the average $Y_i(1)$ for Always-Reporters by calculating the average $Y_i(1)$ for the

Always-Reporters that is implied when we make extreme assumptions about the $Y_i(1)$ values of the If-Treated-Reporters.

To describe this procedure more precisely, we introduce a bit more notation. Let Q be the difference between the shares of non-missing subjects in the treatment group and the control group, divided by the share of non-missing in the treatment group:

$$Q = \frac{\pi(R(1) = 1) - \pi(R(0) = 1)}{\pi(R(1) = 1)}. \tag{7.21}$$

Q is a measure of the "extra" observed subjects in the treatment group versus the control group. As an example, if 75% of the treatment group were non-missing and 50% of the control group were non-missing, Q would equal $(75 - 50)/75 = 1/3$. Let $\hat{Y}(1, q)$ be the value of Y_i at the $100q$ percentile of the distribution of the observed values in the treatment group. For example, $\hat{Y}(1, 0.33)$ refers to the 33rd percentile of the distribution of outcomes among treatment group subjects.

An upper bound for the ATE among Always-Reporters may be estimated using the formula:

$$\hat{E}[Y_i(1) \mid R_i(1) = 1, Y_i(1) > \hat{Y}(1, q)] - \hat{E}[Y_i(0) \mid R_i(0) = 1] = \frac{\sum_{i=1}^{N} 1[Y_i > \hat{Y}(q)] \cdot Y_i \cdot z_i}{\sum_{i=1}^{N} 1[Y_i > \hat{Y}(q)] \cdot z_i} - \frac{\sum_{i=1}^{N} Y_i(1 - z_i)}{\sum_{i=1}^{N} (1 - z_i)}, \tag{7.22}$$

where the $1[\cdot]$ operator takes the value 1 if the condition in the brackets holds and takes the value 0 otherwise. Equation (7.22) says that to find the upper bound of the treatment effect, trim off the lowest $100q$% of the values of $Y_i(1)$, average the remaining values, and then subtract the average untreated outcomes. Intuitively, the upper bound among Always-Reporters is formed by attributing the lowest values of Y_i in the treatment group to the If-Treated-Reporters and trimming those off. The proportion of the Y_i values that are trimmed from the observed distribution in the treatment group is determined by the ratio of Always-Reporters and If-Treated-Reporters in the subject pool. After the lowest values in the treatment group are trimmed off, the proportion of subjects with observed values in the treatment group is equal to the proportion of subjects with observed values in the control group. In order to estimate the lower bound, repeat the procedure replacing the condition $Y_i(1) > \hat{Y}(1, q)$ with $Y_i(1) < \hat{Y}(1, 1 - q)$, which attributes the highest values of $Y_i(1)$ to the If-Treated-Reporters, and trims off those observations.[16]

16 The same procedures apply if the proportion of subjects with non-missing outcomes is larger in the control group than the treatment group. In that case, observations are trimmed from the control group.

In sum, extreme value bounds and trimming bounds differ in several important respects. First, the two sets of bounds apply to different estimands. Extreme value bounds are designed to bracket the sample ATE. Trimming bounds are designed to bracket the ATE among Always-Reporters. Second, trimming bounds require an additional assumption, monotonicity, which is not invoked by the extreme value procedure. Third, extreme value bounds fill in all missing outcomes, while trimming bounds exclude some observed values and ignore missing outcomes. Finally, if the range of possible Y_i values is large, even modest levels of attrition will generate large extreme value bounds. By comparison, trimming bounds are not affected by the logical range of potential outcomes. Unlike extreme value bounds, trimming tends to generate narrow bounds even in the face of substantial attrition.

7.5 Addressing Attrition: An Empirical Example

In recent years governments and NGOs have sought to encourage school attendance or higher academic performance through educational subsidies. Angrist, Bettinger, and Kremer examine the long-term consequences of one such effort, a private school voucher program in Colombia.[17] The PACES program awarded vouchers to low-income high school students throughout the country. These vouchers could be used to pay for private secondary schools. In some places, the vouchers were awarded through a lottery, which produced a random assignment of students to one of two experimental groups: voucher "winners" and "losers."

Angrist et al. seek to measure the effect of these subsidies on long-term educational outcomes. The authors assess the effect of the vouchers on performance on educational tests taken around the time of high school graduation. Like most studies that track long-term outcomes, this one confronts an attrition problem: many subjects leave school and do not have scores on the educational performance tests. Concerned that missingness was especially common among students whose potential scores on the test were low, the authors consider a variety of statistical models that address attrition-related bias. In this section, we describe the data and discuss several strategies for estimating the treatment effect given the attrition problem.

Table 7.4 describes the sample of lottery winners and losers. Of the 3,542 subjects, 2,073 are lottery winners. Covariates include age, sex, and whether administrative records include the subject's phone numbers. As expected given random assignment of vouchers, the three covariates are well balanced across experimental conditions.[18]

17 Angrist, Bettinger, and Kremer 2006.

18 A logistic regression of treatment assignment on the three covariates shows that one cannot reject the null hypothesis that the three covariates do not jointly predict treatment assignment ($p = 0.19$).

TABLE 7.4

Covariate balance across treatment and control groups for the PACES voucher experiment

Covariate	Control group mean	Treatment group mean
Age	12.72	12.63
Male	0.488	0.500
Known Phone Number	0.882	0.890
N	1,469	2,073

Source: Angrist, Bettinger, and Kremer 2006.

Outcomes are assessed using two different measures: (1) registration for the college entrance exam, the ICFES, which is taken by 90% of graduating seniors; and (2) performance on the college entrance exam. Reliable administrative records indicate whether a student registers for the ICFES, and registration data are available for all subjects. Estimating the ATE for the first outcome measure is straightforward. A difference-in-means indicates that lottery winners are much more likely to register for the test ($779/2073 = 37.6\%$) than lottery losers ($444/1469 = 30.2\%$). This difference of 7.4 percentage points is too large to be attributed to sampling variability. The 95% confidence interval ranges from 4.2 to 10.5 percentage points.

Estimating vouchers' average effect on college entrance exam scores is considerably more complicated because roughly two-thirds of the sample never took the test. We saw that missingness is affected by treatment assignment (62.4% in the treatment group versus 69.8% in the control group); further investigation reveals that missingness is also predicted by the three covariates. The regression results in Table 7.5 show that attrition is strongly associated with age and that the attrition rate is significantly higher for men. There is also a bit of asymmetry between the predictors of missingness in treatment and control groups. Administrative records of the student's phone number predict missingness in the treatment group but not the control group. However, a joint test of the significance of the interactions between treatment assignment and each of the three covariates indicates that the variation in coefficients across experimental groups is not greater than what one would expect due to sampling variability (see Chapter 9).

What inferences should one draw from the regression results in Table 7.5? Recall that if attrition is independent of potential outcomes, even a large share of missing outcome data will not bias the treatment effect estimate. Does the pattern of attrition suggest that the MIPO assumption is false? Attrition rates are clearly different for the treatment and control groups, and attrition also seems correlated with pre-treatment variables, yet these facts alone do not establish that the MIPO assumption is invalid. Missingness could be related to covariates and treatment assignment and still be

TABLE 7.5

Regression estimates predicting missingness as a function of covariates, by experimental group

	Control group	Treatment group
Age	0.149 (0.008)	0.167 (0.007)
Male	0.052 (0.022)	0.040 (0.019)
Known phone number	0.076 (0.033)	−0.003 (0.030)
Constant	−1.286	−1.502
N	1,469	2,073

Note: Standard errors in parentheses.
Source: Angrist, Bettinger, and Kremer 2006.

unrelated to potential outcomes $Y_i(1)$ and $Y_i(0)$. MIPO says that $Y_i(z|r = 1)$ and $Y_i(z|r = 0)$ have the same distributions; MIPO does not require that the proportion of subjects for whom, for example, $R_i(0,\boldsymbol{x}) = 1$ equal the proportion of subjects for whom $R_i(1,\boldsymbol{x}) = 0$. That said, when the attrition rates are significantly different across groups or correlated with covariates, questions arise. *Something* is causing the difference in attrition, and any argument in defense of MIPO must now explain why MIPO holds in spite of the observed differences between experimental groups and apparent correlation between missingness and background attributes.

When MIPO seems implausible, a fallback position is to assume missingness independent of outcomes given X. One has to be a bit careful about making this argument about a specific set of covariates when many other possible covariates remain unobserved. In this case, it seems especially dubious to maintain that students' age, sex, and phone number are precisely the X required to render conditional missingness innocuous. At best, age, sex, and phone number are proxies for other background attributes that may be related to missingness and potential outcomes. Presumably, if the authors had covariates other than these three variables, such as achievement scores in middle school, they would have wanted to include them in the conditioning set, $\boldsymbol{X}_i$. In order to lend credence to the assertion that these three variables are sufficient to render conditional missingness independent of outcomes, one would ideally like to test whether other prognostic covariates fail to predict missingness within the subgroups defined by age, sex, and phone number.

Suppose, for purposes of illustration, you were willing to assume missingness independent of outcomes. The unbiased estimator for the effect of the voucher is the difference in average test scores between the observed treatment and control groups. The average exam scores show a modest difference between lottery winners (47.6) and losers (46.9) that is statistically distinguishable from zero (one-tailed $p = 0.02$). This estimated effect is about one-eighth of the standard deviation in the control group. Under the assumption that attrition is independent of potential outcomes, it appears

BOX 7.1

Illustration of Reweighting Regression Analysis to Account for Missingness

Using data from Angrist, Bettinger, and Kremer 2006, this example illustrates a procedure for reweighting data under the assumption of MIPO|X.

```
# Generate a variable ("observed") indicating whether the unit is observed (r_i=1)
observed <- 1 - (read == 0)

# Use logistic regression to predict probabilities of being observed
probobs <-
glm(observed~(vouch0*sex)+(vouch0*phone)+(vouch0*age),family=binomial(link="logit"))
$fitted

# Compare distributions of predicted probabilities across experimental conditions
# Check that there are no zero predicted probabilities in either condition
summary(probobs[vouch0==0])
    Min.   1st Qu.   Median     Mean   3rd Qu.     Max.
0.005258  0.090590  0.295300  0.302200  0.413700  0.887600
summary(probobs[vouch0==1])
    Min.   1st Qu.   Median     Mean   3rd Qu.     Max.
0.006938  0.237700  0.449400  0.375800  0.503700  0.872100

# Generate weights: inverse of predicted probability of being observed
wt <- 1/probobs

# Restrict analysis to observed subjects.
sel_valid <- observed == 1
table(sel_valid)
sel_valid
FALSE  TRUE
 2319  1223

# Coefficients for unweighted regression (restricting analysis to observed subjects)
lm(read~vouch0,subset=sel_valid)$coefficients
(Intercept)      vouch0
 46.9208148   0.6827378

# Coefficients for IPW regression (restricting analysis to observed subjects)
lm(read~vouch0,weights=wt,subset=sel_valid)$coefficients
(Intercept)      vouch0
 46.4378182   0.7230303
```

Data source: Angrist, Bettinger, and Kremer 2006.

that although vouchers caused a substantially greater proportion of students to take the college entrance exam, vouchers caused a relatively modest improvement in academic achievement, as measured by entrance exam scores. Because we assumed MIPO, this estimated ATE of vouchers applies to all subjects, regardless of whether they actually took the exam.

Next, suppose that we were to abandon MIPO in favor of MIPO|X. To illustrate how the MIPO|X assumption might be used to adjust the estimated ATE, we examine the relationship between attrition and the available covariates. To keep the calculations simple, we focus on a single covariate, each student's sex. Suppose that, given sex, attrition is unrelated to potential outcomes. We use observed group averages in order to generate the following estimates:

$$\begin{aligned}\hat{E}[Y_i(1) \mid R_i(1) &= 1, \text{female}] = 47.3,\\ \hat{E}[Y_i(0) \mid R_i(0) &= 1, \text{female}] = 46.6,\\ \hat{E}[Y_i(1) \mid R_i(1) &= 1, \text{male}] = 48.0,\\ \hat{E}[Y_i(0) \mid R_i(0) &= 1, \text{male}] = 47.3.\end{aligned} \tag{7.23}$$

We also use the data to estimate the shares of non-missing subjects within each category of sex and experimental assignment:

$$\begin{aligned}\hat{\pi}(z &= 1, \text{female}) = 40.4\%,\\ \hat{\pi}(z &= 0, \text{female}) = 33.8\%,\\ \hat{\pi}(z &= 1, \text{male}) = 34.8\%,\\ \hat{\pi}(z &= 0, \text{male}) = 26.5\%.\end{aligned} \tag{7.24}$$

As described in the previous section, we construct estimates of $E[Y_i(1)]$ and $E[Y_i(0)]$ for the subject pool by estimating $E[Y_i(1)]$ and $E[Y_i(0)]$ for each gender and then using the overall subject pool proportions to reweight the data. The estimate for the subject pool average of $Y_i(1)$ is:

$$\left(\frac{1}{2073}\right)\left(361 * \frac{48.0}{.348} + 419 * \frac{47.3}{.404}\right) = \left(\frac{1}{2073}\right)[1036 * 48.0 + 1037 * 47.3] = 47.6. \tag{7.25}$$

The second expression illustrates how the weighting scheme works: missing treatment group outcomes are "filled in" with the observed subgroup's average outcome. Although there are more females observed than males in the treatment group (419 to 361), the weighting scheme restores the relative contribution to the overall average of the male ($N = 1{,}036$) and female ($N = 1{,}037$) subjects. Similarly, the estimate for the subject pool average $Y_i(0)$ is:

$$\left(\frac{1}{1469}\right)\left(190 * \frac{47.3}{.265} + 254 * \frac{46.6}{.338}\right) = 46.9. \tag{7.26}$$

As it turns out, despite substantial attrition, the weighted data produce an estimated ATE that is the same as the estimate generated by the difference-in-means analysis. The difference in attrition rates by sex was statistically significant but relatively small, and the difference in test scores for males and female students was also relatively small. After a small reweighting of a small difference, the change in estimates is negligible. We leave for the exercises the question of whether reweighting the data using sex, age, and phone number produces materially different estimates.

An alternative to reweighting is placement of bounds around the ATE. Because using bounds requires fewer assumptions, it reduces the risk of bias—but the answers it provides are less precise. To construct extreme value bounds, we replace the missing values with the worst and best possible test outcomes multiplied by proportion missing, and then add them to the observed averages multiplied by proportion registering for the test. Suppose that the lowest possible test score is 0 and the highest score is 100. The upper bound on the treatment effect estimate is then:

$$[47.6 * .376 + 100 * .624] - [46.9 * .302 + 0 * .698] = 66.1. \tag{7.27}$$

The lower bound is:

$$[47.6 * .376 + 0 * .624] - [46.9 * .302 + 100 * .698] = -66.1. \tag{7.28}$$

These bounds suggest that the average effect of vouchers on test scores could be hugely positive or hugely negative. The wide bounds we obtain reflect the high attrition rates and the large range of the potential test scores. The calculations above dramatize the resulting uncertainty: either 62.4 points are added to form the upper bound or 69.8 points are subtracted to form the lower bound. Without making stronger assumptions about the relationship between potential outcomes and attrition, it is impossible to draw meaningful conclusions about the average treatment effect from these data.

Another approach is to apply trimming bounds. The percentage of the treatment group with outcome measures is approximately 37.6%, while only 30.2% of the control group has outcome measures. Under the assumption of monotonicity, an upper bound for the treatment effect estimate for the Always-Reporters is formed by removing the top $(37.6 - 30.2)/37.6 = 19.6\%$ of the treatment group outcomes, and then calculating the difference in the average outcomes among the remaining treatment and the control group subjects. The lower bound is found by removing the lowest 19.6% of the treatment group outcomes before calculating the difference in treatment and control group averages.

Recall that the difference among the observed cases for treatment and control is: $47.6 - 46.9 = 0.7$. Trimming the treatment group means lopping off subjects who are below the 19.6th percentile or above the 80.4th percentile. The test scores associated with these percentiles are 43 and 52, and the number of observations left after trimming is 626. The average test score for the top 626 subjects is 49.6, and the average for the bottom 626 subjects is 45.7. Using these results, the upper and lower bounds may be calculated:

$$\text{Upper bound: } 49.6 - 46.9 = 2.7, \tag{7.29}$$

$$\text{Lower bound: } 45.7 - 46.9 = -0.8. \tag{7.30}$$

These bounds are quite a bit more informative than the extreme value bounds, producing an interval that is 3.5 units wide as opposed to 132.2 units. Trimming, however, answers a narrower question: What is the average effect of the treatment on Always-Reporters, those who would take the entrance exam regardless of whether they are given a voucher? For this type of student the treatment effect appears to be somewhere between weakly negative and modestly positive.

In sum, the vouchers experiment illustrates the uncertainty that suffuses experimental analysis when experiments encounter substantial rates of attrition. This experiment clearly indicates that the vouchers were effective in inducing students to take a college entrance exam, an outcome for which there is complete data. But as to the question of whether vouchers improved college entrance exam scores, the evidence is murky. An analysis based on the assumption that missingness is independent of potential outcomes suggests the vouchers had a small positive effect. The same conclusion emerges when the data are reweighted assuming that missingness is independent of potential outcomes given the students' sex. Extreme value bounds suggest that the data are in principle consistent with almost any average treatment effect. Trimming bounds suggest that for Always-Reporters, the average treatment effect may be weakly positive, but the evidence is equivocal. The online appendix considers several models that impose much stronger assumptions and further expand the range of possible inferences that may be drawn from this dataset. The broader point, which we revisit in Chapter 11, is that the analysis of experiments that encounter attrition confronts two sorts of uncertainty. The first is statistical uncertainty associated with sampling variability. The second is modeling uncertainty, which reflects that no one knows for sure which assumptions should be invoked.

7.6 Addressing Attrition with Additional Data Collection

This chapter has described methods for analyzing an experiment that suffers attrition. None of the analytic options are particularly attractive. A different approach is to make a new attempt to obtain outcome data from subjects with missing outcomes. Sometimes this data collection effort is launched in an effort to repair an experiment that encounters serious attrition problems; in other cases, supplementary data collection may be planned prior to the launch of the experiment. This section compares two supplementary data collection plans, describes their statistical properties, and considers how a researcher planning an experiment might allocate resources between primary and supplementary data collection.

The key challenge in estimating the average treatment effect is to obtain random samples of $Y_i(0)$ and $Y_i(1)$ from the pool of subjects in order to estimate $E[Y_i(0)]$ and $E[Y_i(1)]$. To keep things simple, suppose that attrition occurs only in the control group. Recall from equation (7.4) that the average untreated outcome is:

$$E[Y_i(0)] = E[R_i(0)]E[Y_i(0) \mid R_i(0) = 1] + \{1 - E[R_i(0)]\}E[Y_i(0) \mid R_i(0) = 0]. \tag{7.31}$$

The problem created by attrition is that although we can estimate $E[R_i(0)]$ and $E[Y_i(0) \mid R_i(0) = 1]$, we have no data for $E[Y_i(0) \mid R_i(0) = 0]$ and therefore cannot estimate $E[Y_i(0)]$ without making additional assumptions. The challenge is to obtain data to estimate $E[Y_i(0) \mid R_i(0) = 0]$.

Consider two alternative data collection strategies. One idea is to attempt to obtain Y_i for *all* subjects who are missing outcome data. This data collection campaign might include locating lost subjects or offering another date for a follow-up interview. In practice, it is often difficult and costly to gather outcome data from all subjects whose outcomes went missing initially. An attractive alternative strategy is to make a much more intensive effort to obtain outcome measures for a *randomly selected subset* of subjects with missing data. In our example, we would try very hard to obtain outcomes for a sample of control group subjects with missing outcomes. Assuming this intensive data collection effort succeeds in producing outcome measurements, we can use the average observed outcome for this sample to form an unbiased estimate of $E[Y_i(0) \mid R_i(0) = 0]$.

Confining one's data collection efforts to a random sample may be much more efficient than trying to obtain outcomes for the entire set of subjects with missing data. Imagine, for example, that $N = 10{,}000$ subjects are divided equally between treatment and control, and potential outcomes $Y_i(0)$ and $Y_i(1)$ are binary (Y_i is observed to be either 0 or 1). For simplicity, again suppose that there is no attrition in the treatment group, but half of the control group is missing. In other words, we observe outcomes for all 5,000 subjects in the treatment group, but 2,500 of the control group subjects have missing outcome data. The two alternative data collection strategies are:

> Plan 1: Try again to obtain outcomes from the 2,500 subjects. Suppose you succeed in half the cases, producing outcome measures for 1,250 more subjects.
>
> Plan 2: Randomly sample 100 observations from the 2,500 missing subjects and try very hard to obtain outcome measures for these. Suppose that you succeed in all 100 cases.

After following Plan 1, the attrition rate has been substantially reduced, but missing data still comprise 25% of the control group. Even with the additional 1,250 observations, estimates of the ATE may be severely biased, and extreme value bounds will be large. Plan 2, on the other hand, provides an unbiased estimator. The average of the random sample of missing outcomes is equal to $E[Y_i(0) \mid R_i(0) = 0]$, the missing piece of $E[Y_i(0)]$ in equation (7.31).

Suppose that we want to estimate the ATE while imposing minimal assumptions. Under Plan 1, we would form the upper bound for $E[Y_i(0)]$ by substituting the extreme value of 1 for the 25% still missing and the lower bound by swapping in 0 for the 25% missing. It is easy to see that extreme value bounds will be very wide and may fail to shed much light on even a large average treatment effect. Under Plan 2, we have a treatment effect estimate and no need for bounds. Setting aside the issue of sampling variability, Plan 2 is clearly superior to Plan 1.

That said, the zero attrition scenario under Plan 2 seems a bit too optimistic. Suppose instead that the second round of data collection resulted in outcome measures for 90 of the 100 randomly selected subjects. To discuss this case, we need to define a new potential outcome, $s_i(z)$, scored 1 if the second-round effort yields an outcome and 0 otherwise. In our example, which concerns missingness among subjects assigned to the control group, we focus on the potential outcome $s_i(0)$, which indicates whether second-round data collection succeeds in measuring an outcome for subject i. We are also interested in the random variable $S_i(0)$, which indicates whether a subject would report an outcome in response to a hypothetical second-round data collection effort. Returning to our running example, we rewrite the target $E[Y_i(0) \mid R_i(0) = 0]$ as a weighted average:

$$E[S_i(0) \mid R_i(0) = 0]E[Y_i(0) \mid R_i(0) = 0, S_i(0) = 1] \\ + \{1 - E[S_i(0) \mid R_i(0) = 0]\}E[Y_i(0) \mid R_i(0) = 0, S_i(0) = 0]. \quad (7.32)$$

Paralleling the earlier decomposition and discussion of $E[Y_i(0)]$, the new data can be used to estimate $E[S_i(0) \mid R_i(0) = 0]$ and $E[Y_i(0) \mid R_i(0) = 0, S_i(0) = 1]$. However, we lack an estimate for $E[Y_i(0) \mid R_i(0) = 0, S_i(0) = 0]$ because the average outcomes among hardcore unmeasurable subjects remain outside our grasp. Fortunately, we can estimate the share of subjects who fall into the hardcore category. In this example, the share is relatively small: $0.5 * (10/100) = 0.05$, or 5% of the entire control group. As a consequence, to calculate the upper and lower bounds for $E[Y_i(0)]$, we swap in the largest and smallest possible outcomes for only 5% of the control group.

The comparison of Plans 1 and 2 highlights the potentially enormous advantages of an intensive effort to reduce attrition among a random sample of subjects with missing data. The drawback of the random sampling approach is sampling variability. A random sample of 100 subjects will produce noisy estimates. Table 7.6 describes how the precision of the estimates and the size of the bounds change under different scenarios. One set of scenarios varies the size of the second-round sample from 100 to 400. Another set of scenarios varies the rate at which outcome data are obtained in the second round. Finally, we vary the severity of the attrition problem by considering several scenarios. The first six rows of the table consider the scenario in which MIPO holds in both rounds of data collection. The next six rows consider scenarios in which missingness is related to potential outcomes in each round of data collection. The final six rows present an intermediate scenario in which missingness is related to potential outcomes in the first round of data collection but not the second. For each scenario, Table 7.6 simulates 100,000 experiments and presents the average estimated ATE and extreme values bounds after the first round and after the second-round followup. The first-round estimates of the ATE are based on a comparison of the 5,000 treated subjects with the control subjects whose outcomes were observed

TABLE 7.6

Simulation of the sampling distribution of point estimates and bounds under second-round sampling (results averaged over 10, 000 simulations, true ATE = 0.10)

		Estimates based only on initial data collection				Estimates based on both rounds of data collection			
Scenario*	N of control follow-ups	Point estimate of ATE	SE	Extreme value lower bounds	Extreme value upper bounds	Point estimate of ATE	SE	Extreme value lower bounds	Extreme value upper bounds
{50,5}	100	0.100	0.013	−0.175	0.325	0.100	0.027	0.086	0.111
{50,5}	400	0.100	0.012	−0.175	0.325	0.100	0.015	0.086	0.111
{50,25}	100	0.100	0.012	−0.175	0.325	0.099	0.026	0.031	0.156
{50,25}	400	0.100	0.012	−0.175	0.325	0.100	0.015	0.031	0.156
{50,50}	100	0.100	0.012	−0.175	0.325	0.101	0.026	−0.037	0.213
{50,50}	400	0.100	0.012	−0.175	0.325	0.101	0.015	−0.037	0.213
{A,B}	100	0.001	0.012	−0.224	0.276	0.091	0.026	0.075	0.105
{A,B}	400	0.001	0.012	−0.224	0.276	0.091	0.015	0.075	0.105
{A,C}	100	0.001	0.012	−0.224	0.276	0.083	0.026	0.041	0.120
{A,C}	400	0.001	0.012	−0.224	0.276	0.083	0.015	0.041	0.120
{A,D}	100	0.001	0.012	−0.225	0.276	0.071	0.026	−0.003	0.139
{A,D}	400	0.001	0.012	−0.225	0.275	0.071	0.015	−0.003	0.139
{A,5}	100	0.001	0.012	−0.225	0.275	0.098	0.025	0.084	0.109
{A,5}	400	0.001	0.012	−0.224	0.276	0.097	0.015	0.084	0.109
{A,25}	100	0.001	0.012	−0.225	0.275	0.085	0.025	0.018	0.143
{A,25}	400	0.001	0.012	−0.225	0.275	0.086	0.015	0.019	0.144
{A,50}	100	0.001	0.012	−0.224	0.276	0.067	0.024	−0.062	0.188
{A,50}	400	0.001	0.012	−0.225	0.276	0.067	0.015	−0.062	0.188

*Scenarios are denoted as follows: The first number or letter indicates the percent missing in the initial round. The second number or letter indicates the number missing in the second round. Scenario A refers to missingness rates of {0.5, 0.5, 0.7, 0.3}, respectively, in the following four strata: (1) $Y_i(1) = 1, Y_i(0) = 0$, (2) $Y_i(1) = 0, Y_i(0) = 1$, (3) $Y_i(1) = 0, Y_i(0) = 0$, and (4) $Y_i(1) = 1, Y_i(0) = 1$. Scenarios B, C, and D refer to second-round missingness rates of {0.05, 0.05, 0.1, 0}, {0.25, 0.25, 0.3, 0}, and {0.5, 0.5, 0.55, 0.45} in the four strata.

based on the initial round of data collection. The second-round estimates of the ATE are based on a comparison of the 5,000 treatment subjects and a weighted average of the control subjects whose outcomes are observed in each round. If the proportion of control subjects who go missing in the first round is x, the average outcomes for the controls in rounds 1 and 2 are weighted $1 - x$ and x, respectively.

As intuition would suggest, the second-round follow-up strategy works best when the second-round sample is large and obtains outcome data for a high proportion of the targeted sample, with second-round success rates that are independent of potential outcomes. The advantages of the second-round sampling strategy are most apparent when missingness is strongly related to potential outcomes. Consider an example in which missingness is related to potential outcomes and second-round efforts recover outcomes for 75 out of 100 randomly targeted subjects (scenario {A,25} in Table 7.6). In this example, the true ATE is 0.10. A naive difference-in-means based on non-missing values generates an average estimate of 0.001 (SE = 0.012). By comparison, the estimator that makes use of the second-round sample generates an average estimate of 0.085 (SE = 0.025), which is less than one standard error away from the true parameter of 0.10. The second-round sampling approach also leads to a dramatic reduction in the extreme value bounds. Using only the first-round data, the bounds extend from −0.225 to 0.275. With the second-round data, the bounds are one-fourth as wide: 0.018 to 0.143.

Of course, even when MIPO holds, the second-round sampling strategy still plays a valuable role, since the researcher is often unsure about the plausibility of MIPO. The second round of data collection narrows the expected range of the extreme value bounds, in effect narrowing the range of uncertainty about the possible consequences of attrition.

Given the attractiveness of a second-round sampling strategy, the remaining question for someone designing a study is how to allocate a fixed budget. In the extreme case, no money is allocated to second-round sampling, and the freed-up resources are invested in increasing the number of treatment and control subjects. Alternatively, the size of the treatment and control group could be reduced to make funds available for a second round of data collection. The specifics of this allocation exercise are a bit complex and so are relegated to the Chapter 7 appendix, but the idea is fairly simple. Suppose the objective were to find the budget allocation that minimized the expected mean square error of the estimated ATE, and suppose that second-round observations cost 10 times more than observations in the initial sample. Allocating more observations to the initial sample reduces variance but fails to address attrition-related bias. Allocating more observations to the second-round sample decreases bias but is expensive. If you believe there is a risk of severe attrition bias, second-round sampling becomes an attractive option despite the cost.

One drawback to second-round sampling, apart from extra expense, is that special care must be taken to ensure that the procedures used to measure outcomes and the context within which measurements are gathered are as similar as possible to the initial round of data collection. If the control group were measured using different standards in the follow-up round, the symmetry between treatment and control groups would be jeopardized. In effect, changing standards of measurement represent a violation of the exclusion restriction, since a factor (measurement procedures) that is systematically correlated with treatment assignment is related to potential outcomes. When assessing the advantages of a second-round sampling approach, one must also consider whether it is feasible to measure outcomes using similar standards in both rounds.

7.7 Two Frequently Asked Questions

Before concluding our discussion of attrition, we address two questions that often arise when experimental researchers encounter missing data. The first question concerns the distinction between missing outcome data and missing data for covariates. What are the consequences of missing covariate data, and how should one address this problem?

Missing data for covariates is typically a much less serious problem than missing data for outcomes. When analyzing a simple randomized experiment using regression, the inclusion of pre-treatment covariates is optional (see Chapter 4). One could obtain consistent estimates regardless of whether one elects to control for covariates. If one chooses to include a covariate X_i for which some of the values are missing, the simplest solution is to:

1. Assign an arbitrary value, such as $x = 0$, for subjects with missing values.
2. Create a new dummy variable scored 1 for subjects who are missing and 0 otherwise.
3. Regress outcomes on the treatment, X_i, and the new dummy variable.

This approach is similar to substituting $\overline{X}$ for missing values of X_i. If X_i is a categorical variable, code missingness as a distinct category when creating dummy variables. For example, if some observations are missing data for the variable sex, create three categories: male, female, and missing. Use two dummy variables to represent sex as a covariate in a regression.

The second question arises when researchers are trying to devise ways of containing the problem of attrition after it has manifested itself. Sometimes rates of attrition in an experimental sample vary from subgroup to subgroup. Given that attrition is such a vexing problem, why not simply confine your data analysis to the subgroups in your sample that do not suffer from attrition? This appealing but flawed strategy will produce biased treatment effect estimates.

Consider the following example. Suppose you want to evaluate a program to help subjects get into better physical condition, and the intervention provided to

TABLE 7.7
Potential outcomes and pattern of attrition for two pairs of subjects

Observation	Pair	$Y_i(0)$	$Y_i(1)$	$r_i(0)$	$r_i(1)$	$Y_i(0)\|r_i(0)$	$Y_i(1)\|r_i(1)$
1	A	4	6	1	1	4	6
2	A	4	2	1	0	4	Missing
3	B	8	8	1	1	8	8
4	B	9	9	1	1	9	9

those assigned to the treatment group is a session with a personal trainer and some reading material. The control group gets only the reading material. Subjects are asked to return after six months for some endurance testing. At the time of initial enrollment in the experiment, Subjects 1 and 2 report that they ordinarily get a minimal amount of exercise. In contrast, Subjects 3 and 4 report that they are very active. The experimenter blocks the subject pool into pairs based on this information, and then randomly assigns one subject in each pair to treatment and the other to control. Outcomes are measured by minutes of jogging on a treadmill at increasing speeds. Table 7.7 shows the potential outcomes and pattern of attrition for the two pairs of subjects (the pairs are Subjects 1 and 2, 3 and 4).

Table 7.7 shows that the intervention has a zero average treatment effect, although the intervention helps some subjects and harms others. Some people start a rigorous exercise regime and thrive, while others find it too hard and give up exercising altogether. The zero treatment effect for the subjects who are already exercising is consistent with the idea that those who enter the experiment in good physical shape are already getting the full benefits of exercise prior to the intervention. Table 7.7 shows that for some patterns of random assignment the experiment produces missing outcome data. If a subject is out of shape, he or she may not want to return for the endurance testing. Suppose that if a subject expects to last fewer than three minutes on the treadmill, the subject does not show up for the evaluation. Focusing on the first pair, Subjects 1 and 2, notice that Subject 2 attrites when assigned to treatment.

Using Table 7.7 we can evaluate the proposed procedure for confining attrition, which holds that if there is attrition in a pair, you can drop the entire pair from the analysis and focus on the portion of the sample that is attrition free. Using the remaining observed outcome data, average the treatment and control outcomes; the difference is the estimated treatment effect. Is this an unbiased estimator?

No. As shown in Table 7.8, the expected value of the estimator is 0.5, not zero. The estimator is upwardly biased because attrition is correlated with the potential outcomes. The first pair {1, 2} is dropped from the data analysis whenever the subject with the negative treatment effect is assigned to the treatment group and is retained whenever the subject with the positive treatment effect is assigned to the treatment

TABLE 7.8

Four possible assignments to treatment and control and the estimated treatment effect associated with each assignment, applying a rule of dropping pairs in which attrition occurs

Subjects assigned to treatment	Subjects assigned to control	Estimated treatment effect	Are any pairs dropped?
{1,3}	{2,4}	$\frac{6+8}{2} - \frac{4+9}{2} = 0.5$	No
{1,4}	{2,3}	$\frac{6+9}{2} - \frac{4+8}{2} = 1.5$	No
{2,3}	{1,4}	$8 - 9 = -1$	First pair is dropped
{2,4}	{1,3}	$9 - 8 = 1$	First pair is dropped

Note: The true ATE is zero, but the average estimated ATE using the above procedure is $(0.5 + 1.5 - 1 + 1)/4 = 0.5$.

group. This asymmetry produces bias because we only hear "good news" about the treatment from this pair. The estimator does not in general produce an unbiased estimator of the treatment effect for the pairs of subjects that experience no attrition. Had the researcher focused on just the second pair from the outset, the treatment effect estimate would have been unbiased. However, because the researcher cannot see Table 7.7, there is no reason why the researcher would elect to discard pair 1 when no attrition is observed within the pair. The broader lesson is that what seems like a sensible procedure—drop subsets of the data that suffer attrition—leads to bias when attrition is a function of potential outcomes.

SUMMARY

Attrition poses a serious threat to all forms of social research, regardless of whether interventions are randomly assigned, but the sting of attrition is felt most acutely by those conducting experiments. Experimental researchers craft randomized experiments in order to obtain an unbiased estimate of a causal effect, only to discover that attrition has introduced the threat of bias. Because attrition concerns unobserved outcomes, it naturally invites controversy. Assumptions are required in order to obtain estimates, and there is inevitably a range of opinions about what assumptions are warranted by theory or evidence.

Some of these controversies can be sidestepped by approaches that strive only to bound average treatment effects. One agnostic approach is to insert extreme values in place of missing values. When attrition rates are large in relation to the anticipated ATE, these bounds are often uncomfortably wide. Another approach is to bound the average treatment effect of a subgroup, those who always report outcomes regardless of whether they are assigned to treatment or control. This approach, which involves trimming outcomes at the top or bottom of the outcome distribution for the experimental group with a lower rate of missingness, makes somewhat stronger assumptions and tends to produce more informative bounds. The price, however, is uncertain interpretation. Is the monotony assumption on which trimming rests valid? If so, how far can we generalize based on the estimated ATE for this subgroup?

Given the many drawbacks of analyzing experiments that experience attrition, researchers should take pains to design studies with attrition in mind. This chapter discussed one design idea—follow-up sampling to measure outcomes on subjects who go missing after the first round—but other suggestions quickly come to mind. At a mundane level, maintain good administrative records (or good relations with those in charge of releasing information) in anticipation of maximizing the chances of obtaining outcome measures. Better still, gather several outcome measures, each drawing information from a different source, so that missing data on one outcome variable may be imputed using information from other parallel measures. An abbreviated list of data sources includes public documents, government records, market research data, direct observation, surveys of the participants, or surveys of the participants' friends or relatives. It is not a bad idea to start an experimental project by first considering the availability of outcome measures.

SUGGESTED READINGS

Allison (2002) and Graham (2009) provide introductions to the vast literature on missing data; Little and Rubin (2002) provide more technical detail. Manski (1995, 2007) discusses the use of worst-case bounds when outcome data are missing, and Lee (2005) discusses trimming. DiNardo, McCrary, and Sanbonmatsu (2006) illustrate different approaches for handling attrition when analyzing experimental data. Additional data collection through double sampling is discussed in Cochran (1977) and Lohr (2010), and an empirical application is presented in Gerber, Green, Kern, and Blattman (2011).

EXERCISES: CHAPTER 7

1. Important concepts:
   (a) Equation (7.1) describes the relationship between potential missingness and observed missingness. Explain the notation used in the expression $r_i = r_i(0)(1 - z_i) + r_i(1)z_i$.
   (b) Explain why the assumption that $Y_i(z) = Y_i(z, r(z) = 1) = Y_i(z, r(z) = 0)$ amounts to an "exclusion restriction."
   (c) What is an "If-Treated-Reporter"?
   (d) What are extreme value bounds?
2. Suppose that $r_i(1) = r_i(0)$ for all subjects in an experiment. In other words, all subjects are either Always-Reporters or Never-Reporters. Show that when the treatment effect is the same for all subjects, the difference-in-means for subjects with observable outcomes shown in equation (7.6) is the same as the overall ATE in equation (7.5).

3. Construct a hypothetical schedule of potential outcomes to illustrate each of these cases:
   (a) The proportion of missing outcomes is expected to be different for the treatment and control groups, yet the difference-in-means estimator is unbiased when applied to observed outcomes in the treatment and control groups.
   (b) The proportion of missing outcomes is expected to be the same for the treatment and control groups, yet the difference-in-means estimator is biased when applied to observed outcomes in the treatment and control groups.
4. Construct a hypothetical schedule of potential outcomes for $Y_i(z)$ and $R_i(z)$ to show that under some random assignments, a researcher may estimate extreme value bounds that do *not* encompass the true ATE.
5. Suppose you were to encounter missingness in the course of conducting an experiment. You look for clues about the causes and consequences of missingness by conducting three lines of investigation: (1) assessing whether rates of missingness differ between treatment and control groups, (2) assessing whether covariates predict which subjects have missing outcomes, and (3) assessing whether the predictive relationship between missingness and covariates differs between treatment and control groups. In what ways would these three lines of investigation inform the analysis and interpretation of your experiment?
6. From the online appendix (http://isps.research.yale.edu/FEDAI), download the data used in the Angrist, Bettinger, and Kremer article.[19] Using the voucher treatment and two covariates (sex and valid phone number), develop a linear regression model that predicts nonmissingness. Use the predicted values from this model to generate inverse probability weights, taking care to verify that predicted values are nonnegative and not greater than 1.0. Run a weighted regression of reading test scores on winning the voucher, using inverse probability scores as weights. Interpret the estimates.
7. Sometimes experimental researchers exclude subjects from their analysis because the subjects (1) appear to understand what hypothesis the experiment is testing, (2) seem not to be taking the experiment seriously, or (3) fail to follow directions. Discuss whether each of these three practices is likely to introduce bias when the researcher compares average outcomes among non-excluded subjects.
8. Ditlmann and Lagunes report the results of an experiment in which Hispanic and non-Hispanic confederates attempted to use a personal check to purchase $10 gift certificates at 217 retail stores.[20] Confederates, who were trained to behave in a similar manner, were randomly assigned to each store. One of the outcome measures is whether the retail clerk asks to see the confederate's photo identification. A second outcome is whether, for those who were asked to present identification, the identification card (which was supplied by the experimenters) was accepted as valid. Suppose the question of interest were: Are clerks more likely to accept the identification card when it is presented by a white or Hispanic shopper? Because some shoppers were never asked to present identification, their outcomes are missing. Define the treatment as 0 if non-Hispanic and 1 if Hispanic. Define the request for identification as 0 if no request is made and 1 if a request is

19 Angrist, Bettinger, and Kremer 2006.

20 Ditlmann and Lagunes 2010.

made. Define the acceptance of identification as 0 if identification is rejected and 1 if it is accepted. The table below shows the number of retailers who requested and/or accepted identification, by experimental condition.

	White Shopper	Hispanic Shopper
No ID Requested	28	17
ID Requested and Accepted	50	68
ID Requested but Rejected	28	26
Total N	106	111

(a) The data seem to suggest that Hispanics who presented identification were more likely to have their IDs accepted than whites who presented identification. Explain why this pattern in the data may give the misleading impression that retailers discriminate in favor of Hispanics.

(b) Use extreme value bounds to fill in the missing outcomes (acceptance or rejection of identification) for those subjects who never presented identification. Interpret your results.

(c) Is the monotonicity assumption on which trimming bounds rest defensible in this application? Calculate the trimming bounds and interpret the results.

9. Suppose a researcher studying a developing country plans to conduct an experiment to assess the effects of providing low-income households with cash grants if they agree to keep their children in school and take them for regular visits to health clinics. The primary outcome of interest is whether children in the treatment group are more likely to complete high school. A random sample of 1,000 households throughout the country is allocated to the treatment group (cash grants), and another sample of 1,000 households is allocated to the control group.

   (a) Suppose that halfway through the project, a civil war breaks out in half of the country. Researchers are prevented from gathering outcomes for 500 treatment and 500 control subjects living in the war zone. What are the implications of this type of attrition for the analysis and interpretation of the experiment?

   (b) Another identical experiment is performed in a different developing country. This time the attrition problem is as follows: households that were offered cash grants are more likely to live at the same address years later, when researchers return in order to measure outcomes. Of the 1,000 households assigned to the treatment group, 900 are found when researchers return to measure outcomes, as opposed to just 700 of the 1,000 households in the control group. What are the implications of this type of attrition for the analysis and interpretation of the experiment?

10. Table 7.6 summarizes the results of a series of simulations. Recall from the discussion of this table in section 7.6 that each simulation considers the accuracy with which conventional and second-round sampling procedures recover the ATE. Based on the results presented in the table, address the following questions.

   (a) The first six rows of the table consider scenarios in which missingness is unrelated to potential outcomes. Each scenario varies the rate of missingness and the number of observations gathered in the second round of data collection. Compare the two estimators (one based on initial data collection only, the other based on both rounds of data collection) in terms of bias, precision, and the width of the extreme value bounds.

(b) The next six rows of the table consider scenarios in which missingness is related to potential outcomes. Again, compare the two estimators in each scenario in terms of bias, precision, and the width of the extreme value bounds.

(c) Perform the same comparisons for the scenarios in the final six rows of the table, in which missingness is related to potential outcomes in the first round but unrelated to potential outcomes in the second round.

(d) Overall, under which scenario does estimation based on both rounds of data analysis have the greatest comparative advantage over estimation based only on the initial round of data collection? Under which scenario does estimation based only on the initial round of data collection have the greatest comparative advantage over estimation based on both rounds?

APPENDIX

A7.1 Optimal Sample Allocation for Second-Round Sampling

Researchers designing experiments must often evaluate alternative ways of allocating limited resources in order to achieve their statistical objectives. This appendix illustrates how to approach the trade-offs that occur when resources are scarce. We examine the design question formally for a special case. The basic approach, which involves checking how each of the feasible designs performs according to a specified objective, is very general. To solve an optimal design problem you must answer a series of questions:

1. What is the objective? Your objective is the formula you will use to compare alternative research designs. One's objective might be, for example, to minimize the expected mean squared error or to minimize the variance among unbiased estimators.
2. What is the budget?
3. What is the set of feasible designs? The term *feasible* describes any design that comes in at or below the budget.
4. Finally, what feasible allocation maximizes the objective? For example, should the researcher spend money to recruit more subjects, or should those resources be allocated to reduce the rate of attrition?

Following the discussion of second-round sampling in section 7.6, we consider the question of how to estimate $E[Y_i(0)]$ when there is attrition from the control group. Suppose that there are two technologies for obtaining outcome measures: the Standard Method and Enhanced Effort. The Standard Method involves a single round of data collection, after which some subjects will have missing outcomes. Enhanced Effort is applied to a random sample of subjects with missing outcome data under the Standard Method. Suppose that each subject assigned to the control group costs C and there is no additional cost associated with the Standard Method. Let α be the proportion of subjects who are *not* missing outcome data when the Standard Method is used to obtain outcome measures. The cost of using Enhanced Effort is an extra amount, C_E, over the standard cost C. To keep this example simple, we assume that when Enhanced Effort is used, 100% of the subjects provide outcome measures.

Let B be the total budget available for estimating $E[Y_i(0)]$. The design question is: "How large should the control group be and how many of the control group subjects should be given Enhanced Effort?" The maximum budget available for any possible design is:

$$B = C \cdot N + C_E \cdot N_E, \tag{A7.1}$$

where N is the number of subjects in the control group, and N_E is the number of subjects given the Enhanced Effort. The number of subjects given Enhanced Effort and the number of subjects in the control group are related in the following way: $N_E \leq (1 - \alpha)N$. The total cost equation quantifies the cost tradeoff between spending money to recruit more control group subjects versus spending extra resources to measure missing outcomes. For example, suppose C_E is three times C. Given a fixed budget, for each subject selected for Enhanced Effort, the size of the control group must be reduced by three subjects in order to stay within the budget.

Recall that we earlier demonstrated that when there is 100% response to a second-round follow-up measurement effort aimed at a random sample of control group subjects with missing outcomes, the average $Y_i(0)$ for the subjects who respond to this Enhanced Effort and for those who responded to the first-round effort can be combined in order to construct an unbiased measure of $E[Y_i(0)]$. Using this result, we now consider the question of how many subjects to allocate to Enhanced Effort.

We assume that the researcher places a priority on obtaining unbiased estimates of the average treatment effect and therefore will allocate at least some resources to Enhanced Effort. In this example, the researcher's objective is to minimize the variance among unbiased estimators. The design problem is therefore to select an allocation to the Standard Method versus Enhanced Effort such that the resulting estimator is unbiased and as precise as possible. The unbiased estimator of $E[Y_i(0)]$ is:

$$\alpha \hat{Y}_S(0) + (1 - \alpha)\hat{Y}_E(0), \tag{A7.2}$$

where $\hat{Y}_S(0)$ refers to the average outcome obtained by the Standard Method, and $\hat{Y}_E(0)$ refers to the average outcome obtained by the Enhanced Method. For simplicity, we ignore sampling covariance between $\hat{Y}_S(0)$ and $\hat{Y}_E(0)$, and we assume that the variance of $Y_i(0)$ may be approximated by σ regardless of a subject's potential missingness. In other words, we assume that regardless of whether $r_i(0) = 0$ or $r_i(0) = 1$, the variance of $Y_i(0)$ remains approximately the same.

For any allocation N and N_E, the variance of the estimator for $E[Y_i(0)]$ equals:

$$\sigma\left(\frac{\alpha}{N} + \frac{(1-\alpha)^2}{N_E}\right). \tag{A7.3}$$

The researcher's problem is to select N and N_E to minimize (A7.3) subject to the budget constraint in equation (A7.1). The cost of a control observation can be normalized to 1 so that B and C_E are expressed in terms of C. The budget may now be written:

$$B = N + C_E \cdot N_E. \tag{A7.4}$$

Because the optimal allocation of N and N_E will exhaust the budget, we can substitute the budget constraint into equation (A7.3):

$$\sigma\left(\frac{\alpha}{N} + \frac{C_E(1-\alpha)^2}{B - N}\right). \tag{A7.5}$$

Equation (A7.5), the function that describes the objective, is now written solely in terms of one choice variable, N. We wish to find the value of N that minimizes the objective function (A7.5) subject to the obvious restrictions that $N > 0$ and $N_E > 0$. Ignoring these inequality constraints for the moment, N ranges from 0 to B. To find the allocation that minimizes (A7.5), search over the range of feasible values for N. (This type of exhaustive, trial-and-error search is called a "grid search.") Notice that when N is selected, N_E is implicitly selected as well, due to the budget constraint. Check that the best value for N is such that the inequalities are satisfied.

An example illustrates the value of carefully considering the allocation of resources to the Standard Method versus Enhanced Effort. Suppose that each observation measured using Enhanced Effort costs 10 times the Standard Method. For this numerical example, we set the attrition rate under the Standard Method at 50%. The budget is sufficient to pay for 1,000 subjects using the Standard Method. Table A7.1 shows the relative variances expected under alternative feasible allocations of subjects to N and N_E. The variance, σ, is a multiplicative constant in the variance expression and so affects the absolute level but not the relative size of the variance across allocations; we therefore set σ to 1 for convenience. The feasible set of allocations satisfies the requirements $N > 0$ and $N_E > 0$ and $N_E \leq (1 - \alpha)N$. The smallest variance is obtained when approximately 300 subjects are assigned to the control group, and of the approximately 150 subjects who have missing data, 70 are randomly selected for Enhanced Effort. Had the researcher instead decided to assemble a larger control group (leaving fewer resources to follow-up), the variance of the estimated $E[Y_i(0)]$ would be higher. For instance, suppose the researcher selected the design with 800 subjects: approximately 400 would be missing outcomes, and 20 of these would be randomly assigned for intensive follow-up. The sampling variance under this design would be more than twice as great. Although the number of subjects with outcome data is larger when the size of the control group is larger, the extra observations help to estimate a quantity that is already being estimated relatively well.

If the N that minimizes the variance is such that more than the bare minimum subjects are allocated to the Standard and Enhanced methods, we say that we have found an "interior solution." For those familiar with calculus, it can be confirmed that the value of N that produces the lowest variance is the value of N such that the first derivative with respect to N of equation (A7.5) equals zero. A useful way to express the optimal allocation is the ratio of subjects allocated to N_E as a share of the total control group subjects N. The ratio of N_E to N that satisfies the zero derivative condition for the design problem addressed in this example is:

$$\frac{N_E}{N} = \frac{1-\alpha}{\sqrt{\alpha C_E}}. \tag{A7.6}$$

Recall that at most $(1 - \alpha)N$ subjects can be assigned to Enhanced Effort. This constraint implies that a smaller number of subjects than the full pool of unmeasured

TABLE A7.1

Results of optimization problem for the case in which enhanced effort costs 10 times as much as standard data collection and the budget is equivalent to the cost of 1,000 standard observations

Allocation to N	100	150	200	250	300	350	400	450	500	550	600	650	700	750	800
Allocation to N_E	90	85	80	75	70	65	60	55	50	45	40	35	30	25	20
Resulting variance of $\hat{E}[Y_i(0)]$	0.007778	0.006275	0.005625	0.005333	0.005238	0.005275	0.005417	0.005657	0.006	0.006465	0.007083	0.007912	0.009048	0.010667	0.013125

Note: The shaded column indicates the allocation that minimizes variance of the resulting estimate.

subjects is assigned for Enhanced Effort whenever $\sqrt{\alpha C_E} > 1$ or $\alpha > 1/C_E$. Inserting the values from our example ($\alpha = 0.5$, $C_E = 10$) into equation (A7.6), we find the optimal ratio of N_E/N to be 0.223, which is approximately equal to the ratio of $70/300$ obtained from the grid search depicted in Table A7.1.

In principle, there may be many values of N at which an arbitrary function's first derivative is to equal zero. Additional conditions should be checked to make sure that the value of N is (at least) a local minimum for the function rather than a maximum: the condition for a local minimum is that the second derivative of the function is positive. When the second derivative of the objective formula is positive for all values of N over the range of N under consideration for the experiment, the N that satisfies the zero first derivative condition is the overall "best" value for N. This condition on the second derivative, which is somewhat stronger than required but sufficient for a global extreme value, is satisfied for the research design problem we are considering. More generally, calculus methods can be used for more complicated cases involving several choice variables. These methods involve taking the first derivative of the augmented objective function with respect to the choice variables, where the new objective function has been modified to include the equality and inequality constraints on the values of the expanded set of choice variables.

CHAPTER 8

Interference between Experimental Units

When defining a potential outcome as $Y_i(1)$ if subject i is treated and $Y_i(0)$ otherwise, we implicitly assume that this subject's potential outcomes are influenced solely by whether it receives the treatment. Treatments administered to other units are irrelevant. Formally, if we use the vector $\boldsymbol{d}$ to list the treatment status of each subject ($d = 0$ or $d = 1$), the non-interference assumption says that $Y_i(\boldsymbol{d}) = Y_i(d)$. In other words, potential outcomes for subject i respond only to the subject's own treatment status, d, not the treatments administered to other subjects.[1] In cases where noncompliance causes the vector of assignments $\boldsymbol{z}$ to diverge from the vector of actual treatments $\boldsymbol{d}$, non-interference implies that each subject is unaffected by the treatments and assignments of others: $Y_i(\boldsymbol{z}, \boldsymbol{d}) = Y_i(z, d)$.

The stipulation that potential outcomes for a unit reflect only the treatment status of that unit is known by different names in the literature on experimentation. Sometimes this assumption is called the Individualistic Treatment Response[2] or the Stable Unit Treatment Value Assumption (SUTVA).[3] The latter term is more common but, unfortunately, its wording does not do much to convey the intuition of "no interference between units," which is the key idea that must reside in a researcher's working memory. Treatments are assumed to be "stable" or "non-interfering" in the technical sense that potential outcomes are unaffected by how the randomization happens to come out. Non-interference is a substantive assumption about how units behave when treatments are allocated—in essence, it declares that the schedule of potential outcomes is the true schedule; there are no lurking potential outcomes

1 This version of the non-interference assumption might be described as "global non-interference," in the sense that it applies to any allocation of m treatments to N subjects, where m may be as small as 0 or as large as N. A weaker requirement of "local non-interference" holds that potential outcomes reflect the subject's own receipt of treatment provided that exactly m treatments are administered to N subjects. As the examples discussed below suggest, the assumption of local non-interference may be no longer credible when applied to interventions in which higher or lower proportions of subjects are treated.

2 Manski 2012.

3 Rubin 1990.

What you will learn from this chapter:

1. How interference between units affects the way in which potential outcomes are defined.
2. How interference may lead to biased inferences when one compares treatment and control groups.
3. Experimental designs for detecting the effects of interference.
4. The complications that interference presents for spatial analysis.
5. The connection between the assumption of non-interference and the assumptions necessary for unbiased inference from within-subjects design.

that emerge based on who gets what treatment. As discussed in Chapter 2, non-interference is one of the core assumptions needed to establish the unbiasedness of the difference-in-means estimator.

The non-interference assumption also helps clarify what we mean when we define a treatment effect as $Y_i(1) - Y_i(0)$. By way of illustration, consider a simple example of three sales agents working at a retail store. At the end of July, the store plans to introduce a new system for recognizing its employees and will honor one of the three sales agents as "Salesperson of the Month." Potential outcomes in terms of each agent's August sales are shown in Table 8.1. The last column shows the potential outcomes if no one is honored. Compared to this benchmark, the first sales agent, Mary, will increase her productivity if honored. Peter's mediocre productivity is unaffected regardless of who is honored. Limor is unaffected by the honor but like Mary, becomes demoralized if Peter is honored, in which case her productivity plummets.

A researcher is asked to evaluate how workers respond to being honored and proposes an experiment in which the honor is assigned at random to one of the three sales agents. The researcher defines the causal estimand as the difference between two potential outcomes: $Y_i(1)$ if subject i is honored and $Y_i(0)$ if subject i is not honored. Here is the catch: the researcher, unaware of the schedule of potential outcomes, does

TABLE 8.1

Sales made by each agent, depending on who is awarded "Salesperson of the Month" in the prior month

Sales agent	Outcomes if Mary is honored	Outcomes if Peter is honored	Outcomes if Limor is honored	Outcomes if nobody is honored
Mary	100	50	70	70
Peter	50	50	50	50
Limor	90	50	90	90

not realize that each worker's $Y_i(0)$ is unstable in the sense that it varies depending on who receives the honor. Instead, the researcher supposes that $Y_i(0)$ simply refers to the potential outcome in the last column, where no one is honored. Using that definition, the average treatment effect is $((100 - 70) + (50 - 50) + (90 - 90))/3 = 10$. If the store chooses the winner by lottery, however, the three randomizations will produce apparent treatment effects of $(100 - (50 + 90)/2) = 30, (50 - (50 + 50)/2) = 0$, and $(90 - (70 + 50)/2) = 30$, which average to 20. This example illustrates how interference may lead to bias. When defining the estimand, the researcher ignores the fact that potential outcomes for observation i depend on whether treatment is administered to observation j. The ATE is exaggerated in this case because the average $Y_i(0)$ value when Peter is honored is lower than the average $Y_i(0)$ value when no one is honored.

In the social and health sciences, it is not unusual to encounter situations in which an observation's potential outcomes vary not only according to the treatment it receives but also according to the treatments that other observations receive. Examples of social phenomena that cause the treatment of one unit to have repercussions for other units include:

- Contagion: The effect of being vaccinated on one's probability of contracting a disease depends on whether others are vaccinated. The causal effect of vaccination is likely to be small if one is surrounded by others who are vaccinated, large if one is surrounded by unvaccinated people.
- Displacement: Police interventions designed to suppress crime in one location may displace criminal activity to nearby locations. A neighborhood's potential outcomes in terms of crime rates may vary depending on its proximity to the intervention. An "untreated" location may experience a surge in crime if it is adjacent to a treated neighborhood. The causal effect of police intervention would be exaggerated if one were to compare treated locations to adjacent untreated locations, because the nearby untreated locations will experience displaced criminal activity.
- Communication: Interventions that convey information about commercial products, entertainment, or political causes may spread from individuals who receive the treatment to others who are nominally untreated. The causal effect of direct treatment may be substantial, but as word spreads from the treated group to the untreated group, the apparent effect of the information may seem negligible.
- Social comparison: An intervention that offers housing assistance to a treatment group may change the way in which those in the control group evaluate their own housing conditions. Potential outcomes in the control group, in other words, decrease when the treatment group is treated.
- Deterrence: Policy interventions are sometimes designed to "send a message" about what the government intends to do or what it has the capacity to do. For example, a federal agency may launch a series of random audits of local

governments to deter corruption in the treated locations as well as in untreated locations that learn of the audit policy.

- Persistence and memory: Within-subjects experiments, in which outcomes for a given unit are tracked over time, measure the effects of introducing a stimulus on outcomes in subsequent time periods. An individual's potential outcomes at each point in time are often more complicated than one response if treated and another if not treated. A subject may remember past interventions and may respond to future interventions differently.

Social phenomena are not the only sources of interference. Potential outcomes for subject *i* may be affected by the treatment of subject *j* when an intervention redeploys resources from the control group to the treatment group. For example, an experimental curriculum may cause teachers to spend more time with the treatment group and less time with the control group.[4]

This long list of possible sources of interference is a daunting reminder that researchers must pay special attention to interference when positing potential outcomes, defining the ATE, and designing an experiment to estimate it. One of the aims of this chapter is to introduce readers to potential outcomes that go beyond the simple framework of $Y_i(1)$ if unit i is treated and $Y_i(0)$ otherwise. More nuanced schedules of potential outcomes demand more complex experimental designs. A second aim of this chapter is to introduce design principles that accommodate certain forms of spillover across units. A final aim of this chapter is to show how non-interference plays an important role in the analysis of experiments that track subjects over time. In order to extract information from experimental data, non-interference stipulations must be made, and the purpose of this chapter is to call attention to what those stipulations are and how they might themselves become the object of experimental inquiry.

8.1 Identifying Causal Effects in the Presence of Localized Spillover

In order to illustrate the nature and consequences of interference as well as experimental designs that address the issue of spillover, we start with an example that commonly arises in voter mobilization research. Consider a political campaign that

4 An experimental curriculum that diverts resources from the control group to the treatment group could be interpreted as an excludability violation insofar as the difference between treatment and control outcomes, in expectation, reflects both the treatment effect (the curriculum) and a lurking variable, the withdrawal of resources from the control group. The reason this example is used to illustrate interference is that the extent to which resources are diverted from the control group depends on the outcome of the random assignment; potential outcomes may vary depending on which students are assigned to each experimental condition.

attempts to mobilize voters by sending them direct mail. Suppose that each registered voter on the target list resides at the same address with one other voter on the list. We'll refer to these as two-voter households. Imagine that the set of two-voter households is divided into the following randomly assigned groups: in 5,000 households, both voters are targeted for mail; in 5,000 households, neither voter is targeted for mail; and in 10,000 households, exactly one voter is randomly targeted for mail. Thus, 10,000 people are exposed to mail along with their housemates; 10,000 people receive no mail and neither do their housemates; 10,000 people receive mail when their housemates do not; and 10,000 people receive no mail themselves, but their housemates receive mail. This type of experiment is termed a *multi-level design* because there are in effect two levels of randomization. Households are selected for treatment or control; if a household is selected for treatment, one or both individuals are randomly selected. More elaborate designs of this kind include additional levels, such as the precinct or zip code in which households are situated.[5]

Suppose that potential outcomes are a function of whether one receives mail oneself *and* whether one's housemate receives mail. We need to expand our notational system to take account of the fact that potential outcomes now respond to two different inputs. In order to avoid notational clutter, we dispense with the subscript i for each individual. For each voter, we have four potential outcomes: Y_{00}, Y_{01}, Y_{10}, Y_{11}, where the first subscript refers to whether the housemate is treated, and the second subscript refers to whether the voter is treated. Because potential outcomes are explicitly defined only for within-household spillovers, the operative non-interference assumption is that subjects' potential outcomes are unaffected by treatments administered to those *outside* their own household.

Based on these four potential outcomes, we can define several interesting causal effects. The difference between Y_{01} and Y_{00} defines the causal effect of being targeted oneself for mail while one's housemate receives nothing. When defining the effect of mail on the recipient, it is important to specify what the housemate receives, because another way of defining the effect of mail on the recipient is to compare Y_{11} and Y_{10}, which indicates the influence of mail on those whose housemates receive mail. Similarly, intra-household spillover effects can be defined in two ways. The difference between Y_{10} and Y_{00} indicates the spillover effect on those who receive no mail but whose housemates receive mail. A related estimand compares Y_{11} and Y_{01}, which again reflects spillover effects, this time among those who receive mail themselves. Notice that the latter two estimands do not pinpoint how spillovers occur; they simply summarize the cumulative effect of communication between housemates, exposure to the housemate's mail, lowered transaction costs of voting if the housemate is traveling to the polls, and the like.

5 See Sinclair, McConnell, and Green 2012 for an example of a multilevel experiment in which random assignment determines which postal codes are treated, which households are treated within postal codes, and which individuals are treated within households.

Our experimental design reveals one of the four potential outcomes for each voter. Because voters are randomly assigned, the average potential outcome in each experimental group can be used to identify each of the average causal effects defined above. In this example, the non-interference assumption is *not* violated by within-household spillovers because the experimental design accommodates the expanded schedule of potential outcomes. Although non-interference is ordinarily summarized in shorthand form as "no spillovers," a more accurate shorthand is "no unmodeled spillovers."[6] In this example, non-interference would be violated if voters were affected by mail sent to households other than their own, because the identification strategy presupposes no spillovers of this type.[7]

Be careful not to confuse non-interference with other concerns. One complication that arises when analyzing a multi-level experiment is how to estimate standard errors, hypothesis tests, and confidence intervals. Under simple random assignment, standard errors, hypothesis tests, and confidence intervals are calculated based on the assumption that each of the subjects is assigned independently to treatment or control. Here, housemates are assigned as pairs, and housemates may share similar potential outcomes. When simulating the sampling distribution of the estimated treatment effect using the methods introduced in Chapter 3, a researcher must take clustered assignment into account. This issue, however, is distinct from non-interference. Voting patterns may be correlated within households even when there are no spillover effects—for instance, both household members may share similar socioeconomic backgrounds and exposure to the same campaign advertisements and local organizing activity. The converse is true as well: spillovers could occur even in households where pairs of voters have uncorrelated potential outcomes. Non-interference and correlated potential outcomes are different and have different implications for estimation. Violations of non-interference lead to biased estimates, while correlated potential outcomes affect standard errors under cluster random assignment when clusters contain the same number of subjects.

Biased estimates due to interference are a much bigger concern and our focus here. In order to see how bias is introduced when spillovers are ignored, imagine that the experiment described above were conducted, but the researcher takes notice only of whether a given subject is targeted directly by the campaign when defining potential outcomes. The researcher defines the estimand as $\overline{Y}_{01} - \overline{Y}_{00}$, which is assumed to be the same as $\overline{Y}_{01} - \overline{Y}_{10}$ since spillovers are ignored. Instead of analyzing four distinct groups, the researcher combines those who receive mail (whose observed

6 One way to check the adequacy of an experimental research design is to compare the number of assigned experimental groups to the number of columns in the posited schedule of potential outcomes. There should be at least as many assigned groups as columns of potential outcomes.

7 Formally, this less restrictive version of the non-interference assumption would require that potential outcomes respond only to treatments d, which now refers to the four ways in which mail is distributed to the subject and the subject's housemate. Potential outcomes respond only to these treatments and not treatments administered outside the household, such that $Y_i(\mathbf{d}) = Y_i(d)$.

average rates of voting are $\hat{\overline{Y}}_{11}$ and $\hat{\overline{Y}}_{01}$, depending on whether their housemate receives mail) and those who do not (whose observed average rates of voting are $\hat{\overline{Y}}_{10}$ and $\hat{\overline{Y}}_{00}$, again depending on whether their housemate receives mail) in an effort to estimate the "effect" of mail, $\overline{Y}_{01} - \overline{Y}_{00}$. What happens when these two groups are compared? When there are actually no within-household spillovers, an experiment that involves randomly assigned groups of equal size generates an expected ATE of:

$$\frac{\overline{Y}_{01} + \overline{Y}_{11}}{2} - \frac{\overline{Y}_{00} + \overline{Y}_{10}}{2} = \overline{Y}_{01} - \overline{Y}_{00}, \tag{8.1}$$

which is unbiased. However, in the presence of spillovers, this naïve estimator may be upwardly or downwardly biased. The naive estimator tends to overestimate $\overline{Y}_{01} - \overline{Y}_{00}$ when $\overline{Y}_{11} - \overline{Y}_{01} > \overline{Y}_{10} - \overline{Y}_{00}$. In other words, when the spillover effects on those who receive mail themselves are greater than the spillover effects among those who receive no mail, ignoring household spillovers will lead to overestimates of the direct effect of mail on those whose housemates receive nothing. When the inequality is reversed, failure to account for spillovers leads to underestimates. In the absence of a multi-level design, the researcher might speculate that underestimates are more likely, since untreated voters are more likely to be influenced by their housemate's treatment. The attractive feature of a multilevel design is that it provides empirical estimates of quantities that would otherwise be the object of speculation.

Giving careful consideration to the non-interference assumption during the design phase of an experiment is helpful because it forces us to define precisely what we mean by the causal effect of interest. Suppose we increase police patrols in an effort to deter crime in targeted neighborhoods. Presumably, the estimand of interest is the difference between two potential outcomes in a given location: crime rates under the current mode of policing and crime rates in the wake of increased police presence. When displacement of crime is ignored, a comparison of treated and untreated areas no longer reveals the estimand of interest. Instead, the researcher answers a different and perhaps unintended causal question: how do crime rates change when an area that would otherwise receive displaced crime due to increased police surveillance of neighboring areas instead receives increased police surveillance?

Being alert to these issues during the design phase of an experimental project may enable the researcher to craft the experiment in ways that allow for the detection and measurement of spillover effects.[8] Multi-level designs shed light on spillovers by varying the degree of firsthand and secondhand exposure to the treatment. In the vaccination example, one could vary whether individuals receive a vaccine as well as varying rates of vaccination in different areas. Below, we discuss the design and analysis of experiments in which proximity is believed to determine the degree of spillover.

8 The appendix at http://isps.research.yale.edu/FEDAI presents an example of the complications that arise when one models spillovers in the context of an experiment that encounters noncompliance.

8.2 Spatial Spillover

To this point we have weakened the non-interference assumption just slightly to permit spillovers that are confined to voters residing at the same address. Instances where spillovers are confined to specific settings make for challenging but manageable research projects. Opportunities to study spillovers of this kind frequently arise when randomly assigned programs are targeted at specific social groups. For example, when girls are targeted for educational interventions in randomly selected schools, boys in the same schools may be affected indirectly.[9] When government subsidies encourage poor families in randomly selected villages to keep their children in grade school, educational outcomes change among children in the same villages whose families are ineligible for subsidies.[10] In each instance, non-interference is relaxed to allow for spillovers within households, schools, or villages, but spillovers across these boundaries are ignored.

In many applications, spillovers are not obviously confined to entities such as households, schools, or villages. A typical application involves treatments administered in various locations. The researcher wants to know whether proximity to treated locations affects outcomes among observations that are not treated directly themselves. Classic examples include health interventions designed to prevent some kind of ailment that can be transmitted from one person to another. The intervention is directed at randomly selected locations (e.g., villages), and researchers look not only at the direct effects of the treatment but also at whether untreated units are affected by their proximity to the nearest treated unit or the number of treated units within a certain radius.

At first glance, analyzing spatial spillover effects in the context of a randomized intervention seems straightforward. First, develop some sort of metric for proximity. Next, deploy treatments at randomly chosen locations, and compare the outcomes of units in the immediate vicinity of treatment to those farther away. Unfortunately, neither measurement nor estimation turns out to be straightforward.

Developing measures of proximity presupposes a model of how spillover effects are transmitted. If we are studying the transmission of social norms, Euclidean distance might be a good way to measure proximity if people are walking on foot but a poor way if they are communicating via cell phones. Sometimes physical proximity makes good sense as a causal explanation provided it is scaled in appropriate units. If the intervention is police surveillance of certain areas, a researcher may reason that what matters is not how far in miles the nearest police responder is but how far in minutes. In some cases, physical distance is just a shorthand way of characterizing the quantity of exposure. A relatively agnostic approach is to define proximity

9 Kremer, Miguel, and Thornton 2009.
10 Lalive and Cattaneo 2009.

according to whether a subject is located within a certain radius of a treated subject. Specifying this radius, however, is another source of uncertainty.[11]

Even when researchers are fairly confident about the way in which they have measured proximity, estimation presents some formidable challenges. A common mistake is to assume that exposure to spillover effects is determined by simple random assignment because treatments are determined by simple random assignment. Typically, units in a spatial experiment have very different probabilities of exposure to spillovers. Just as failure to account for varying assignment probabilities leads to biased estimation in the context of blocked designs (see Chapters 3 and 4), failure to account for varying probabilities of secondhand treatment leads to bias here.

In order to see how estimation problems arise, consider the very simple spatial arrangement depicted in Figure 8.1: locations A through F are evenly spaced on a number line ranging from 1 to 6. (In this example, we assume that the proper distance metric is known, so as to distinguish the estimation problem from the measurement problem that we just discussed.) Experimental subjects reside at locations A, B, C, D, and F (E is unoccupied). Suppose that just one of the locations is slated to receive treatment. For concreteness, imagine that the observations are rural villages, the intervention is the construction of a small clinic, and the outcome is a measure of each village's level of health. The potential outcomes listed in Table 8.2 show how each observation would respond to the location of the intervention. There are three types of potential outcomes: Y_{01} refers to the health level that would result if the village received the clinic; Y_{10} refers to the health level that would result if an adjacent village receives the clinic; and Y_{00} refers to the health level that would occur if neither the village nor its adjacent neighbors receive the treatment. Notice that the non-interference assumption now rules out spillover effects from non-adjacent locations.

FIGURE 8.1

Spatial location of five villages along a line

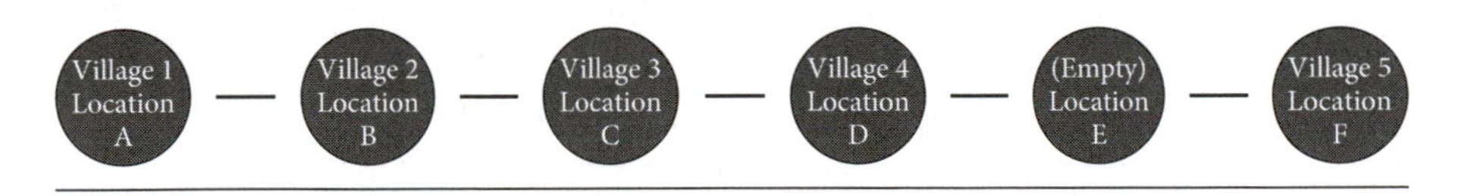

11 Ideally, one would let the data determine the rate at which the spillover effect decays with distance and the rate at which exposure to multiple treated units amplifies the spillover effect. The problem is that flexible models tend to generate imprecise estimates. In order to obtain results that are precise enough to be informative, the researcher may have to impose some structure on the specification. There is a fine line between introducing reasonable assumptions and imposing ad hoc stipulations. Readers of work in this genre must be attentive to the modeling decisions that drive the results and should look for evidence that conclusions about spillovers are robust to different sets of modeling assumptions.

TABLE 8.2

Five villages' potential health outcomes in response to the creation of a clinic

Village	Untreated (Y_{00})	Adjacent village treated (Y_{10})	Treated (Y_{01})
1	0	2	0
2	6	2	10
3	0	4	4
4	6	6	6
5	6	NA	3

NA indicates that according to the experimental design, Village 5 can never be adjacent to a treated village.

Table 8.2 reveals two important features of this example. First, some observations never manifest certain types of potential outcomes. For example, the subject at location F can never manifest a Y_{10} outcome because it can never be adjacent to a treated unit; because this potential outcome can never be observed, we exclude this subject from the definition of the average treatment effect $E[Y_{10} - Y_{00}]$—in effect, the estimand becomes the ATE among villages that could express the potential outcomes Y_{10} and Y_{00}. Second, the probability of assignment to each treatment condition varies from one observation to the next. As summarized in Table 8.3, the village at location A has a 0.20 probability of being exposed to spillovers from an adjacent treated location, whereas the village at location B has a 0.40 probability. As explained in Chapter 3, the appropriate way to handle different probabilities of exposure is to weight each observation by the inverse of the probability that it entered into its assigned experimental condition (which forces us to exclude subjects with zero probability).

For example, when the treatment occurs at location D, Villages 1, 2, and 5 express their Y_{00} potential outcomes, Village 4 expresses a Y_{01} potential outcome, and Village 3

TABLE 8.3

Potential outcomes expressed based on the location of the clinic and the probability that each village expresses each type of potential outcome

	Assigned treatment location							
Village	A	B	C	D	F	Pr(assignment to control)	Pr(assignment to spillover)	Pr(assignment to treatment)
1	Y_{01}	Y_{10}	Y_{00}	Y_{00}	Y_{00}	0.6	0.2	0.2
2	Y_{10}	Y_{01}	Y_{10}	Y_{00}	Y_{00}	0.4	0.4	0.2
3	Y_{00}	Y_{10}	Y_{01}	Y_{10}	Y_{00}	0.4	0.4	0.2
4	Y_{00}	Y_{00}	Y_{10}	Y_{01}	Y_{00}	0.6	0.2	0.2
5	Y_{00}	Y_{00}	Y_{00}	Y_{00}	Y_{01}	0.8	0	0.2

expresses a Y_{10} potential outcome. The weighted difference-in-means estimator of $E[Y_{01} - Y_{00}]$ adjusts for the fact that the probabilities of being untreated are 0.60, 0.40, and 0.80 for Villages 1, 2, and 5, respectively:[12]

$$\hat{E}[Y_{01} - Y_{00}] = \frac{\frac{6}{0.2}}{\frac{1}{0.2}} - \frac{\frac{0}{0.6} + \frac{6}{0.4} + \frac{6}{0.8}}{\frac{1}{0.6} + \frac{1}{0.4} + \frac{1}{0.8}} = 1.85. \quad (8.2)$$

The calculation in equation (8.2) excludes Village 3 because this village expresses the potential outcome Y_{10}. In addition to excluding Village 4, which is directly treated, the weighted difference-in-means estimate of $E[Y_{10} - Y_{00}]$ excludes observation 5, because it can never express a Y_{10} potential outcome:

$$\hat{E}[Y_{10} - Y_{00}] = \frac{\frac{4}{0.4}}{\frac{1}{0.4}} - \frac{\frac{0}{0.6} + \frac{6}{0.4}}{\frac{1}{0.6} + \frac{1}{0.4}} = 0.40. \quad (8.3)$$

The weighted difference-in-means estimator is not unbiased in small samples, but the bias in this example is fairly tame.[13] The true average effect of receiving the treatment at one's own location is defined for all five units: $E[Y_{01} - Y_{00}] = 1$. The average weighted difference-in-means across all possible randomizations is 0.81. The true average effect of spillovers from adjacent units is defined for four units at locations A, B, C, D, for which $E[Y_{10} - Y_{00}] = 0.5$. The average weighted difference-in-means across all possible randomizations is 0.33, which, again, is somewhat biased.

By comparison, the *unweighted* difference-in-means is severely biased. Its average estimate of $E[Y_{10} - Y_{00}]$ across all randomizations is −1.0, which suggests that spillovers have a negative effect, when in fact their effect is positive. The same goes for unweighted regression. Regressing outcomes on indicator variables for being treated and for being adjacent to a treated village produces the same estimates as unweighted difference-in-means. Although very elementary, this example has important practical implications. Even when we correctly specify the distance that spillover effects travel, our estimates may

12 Equation (8.2) divides by the sum of the weights, whereas the corresponding equations in Chapter 3 divided by weights that, by construction, summed to N.

13 This bias arises in small samples because the weighted N in the denominator of the weighted average varies across randomizations. An unbiased but imprecise estimator of the average spillover effect uses a weighted difference-in-totals:

$$\frac{1}{N}\left(\sum_{i=1}^{N} \frac{d_{10,i} Y_{10,i}}{p_{10,i}} - \sum_{i=1}^{N} \frac{d_{00,i} Y_{00,i}}{p_{00,i}}\right),$$

where $p_{10,i}$ refers to the probability that subject i is exposed to spillovers and $p_{00,i}$ refers to the probability of being in the control group. This estimator is similar to the weighted difference-in-means estimator but divides by N rather than the sum of the weights in each group. Applying this estimator to the data used in equation (8.3) yields an estimate of $1/4 * [4/0.4 - (0/0.6 + 6/0.4)] = -1.25$. An analogous estimator may be used to estimate the effect of receiving the treatment directly. Applying this estimator to the data used in equation (8.2) yields an estimate of 1.50. For more on this issue and how it plays out in the context of clustered assignment, see Middleton and Aronow 2011 and Samii and Aronow 2012.

be severely biased if we fail to take into account the probability that each unit is exposed to spillovers. Bias in the unweighted difference-in-means estimator stems from the same problem discussed in Chapters 2, 3, and 4: subjects' potential outcomes (see Table 8.2) are related to the probability that they receive the spillover treatment (see Table 8.3).

8.2.1 Using Nonexperimental Units to Investigate Spillovers

The random deployment of treatments sets the stage for a downstream analysis of spillover effects on subjects outside the experiment. Returning to our clinic example, imagine a set of locations to the left or right of the area depicted in Figure 8.1. These units in surrounding areas are nonexperimental in the sense that they were not eligible to receive the treatment themselves. These surrounding units are, however, subject to spillovers, and since the treatments are randomly assigned, exposure to spillovers is randomly assigned. If the surrounding units are sufficiently numerous, they may provide a more powerful test of spillovers than the experimental units themselves.

The statistical procedure for studying spillover effects directed at surrounding units is less complicated than the one described in the previous section. Because the surrounding units are not eligible for treatment, they permit us to estimate only the spillover effect, not the immediate treatment effect. In other respects, the estimation approach is the same. Our estimator must take into account the different probabilities of exposure to spillovers, which in turn requires us to stipulate how proximity is defined and measured.

When outcome measures are readily available, failure to investigate the effects of spillovers on nonexperimental units is like leaving money on the table. Regardless of whether the average spillover effect turns out to be large or small, the estimates provide useful substantive and methodological insights. The discovery of spillovers encourages follow-up research designed to investigate the mechanisms by which effects are transmitted and cautions researchers to design experiments in ways that are robust to non-interference violations. Conversely, evidence suggesting that spillover effects are negligible gives experimenters more leeway in designing studies because interference is no longer a pressing concern.

8.3 An Example of Spatial Spillovers in Two Dimensions

This section illustrates the estimation of spillovers for both experimental and non-experimental units when locations are arrayed in two-dimensional space. Our example is patterned after experiments that have attempted to gauge the effects of increasing police presence in "hotspots," or areas with high levels of criminal activity.[14] We begin by positing a full schedule of potential outcomes and illustrate how different modeling decisions affect the sampling distribution of the estimates.

We first analyze a set of experimental hotspots before attempting to detect spillovers in locations that were not part of the initial randomization. Consider an experiment conducted on 30 hotspots, 10 of which receive additional police patrols; the remaining 20 hotspots in the control group receive the usual amount of policing. The goal of the experiment is to assess the effects of the extra patrols on crime, in both the immediate vicinity of the patrol and the surrounding area. The researcher records the geographic location of each hotspot and measures the number of crimes in the month following the intervention.

Figure 8.2 displays the spatial arrangement of the 30 hotspots. The large circles indicate the half-kilometer radius surrounding each hotspot. The hypothetical schedule of potential outcomes in Table 8.4 includes columns for Y_{00} (the hotspot is not

FIGURE 8.2

Spatial arrangement of hotspots and spillover zones

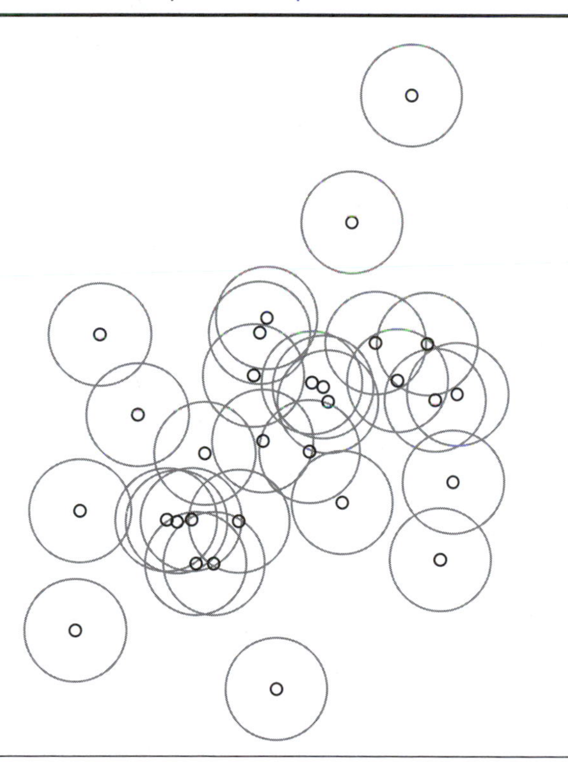

14 Examples of this type of experiment may be found in Sherman and Weisburd 1995; Weisburd and Green 1995; Ratcliffe et al. 2011; and Braga and Bond 2008.

TABLE 8.4

Potential and observed outcomes following a hypothetical police intervention in randomly selected locations

Hotspot	No. Hotspots within 500 meters	No. Hotspots within 750 meters	Y_{00}	Y_{01}	Y_{10}	Y_{11}	Assignment	Exposure	Y
1	1	2	30	25	35	23	0	00	30
2	0	0	10	5	15	3	0	00	10
3	3	5	60	55	65	53	0	10	65
4	0	0	10	5	15	3	1	01	5
5	3	6	70	65	75	63	0	10	75
6	0	0	10	5	15	3	0	00	10
7	0	1	20	15	25	13	0	00	20
8	1	3	40	35	45	33	0	10	45
9	0	0	10	5	15	3	1	01	5
10	0	5	60	55	65	53	1	01	55
11	0	0	10	5	15	3	0	00	10
12	0	0	10	5	15	3	0	00	10
13	1	5	60	55	65	53	1	11	53
14	3	6	70	65	75	63	0	10	75
15	2	6	70	65	75	63	0	00	70
16	1	5	60	55	65	53	0	00	60
17	2	7	80	75	85	73	0	00	80
18	2	6	70	65	75	63	0	00	70
19	2	3	40	35	45	33	1	01	35
20	5	6	70	65	75	63	0	10	75
21	1	5	60	55	65	53	1	11	53
22	2	3	40	35	45	33	0	00	40
23	2	6	70	65	75	63	0	00	70
24	3	5	60	55	65	53	1	01	55
25	0	0	10	5	15	3	1	01	5
26	1	4	50	45	55	43	1	01	45
27	0	0	10	5	15	3	1	01	5
28	2	6	70	65	75	63	0	00	70
29	0	0	10	5	15	3	0	00	10
30	1	5	60	55	65	53	0	00	60

treated, and there are no treated hotspots within 500 meters), Y_{01} (the hotspot is treated, and there are no treated hotspots within 500 meters), Y_{10} (the hotspot is not treated, but there is at least one treated hotspot within 500 meters), and Y_{11} (the hotspot is treated, and there is at least one treated hotspot within 500 meters). In our stylized example, these potential outcomes are crime rates generated by the following equations:

$$\begin{aligned} Y_{00} &= 10 + \mu_i, \\ Y_{01} &= 5 + \mu_i, \\ Y_{10} &= 15 + \mu_i, \\ Y_{11} &= 3 + \mu_i, \end{aligned} \tag{8.4}$$

where $\mu_i = 6$ (number of hotspots within 750 meters). You can think of μ_i as an unmeasured feature of each location that contributes to its crime rate. The motivation behind the four intercepts is that policing suppresses crime in the immediate area and displaces it to surrounding areas that are not themselves under heightened police surveillance. The means for the potential outcomes listed in Table 8.4 for Y_{00}, Y_{01}, Y_{10}, and Y_{11} are 43.3, 38.3, 48.3, and 36.3, respectively. In keeping with the intercepts in equation (8.4), police surveillance of the immediate area on average decreases crime, $E[Y_{01} - Y_{00}] = -5$; surveillance of an adjacent area increases crime, $E[Y_{10} - Y_{00}] = 5$; and the combination of immediate and neighboring surveillance decreases crime, $E[Y_{11} - Y_{00}] = -7$.

The last column in Table 8.4 illustrates the outcomes that would be observed based on a random assignment of 10 hotspots to the treatment group. The spatial pattern produced by this assignment is illustrated in Figure 8.3. The black dots indicate the treated hotspots, the grey dots are untreated hotspots that fall within a half kilometer radius of a treated hotspot, and the hollow dots are hotspots that are untreated and unaffected by the treatment of other hotspots.

How should we analyze these data? Let's first consider two ill-advised approaches. One is to ignore spillover and compare average outcomes between hotspots that are treated directly (those manifesting Y_{01} or Y_{11} potential outcomes) and hotspots that are untreated (those manifesting Y_{10} or Y_{00} potential outcomes) in order to recover the effect of police presence in the immediate vicinity. Not surprisingly, this approach is biased because it ignores spillovers. The average estimate that emerges from 100,000 simulated replications of this experiment is −7.3, which is not the true average effect, $E[Y_{01} - Y_{00}] = -5$. Another biased approach is to compare the unweighted sample estimates of $\overline{Y}_{10}$ (using the average outcome among untreated locations that were near a treated location) and $\overline{Y}_{00}$ (using the average outcome among untreated locations that were not near a treated location). On its face, this comparison seems sensible, but it ignores the fact that the spatial arrangement of the hotspots is related to potential outcomes and the probability of being assigned to either the untreated or spillover conditions varies across hotspots. The average estimate that emerges from 100,000 replications of this experiment is −3.8, which is far from $E[Y_{10} - Y_{00}] = 5$.

FIGURE 8.3
Spatial arrangement of treatment assignments

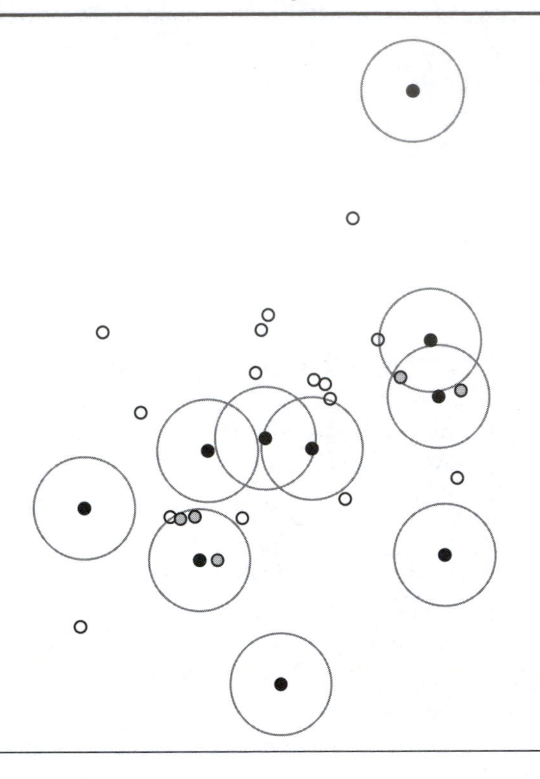

A more sensible estimation approach begins with the realization that spatial experiments expose clusters of subjects to treatment with varying probabilities. Although simple random assignment is used to assign treatment, subjects in different locations have different probabilities of being treated by spillovers. Some subjects are close to many potentially treated locations and therefore have a high probability of assignment to spillovers; other subjects are in outlying areas with low probability of exposure to spillovers. Clustering occurs because neighboring hotspots tend to be assigned as pairs or triplets to the same spillover (see Figure 8.3). Let's grapple first with varying probabilities of exposure to spillovers. Because the process of assigning subjects to experimental conditions is easily repeated, we can calculate the probability that each hotspot ends up in one of the four experimental groups. Table 8.5 reports the probabilities based on 100,000 random assignments. Table 8.5 reveals that 11 of the 30 observations have zero probability of being exposed to spillovers; these hotspots shed no light on spillover effects and can identify only the immediate effects of police presence. Across 100,000 simulated experiments, the average difference between the sample average of $\overline{Y}_{01}$ and $\overline{Y}_{00}$ within the subgroup

TABLE 8.5

Probabilities of assignment to each type of exposure (based on 5,000,000 simulated random assignments)

Hotspot	No. Hotspots within 500 meters	Prob(00)	Prob(01)	Prob(10)	Prob(11)
1	1	0.436	0.230	0.230	0.104
2	0	0.667	0.333	0.000	0.000
3	3	0.177	0.104	0.490	0.229
4	0	0.667	0.333	0.000	0.000
5	3	0.177	0.104	0.490	0.229
6	0	0.667	0.333	0.000	0.000
7	0	0.667	0.333	0.000	0.000
8	1	0.437	0.230	0.230	0.103
9	0	0.667	0.333	0.000	0.000
10	0	0.667	0.333	0.000	0.000
11	0	0.667	0.333	0.000	0.000
12	0	0.667	0.333	0.000	0.000
13	1	0.437	0.230	0.230	0.104
14	3	0.177	0.104	0.490	0.229
15	2	0.281	0.156	0.386	0.177
16	1	0.437	0.230	0.230	0.103
17	2	0.281	0.156	0.386	0.177
18	2	0.281	0.156	0.386	0.177
19	2	0.281	0.156	0.386	0.177
20	5	0.065	0.043	0.601	0.290
21	1	0.437	0.230	0.230	0.104
22	2	0.281	0.156	0.386	0.177
23	2	0.281	0.156	0.386	0.177
24	3	0.177	0.104	0.490	0.229
25	0	0.667	0.333	0.000	0.000
26	1	0.437	0.230	0.230	0.103
27	0	0.667	0.333	0.000	0.000
28	2	0.281	0.156	0.386	0.177
29	0	0.667	0.333	0.000	0.000
30	1	0.437	0.229	0.230	0.104

that could never be exposed to spillovers is −4.9, which is close to the true ATE of −5 for these 11 observations.

The remaining 19 hotspots may be classified into four categories based on their probabilities of receiving each treatment. In order to adjust for different probabilities of treatment, we use inverse probability weighting (see Chapter 3). For example, if Hotspot #1 were assigned to the direct treatment (01) condition, it would be weighted by a factor of $1/0.230 = 4.348$, because it has a probability of 0.230 of being assigned to this condition. Weighting the 19 hotspots in this way and estimating treatment effects for 100,000 simulated experiments yields an average estimate of −4.2 when we compare sample estimates of $\overline{Y}_{01}$ and $\overline{Y}_{00}$. Comparing sample estimates $\overline{Y}_{10}$ and $\overline{Y}_{00}$ gives an average estimate of 7.0, which is higher than the true parameter of 5.0; comparing $\overline{Y}_{11}$ and $\overline{Y}_{00}$ gives an average estimate of −3.8, which is higher than the true parameter of −7.0. This biased pattern of results occurs in part because the assignment of spillovers is clustered, and as noted in Chapter 3, clustered assignment can produce biased results when the number of clusters is small. Another source of bias is the fact that the weights used by our weighted estimator vary depending on which units are assigned to treatment. All else being equal, these biases diminish as the number of observations increases, and here the bias largely disappears when we change the example to include 300 hotspots, 100 of which are assigned to the treatment group. In this expanded example, comparing sample estimates of $\overline{Y}_{01}$ and $\overline{Y}_{00}$ yields an average estimate of −4.9; comparing sample estimates of $\overline{Y}_{10}$ and $\overline{Y}_{00}$ yields an average estimate of 5.2, and comparing sample estimates of $\overline{Y}_{11}$ and $\overline{Y}_{00}$ yields an average estimate of −6.7. These estimates are close to the true parameters used to generate the simulation. Increased sample size not only improves precision but also facilitates unbiased estimation.[15]

Due to the logistical challenges of orchestrating experiments that target locations, it may be impossible to expand the number of experimental locations. Fortunately, units that are ineligible for treatment provide "free" supplementary results. In order to see the value of these additional observations, consider Figure 8.4, which shows 100 locations (marked with an X) in relation to the 30 hotspots (the small circles) that comprise the experimental sample. The black dots indicate treated hotspots, and the larger circles indicate the range of spillovers from these treated hotspots. To simulate the potential outcomes for these supplementary observations, we assume that locations that lie farther than 500 meters from a treated location have potential outcomes $Y_{00} = \textit{Number of hotspots within 750 meters}$, while those locations that lie within 500 meters of at least one treated hotspot have outcomes $Y_{10} = \textit{Number of hotspots within 750 meters} + 5$.

15 Estimation of confidence intervals may be done in the usual manner. Create a full schedule of potential outcomes by stipulating that the effects of direct and spillover treatments are equal to the estimated average treatment effects. Repeat the random assignment 100,000 times, and report the 0.025 and 0.975 quantiles.

FIGURE 8.4

Spatial arrangement of nonexperimental locations in relation to spillover zones

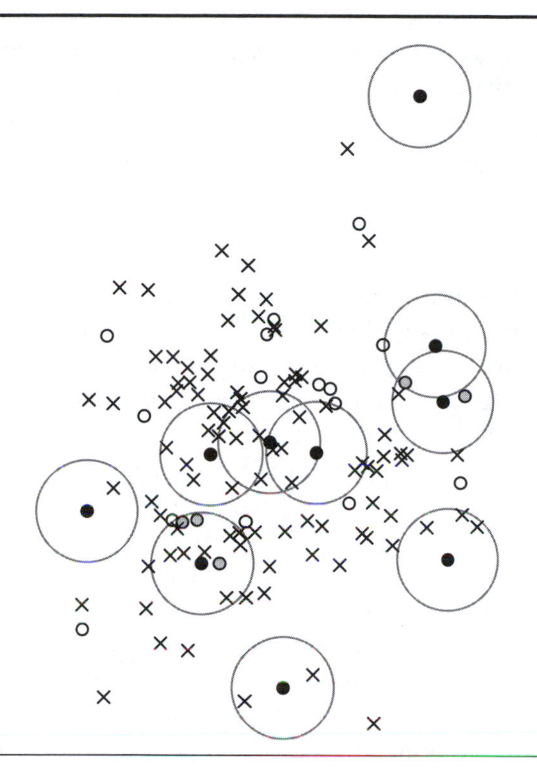

In order to estimate the effect of spillover for these 100 supplementary units, we first determine the probability of exposure to spillovers by repeating the randomization 100,000 times and recording each location's probability of being assigned to the 00 or 10 condition.[16] These probabilities are then used to weight each of the observations, and the weighted averages, $\hat{\bar{Y}}_{10}$ and $\hat{\bar{Y}}_{00}$, are used to estimate the average spillover effect. Replicating this experiment 100,000 times reveals that this difference-in-means is nearly unbiased, yielding an average estimate of 5.2. In this example, nonexperimental locations make a useful contribution to the study of spillovers. The large number of 100 supplementary locations means that clustering does not cause much bias, and the spillover effect is estimated with reasonable precision (the standard error across 100,000 experiments is 1.1).

16 Some locations have no proximal hotspots and therefore have no chance of being treated. These units must be excluded from the analysis. In effect, the estimand is the average effect of spillovers for those locations that are potentially subject to them.

Notice that in analyzing this experiment, we estimated average treatment effects separately for three distinct sets of observations (experimental locations that cannot be exposed to spillovers, experimental locations that may be exposed to spillovers, and nonexperimental locations that may be exposed to spillovers). This approach serves two purposes. First, estimating average effects by subgroup provides a more detailed picture of the results. Second, it prevents us from accidentally introducing modeling assumptions when estimating and interpreting treatment effects for different groups. The average spillover effect for experimental units, for example, may be different from the average spillover effect for nonexperimental units. We can always combine our estimates in order to estimate the ATE for the two groups combined or for the population from which both groups are drawn (see Chapter 11), but that requires an explicit modeling decision.

This example raises two further questions. First, what happens when we posit the wrong model of spillovers? Suppose, for example, that we assume that the true spillover radius is 250 meters rather than 500 meters. Or suppose that we thought that the spillovers were a function of the distance to the nearest treated unit or some complicated weighted average of distance to all treated units. Our exploration of various models suggests that applying the wrong model of spillovers to these data produces distorted results, but the basic pattern of spillovers tends to be recovered.[17] The broader answer, however, is that model specification is an important source of uncertainty, and researchers should present results from a variety of different models so that readers can appreciate the extent to which substantive conclusions turn on modeling decisions.

Second, rather than using simple random assignment, how might a researcher design this experiment to better gauge the immediate and spillover effects of police patrols? The answer to this question depends in part on how agnostic the researcher wishes to be about the underlying model of spillovers. If the aim is to make relatively few modeling assumptions, one attractive approach is to focus attention on pairs of adjacent hotspots, preferably ones that are relatively isolated from other hotspots. These pairs could be assigned to 00, 01, 10, and 11 conditions, akin to the direct mail and voting example in section 8.1. The paired design solves the problem of bias due to clustering, because each of the clusters contains the same number of units. Another design recommendation is to use hypothetical randomizations to predict the probability that experimental or supplementary locations will be exposed to proximal treated units. This exercise provides a sense of how clustering is likely to play out in your application and may suggest ways of restricting the set of admissible randomizations so that the resulting distribution of 00, 01, 10, and 11 treatments is as informative as possible. Where data on outcomes exist for each unit over time, you can assess the precision with which immediate and spillover effects are likely to be estimated by conducting mock interventions and assessing the sampling distribution of your

17 Readers may review these models and results at http://isps.research.yale.edu/FEDAI.

estimates. None of these design strategies eliminates the guesswork associated with spatial spillovers, but at least you will be better prepared to grapple with the estimation challenges that arise.

8.4 Within-Subjects Design and Time-Series Experiments

Within-subjects experimentation refers to studies where a single person or entity is tracked over time and random assignment determines *when* a treatment is administered. For example, a series of public health experiments have placed signs at subway entrances encouraging patrons to walk upstairs rather than ride the escalators. Signage is posted or removed each month, and outcomes are measured by the number of people using the stairway during each monthly period.[18] In this example, the subject is the location (the stairway) as assessed in each period, so six months furnish six observations. The subscript i now refers to each period in which outcomes are measured. In contrast to between-subjects design, which tests the effect of an intervention by assigning N subjects to treatment or control and comparing outcomes in each group, within-subjects design assigns a single subject to periods of treatment or control and compares outcomes under each condition.

The allure of within-subjects designs is their capacity to generate precise treatment estimates with a single subject. Precision derives in part from the fact that individuals or entities are compared to themselves, which means that background attributes hold constant. Another feature of most within-subjects experiments is that over-time comparisons are made under controlled conditions, so that the principal source of over-time variation is the treatment. Consider a simple example of a physiological experiment testing the relationship between the amount of light and the extent of pupil dilation. A within-subjects design would place subjects in a dim room and randomly introduce varying amounts of light, measuring the eye's reaction. Sometimes the aim of experimenting on a particular person or place is to learn about the ATE for that subject. For example, you might be interested in whether embarking on a program of diet and exercise will lower your cholesterol; the ATE among some broader collection of people is a secondary concern. Or you might be interested in a policy intervention's effects on your country of origin, not a collection of countries. On the other hand, sometimes the primary research aim is to generalize to a broader class of subjects, in which case the experiment may need to be replicated on subjects that are representative of some broader population.

Outside of psychology, social scientists rarely use within-subjects designs. Instead, researchers who track subjects over time tend to rely on *interrupted*

---

18 Brownell, Stunkard, and Albaum 1980.

time-series, which record outcomes for a single unit before and after an intervention occurs. The timing of the intervention is not determined randomly. For example, a study of a one-month police "crackdown" in a particular section of a large city tracked robberies and burglaries in that area before, during, and after the intervention.[19] This type of non-randomized design is often criticized on the grounds that the timing of the intervention might coincide with trends in the data having nothing to do with the treatment itself. Police crackdowns might occur during spikes in crime, and "regression to the mean" alone might produce an apparent treatment effect even if the treatment had no effect. Or police crackdowns might occur in anticipation of an ever-worsening crime wave, which might lead the researcher to mistakenly conclude that the crackdown exacerbated criminal activity.

Random assignment overcomes some, but not all, of the complications associated with over-time comparisons. In order to appreciate the challenges of extracting causal inferences when interventions are timed randomly, let's take a closer look at the simplest possible research design: an experiment in which a single subject is observed over two periods. In the period prior to the study, the subject is untreated. A coin flip determines whether the subject receives the treatment in period 1 or period 2.

Because potential outcomes could conceivably depend on when interventions occur (past, present, or future), we must again expand our notation to reflect possible treatment sequences. For our two-period example, the relevant potential outcomes are denoted $Y_{t-1,t,t+1}$, where the $t-1$ subscript refers to whether a treatment is administered in the preceding time period, t refers to the current time period, and $t+1$ refers to the next time period. In order to reduce clutter, we omit the i subscript, but each potential outcome should be understood to refer to a specific point in time. This notation allows us to use Y_{010} in period i, for example, to refer to that period's potential outcome if the subject were untreated in the preceding period, treated in the current period, and untreated in the next period. The sole virtue of this otherwise cumbersome notation is that it helps illuminate the assumptions that must be invoked when drawing inferences from within-subjects experiments.

In order to illustrate the notation and the identification problem that a within-subjects design presents, Table 8.6 displays a schedule of potential outcomes under each possible treatment state. For concreteness, suppose the outcome is pupil dilation (in millimeters), and the treatment is 1 if the subject is given eye drops to induce dilation. Because a subject is treated just once, the relevant potential outcomes are Y_{000} (untreated in all periods), Y_{001} (untreated until the subsequent period), Y_{010} (treated in the current period), and Y_{100} (treated in the previous period and untreated thereafter). The example assumes that Y_{000} drifts upward over time as the subject becomes accustomed to laboratory surroundings. In Period 1, Y_{000} is 3 but rises to 4 in Period 2. During the period in which the subject receives the eye drops, pupil dilation

19 Novak, Hartman, Holsinger, and Turner 1999.

TABLE 8.6

Potential outcomes for a two-period within-subjects experiment

	Y_{000}	Y_{001}	Y_{010}	Y_{100}
Period 1	3	3	7	NA
Period 2	4	NA	7	4

Entries are pupil dilation widths, in millimeters. The entries marked NA indicate outcomes that can never be observed because the subject cannot be treated before Period 1 or after Period 2.

increases; Y_{010} is 7 in Period 1 and Period 2. The average treatment effect of the eye drops during the period in which they are administered is $E[Y_{010} - Y_{000}] = 3.5$.

The question is whether this ATE can be recovered from a within-subjects experiment in which a coin flip determines when the treatment is administered. Our subject is tracked over two periods: what do we learn from comparing the outcome in the treated period to the outcome in the untreated period? If the random assignment places Period 1 into control and Period 2 into treatment, we observe the potential outcomes Y_{001} in Period 1 and Y_{010} in Period 2. If random assignment places Period 1 into treatment and Period 2 into control, we observe the potential outcomes Y_{010} in Period 1 and Y_{100} in Period 2. Neither random allocation provides information about Y_{000}.

In order to identify the ATE defined above, we will need to make two additional assumptions. The *no-anticipation assumption* says that potential outcomes are unaffected by treatments that are administered in the future, implying $Y_{001} = Y_{000}$. Substantively, this assumption means that pupil dilation is not caused by the prospect of receiving eye drops or exposure to the preparatory work that goes with it (such as medical staff dimming lights ahead of time.) A second assumption, *no persistence*, requires that potential outcomes in one period are unaffected by treatments administered in prior periods. This assumption implies that $Y_{100} = Y_{000}$. Sometimes experimental procedures make this strong assumption palatable. When conducting physiological experiments on heart rate or pupil dilation, experimenters often include "washout periods" between experimental sessions so as to allow the previous period's effects to dissipate. This design strategy is analogous to the use of buffer zones separating spatially proximal units in order to dampen geographic spillovers.

These assumptions are the within-subjects equivalents of the usual non-interference assumption, which holds that one subject is unaffected by the treatments administered to other subjects. In a within-subjects experiment, a given subject is observed at multiple points in time, effectively becoming multiple subjects. The no-anticipation assumption says that current potential outcomes are unaffected by future treated states, and the no-persistence assumption says that treatments administered in one period do not spill over to the next. When both assumptions hold, the within-subjects design provides unbiased estimates of the ATE. Referring to the potential outcomes in Table 8.6, we see that the two random allocations generate estimates of 4 and 3, which on average recover the true ATE.

Ordinarily, we look to random assignment in order to relax strong assumptions, but the random assignment assumption, which plays such a crucial role in between-subjects designs, plays a limited role here. By specifying a random point at which the intervention occurs, we rule out the possibility that the timing of the intervention is systematically related to potential outcomes in different periods. The more time periods and interventions, the less likely it is that a randomly assigned pattern of treatment assignment will coincide with over-time trends in potential outcomes. Random assignment might also help prevent the subject from anticipating when a treatment will be administered. But random assignment does little to shore up the shakiest assumption, the stipulation that treatments have no anticipated or persistent effects. In a between-subjects design, the control group provides a glimpse of what the treatment group would have looked like in its untreated state. In a within-subjects design, the untreated outcome after a treatment occurs may not resemble the untreated potential outcome that would have been observed in the absence of current or previous treatment.

In sum, despite random assignment, within-subjects designs stand or fall on supplementary non-interference-type assumptions. Researchers invoking these assumptions should carefully consider whether they are justified theoretically and empirically. Do other between-subjects experiments in related domains reveal persistent effects of this treatment? If so, does the experimental design allow for a long enough wash-out period to negate this source of bias? With regard to the no-anticipation assumption, is there reason to believe that potential outcomes in one period could be affected by what a subject expects to encounter in subsequent periods? These questions cannot be answered in the abstract; they depend on the details of a particular application. Within-subjects studies place a premium on aspects of experimental design other than randomization, such as the experimenter's ability to minimize the intrusion of external factors or to introduce washout periods after treatments. Perhaps these requirements account for the fact that within-subjects designs are more common in the physical sciences than the social sciences. When social experiments track subjects over time, they tend to blend elements of between-subjects and within-subjects design, as illustrated in the next section.

8.5 Waitlist Designs (Also Known as Stepped-Wedge Designs)

Experiments that track subjects over time as they move from untreated to treated conditions combine elements of within-subjects and between-subjects designs. At a given point in time, subjects are compared between experimental conditions; over time, a given subject contributes a series of outcomes and in that sense is counted as multiple observations.

Waitlist designs play a valuable scientific and diplomatic role in field experimentation. Their scientific value comes from their ability to track treatment effects among several subjects as they play out over time. Waitlists play a diplomatic role because they overcome the problem of withholding treatment from a control group. In this design, every subject is treated eventually; random assignment determines when they receive treatment. Waitlist designs are sometimes called stepped-wedge designs, due to the visual pattern of treatment assignments over time. As subjects gradually move from control to treatment, the chart of treatment assignments looks like a series of steps.

In order to illustrate the special features of waitlist designs, we present a stylized version of an experiment that assessed the effects of televised campaign advertisements on support for a gubernatorial candidate.[20] In this example, ads are aired during three successive weeks, and outcomes (support for the gubernatorial candidate as gauged by opinion polls) are assessed at the end of each week. Two media markets are randomly assigned to air ads for three weeks starting in week 1. Two markets air ads for two weeks starting in week 2. Two markets air ads for one week starting in week 3. And two media markets air no ads at all. Given eight media markets and three starting periods, the number of possible random allocations is:

$$\frac{8!}{2!6!} \cdot \frac{6!}{2!4!} \cdot \frac{4!}{2!2!} = 2{,}520. \quad (8.5)$$

Suppose that the schedule of potential outcomes reflects two types of treatments: whether a media market airs an ad during the current period and whether a media market airs an ad during the preceding period. Adapting the notation from the previous section, we assume that there are just three relevant potential outcomes: Y_{00} (untreated during preceding and current periods), Y_{01} (untreated during the preceding period but treated during the current period), and Y_{11} (treated in both the preceding and current periods). Given the design, we never observe the potential outcome Y_{10} because media markets never cease to run ads once they start. By assuming that potential outcomes respond only to ads aired in the current and preceding weeks, we invoke the no-anticipation assumption from section 8.5, allowing us to ignore treatments that are allocated after the current period. We also ignore treatments that were allocated two weeks prior, which is tantamount to assuming that advertising's effects "wash out" entirely after two weeks. These are important substantive assumptions that will later affect the way we analyze the results.

The hypothetical schedule of outcomes for this experiment is presented in Table 8.7. Each media market has nine columns of potential outcomes, reflecting the type of treatment that could be received and the timing of each treatment. One column is empty because no Y_{11} values can ever be observed in the first week.

20 Gerber, Gimpel, Green, and Shaw 2011.

TABLE 8.7

Potential outcomes from the advertising waitlist experiment

Y_{00}		Week	
Market	**1**	**2**	**3**
1	4	5	3
2	7	5	4
3	1	2	4
4	4	3	2
5	3	3	3
6	8	4	3
7	2	3	4
8	3	1	2

Y_{01}		Week	
Market	**1**	**2**	**3**
1	7	9	4
2	8	7	7
3	1	2	8
4	4	7	10
5	4	3	2
6	10	6	9
7	2	7	6
8	5	1	2

Y_{11}		Week	
Market	**1**	**2**	**3**
1	NA	9	4
2	NA	8	7
3	NA	2	10
4	NA	8	10
5	NA	3	2
6	NA	8	10
7	NA	7	6
8	NA	2	3

Substantively, the outcomes depicted in the table may be thought of as the gubernatorial candidate's lead in the polls. The table has been crafted so that $E[Y_{01} - Y_{00}]$, the immediate average treatment effect, is 2 and $E[Y_{11} - Y_{00}]$, the combination of immediate and lagged average treatment effects, is 3.

In order to estimate these average treatment effects, we randomly assign the eight media markets to the four experimental groups. Table 8.8 shows the randomly assigned treatments and the resulting observed outcomes. How shall we analyze the results? As usual, we first consider an ill-advised approach. Suppose one were to naively compare average outcomes of treated units to the average outcomes of untreated units. This approach is biased because it ignores the fact that the probability

TABLE 8.8

Advertising waitlist experiment's random assignments and observed outcomes

Assigned treatment condition

	Week		
Market	1	2	3
1	01	11	11
2	00	00	01
3	00	01	11
4	00	00	01
5	00	00	00
6	01	11	11
7	00	00	00
8	00	01	11

Observed outcomes

	Week		
Market	1	2	3
1	7	9	4
2	7	5	7
3	1	2	10
4	4	3	10
5	3	3	3
6	10	8	10
7	2	3	4
8	3	1	3

of assignment to treatment varies from week to week (media markets are much more likely to be assigned to treatment in the final week than in the first week).[21] It is also prone to bias because it ignores lagged effects, treating Y_{11} and Y_{01} as though they were identical.

A better approach begins by calculating the probability of being assigned to each treatment during each period. These probabilities are displayed in Table 8.9. In order to obtain unbiased estimates of the two average treatment effects, one applies inverse probability weights as in Chapter 3.[22] Treated units are weighted by the inverse of the probability of being treated; control units are weighted by the inverse of the probability of being in control. For the random assignment in Table 8.8, calculation of the immediate effect is as follows:

$$\hat{E}[Y_{01} - Y_{00}] = \frac{\frac{7+10}{0.25} + \frac{2+1}{0.25} + \frac{7+10}{0.25}}{\frac{2}{0.25} + \frac{2}{0.25} + \frac{2}{0.25}} - \frac{\frac{7+1+4+3+2+3}{0.75} + \frac{5+3+3+3}{0.50} + \frac{3+4}{0.25}}{\frac{6}{0.75} + \frac{4}{0.50} + \frac{2}{0.25}} = 2.72. \quad (8.6)$$

In order to estimate the combined immediate and lagged effect, we must restrict our attention to the second and third weeks, because this type of treatment cannot occur in the first week.

$$\hat{E}[Y_{11} - Y_{00}] = \frac{\frac{9+8}{0.25} + \frac{4+10+10+3}{0.50}}{\frac{2}{0.25} + \frac{4}{0.50}} - \frac{\frac{5+3+3+3}{0.50} + \frac{3+4}{0.25}}{\frac{4}{0.50} + \frac{2}{0.25}} = 4.13. \quad (8.7)$$

TABLE 8.9

Probabilities of assignment to treatment condition, by week

	Week		
Treatment Condition	**1**	**2**	**3**
Pr(00)	0.75	0.50	0.25
Pr(01)	0.25	0.25	0.25
Pr(11)	0	0.25	0.50

21 Across all possible random assignments, this estimator on average yields an estimate of 0.58, which is far from the actual treatment effects of 2 or 3.

22 In Chapter 3, inverse probability weighting was unbiased in the case of blocked random assignment because the sum of the weights was always the same, regardless of which units were assigned to treatment. Here, the situation is analogous; the stepped wedge design implies a fixed sum of weights, regardless of which units are assigned to treatment. By contrast, under clustered assignment, the sum of the weights may vary depending on which clusters are assigned to treatment.

We could generate a comparable $\hat{E}[Y_{01} - Y_{00}]$ by restricting our attention to weeks 2 and 3. Similarly, we could calculate $\hat{E}[Y_{11} - Y_{00}]$ without imposing the assumption that treatment effects disappear after two weeks by restricting our attention to week 2. We revisit these calculations in exercise 8.11.

The remaining task is to estimate confidence intervals for these two ATEs. Using our estimates of 2.72 and 4.13, we may use the observed data in Table 8.8 to fill in the implied schedule of potential outcomes under the assumption of constant treatment effects. In order to construct our confidence intervals, we form the sampling distribution of the estimated ATEs across all 2,520 possible random allocations. The 0.025 and 0.975 quantiles of these sampling distributions give 95% intervals ranging from (0.23, 5.51) for the immediate treatment effect and (0.89, 7.40) for the immediate and lagged treatment effect. This example illustrates an attractive property of waitlist designs, namely, their ability to extract statistically meaningful estimates from relatively small numbers of subjects. Bear in mind, however, that the power of the design derives in large part from the substantive assumptions about non-interference that are built into the model of potential outcomes.

SUMMARY

Coming to grips with the non-interference assumption forces experimental researchers to specify how potential outcomes respond to all possible random assignments and, in turn, how treatment effects will be defined. Debates about interference inevitably come down to judgment calls—assuming that one can only observe one outcome per subject, the question of how many distinct potential outcomes there may be for each subject is inevitably a matter of speculation. Perhaps for that reason, researchers have a tendency to let knotty problems of interference recede into the background. The message of this chapter is that this tendency should be resisted. Even if a researcher cannot answer every interference-related concern, there is value in carefully defining the estimand and assessing whether the experimental design identifies the parameter of interest.

When evaluating whether an experiment violates the non-interference assumption, think about the ways in which subjects' potential outcomes might by affected by whether other subjects are treated. For example, are subjects likely to discuss or transmit the treatment? Are treatments slated for the treatment group likely to withdraw resources that would otherwise flow to the control group? Are subjects at different points in time counted as distinct observations? Because interference is a possible source of bias, uncertainty about the nature and extent of interference diminishes the credibility of experimental findings.

In order to address these concerns empirically, researchers turn to experimental designs that relax the assumption that subject i is unaffected by whether others are treated. These designs share a common feature: they randomly assign varying degrees of secondhand exposure. Experimental designs that randomize treatments at multiple levels do not altogether eliminate assumptions about spillovers or displacement; rather, they invoke certain assumptions that restrict the way in which spillover effects are transmitted. For example, the model of within-household spillovers presented in section 8.1 stipulated that secondhand effects do not travel across households. The model of spatial spillovers discussed in section 8.3 assumed that spillovers decay as they radiate geographically. The model of temporal spillovers discussed in section 8.5 assumed that spillovers dissipate after two weeks. The challenge is to design ever more flexible experiments that are less reliant on potentially fallible modeling assumptions.

A further challenge is statistical. When subjects are clustered geographically or along other dimensions such as proximity in social networks, exposure to spillovers tends to vary, even when subjects are randomly assigned to treatment. If subjects have different probabilities of exposure to spillovers and these probabilities are related to potential outcomes, data analysis requires special care. Simply comparing average outcomes among those exposed to spillovers and those not exposed is prone to bias. A better approach is to use the randomization procedure to simulate the probabilities of exposure to spillovers and to compare weighted means. Even this method is prone to bias when subjects fall into a small number of geographic clusters that are all exposed to spillover effects together. These complications suggest that researchers should be on the lookout for research opportunities that facilitate unbiased and precise estimation, such as a sample that comprises pairs of geographically proximal units.

On the bright side, the payoff from conducting a well-executed study of spillover effects is often considerable. Secondhand influence has enormous practical implications. If the effects of a treatment are found to resonate through social networks, their cumulative effects may be many times greater than their effects on those who receive the treatment directly. There is tremendous theoretical interest in the extent of interpersonal transmission of effects and, more generally, in the conditions under which effects are transmitted. A great deal hangs in the balance methodologically as well. A finding suggesting a lack of spillover or displacement should not be dismissed as a mere null finding; if true, the lack of interference between units greatly simplifies the task of estimating causal effects.

Finally, the non-interference assumption should be borne in mind whenever one draws generalizations from experimental results. To return again to the crime displacement example, an intervention that intensifies policing in treatment areas may push crime into control precincts, producing an apparent treatment effect. If the objective is to measure the average difference between treated and untreated potential outcomes, this experiment yields a biased result. But suppose your aim is to estimate an ATE defined as follows: the average potential outcome if precincts were treated with heightened police patrols minus the average potential outcome if neighboring precincts were treated with heightened police patrols. Redefining the estimand in this way means that interference is no longer a source of bias. Whether this estimand is useful depends on the broader scientific and practical questions one seeks to address. One should be cautious about concluding from a study of this kind that intensified policing reduces crime, because the evidence is also consistent with the conclusion that intensified policing merely displaces crime. On the other hand, if one lives in a neighborhood that may or may not receive intensified police protection, the experiment provides an instructive answer to the question of how your crime rates are likely to differ from your neighbors' if you receive the treatment and they do not.

SUGGESTED READINGS

Technical discussions of the estimation problems posed by interference may be found in Aronow and Samii (2012). Multilevel designs are discussed by Hong and Raudenbush (2006) and Hudgens and Halloran (2008). For an accessible discussion of non-interference as applied to within-subjects designs, see Rubin (2001). For a discussion of within-subjects design from a randomization inference perspective, see Dugard, File, and Todman (2012). Hussey and Hughes (2007) consider various modeling approaches for analyzing stepped-wedge designs that involve clustered assignment.

EXERCISES: CHAPTER 8

1. Important concepts:
   (a) Interpret the expression $Y_i(\boldsymbol{d}) = Y_i(d)$ and explain how it conveys the non-interference assumption.
   (b) Why are experiments that involve possible spatial spillover effects (such as the example described in section 8.4) said to involve "implicit" clustered assignment?
   (c) In what ways might a within-subjects design violate the non-interference assumption?
   (d) What are the attractive properties of a waitlist (or stepped-wedge) design?
2. National surveys indicate that college roommates tend to have correlated weights. The more one roommate weighs at the end of the freshman year, the more the other freshman roommate weighs. On the other hand, researchers studying housing arrangements in which roommates are randomly paired together find no correlation between two roommates' weights at the end of their freshman year. Explain how these two facts can be reconciled.
3. Sometimes researchers are reluctant to randomly assign individual students in elementary classrooms because they are concerned that treatments administered to some students are likely to spill over to untreated students in the same classroom. In an attempt to get around possible violations of the non-interference assumption, they assign classrooms as clusters to treatment and control, and administer the treatment to all students in a classroom.
   (a) State the non-interference assumption as it applies to the proposed clustered design.
   (b) What causal estimand does the clustered design identify? Does this causal estimand include or exclude spillovers within classrooms?

4. Recall from Chapter 3 (exercise 9) the field experiment conducted by Camerer in which he selected pairs of similar horses running in the same race and randomly placed large wagers on one of them to see if his bets affected the amount of money that other bettors placed on both horses.
   (a) Define the potential outcomes in Camerer's study. What non-interference assumption is invoked?
   (b) What is the causal parameter that this study identifies?
5. In their study of spillover effects, Sinclair, McConnell, and Green sent mailings to randomly selected households encouraging them to vote in an upcoming special election.[23] The mailings used a form of "social pressure," disclosing whether the targeted individual had voted in previous elections. Because this type of mail had proven to increase turnout by approximately 4–5 percentage points in previous experiments, Sinclair, McConnell, and Green used it to study whether treatment effects are transmitted across households. Employing a multi-level design, they randomly assigned all, half, or none of the members of each nine-digit zip code to receive mail. For purposes of this example, we focus only on households with one registered voter. The outcome variable is voter turnout as measured by the registrar of voters. The results are as follows. Among registered voters in untreated zip codes, 1,021 of 6,217 cast ballots. Among untreated voters in zip codes where half of the households received mail, 526 of 3,316 registered voters cast ballots. Among treated voters in zip codes where half of the households received mail, 620 of 2,949 voted. Finally, among treated voters in zip codes where every household received mail, turnout was 1,316 of 6,377.
   (a) Using potential outcomes, define the treatment effect of receiving mail addressed to subject i.
   (b) Define the "spillover" treatment effect of being in a zip code where varying fractions of households are treated.
   (c) Propose an estimator for estimating the firsthand and secondhand treatment effects. Show that the estimator is unbiased, explaining the assumptions required to reach this conclusion.
   (d) Based on these data, what do you infer about the magnitude of the mailing's direct and indirect effects?
6. Using the potential outcomes from the clinic example in Table 8.2, calculate the following estimates.
   (a) Estimate $E[Y_{01} - Y_{00}]$ for the random assignment that places the treatment at location A.
   (b) Estimate $E[Y_{10} - Y_{00}]$ for the random assignment that places the treatment at location A, restricting the sample to the set of villages that have a non-zero probability of expressing both of these potential outcomes.
   (c) In order to make a more direct comparison between these two treatment effects, estimate $E[Y_{01} - Y_{00}]$, restricting the sample to the same set of villages as in part (b).
7. Lab experiments sometimes pair subjects together and have them play against one another in games where each subject is rewarded financially according to the game's outcome. One such game involves making monetary contributions to a public good (e.g., preserving

23 Sinclair, McConnell, and Green 2012.

the environment); the game can be arranged such that each player gains financially if both of them make a contribution, but each player is better off still if they contribute nothing while their partner in the game makes a contribution. The treatment is whether the pair of players is allowed to communicate prior to deciding whether to contribute. Suppose that a lab experimenter recruits four subjects and assigns them randomly as pairs to play this game. The outcome is whether each player makes a contribution: Y_i is 1 if the player contributes and 0 otherwise. Each player has three potential outcomes: Y_{0i} is the outcome if players are prevented from communicating, Y_{1i} is the outcome if a player communicates with another player who is "persuasive," and Y_{2i} is the outcome if a player communicates with another player who is "unpersuasive." The table below shows the schedule of potential outcomes for four players, two of whom are persuasive and two of whom are unpersuasive.

Subject	Type	Y_{0i}	Y_{1i}	Y_{2i}
1	Persuasive	0	1	0
2	Persuasive	1	1	0
3	Unpersuasive	0	0	0
4	Unpersuasive	1	1	1

(a) Calculate the average treatment effect of $Y_{1i} - Y_{0i}$. Calculate the average treatment effect of $Y_{2i} - Y_{0i}$.

(b) How many random pairings are possible with four subjects?

(c) Suppose that the experimenter ignores the distinction between Y_{1i} and Y_{2i} and considers only two treatment conditions: the control condition prevents communication between pairs of players, and the treatment condition allows communication. Call the observed outcomes in the communication condition Y^*_{1i}. Across all possible random pairings of subjects, what is the average difference-in-means estimate when the average Y^*_{1i} is compared to the average Y_{0i}? Does this number correspond to either of the two estimands defined in part (a)? Does it correspond to the average of these two estimands?

(d) What is the probability that a persuasive subject is treated by communicating with an unpersuasive subject? What is the probability that an unpersuasive subject is treated by communicating with an unpersuasive subject?

(e) Briefly summarize why a violation of the non-interference assumption leads to biased difference-in-means estimates in this example.

(f) Would bias be eliminated if the experimenter replicated this study (with four subjects) each day and averaged the results over a series of 100 daily studies?

(g) Would bias be eliminated if the experimenter assembled 400 subjects at the same time (imagine 100 subjects for each of the four potential outcomes profiles in the table) and assigned them to pairs? Hint: Answer the question based on the intuition suggested by part (d).

8. Concerns about interference between units sometimes arise in survey experiments. For example, surveys sometimes administer a series of "vignettes" involving people with different attributes. A respondent might be told about a low-income person who is

randomly described as white or black; after hearing the description, the respondent is asked to rate whether this person deserves public assistance. The respondent is then presented with a vignette about a second person, again randomly described as white or black, and asked about whether this person deserves public assistance. This design creates four experimental groups: (a) two vignettes about blacks, (b) two vignettes about whites, (c) a black vignette followed by a white vignette, and (d) a white vignette followed by a black vignette. Each respondent provides two ratings.

(a) Propose a model of potential outcomes that reflects the ways that subjects might respond to the treatments and the sequences in which they are presented.

(b) Using your model of potential outcomes, define the ATE or ATEs that a researcher might seek to estimate.

(c) Suggest an identification strategy for estimating this (these) causal estimand(s) using the data from this experiment.

(d) Suppose a researcher analyzing this experiment estimates the average "race effect" by comparing the average evaluation of the white recipient to the average evaluation of the black recipient. Is this a sound approach?

9. Use data from the hotspots experiment in Table 8.4 (these data are also available at http://isps.research.yale.edu/FEDAI) and the probabilities that each unit is exposed to immediate or spillover treatments (Table 8.5) to answer the following questions:

(a) For the subset of 11 hotspot locations that lie outside the range of possible spillovers, calculate $E[Y_{01} - Y_{00}]$, the ATE of immediate police surveillance.

(b) For the remaining 19 hotspot locations that lie within the range of possible spillovers, calculate $E[Y_{01} - Y_{00}]$, $E[Y_{10} - Y_{00}]$, and $E[Y_{11} - Y_{00}]$.

(c) Use the data at http://isps.research.yale.edu/FEDAI to estimate the average effect of spillover on nonexperimental units. Note that your estimator must make use of the probability that each unit lies within 500 meters of a treated experimental unit; exclude from your analysis any units that have zero probability of experiencing spillovers.

10. A doctoral student conducted an experiment in which she randomly varied whether she ran or walked 40 minutes each morning.[24] In the middle of each afternoon over a period of 26 days, she measured the following outcome variables: (1) her weight (minus a constant, for privacy's sake), (2) her score in a game of Tetris, (3) her mood on a 0–5 scale, with 5 being the most pleasant, (4) her energy level on a 0–5 scale, with 5 being the most energetic, and (5) whether she answered correctly a randomly selected problem from the math section of the GRE. Outcomes are missing for days 13 and 17. The data are listed below.

Day	Run	Weight	Tetris	Mood	Energy	Appetite	GRE
1	1	21	11092	3	3	0	1
2	1	21	14745	3	1	2	0
3	0	20	11558	3	3	0	1
4	0	21	11747	3	1	1	1

24 Hough 2010.

5	0	21	14319	2	3	3	1
6	1	19	7126	3	2	0	1
7	0	20	16067	3	4	0	0
8	0	20	3939	3	2	0	1
9	1	21	28230	4	2	0	0
10	0	21	17396	4	4	1	1
11	1	20	36152	1	4	0	0
12	0	20	16567	4	4	1	1
13	0	20					
14	1	18	11853	4	2	0	1
15	1	18	20433	4	2	2	1
16	1	18	20701	3	4	0	0
17	0	20					1
18	1	19	17509	3	3	1	1
19	0	21	9779	3	3	1	0
20	0	22	18598	3	3	1	1
21	1	20	36665	2	3	0	1
22	0	21	8094	4	3	1	1
23	1	19	48769	2	5	0	0
24	1	20	22601	4	4	1	1
25	1	19	37950	4	4	0	1
26	1	20	56047	4	4	0	1

(a) Suppose you were seeking to estimate the average effect of running on her Tetris score. Explain the assumptions needed to identify this causal effect based on this within-subjects design. Are these assumptions plausible in this instance? What special concerns arise due to the fact that the subject was conducting the study, undergoing the treatments, and measuring her own outcomes?

(b) Estimate the effect of running on Tetris score. Use randomization inference to test the sharp null hypothesis that running has no immediate or lagged effect on Tetris scores.

(c) One way to lend credibility to within-subjects results is to verify the no-anticipation assumption. Use the variable Run to predict the Tetris score *on the preceding day*. Presumably, the true effect is zero. Does randomization inference confirm this prediction?

(d) If Tetris responds to exercise, one might suppose that energy levels and GRE scores would as well. Are these hypotheses borne out by the data?

11. Return to the stepped-wedge advertising example in section 8.6 and the schedule of assigned treatments in Table 8.8.

(a) Estimate $E[Y_{01} - Y_{00}]$ by restricting your attention to weeks 2 and 3. How does this estimate compare to the estimate of $E[Y_{11} - Y_{00}]$ presented in the text, which is also identified using observations from weeks 2 and 3?

(b) Estimate $E[Y_{11} - Y_{00}]$ without imposing the assumption that treatment effects disappear after two weeks by restricting your attention to week 2.

CHAPTER 9

Heterogeneous Treatment Effects

To this point, we have focused primarily on the challenges of estimating average treatment effects. Rarely is it plausible to suppose that every observation responds to an intervention in exactly the same way. Whether interventions occur in the realm of economics, politics, criminology, education, or public health, it is difficult to think of an experiment whose treatment could be expected to exert the same effect on every individual, organization, or region. In this chapter, we move beyond *average* effects and investigate *variability* in treatment effects.

Understanding variation in treatment effects can be of enormous practical and scientific value. From a practical standpoint, those deciding whether to implement a program or policy want to know which individuals will be most responsive and under what conditions. An advertising campaign that appeals to teenagers may drive away older consumers. Job training programs that increase earnings among those who volunteer for them may have no effect on those who are required to undergo job training in order to maintain their eligibility for public assistance. Government efforts to increase compliance with tax laws may succeed or fail depending on whether taxpayers are in professions that allow them to conceal their income.

From a scientific standpoint, the study of variation in treatment effects can provide important clues about why a treatment does or does not work. For example, experiments have suggested that the effects of policies that allow parents to select where their children will attend public school hinge on parents' values. Experimental data show that allowing greater choice benefits children whose parents prioritize academic quality when selecting schools; the policy does not improve educational outcomes among children whose parents select schools based on proximity to home.[1] This finding implies, and subsequent research confirms, that one way to improve educational outcomes among children whose parents would ordinarily choose schools based on proximity is to provide information that encourages parents to place greater

1 Hastings, Kane, and Staiger 2006.

What you will learn from this chapter:

1. Methods of detecting heterogeneous treatment effects that make minimal modeling assumptions.
2. How to model heterogeneous treatment effects using regression, and how to interpret the results.
3. The advantages of multi-factor experiments that shed light on the conditions under which a treatment effect is strong or weak.

weight on academic quality when selecting schools.[2] Some of the most intriguing findings in social science involve discoveries that indicate the special combination of ingredients that make for a large or small treatment effect.

When comparing treatment effects across subgroups and settings, the challenge is to interpret the results in a rigorous scientific manner, so that conclusions are not driven by sampling variability or ad hoc assumptions. Often, the researcher senses that treatment effect variation lies buried in the data but lacks sufficient precision to pinpoint the source of variation reliably. One of the hardest things to admit as a researcher is that a study's conclusions remain murky and tentative. The investigation of treatment effect variation has acquired a shady reputation because researchers have latched onto findings that later proved to be ephemeral when subjected to further experimental testing. This chapter encourages exploration while urging researchers to interpret their discoveries with caution until confirmed by follow-up experiments.

This chapter discusses a series of approaches for investigating treatment effect variation. The starting point is a review of the inherent problems of inference that a researcher confronts when attempting to study heterogeneous effects. We consider what can be learned from statistical techniques that make minimal assumptions when searching for heterogeneity. These techniques may serve as a useful initial step, giving the researcher a preliminary sense of whether a more theory-guided investigation of heterogeneity is likely to prove fruitful. We next discuss two of these more structured approaches. One approach involves the study of how treatment effects vary across different values of the covariates, which is sometimes termed the study of *treatment-by-covariate interactions* or *subgroup analysis*. Another approach deploys designs that simultaneously vary treatments and the experimental context in which they are deployed or received. We show how data generated by either approach may be analyzed using the tools discussed in previous chapters, noting some of the modeling uncertainties that attend the investigation of treatment effect heterogeneity.

2 Hastings and Weinstein 2008.

9.1 Limits to What Experimental Data Tell Us about Treatment Effect Heterogeneity

Suppose we conduct an experiment involving N subjects, m of whom are randomly assigned to the treatment condition. For each subject, the treatment effect is defined as the difference between the treated and untreated potential outcomes, $\tau_i \equiv Y_i(1) - Y_i(0)$. Suppose that noncompliance is not an issue; no one in the control group is treated, and everyone in the treatment group receives the treatment. We observe the $Y_i(0)$ potential outcomes for the control group and the $Y_i(1)$ potential outcomes for the treatment group.

Treatment effect heterogeneity refers to the variance of the treatment effect τ_i across subjects. We seek to estimate $\text{Var}(\tau_i)$ and, in particular, to test whether $\text{Var}(\tau_i) > 0$. If we find evidence of treatment effect heterogeneity, our next step is to investigate the conditions under which treatment effects are large or small.

The study of heterogeneity presents more formidable estimation problems than the study of average treatment effects. The variance in treatment effects may be expressed in terms of the variances of the potential outcomes and their covariance:

$$\text{Var}(\tau_i) = \text{Var}(Y_i(1) - Y_i(0)) = \text{Var}(Y_i(1)) + \text{Var}(Y_i(0)) - 2\text{Cov}(Y_i(1), Y_i(0)). \quad (9.1)$$

The empirical problem is that an experiment does not furnish the information necessary to estimate each of the components of $\text{Var}(\tau_i)$. The data from an experiment shed light on the *marginal distribution* of potential outcomes. The observed values of $Y_i(0)$ from the control group enable us to estimate the mean of the overall distribution of $Y_i(0)$, its variability, skewness, and higher moments. The same can be said of the observed values of $Y_i(1)$ from the treatment group; we can use these m observations to make inferences about the distribution of $Y_i(1)$ values across all N subjects. In terms of equation (9.1), our data provide sample estimates of $\text{Var}(Y_i(0))$ and $\text{Var}(Y_i(1))$. On the other hand, because we never observe both the $Y_i(0)$ and $Y_i(1)$ potential outcomes for any subject, we lack any direct information about the *joint distribution* of these potential outcomes. We do not know, for example, whether high values of $Y_i(0)$ tend to coincide with high or low values of $Y_i(1)$. We therefore cannot estimate $\text{Cov}(Y_i(0), Y_i(1))$.

A simple example underscores the consequences of our inability to link the observed $Y_i(0)$ values to the unobserved $Y_i(1)$ values and vice versa. Suppose our experiment had six experimental subjects, and we observed outcomes {1, 2, 3} in the control group and {4, 5, 6} in the treatment group. The estimated ATE is 3. But how variable is the treatment effect across subjects? Reasoning that the $Y_i(0)$ values in the control group have the same distribution as the $Y_i(0)$ values in the treatment group, we could pair each of the observed $Y_i(1)$ values with one of the observed $Y_i(0)$ values.

But which values should be paired together? There are three possible $Y_i(0)$ values for each $Y_i(1)$ value. If we assembled the potential outcomes such that the pairings were {(1,4), (2,5), (3,6)}, the treatment effect would be 3 for all subjects and $\text{Var}(\tau_i) = 0$. On the other hand, if we arranged the pairs {(1,6), (2,4), (3,5)}, the treatment effects would be 5, 2, and 2, implying that $\text{Var}(\tau_i) = 2$. Without knowing how to arrange the pairing of potential values, we cannot calculate $\text{Var}(\tau_i)$.

9.2 Bounding Var(τ_i) and Testing for Heterogeneity

Although experimental data do not enable us to estimate the value of $\text{Var}(\tau_i)$, we can use this arbitrary pairing procedure to estimate bounds suggesting how large or small $\text{Var}(\tau_i)$ may be. Establishing bounds involves pairing values of $Y_i(0)$ and $Y_i(1)$ such that the implied $\text{Cov}(Y_i(0), Y_i(1))$ is as large or as small as possible. When there are equal numbers of observations in the treatment and control groups, these pairings are easily arranged. First, sort the observed values of $Y_i(0)$ in ascending order. Next, sort the observed values of $Y_i(1)$ in ascending order. Pair the first sorted $Y_i(0)$ value with the first sorted $Y_i(1)$ value, the second sorted $Y_i(0)$ value with the second sorted $Y_i(1)$ value, and so on. The differences between these paired potential outcomes generate estimates of τ_i when the sample covariance $\widehat{\text{Cov}}(Y_i(0), Y_i(1))$ is as large as possible. The variance of these $\hat{\tau}_i$ provides a lower bound for $\text{Var}(\tau_i)$. A lower bound that is substantially greater than zero suggests treatment effect heterogeneity. To estimate an upper bound for $\text{Var}(\tau_i)$, reverse the procedure: pair observed values of $Y_i(0)$ sorted in ascending order with observed values of $Y_i(1)$ sorted in descending order. The variance of the implied $\hat{\tau}_i$ provides an upper bound for $\text{Var}(\tau_i)$. These upper and lower bounds are estimates and therefore subject to sampling variability, but they indicate what the data at hand tell us about $\text{Var}(\tau_i)$.[3]

When does this sorting method produce bounds that are narrow enough to be useful? This method tends to be most useful when variation in $Y_i(0)$ is low. Consider the limiting case in which all of the $Y_i(0)$ values are identical: the covariance between $Y_i(0)$ and $Y_i(1)$ will be zero regardless of how the values are sorted. The practical implication is that this approach works well when researchers pre-screen subjects in an effort to come up with relatively homogeneous $Y_i(0)$ values. The success of pre-screening can be assessed by examining the variance of Y_i outcomes in the control group. Another option is to administer a pre-test (X_i) that is highly predictive

3 Because the values of $Y_i(0)$ and $Y_i(1)$ that we actually observe are random samples of all $Y_i(0)$ and $Y_i(1)$ for all subjects, the bounds are subject to sampling variability and do not test the null hypothesis of no heterogeneity. Heckman, Smith, and Clements (1997) and Djebbari and Smith (2008) propose a simulation-based test of the null hypothesis that the lower bound on the variance of the treatment effect is zero. We discuss a simpler test that compares variances in the treatment and control groups.

BOX 9.1

Estimating Bounds on the Amount of Treatment Effect Heterogeneity

Treatment effect heterogeneity may be summarized in terms of the variance of the subject-level treatment effect, τ_i. As equation (9.1) suggests, estimating $\text{Var}(\tau_i)$ requires an estimate of the covariance between $Y_i(0)$ and $Y_i(1)$, but this covariance cannot be estimated directly because we only observe one of the two potential outcomes for each subject.

Heckman, Smith, and Clements (1997) propose a method for estimating bounds for $\text{Var}(\tau_i)$. Their method pairs the $Y_i(0)$ values observed in the control group with the $Y_i(1)$ values observed in the treatment group in order to estimate the maximum and minimum covariance between $Y_i(0)$ and $Y_i(1)$. The method is easy to implement when the treatment and control groups contain the same number of subjects. Suppose you observe outcomes {1, 2, 3} in the control group and {4, 5, 6} in the treatment group. In order to estimate $\text{Var}(\tau_i)$ assuming the maximum covariance between $Y_i(0)$ and $Y_i(1)$, sort the $Y_i(0)$ observed in the control group in ascending order, and sort the $Y_i(1)$ observed in the treatment group in ascending order. Pairing the treatment and control outcomes {(1,4), (2,5), (3,6)} allows us to estimate τ_i for each subject and a lower bound for $\text{Var}(\tau_i)$. In order to estimate the upper bound, sort the $Y_i(0)$ values in ascending order and the $Y_i(1)$ values in descending order. The resulting pairings {(1,6), (2,5), (3,4)} allow us to estimate the upper bound for $\text{Var}(\tau_i)$. If the number of subjects in treatment and control differ, pair each percentile within the treatment distribution to its corresponding percentile within the control distribution.

of untreated potential outcomes ($Y_i(0)$). By rescaling the outcome variable so that it represents a change score ($Y_i - X_i$), one reduces the variability in outcomes and therefore narrows the range of possible values of $\text{Cov}(Y_i(0) - X_i, Y_i(1) - X_i)$. The result can be a marked improvement in the tightness of the bounds.

Another statistical technique for detecting heterogeneity is to test the null hypothesis that $\text{Var}(\tau_i) = 0$. What sort of evidence would lead us to reject this hypothesis? Consider the implications of comparing $\text{Var}(Y_i(1))$ to $\text{Var}(Y_i(0))$, both of which can be estimated from experimental data. Because

$$\text{Var}(Y_i(1)) = \text{Var}(Y_i(0) + \tau_i) = \text{Var}(Y_i(0)) + \text{Var}(\tau_i) + 2\text{Cov}(Y_i(0), \tau_i), \quad (9.2)$$

the equality $\text{Var}(Y_i(1)) = \text{Var}(Y_i(0))$ holds when:

$$\text{Var}(\tau_i) = -2\text{Cov}(Y_i(0), \tau_i). \quad (9.3)$$

Under the null hypothesis that τ_i is constant across subjects, both sides of equation (9.3) are zero, since the covariance between a variable and a constant is zero.[4] Thus, rejecting the hypothesis $\text{Var}(Y_i(1)) = \text{Var}(Y_i(0))$ means rejecting the hypothesis that $\text{Var}(\tau_i) = 0$.

In order to illustrate the procedure for testing the hypothesis that $\text{Var}(Y_i(1)) = \text{Var}(Y_i(0))$, let's revisit the teacher incentives experiment discussed in Chapter 4.[5] Recall that elementary school teachers in treatment schools were paid bonuses according to their students' academic performance on standardized tests. In the actual experiment, 100 schools were assigned to the treatment group and 100 to the control group. Between the baseline test (prior to the intervention) and the one year follow-up test, the average school in the treatment group showed an improvement of 11.70 points, as compared to 8.20 points in the control group. This estimated ATE is far greater than we would expect by chance ($p < 0.01$). Our focus here is whether treatment effects vary, as suggested by a visual inspection of the distribution of scores in each group (Figure 9.1). The dispersion in change scores appears to differ by experimental group: the treatment group's variance is 91.20, as compared to 59.29 in the control group. In order to test the null hypothesis that $\text{Var}(\tau_i) = 0 \mid \widehat{\text{ATE}}$, we calculate the p-value of the observed absolute difference-in-variances of $|91.20 - 59.29| = 31.91$. The procedure is as follows. We first create the full schedule of potential outcomes by assuming a constant treatment effect equal to the estimated ATE of $11.70 - 8.20 = 3.50$. Next, we perform 100,000 random assignments and record the estimated absolute difference-in-variances for each. Only 8,766 of the 100,000 estimates are greater than or equal to the observed estimate of 31.91, implying a p-value of 0.088. We cannot reject the null hypothesis that the ATE is 3.50 for all subjects at the $p < 0.05$ level, but this borderline p-value nevertheless hints that offering cash incentives to teachers may produce effects that vary from school to school.

The virtue of using bounds and randomization inference to explore treatment effect heterogeneity is that these methods impose few assumptions and require no additional data. Unfortunately, in practice, this minimalist approach suffers from two limitations. First, bounds on the variance of τ_i tend to be wide, and tests of equal variances tend to lack power. Because the researcher supplies no theoretical or empirical guidance about the subgroups that harbor different treatment effects, this assessment of heterogeneity may be unable to furnish a decisive answer, as we saw in the previous example. For this reason, estimating the bounds of $\text{Var}(\tau_i)$ and testing whether $\text{Var}(\tau_i) = 0$ may be regarded as a preliminary step en route to more structured assessments of treatment effect heterogeneity. A significant test statistic in this preliminary phase encourages further investigation of heterogeneity, but even a

4 The condition may also be satisfied in special cases where the covariance between $Y_i(0)$ and τ_i is negative, so equality between $\text{Var}(Y_i(1))$ and $\text{Var}(Y_i(0))$ does not rule out treatment effect heterogeneity.

5 Muralidharan and Sundararaman 2011.

FIGURE 9.1

Distribution of outcomes for treatment and control groups in the teacher incentives experiment

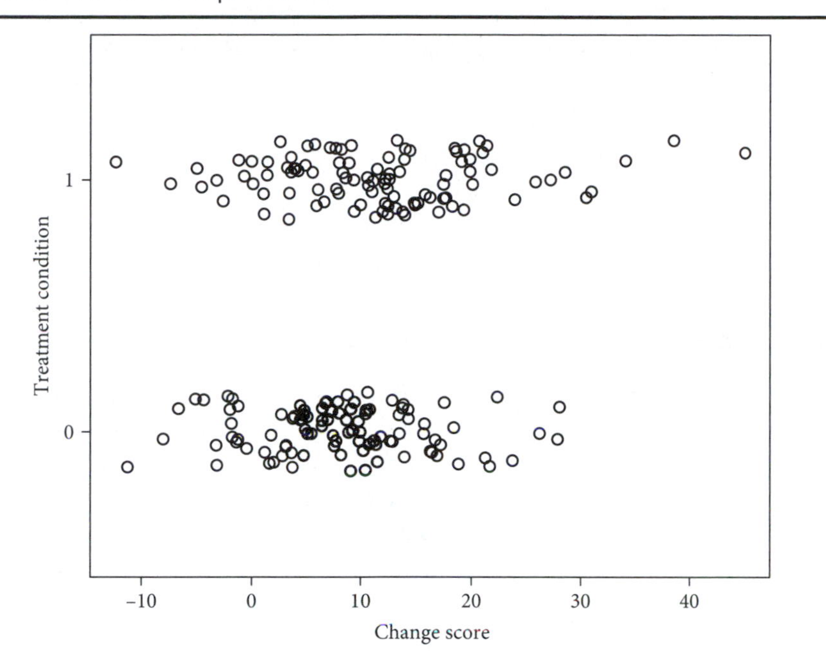

Source: Muralidharan and Sundararaman 2011. The plotted circles have been jittered to make it easier to see each observation.

marginal test statistic may lead to important discoveries when more powerful tests are conducted.

Second, the methods presented above are suited to continuous outcomes, such as tipping rates, income, or test scores. Binary outcomes, such as voting, must be analyzed in a different manner. Treatment effects are almost certainly heterogeneous when outcomes are binary—if a treatment increases voter turnout by 5 percentage points, it must be the case that some (but not all) of the subjects were induced to vote. Because binary outcomes contain so little information, placing bounds on the degree of heterogeneity often leads to uninformative results unless one imposes additional assumptions, such as the stipulation that all treatment effects are greater than (or less than) zero.[6] The special considerations that arise in the analysis of binary outcomes are discussed in the online appendix (http://isps.research.yale.edu/FEDAI). Those working with binary outcomes will typically need to use the exploratory and design-based approaches discussed in the next section.

6 This assumption is known as *monotone treatment response* (Manski 1997).

9.3 Two Approaches to the Exploration of Heterogeneity: Covariates and Design

Suppose that a researcher's theoretical intuitions or preliminary hypothesis tests suggest the presence of heterogeneous effects. This section discusses two ways of investigating the conditions under which τ_i varies. The first method partitions the subjects into subgroups in order to examine *treatment-by-covariate interactions*, or variation in ATEs from subgroup to subgroup. The second method introduces additional interventions in order to assess *treatment-by-treatment interactions*, or variation in ATEs across other randomly assigned treatment conditions.

9.3.1 Assessing Treatment-by-Covariate Interactions

Experimental researchers routinely use covariates to assess hypotheses about when and why treatment effects vary. These covariates partition subjects into groups based on subjects' individual attributes or on attributes of the context in which an experiment occurs. In effect, the researcher conducts a miniature experiment within each subgroup. The ATE within a subgroup is called a *conditional average treatment effect*, or CATE. When researchers speak of interaction effects between a treatment and one or more covariates, they are referring to the difference between CATEs.

The study of interactions is often guided by theoretical intuitions about when a treatment is likely to be most effective. To draw on examples from previous chapters, one might partition Indian villages according to the amount of money they historically allocated to water sanitation in order to test whether the budgetary effects of randomly appointing women to village leadership roles are greater where women villagers' concerns were neglected in the past. This analysis, which grows out of theoretical arguments about the expression of pent-up villager demands for policy change,

BOX 9.2

Definition: Interaction Effect

An interaction effect refers to the change in treatment effect that occurs from one subgroup to the next. Subgroups may be defined using covariates or randomly assigned factors. For example, if in a study of voter mobilization the treatment were door-to-door canvassing, the subgroups could be defined in terms of covariates such as voters' age or party registration, or they could be defined based on another randomly assigned factor, such as whether the voter is also sent mail.

BOX 9.3

Definition: Conditional Average Treatment Effect (CATE)

The CATE is the ATE for a defined subset of the subjects. For example, the ATE among middle-aged male professors would be a CATE.

compares two CATEs: one for villages where past allocations to water sanitation were low and another where they were high. Similarly, theoretical arguments about attitude change might inspire subgroup comparisons when interventions expose people to new ideas and information. Covariates measured by the baseline survey of Pakistani Muslims may be used to test whether the Hajj's effects on attitudes toward people from other countries are most pronounced among subjects who previously had little exposure to other peoples and cultures. Again, the analysis compares the CATE among those subjects who previously had little exposure to the CATE among those who had higher levels of exposure.[7] So long as partitions are based on pre-treatment covariates, these investigations provide an unbiased assessment of the difference between two CATEs.

In order to see how this type of statistical analysis plays out in practice, let's return to the teacher incentives experiment that we analyzed in section 9.2. What individual or contextual attributes might contribute to variation in τ_i? The researchers who conducted this study considered two classes of covariates, the attributes of students and the attributes of their teachers.

Interestingly, the researchers find little evidence that CATEs vary by student characteristics, such as sex, ethnicity, or parents' affluence or literacy. Consider, for example, what happens when we partition schools by the average level of parent literacy.[8] Among the 100 schools with pre-treatment literacy scores below the median, the estimated CATE is $11.14 - 7.83 = 3.31$. Among the 100 schools with literacy scores above the median, the estimated CATE is $12.26 - 8.57 = 3.69$. Lacking a clear expectation about which CATE should be greater, we conduct a two-tailed test to assess whether the *difference* in CATEs could have occurred by chance.

7 Because the Hajj experiment involves two-sided noncompliance, there is further reason to suspect treatment effect heterogeneity. The effect of assigned treatment (the ITT) is heterogeneous so long as the ATE among Compliers is nonzero, since (under the excludability assumption) the ITT among Always-Takers and Never-Takers is zero.

8 This analysis could also be conducted at the individual level, conditioning on each student's own parents' literacy. Because the experiment used school-level clustered assignment and the number of students in each literacy category varies by school, this analysis requires a more complex hypothesis test that takes clustered assignment into account. For ease of exposition, we use school-level data and partition the sample by school-level attributes.

The null hypothesis is that the CATEs in both groups are equal to the estimated ATE. In order to test this hypothesis using randomization inference, we form the full schedule of potential outcomes by assuming a constant ATE of 3.50 and assign subjects randomly to treatment and control groups. We calculate the difference between the two estimated CATEs and repeat the exercise 100,000 times to see how often one obtains an observed difference at least as large as $|3.69 - 3.31| = 0.38$. In this case, the observed CATEs are not significantly different, as $p = 0.88$.

This hypothesis test may also be conducted using regression. The F-statistic compares the sum of squared residuals from two models, a null model that assumes that both subgroups share the same ATE and an alternative model that allows the CATEs to differ. The null model is:

$$Y_i = a + bI_i + cP_i + u_i, \tag{9.4}$$

where I_i is an indicator variable scored one if the school receives the incentives treatment and where P_i is an indicator variable scored one if the school's level of parental literacy is above the median. This null model posits just a single ATE, represented by the parameter b. The alternative model posits two CATEs, one for schools with low levels of parental literacy and another for schools with high levels. The two-CATEs hypothesis is, in effect, the regression model in equation (9.4) with an additional parameter:

$$Y_i = a + bI_i + cP_i + dI_iP_i + u_i. \tag{9.5}$$

The two CATEs become apparent when we condition on parental literacy. When parental literacy is low ($P_i = 0$), the CATE is b:

$$Y_i = a + bI_i + u_i. \tag{9.6}$$

When parental literacy is high ($P_i = 1$), the CATE is $b + d$:

$$Y_i = a + bI_i + c + dI_i + u_i = (a + c) + (b + d)I_i + u_i. \tag{9.7}$$

The parameter d is called the *interaction effect* because it represents the change in CATEs that occurs when parental literacy changes. We now have two "nested" models because the null model in equation (9.4) is a special case of equation (9.5), where d is assumed to be zero.[9] The F-statistic compares the sum of squared residuals from these two nested models:

$$F = \frac{\dfrac{\textit{SSR under the null hypothesis} - \textit{SSR under the alternative hypothesis}}{\textit{Number of parameters in the alternative model} - \textit{Number of parameters in the null model}}}{\dfrac{\textit{SSR under the alternative hypothesis}}{N - \textit{Number of parameters in the alternative model}}}. \tag{9.8}$$

9 The null model, which imposes the restriction that $d = 0$, is sometimes called the restricted model. The alternative model, which treats d as a parameter to be estimated, is called the unrestricted model.

Recall that the sum of squared residuals is a measure of a regression model's predictive accuracy; the smaller the SSR, the better the model fit. The numerator in equation (9.8) is zero when the sum of squares from the alternative regression model, which allows for interactions, is exactly the same as the sum of squares under the null model, which ignores interactions. If, by allowing for different ATEs in different subgroups, the alternative model reduces the SSR, the numerator in equation (9.8) will be positive. In this example,

$$\hat{F} = \frac{\frac{14856.1-14854.4}{4-3}}{\frac{14854.4}{200-4}} = 0.02. \tag{9.9}$$

In order to assess the p-value of this statistic, we generate the full schedule of potential outcomes assuming that $\hat{b} = 3.50$ from the null model is the treatment effect for all schools. We repeat the random assignment, estimate the null and alternative regressions, and calculate the F-statistic. When this procedure is replicated 100,000 times, we find that 88% of the F-statistics generated under the null model are greater than the observed F-statistic of 0.02, implying that $p = 0.88$.[10] The F-test generated the same p-value as testing the difference-in-CATEs. The advantage of using regression is that it easily extends to tests involving several categories or variables, as illustrated below.

The same statistical approach may also be used to assess whether treatment effects vary according to teacher characteristics, such as their years of teaching experience, education, salary, and teaching style (as assessed prior to the intervention). The results suggest that treatment effect heterogeneity is a function of teacher attributes. Comparing CATEs for high and low values of each variable, the researchers find that treatment effects are largest for teachers who have lower salaries, fewer years of experience, higher levels of education, and more active teaching styles.[11] This discovery is potentially important for policy makers and social scientists because it provides some hints about possible ways of restructuring incentive pay to enhance its effects; the lack of heterogeneous effects by student characteristics suggests that incentives may help a broad cross-section of the student population.

9.3.2 Caution Is Required When Interpreting Treatment-by-Covariate Interactions

After carefully investigating subgroup differences in order to better understand the policy implications of their findings, the researchers who conducted the teacher incentives study nevertheless remain cautious when interpreting their results. Stepping back from

10 A conventional F-test is based on the F distribution, whose shape depends on three numbers, N, the number of interactions in the alternative model, and the total number of parameters in the alternative model. These numbers, which are the denominators in the upper and lower parts of equation (9.8), have no effect on randomization tests because they remain the same across all simulated randomizations. For more on the use of randomization inference for the testing of nested hypotheses, see Edgington and Onghena 2007.

11 Muralidharan and Sundararaman 2011, Table 5.

this exemplary study and thinking about subgroup analysis more generally, why is caution warranted?

One reason for caution is the *multiple comparisons problem*. A dataset with an extensive set of covariates allows a researcher to examine a large number of subgroup partitions. If the researcher compares enough subgroups, odds are good that statistically significant interactions with at least one covariate will pop up merely by chance. For example, suppose a researcher assesses whether each of 20 covariates interacts with the treatment. For the sake of illustration, suppose the covariates are uncorrelated with one another and that the treatment has an effect of zero for all subjects. The probability of finding at least one covariate that significantly interacts with the treatment at the 0.05 significance level is $1 - (1 - 0.05)^{20} = 0.642$. Perhaps more striking is the fact that the probability of finding at least one covariate that significantly interacts with the treatment at the 0.01 significance level is $1 - (1 - 0.01)^{20} = 0.182$. In other words, roughly one in six researchers who evaluate treatment-by-covariate interactions with 20 different covariates will find what would ordinarily be considered very convincing evidence of at least one interaction, even though the treatment effect is actually zero for every subject.

The practical implication is that when evaluating difference in CATEs, you should keep track of the number of hypothesis tests you conduct. A convenient way to guard against false discoveries is to use the Bonferroni correction: if you conduct h hypothesis tests, divide your target p-value by h. For example, if you conduct 20 subgroup analyses at the 0.05 significance level, only reject the null hypothesis if your p-value for any particular comparison is below 0.0025. Had we used a Bonferroni correction in our previous calculations, our stock of statistically significant red

BOX 9.4

Definition: Bonferroni Correction

The more significance tests one performs, the greater the likelihood of falsely rejecting the null hypothesis at least once. The Bonferroni correction reduces the target p-value in proportion to the number of significance tests that the researcher conducts. If the researcher seeks to test hypotheses at the α significance level and conducts q tests, the target p-value becomes α/q. This correction ensures that the probability of rejecting the null for any given test is no greater than α. For example, if a researcher conducts 10 tests at the $\alpha = 0.05$ level, the target p-value should be 0.005 for each test. The Bonferroni correction is easy to calculate and errs on the side of being conservative. For other methods of correcting for multiple comparisons, see Shaffer (1995).

herrings would have returned to appropriate levels. For example, the probability of detecting at least one interaction that is significant at the $0.01/20 = 0.0005$ level in 20 attempts is $1 - (1 - (0.01/20))^{20} = 0.01$.

The multiple comparisons problem is at root an accounting issue; in order to properly gauge an estimated interaction's level of statistical significance, we need to know something about the procedure by which this apparent interaction was discovered. Sometimes the subgroup comparisons are specified in advance in the form of a planning document (see Chapter 13), in which case correcting for multiple comparisons is straightforward.[12] Interpretation becomes more problematic in cases where the number of comparisons is unknown. In this case, a reasonable approach is to regard hypothesis tests with skepticism pending replication by another study.

Another way to avoid the multiple comparisons problem is to rephrase the null hypothesis. Instead of asking whether partitioning a single covariate into two groups leads to significantly different CATEs, one might ask whether the groups formed by partitioning the sample based on several covariates leads to an array of CATEs that vary more than one would expect by chance. For example, a researcher could partition teacher specialty into four categories (math, language, social studies, other) and further partition teachers according to gender, thereby creating $4 \times 2 = 8$ subgroups. One could test whether the F statistic from a regression of outcomes on indicator variables for each subgroup and each subgroup's treatment is significantly greater than one would expect by chance if all eight CATEs were equal to the estimated ATE. This method extends easily to accommodate additional covariates. Continuous covariates such as age or years of teaching experience can either be recoded into categories or modeled using regression with interaction terms (see below). When testing the joint significance of all interactions, rejecting the null hypothesis implies that CATEs vary more than can be explained by sampling variability. This test may be much more powerful than the agnostic hypothesis test of section 9.2, especially if the covariates used in the analysis reflect theoretical expectations about the conditions under which treatment effects are likely to be strong or weak.

Even when the statistical results from hypothesis tests strongly suggest heterogeneity, a more fundamental limitation of subgroup analysis is that it is essentially nonexperimental in character. Suppose that the statistical evidence in the teacher incentives study left no doubt that treatment effects were larger when students were

---

12 An alternative, simulation-based approach to the Bonferroni correction uses the following steps: (1) generate the full schedule of potential outcomes assuming constant treatment effects equal to the observed estimates, (2) generate 100,000 random allocations in order to form the sampling distribution of the difference in CATEs for each variable, and (3) search across the range of observed p-values until you find one that makes the probability of obtaining at least one significant difference in CATEs 5%. In other words, this p-value will be strict enough to allow only 5,000 of the simulated samples to generate one or more significant differences in CATEs. The online appendix (http://isps.research.yale.edu/FEDAI) provides an example using R.

taught by highly educated teachers. Does this finding imply that treatment effects would rise if teachers with a high school diploma were replaced by teachers who have earned a higher degree? Not necessarily. It could be that educational attainment is merely a marker for other factors, such as the value that the surrounding community places on education. Teachers' education *predicts* the size of the treatment effect, but changing teachers' level of education may not *cause* an increase in the size of the conditional average treatment effect.

Subgroup analysis should be thought of as an exploratory or descriptive exercise. Descriptions can be useful: the administrator, firm, or campaign that wants to know where to direct an intervention in order to maximize its effects would benefit from knowing which subgroups tend to be most responsive to the treatment. If the aim is simply to predict when treatment effects will be large, the researcher need not have a correctly specified causal model that explains treatment effects. Any prognostic variables will do. Noticing that treatment effects tend to be large in some groups and absent from others can provide important clues about why treatments work. But resist the temptation to think that subgroup differences establish the causal effect of randomly varying one's subgroup attributes. The researcher may sift through correlations between CATEs and various individual and contextual factors but must stop short of inferring that randomly changing the values of the covariates via some sort of intervention would lead to a change in treatment effect.

The following example exposes the risks of extracting causal inferences from subgroup analysis. Gerber, Green, and Larimer conducted a large-scale experiment in 2006 in which they sent registered voters mail that showed whether they cast ballots in a 2004 election.[13] (More information about this study appears in Chapter 10.) Those who were sent this mailing voted at a rate of 34.5%, as compared to the control group, which voted at a rate of 29.7%. The estimated ATE is 4.8 percentage points with a standard error of just 0.3 percentage points. It turns out that the estimated CATE is especially large when one restricts attention to those in the treatment and control groups who actually voted in the 2004 election. Their CATE is estimated to be 5.9 percentage points, which is significantly ($p < .01$) greater than the 4.1 percentage point CATE estimate among those who did not vote in 2004. On a purely statistical level, this interaction is clear, but what does it mean substantively? Does this result imply that receiving notice that you voted in a past election has a stronger effect than receiving notice that you did not vote? A follow-up experiment conducted in 2007 suggests not. Gerber, Green, and Larimer show that randomly varying whether voters are shown an election in which they voted or an election in which they abstained generated precisely the opposite effect: those shown that they did not vote in a past election were significantly *more* likely to vote than those shown that they did vote in the past.[14] The analysis of subgroups conflates the content of the mailings with

13 Gerber, Green, and Larimer 2008.

14 Gerber, Green, and Larimer 2010.

the types of people who receive them. The analysis of treatment-by-covariate interactions may be credited with calling attention to an interesting puzzle, but a direct experimental test is required in order to distinguish the attributes of the treatment from the attributes of those who receive it.

9.3.3 Assessing Treatment-by-Treatment Interactions

The basic limitations of subgroup analysis can be overcome with more elaborate experimental designs that manipulate both the treatment as well as the personal or contextual factors that are thought to affect the size of the treatment effect. In field experiments, a common approach is to use a design in which a treatment factor is "crossed" with a factor that varies the context in which the treatment is delivered. With ample subjects and the resources to treat them in various ways, a researcher could implement a factorial design in which at least some subjects receive every possible combination of treatments.

Three examples give a flavor of how these designs look in practice. Gerber and Green's voter mobilization study involved canvassing, calling, or mailing New Haven residents in order to encourage them to vote.[15] Rather than study subgroups that ordinarily receive greater or lesser amounts of campaign communication, this study crossed these three forms of campaign activity (door-to-door canvassing, phone calls, and direct mail) in order to see whether the effects of exposure to one form of

BOX 9.5

Definition: Factorial Design

Suppose that an experiment involves two or more factors, each with two or more experimental conditions. A factorial design allocates subjects at random to every combination of experimental conditions. For example, if factor one consists of conditions A and B, and factor two consists of conditions C, D, and E, a factorial design would assign subjects to treatments {AC, AD, AE, BC, BD, BE}. Factorial designs allow researchers to study the way in which the treatment effect of one variable changes depending on the levels of other randomly assigned factors. When the number of factors and conditions is large, it may be impractical to assign subjects to every combination of conditions. Designs that omit some combinations are termed "fractional" factorial designs. The design of fractional factorial experiments is itself the subject of entire reference texts. A classic text is Cochran and Cox (1957).

15 Gerber and Green 2000.

communication were enhanced or diminished by random exposure to other campaign messages. Another example is Olken's study of corruption monitoring in Indonesia.[16] One treatment, audits of road construction projects, was crossed with another treatment, invitation of villagers to attend public meetings or provide anonymous information, in order to gauge whether the effectiveness of official audits varied by the level of villager involvement. A final example is a discrimination experiment by Rosen, who emailed state legislators to find out how they would respond to a request for help from a putative constituent seeking to find out about immigration rules.[17] This study randomly varied the apparent ethnicity of the email's author; in one condition, the author was Colin Smith, and in the other, José Ramirez. For each putative letter writer, the quality of the email's prose was manipulated; in one condition, it contained grammatical errors, and in another condition it did not. This design allows the researcher to investigate whether ethnic discrimination depends on the letter writer's ability to communicate, which subtly conveys information about education and social class.

Let's take a closer look at the letter-writing experiment, which uses a simple 2×2 design. Table 9.1 shows legislators' response rates to each of the four experimental conditions.[18] When both letter-writers use good grammar, Colin is more likely to receive a reply than José (52% versus 37%). This pattern seems to suggest that legislators and their staff discriminate against the minority letter writer. However, when the experimental letter is poorly written, the pattern is reversed, and legislators seem to favor José (34%) over Colin (29%).[19]

TABLE 9.1

Percentage of replies to a request made to state legislators, by the putative ethnicity of the constituent and the quality of the prose used in the request

	Colin		José	
	Good grammar	Bad grammar	Good grammar	Bad grammar
% Received Reply	52	29	37	34
(N)	(100)	(100)	(100)	(100)

Source: Rosen 2010.

16 Olken 2007.

17 Rosen 2010. Rosen's research is patterned after Butler and Broockman 2011, which compared legislators' responses to putatively black or white constituents.

18 Our analysis focuses on a subset of Rosen's data in order to simplify the presentation of results.

19 The same four numbers may also be used to address the question: Which author is most severely punished for poor writing? The quality of prose strongly affects the likelihood of a reply when Colin is the author (52% versus 29%), but José's prose is relatively inconsequential (37% versus 34%).

This example illustrates some attractive features of factorial design. The deployment of a wide array of treatments is especially useful at the beginning of a research program, when the aim is to explore various treatment combinations as a prelude to further investigation of interventions that appear to be especially effective. In this case, the use of varying treatments sheds light on the conditions under which legislators discriminate in favor of Colin. Apparently, Colin Smith only elicits a substantially different response when he signals a high degree of linguistic competence (and by implication, education and affluence). Had the quality of prose been uniformly high, we might have been led to believe that non-Hispanics were categorically favored over Hispanics. The findings generated by this study suggest further investigation of whether legislators respond according to whether they perceive the constituent to be part of an influential segment of society.

Apart from requiring the researcher to implement a more complex experiment, the use of factorial design involves relatively little downside risk. When the two factors (e.g., ethnicity of the letter writer and writing quality) are uncorrelated, the researcher can always ignore one factor if it proves to be inconsequential. For example, if one were to conclude that writing quality had no effect on outcomes, one could compare all letters sent by Colin to all letters sent by José. If the different versions of the treatment do appear to have different effects, remember that the average effect of your treatment is a weighted average of the ATEs for each version. This point should be kept in mind when drawing generalizations based on a multi-factor experiment. In this case, the average effect of the letter in this study reflects the fact that half of the letters from each author were sent with grammatical errors.

One further caveat about factorial design concerns experiments encountering noncompliance: estimates may be imprecise if there are very few Compliers in each treatment group. It may be impossible to estimate the combined effect of multiple treatments in the face of noncompliance, because the assignment to multiple treatments results in very few people who in fact receive all treatments.

9.4 Using Regression to Model Treatment Effect Heterogeneity

In Chapter 4, we expressed each subject's potential outcomes Y_i as a linear regression equation:

$$Y_i = Y_i(0)(1 - d_i) + Y_i(1)d_i = a + bd_i + u_i. \tag{9.10}$$

Here, the intercept $a = \mu_{Y(0)}$, or the average value of the untreated potential outcomes. The slope $b = \mu_{Y(1)} - \mu_{Y(0)}$ represents the ATE. The independent variable is the treatment, d_i. The unobserved part of the regression model is the disturbance term $u_i = Y_i(0) - \mu_{Y(0)} + [(Y_i(1) - \mu_{Y(1)}) - (Y_i(0) - \mu_{Y(0)})]d_i$. This model allows

BOX 9.6

Definition: Systematic and Idiosyncratic Heterogeneity

Systematic heterogeneity is variation in τ_i that can be predicted by observed covariates or other experimental factors.

Idiosyncratic heterogeneity is residual variation in τ_i that is not predicted by covariates or other experimental factors.

for heterogeneous treatment effects but makes no attempt to explain why heterogeneity arises. The so-called *idiosyncratic heterogeneity* in treatment effects is simply relegated to the unobserved disturbance term.

Modeling heterogeneity involves specifying how observed variables contribute to higher or lower average treatment effects. By introducing treatment-by-covariate or treatment-by-treatment interactions to the right hand side of the equation, the expanded regression model removes *systematic heterogeneity* from the undifferentiated pool of heterogeneity, calling what remains idiosyncratic. Let's start with a simple example of a treatment-by-treatment interaction and build up to a more complicated model that includes both types of interactions.

In the letter-writing experiment, there are four experimental conditions, and one way to write down a regression model is to use indicator variables for each type of letter:

$$Y_i = b_1 L_{i\,Good\,Grammar}^{Non-Hispanic} + b_2 L_{i\,Good\,Grammar}^{Hispanic} + b_3 L_{i\,Bad\,Grammar}^{Non-Hispanic} + b_4 L_{i\,Bad\,Grammar}^{Hispanic} + u_i. \quad (9.11)$$

The regression estimates of the four coefficients are simply the mean outcomes in each of the four experimental groups. This regression may be expressed in an equivalent manner by redefining the right-hand-side variables. Let J_i be an indicator variable scored 1 if the writer is José Ramirez and 0 if the writer is Colin Smith, and let G_i be an indicator variable scored 1 if the letter is written with grammatical errors and 0 otherwise.

$$Y_i = a + bJ_i + cG_i + d(J_iG_i) + u_i. \quad (9.12)$$

Equations (9.11) and (9.12) look different but are actually equivalent.[20] One can move between the two equations by relabeling the parameters: $b_1 = a$, $b_2 - b_1 = b$,

20 This model may also be written in terms of potential outcomes. The treatment is Hispanic authorship, and $a = \mu_{Y(0)}^{G=0}$, or the average value of the untreated potential outcomes when the letter is well-written, and $b = \mu_{Y(1)}^{G=0} - \mu_{Y(0)}^{G=0}$ represents the ATE when the letter is well-written. The coefficient $c = \mu_{Y(0)}^{G=1} - \mu_{Y(0)}^{G=0}$ is the difference in intercepts between poorly written and well-written letters. The coefficient $d = \left(\mu_{Y(1)}^{G=1} - \mu_{Y(0)}^{G=1}\right) - \left(\mu_{Y(1)}^{G=0} - \mu_{Y(0)}^{G=0}\right)$ is the difference between the ATE for poorly written and well-written letters.

$b_3 - b_1 = c$, and $(b_4 - b_3) - (b_2 - b_1) = d$. In equation (9.12), b is the average effect of signing a letter with José Ramirez's name rather than Colin Smith's name *given that the letter is well-written*. This interpretation of b is easy to see when we rewrite equation (9.12) assuming that the letter is well-written ($G_i = 0$):

$$Y_i = a + bJ_i + c \times 0 + dJ_i \times 0 + u_i = a + bJ_i + u_i. \quad (9.13)$$

Similarly, we interpret c as the average effect of composing a letter with grammatical errors *given that the letter is written by Colin Smith*. To see this interpretation more clearly, rewrite equation (9.12) assuming that $J_i = 0$:

$$Y_i = a + b \times 0 + cG_i + d \times 0 \times G_i + u_i = a + cG_i + u_i. \quad (9.14)$$

The interaction effect d dictates the way that b changes as G_i changes from 0 to 1 (and, equivalently, the way that c changes as J_i changes from 0 to 1). For example, when $G_i = 1$,

$$Y_i = a + bJ_i + c + dJ_i + u_i = (a + c) + (b + d)J_i + u_i, \quad (9.15)$$

and the average effect of J_i becomes $(b + d)$. Comparing equations (9.13) and (9.15) reveals that the ATE of J_i changes by d as G_i changes from 0 to 1. The same steps reveal that the ATE of G_i changes by d as J_i changes from 0 to 1.

Applying the regression model in equation (9.12) to the letter writing data recovers exactly the four percentages in Table 9.1. The estimates are:

$$\hat{Y}_i = \hat{a} + \hat{b}J_i + \hat{c}G_i + \hat{d}(J_iG_i) = 0.52 + (-0.15)J_i + (-0.23)G_i + (0.20)(J_iG_i). \quad (9.16)$$

One advantage of the regression framework is that it facilitates hypothesis testing. One hypothesis of interest is whether the ATE of authorship when grammar is poor is different from the ATE of authorship when grammar is good. This hypothesis may be evaluated by assessing whether the interaction parameter d is different from zero. Under the null hypothesis that d is zero, the regression model contains only the coefficients a, b, and c. In order to test the adequacy of the null model, we calculate an F-statistic, which compares the sum of squared residuals from two nested regression models, equation (9.12) and a null model in which d is assumed to be zero. The estimated F-statistic is:

$$\hat{F} = \frac{\dfrac{SSR_{Null} - SSR_{Alternative}}{Parameters_{Alternative} - Parameters_{Null}}}{\dfrac{SSR_{Alternative}}{N - Parameters_{Alternative}}} = \frac{\dfrac{92.30-91.30}{4-3}}{\dfrac{91.30}{400-4}} = 4.34. \quad (9.17)$$

In order to assess the p-value of this statistic, we use randomization inference. We generate a full schedule of potential outcomes assuming the constant ATEs ($\hat{b}$ and $\hat{c}$)

estimated under the null model when d is assumed to be zero. We repeat the randomization, estimate the regression model in equation (9.12), and calculate the F-statistic comparing it to the null model in which d is assumed to be zero. The p-value is the proportion of simulated experiments that generate an F-statistic larger than the observed value, 4.34. Our test reveals that 3,706 of the 100,000 simulations generated an F-statistic greater than or equal to 4.34, implying a p-value of 0.037. We therefore reject the null hypothesis that the ATEs are equal with $p < 0.05$. Evidently, whether a letter contains grammatical errors changes the effect of the letter writer's ethnicity, and the letter writer's ethnicity changes the effect of grammatical errors.

This regression model may be expanded to accommodate treatment-by-covariate interactions, with the usual proviso that treatment-by-covariate interactions may have no clear causal interpretation. The letter-writing experiment was directed at 200 Hispanic legislators and 200 non-Hispanic legislators, and intuition suggests that Hispanic legislators would be less likely to discriminate in favor of Colin Smith. Letting H_i be an indicator variable scored 1 for Hispanic legislators and 0 otherwise, we expand equation (9.12) to include four additional parameters:

$$Y_i = a + bJ_i + cG_i + d(J_iG_i) + eH_i + f(J_iH_i) + g(G_iH_i) + h(J_iG_iH_i) + u_i. \tag{9.18}$$

When this model is estimated using the letter-writing data, the eight parameters in this regression model exactly reproduce the eight percentages in Table 9.2:

$$\hat{Y}_i = 0.62 - 0.22J_i - 0.28G_i + 0.24(J_iG_i) - 0.20H_i + 0.14(J_iH_i) + 0.10(G_iH_i) - 0.08(J_iG_iH_i). \tag{9.19}$$

In order to move from regression results to more readily interpretable CATEs, hold one of the treatments constant at 0 or 1. For example, when letters contain no grammatical errors, Hispanic legislators are much less prone to discriminate against José Ramirez than are non-Hispanic legislators. For Hispanic legislators, the estimated CATE of authorship is $-0.22 + 0.14 = -0.08$, as compared to -0.22 for non-Hispanic legislators. When the letters contain grammatical errors, the interaction is much weaker. For Hispanic legislators, the estimated CATE is $-0.22 + 0.24 + 0.14 - 0.08 = 0.08$, as compared to $-0.22 + 0.24 = 0.02$ for non-Hispanic legislators.

If we want to test whether Hispanic and non-Hispanic legislators are affected differently by the letter writer, the quality of the prose, or the combination of the two, we may test the joint significance of f, g, and h. The null model posits no interactions and constrains these three parameters to be zero:

$$Y_i = a + bJ_i + cG_i + d(J_iG_i) + eH_i + u_i. \tag{9.20}$$

Notice that the null model still includes the parameter e, because H_i functions as an ordinary covariate, improving precision by accounting for some of the variability in outcomes. (As usual, the parameter e has no causal interpretation, as H_i is not randomly assigned.) The null model, in other words, allows Hispanic legislators to

TABLE 9.2

Percentage of replies to a request made to state legislators by the putative ethnicity of the constituent, the quality of the prose used in the request, and the ethnicity of the legislator

	Non-Hispanic legislator				Hispanic legislator			
	Colin		José		Colin		José	
	Good grammar	Bad grammar	Good grammar	Bad grammar	Good grammar	Bad grammar	Good grammar	Bad grammar
% Received reply	62	34	40	36	42	24	34	32
(N)	(50)	(50)	(50)	(50)	(50)	(50)	(50)	(50)

Source: Rosen 2010.

respond to letters at different rates but assumes that they respond to variations in authorship and prose in the same manner as non-Hispanic legislators. The F-test comparing the nested models in equations (9.18) and (9.20) is 0.55, which has a p-value of 0.65. We cannot reject the hypothesis that Hispanic and non-Hispanic legislators (and their staff) respond to the two treatments in the same way.

Although we cannot reject the null hypothesis of no interaction with legislator ethnicity, our ability to detect interactions is limited, given the relatively small number of Hispanic and non-Hispanic legislators in each experiment condition. As so often happens when comparing CATEs, we cannot reject the null hypothesis of no treatment-by-covariate interactions, yet our estimated interactions have large standard errors, leaving open the possibility that substantively important interactions went undetected due to sampling variability. The level of statistical uncertainty is apparent when we estimate the 95% confidence interval associated with the parameter h in equation (9.18): $(-0.451, 0.290)$. The three-way interaction between author ethnicity, quality of prose, and legislator ethnicity could be strongly positive or negative.

This uncertainty worsens as we add covariates such as the legislators' political party and include their interactions with the two treatments. Each additional covariate effectively partitions our original sample into smaller and smaller subgroups. One can recover some of the loss in statistical power by introducing modeling assumptions, such as the lack of interaction between certain factors and covariates. When interactions are omitted, the correlation among right-hand-side variables diminishes, which in turn reduces sampling variability. The question is whether these modeling assumptions are specified in advance of seeing the results (i.e., in a planning document) or whether they are suggested by the pattern of outcomes. Post hoc specification decisions are a reasonable part of exploratory data analysis, but the conclusions that one draws should be regarded as provisional pending replication by another experiment.

9.5 Automating the Search for Interactions

Regression is a useful device for estimating interactions—assuming that we have a model in mind as we approach the data. In that case, we estimate a single regression or compare two nested models, and that's that. In principle, regression could be used to examine interactions between several treatments and covariates, but in practice, the complexity of managing and interpreting large numbers of interactions becomes overwhelming. And the more discretion researchers have when adding or dropping variables, the farther they stray from generating results that are reproducible or have known sampling distributions.

One possible solution is to keep things simple. Very few experimental research agendas in the social sciences are so well-developed that the active research frontier concerns the interaction among three different variables. Indeed, few experimental literatures have generated replicable interactions between two variables. One can go a long way focusing on one or two substantively interesting interactions, showing them to be robust and reproducible.

Another approach is to automate the search for interactions. Developments in the field of machine learning coupled with increasingly powerful computers have generated new classes of algorithms that methodically comb through vast numbers of possible interactions. A full description of these methods goes beyond the scope of this chapter, but the basic idea is fairly straightforward. The computer is instructed to use a set of covariates to partition the sample into subgroups. By repeatedly splitting the sample into subgroups, the method offers a very flexible way to explore a vast number of high-order interactions. In order to further prevent false discoveries, this automated search is replicated a large number of times using different variables and criteria for subdividing the sample. The researcher plays a minimal role in the process because the tuning parameters that guide the search tend to play a minor role in shaping what the computer finds. Are interactions detected using automated search methods more likely to be confirmed by follow-up experiments than interactions detected by researcher-guided analysis? This remains an open question and important area for future research.

SUMMARY

On the surface, the study of heterogeneous treatment effects seems straightforward. Partition the subjects according to some set of characteristics and compare different subgroups' apparent treatment effects. On closer examination, the investigation of treatment effect variation presents an array of statistical and conceptual challenges:

1. Experimental results do not allow us to identify the degree of treatment effect heterogeneity because we do not observe the joint distribution of $Y_i(0)$ and $Y_i(1)$. Because the covariance between $Y_i(0)$ and $Y_i(1)$ cannot be inferred from the data, we cannot estimate $\text{Var}(\tau_i)$. We may instead estimate bounds for $\text{Var}(\tau_i)$ by calculating the largest and smallest covariances implied by the data. Another strategy is to test the null hypothesis that $\text{Var}(\tau_i) = 0$ by comparing the observed variances of $Y_i(0)$ and $Y_i(1)$. The hypothesis that $\text{Var}(\tau_i) = 0$ implies that $\text{Var}(Y_i(0)) = \text{Var}(Y_i(1))$, and so observing markedly different variances suggests that treatment effects are heterogeneous. Unfortunately, in many applications, these bounds fail to give a clear indication of whether $\text{Var}(\tau_i) > 0$, and this test of $\text{Var}(\tau_i) = 0$ lacks power.
2. Additional power is achieved when researchers partition subjects based on covariates or experimental treatments. The conditional average treatment effects (CATEs) within these subgroups may be compared using randomization inference. Regression provides a flexible framework for estimating interaction effects and testing their significance. However, when testing a series of interaction hypotheses, researchers confront the multiple comparisons problem: even if there were in fact no interactions, the probability that at least one estimated interaction proves significant rises as the number of tests increases. One way to address the multiple comparisons problem is to adjust the size of each test using the Bonferroni correction. Another way is to assess the joint significance of all interactions considered together using the F-test to compare nested models.
3. When assessing treatment-by-covariate interactions, researchers find themselves on the border between experimental and nonexperimental research. The experimental treatments are randomly assigned, but the covariates with which they interact are not. When CATEs are found to vary depending on the value of a covariate, interpretation remains ambiguous. The subgroups defined by the covariate may have different potential outcomes. Treatment-by-covariate interactions may provide useful descriptive information about which types of subjects are most responsive to treatment, but the theoretical question of whether these interactions are causal requires an experimental design that randomly varies what are believed to be the relevant subject attributes or contextual characteristics.
4. Multi-factor experiments have the potential to shed light on practical questions (What combinations of treatments are most effective?) and theoretical questions (Under what conditions are treatment effects large or small?). Although potentially valuable, experiments that test treatment-by-treatment interactions may present practical challenges. In field settings, it may be difficult to manipulate subject attributes (e.g., education) and even more difficult to do so in ways that isolate specific theories about why interactions occur (e.g., exposure to a broad spectrum of ideas and viewpoints). Somewhat easier, although by no means

trivial, is the task of manipulating the context in which treatments are deployed. In the examples discussed above, context is manipulated by randomly varying the information that people receive, the number of times they are contacted by campaigns, or their opportunity to air grievances at public meetings.

5. Even well-designed experiments may run short on statistical power when assessing interactions, especially when the number of possible interactions is large. Power may be enhanced by selectively excluding certain interactions from consideration. Ideally, these modeling decisions are guided by a planning document that specifies ex ante which interactions are to be tested. In the absence of a planning document, the challenge is to explore and report interactions in a systematic manner, taking into account modeling uncertainty when reporting statistical tests and confidence intervals. Machine learning algorithms that automate the search for interactions are an attractive option, reducing the role of discretion and allowing for a more rigorous assessment of statistical uncertainty.

Given the challenges of investigating treatment effect heterogeneity, the advice of this chapter is to be methodical and cautious. In the absence of a planning document that spells out the interactions to be tested, begin by assessing whether methods that make relatively few assumptions, such as bounds or simple hypothesis tests, shed any light on the question. If the answer is yes (or is sufficiently uncertain to be taken as a tentative yes), use your theoretical intuitions to come up with testable propositions about interaction effects. If possible, build additional factors into your experimental design, because random manipulation of subject attributes and contextual characteristics facilitates causal inference. Less informative but still valuable is the investigation of treatment-by-covariate interactions. As you analyze the results, keep track of the number of interactions to be tested, and err on the side of skepticism by assessing whether the p-value withstands a Bonferroni correction. Better to be a bit less forceful in touting one's discovery of an interaction, lest the next study reveal your finding to be a chimera of sampling variability.

SUGGESTED READINGS

Rosenbaum (2010: section 2.4) provides a good introduction to the topic of treatment effect heterogeneity, and Abbring and Heckman (2007) review the literature. Heckman, Smith, and Clements (1997) describe how to estimate bounds for treatment effect heterogeneity, and Djebbari and Smith (2008) apply this technique to the PROGRESA social assistance program in Mexico. Crump et al. (2008) derive a test for treatment effect heterogeneity that exploits the presence of covariates. Bitler, Gelbach, and Hoynes (2006) show that conclusions about the effectiveness of social programs can be misleading when researchers ignore possible treatment effect heterogeneity. Byar (1985), Pocock et al. (2002), and Rothwell (2005) discuss issues of multiple testing and subgroup analyses. Dehejia (2005) and Imai and Strauss (2011) consider the decision problem that arises from heterogeneous treatment effects and formulate strategies to maximize the cost-effectiveness of policy interventions. Green and Kern (2011) provide an introduction to Bayesian additive regression trees, one of several automated approaches to the investigation of heterogeneous treatment effects.

EXERCISES: CHAPTER 9

1. Important concepts:
   (a) Define CATE. Is a Complier average causal effect (CACE) an example of a CATE?
   (b) What is an interaction effect?
   (c) Describe the multiple comparisons problem and the Bonferroni correction.
2. The standard error formula given in equation (3.4) suggests that, all else being equal, reducing variance in $Y_i(0)$ helps reduce sampling uncertainty. Referring to the procedure outlined in section 9.2, explain why the same principle applies to estimating bounds on treatment effect heterogeneity.
3. One way to reduce variance in $Y_i(0)$ is to block on a prognostic covariate. When blocking is used, the joint distribution of $Y_i(0)$ and $Y_i(1)$ is simulated within blocks using the bounding procedure described in section 9.2. Using the schedule of potential outcomes below, show how the maximum and minimum values of the covariance of $Y_i(0)$ and $Y_i(1)$ compare to the maximum and minimum values of the covariance of $Y_i(0)$ and $Y_i(1)$ for the dataset as a whole (i.e., had blocking not been used).

Block	Subject	$Y_i(0)$	$Y_i(1)$
A	A-1	0	2
A	A-2	1	5
A	A-3	1	3
A	A-4	2	1
B	B-1	2	3
B	B-2	3	3
B	B-3	4	9
B	B-4	4	7

4. Suppose that a researcher compares the CATE among two subgroups, men and women. Among men ($N = 100$), the ATE is estimated to be 8.0 with a standard error of 3.0, which is significant at $p < 0.05$. Among women ($N = 25$), the CATE is estimated to be 7.0 with an estimated standard error of 6.0, which is not significant, even at the 10% significance level. Critically evaluate the researcher's claim that "the treatment only works for men; for women, the effect is statistically indistinguishable from zero." In formulating your answer, address the distinction between testing whether a single CATE is different from zero and testing whether two CATEs are different from each other.
5. The table below shows hypothetical potential outcomes for an experiment in which low-income subjects in a developing country are randomly assigned to receive (i) loans to aid their small businesses; (ii) business training to improve their accounting, hiring, and inventory-management skills; (iii) both; or (iv) neither. The outcome measure is business income during the subsequent year. The table also includes a pre-treatment covariate, an indicator scored 1 if the subject was judged to be proficient in these basic business skills.

Subject	$Y_i(loan)$	$Y_i(training)$	$Y_i(both)$	$Y_i(neither)$	Prior business skills
1	2	2	3	2	0
2	2	3	2	1	0
3	5	6	6	4	1
4	3	1	5	1	1
5	4	4	5	0	0
6	10	8	11	10	1
7	1	3	3	1	0
8	5	5	5	5	1
Average	4	4	5	3	0.5

(a) What is the ATE of the loan if all subjects were also to receive training?
(b) What is the ATE of the loan if no subjects receive training?
(c) What is the ATE of the training if all subjects also receive a loan?
(d) What is the ATE of the training if no subjects receive a loan?
(e) Suppose subjects were randomly assigned to one of the four experimental treatments in equal proportions. Use the table above to fill in the expected values of the four regression coefficients for the model and interpret the results:

$$Y_i = \alpha_0 + \alpha_1 Loan_i + \alpha_2 Training_i + \alpha_3 (Loan_i \cdot Training_i) + e_i.$$

(f) Suppose a researcher were to implement a block randomized experiment, such that two subjects with business skills are assigned to receive loans, and two subjects without business skills are assigned to receive loans, and the rest are assigned to control. No subjects are assigned to receive training. The researcher estimates the model

$$Y_i = \gamma_0 + \gamma_1 Loan_i + \gamma_2 Skills_i + \gamma_3 (Loan_i \cdot Skills_i) + e_i.$$

Over all 36 possible random assignments, the average estimated regression is as follows:

$$Y_i = 1.00 + 1.25 Loan_i + 4.00 Skills_i - 0.50 (Loan_i \cdot Skills_i).$$

Interpret the results and contrast them with the results from part (e). (Hint: the block randomized design does not affect the interpretation. Focus on the distinction between treatment-by-treatment and treatment-by-covariate interactions.)

6. Rind and Bordia studied the tipping behavior of lunchtime patrons of an "upscale Philadelphia restaurant" who were randomly assigned to four experimental groups.[21] One factor was server sex (male or female), and a second factor was whether the server draws a "happy face" on the back of the bill presented to customers.[22] Download the data located at http://isps.research.yale.edu/FEDAI.

21 Rind and Bordia 1996.

22 The authors took steps to ensure the blindness of the servers to the happy face condition, which was determined only moments before the bill was delivered. The authors also instructed waitstaff to deliver bills and walk away, so that there would be no additional interaction with customers. It is not clear whether the sex of the server was randomly assigned.

(a) Suppose you ignored the sex of the server and simply analyzed whether the happy face treatment has heterogeneous effects. Use randomization inference to test whether $\text{Var}(\tau_i) = 0$ by testing whether $\text{Var}(Y_i(1)) = \text{Var}(Y_i(0))$. Construct the full schedule of potential outcomes by assuming that the treatment effect is equal to the observed difference-in-means between $Y_i(1)$ and $Y_i(0)$. Interpret your results.

(b) Write down a regression model that depicts the effect of the sex of the waitstaff, whether they write a happy face on the bill, and the interaction of these factors.

(c) Estimate the regression model in (b) and test the interaction between waitstaff sex and the happy face treatment. Is the interaction significant?

7. In their 2004 study of racial discrimination in employment markets, Bertrand and Mullainathan sent resumes with varying characteristics to firms advertising job openings. Some firms were sent resumes with putative African American names, while other firms received resumes with putatively Caucasian names. The researchers also varied other attributes of the resume, such as whether the resume was judged to be of high or low quality (based on labor market experience, career profile, gaps in employment, and skills listed).[23] The table below shows the rate at which applicants were called back by employers, by the city in which the experiment took place and by the randomly assigned attributes of their applications.

	Boston				Chicago			
	Low-quality resume		High-quality resume		Low-quality resume		High-quality resume	
	Black	White	Black	White	Black	White	Black	White
% Received call from employer	7.01	10.15	8.50	13.12	5.52	7.16	5.28	8.94
(N)	(542)	(542)	(541)	(541)	(670)	(670)	(682)	(682)

(a) For each city, interpret the apparent treatment effects of race and resume quality on the probability of receiving a follow-up call.

(b) Propose a regression model that assesses the effects of the treatments, interaction between them, and interactions between the treatments and the covariate, city.

(c) Estimate the parameters in your regression model. Interpret the results. (This can be done by hand based on the percentages given in the table.)

8. In Chapter 3, we analyzed data from Clingingsmith, Khwaja, and Kremer's study of Pakistani Muslims who participated in a lottery to obtain a visa for the pilgrimage to Mecca.[24] By comparing lottery winners to lottery losers, the authors are able to estimate the effects of the pilgrimage on various attitudes, including views about people from other countries. Winners and losers were asked to rate the Saudi, Indonesian, Turkish, African, European, and Chinese people on a five-point scale ranging from very negative (-2) to very positive ($+2$). Adding the responses to all six items creates an index ranging from -12 to $+12$. The key results are presented in the table below.

23 Bertrand and Mullainathan 2004, p. 994.

24 Clingingsmith, Khwaja, and Kremer 2009.

	Control group	Treatment group
N	448	510
Mean	1.868	2.343
Variance	5.793	6.902
Absolute difference in variances	1.109	

(a) Explain the meaning of "absolute difference in variances."

(b) Describe how one could use randomization inference to test the null hypothesis of constant treatment effects.

(c) Assume that researchers applied the method you proposed in part (b) and simulated 100,000 random assignments, each time calculating the absolute difference in variances; they find that 25,220 of these differences are as large or larger than 1.109, the absolute difference in variances observed in the original sample. Calculate the p-value implied by these results. What do you conclude about treatment effect heterogeneity in this example?

(d) Suppose that this experiment were partitioned into subgroups defined according to whether the subjects had travelled abroad in the past. Suppose that the CATE among those who had previously travelled abroad were 0 and that the CATE among those who had not travelled abroad were 1.0. Suppose this difference in CATEs were significant at $p < .05$. Does this result imply that randomly encouraging people to travel abroad eliminates the Hajj's effect?

9. An example of a two-factor design that encounters one-sided noncompliance may be found in Fieldhouse et al.'s study of voter mobilization in the United Kingdom.[25] In this study, the first factor is whether each voter was mailed a letter encouraging him or her to vote in the upcoming election. The second factor is whether each voter was called with an encouragement to vote. Noncompliance occurs in the case of phone calls, as some targeted voters cannot be reached when called. The experimental design consists of four groups: a control group, a mail-only group, a phone-only group, and a group targeted for both mail and phone. The following table shows the results by assigned experimental group.

	Control	Mail Only	Phone Only	Mail and Phone
N	5,179	4,367	3,466	2,287
Number Contacted by Phone	0	0	2,003	1,363
Among those Assigned to this Experimental Group, Percent who Voted	39.7%	40.3%	39.7%	41.8%
Among those Contacted by Phone, Percent who Voted	Not Applicable	Not Applicable	46.5%	46.8%

(a) Show that, under certain assumptions, this experimental design allows one to identify the following parameters: (i) the ATE of mail, (ii) the Complier average causal

25 Fieldhouse et al. 2010.

effect (CACE) of phone calls, (iii) the CATE of mail among those who comply with the phone call treatment, (iv) the CATE of mail among those who do not comply with the phone call treatment, and (v) the CACE of phone calls among those who receive mail.

(b) Using the identification strategies you laid out in part (a), estimate each of the five parameters using the results in the table.

(c) In Chapters 5 and 6, we discussed the use of instrumental variables regression to estimate CACEs when experiments involve noncompliance. Here, we can apply instrumental variables regression to a factorial experiment in which one factor encounters noncompliance. With the replication dataset at http://isps.research.yale.edu/FEDAI, use instrumental variables regression to estimate the parameters of the Vote equation in the following three-equation regression model:

$$Phone\_Contact_i = \alpha_0 + \alpha_1 Mail_i + \alpha_2 Phone\_Assign_i + \alpha_3 (Phone\_Assign_i \cdot Mail_i) + e_i$$

$$Phone\_Contact_i \cdot Mail_i = \gamma_0 + \gamma_1 Mail_i + \gamma_2 Phone\_Assign_i + \gamma_3 (Phone\_Assign_i \cdot Mail_i) + \varepsilon_i$$

$$Vote_i = \beta_0 + \beta_1 Mail_i + \beta_2 Phone\_Contact_i + \beta_3 (Phone\_Contact_i \cdot Mail_i) + u_i$$

Interpret the regression estimates in light of the five parameters you estimated in part (b). Which causal parameters does instrumental variables regression estimate or fail to estimate?

CHAPTER 10

Mediation

Some of the most interesting and useful discoveries in science occur when researchers discover *intervening* or *mediating* variables that transmit the influence of an experimental intervention. In one of the first experiments ever conducted, the introduction of limes into the diet of seafarers in the eighteenth century dramatically reduced the incidence of scurvy, and eventually twentieth-century scientists figured out that the key mediating ingredient was vitamin C. The experimental treatment was a dietary supplement of limes; the mediating or intervening variable was intake of vitamin C. Equipped with knowledge about how an experimental treatment works, researchers may devise other, possibly more efficient ways of achieving the same effect. Modern seafarers can prevent scurvy with limes or simply with vitamin C tablets.

Arresting examples of mediators abound in the physical and life sciences. Indeed, not only do scientists know that vitamin C mediates the causal relationship between limes and scurvy, they also understand the biochemical process by which vitamin C counteracts the onset of scurvy. In other words, mediators themselves have mediators. Physical and life scientists continually seek to pinpoint ever more specific explanatory agents.

Social scientists, too, are eager to isolate causal mechanisms. When an experiment indicates that a treatment influences an outcome, researchers immediately express curiosity about the channels through which an experimental treatment transmits its influence. It is quite common for authors of research articles to both demonstrate an experimental effect and offer hypotheses about how the effect is transmitted through mediating variables. An especially interesting example of mediation analysis is Rikhil Bhavnani's study of local government representation in India.[1] A randomly selected portion of local council seats are reserved for women candidates, and only

1 Bhavnani 2009.

What you will learn from this chapter:

1. The definition of mediation and how researchers try to assess causal pathways using regression.
2. Why regression-based models of mediation tend to exaggerate the degree to which a proposed mediator explains the effect of a random intervention.
3. How the attempt to partition an experimental effect into "direct" and "indirect" causal pathways breaks down when causal effects are heterogeneous.
4. The statistical connections between mediation analysis and encouragement designs.
5. The inherent challenges of detecting mediation effects even when both treatments and mediators are experimentally manipulated.
6. An alternative approach, implicit mediation analysis, that involves adding or subtracting elements from random interventions.

women may run for office in these selected constituencies. Bhavnani shows that constituencies where women held reserved seats in 1997 were more likely to elect women representatives in 2002, after reservations were no longer in effect. Why did reservations in one election improve the subsequent electoral fortunes of female candidates? Bhavnani considers many possible mediating factors. First, reservations create a cohort of female incumbents whose experience in office may make them more appealing to voters. Second, reservations give voters an opportunity to change their views about women and, in particular, to learn that women make capable representatives. Third, having a woman representative may increase voter participation, and a surge of new voters might improve the chances of electing a woman after reservations expire. Each of these hypotheses posits a mediator that is influenced by reservations and in turn influences the likelihood that a woman is elected after reservations expire.

In abstract terms, mediation analysis starts with an average causal effect of an intended treatment Z_i on an outcome Y_i.[2] For our purposes, we will assume that this intent-to-treat effect is demonstrated convincingly by an experiment in which Z_i is randomly assigned. The aim of mediation analysis is to identify the pathways through which Z_i transmits its influence to Y_i. The researcher endeavors to determine whether Z_i induced a change in a mediating variable M_i, and whether a Z_i-induced change in M_i produced a change in Y_i. In applications such as the election of female leaders in

2 In order to make our notation match other authors' treatments of mediation, we use capital letters to denote assigned or actual treatment.

India, the scope of the investigation usually expands to encompass multiple mediators (M_{1i} = the number and quality of women candidates, M_{2i} = attitudes about women as leaders, and M_{3i} = the mobilization of new voters). The success of this research enterprise is usually judged according to whether the mediators account for all of the influence that Z_i exerts on Y_i. For example, if the creation of a cohort of female incumbents were found to be the sole reason behind the effect of reservations, the implication would be that reservations affect the subsequent election of women only if they generate female incumbents seeking reelection. If incumbents were for some reason barred from running for reelection, reservations would not increase the share of women elected subsequently. Or, to return to the scurvy example, limes have no effect unless they contain vitamin C.

As a research activity, the analysis of mediation is truly an enormous enterprise, suffusing thousands of social science research articles dating back a half-century. Judging from the frequency with which mediation analysis is conducted, one might not guess how difficult it is to extract reliable inferences about mediation from experimental data or how rare it is to encounter a convincing demonstration in the social sciences. Many researchers seem unaware of the strong assumptions that underlie the mediation analyses they read or conduct. For that reason, the chapter starts with a critique of common statistical practice. We show that the regression models commonly used to demonstrate mediation rely on implausible assumptions that have no connection to experimental design. Using both regression terminology and potential outcomes notation, we explain how violation of these assumptions leads to biased inferences about mediation. Next, we consider experimental designs that try to address issues of mediation by manipulating both treatments and mediators. Even here, formidable conceptual and practical challenges confront the researcher who tries to manipulate mediators experimentally, especially in field settings. Researchers must devise ways of manipulating specific mediators (e.g., the amount of vitamin C) without inadvertently manipulating other mediators (e.g., other vitamins or overall caloric intake). Rarely, if ever, do experimental designs in the social sciences achieve this level of specificity and precision.

Finally, we consider a different approach, implicit mediation analysis. This line of research scales back the ambitions of conventional mediation analysis. Instead of attempting to estimate the channels through which Z_i transmits its influence using a statistical model, implicit mediation analysis takes a design-based approach. The researcher conducts an experiment with an array of treatments in order to investigate how adding or subtracting different ingredients to or from Z_i alters its effects. This general approach of experimenting with a variety of treatments not only has the capacity to shed light on causal mechanisms, it also has the practical and theoretical benefit of refocusing social science energies on the search for especially effective interventions.

10.1 Regression-Based Approaches to Mediation

In Chapter 4, we defined a covariate as a variable whose potential outcomes are unaffected by the treatment. Recall that we even used the term *pre-treatment covariate* to underscore the assumption that a covariate is causally prior to the treatment. In this chapter, we discuss a type of variable that is sometimes considered a covariate but is really an experimental outcome. A mediating variable is caused by an intended treatment (Z_i) and in turn causes the outcome (Y_i).[3] In other words, the assigned treatment (Z_i) affects the mediator (M_i), and either or both Z_i and M_i affect Y_i.

Before jumping into a discussion of how this three-variable system might be described using potential outcomes notation, we first depict it as a regression model. This starting point has not only the advantage of being easier to understand, it also is the model that the vast majority of researchers rely on when discussing mediation.

Most regression-based analyses of mediation rely on some form of a three-equation system:

$$M_i = \alpha_1 + aZ_i + e_{1i}, \quad (10.1)$$

$$Y_i = \alpha_2 + cZ_i + e_{2i}, \quad (10.2)$$

$$Y_i = \alpha_3 + dZ_i + bM_i + e_{3i} \quad (10.3)$$

Here, Y_i is the outcome of interest, Z_i is an assigned treatment, M_i is a mediator of the treatment, and α_1, α_2, and α_3 are intercepts. The variables e_{1i}, e_{2i}, and e_{3i} are unobserved disturbance terms that represent the cumulative effect of omitted variables. Because Z_i is randomly assigned, it is statistically independent of the disturbances e_{1i}, e_{2i}, and e_{3i}. Equations (10.1) and (10.2) are therefore experimental in character, and estimators such as regression give unbiased estimates of the average effect of Z_i on the outcome variable in each equation. M_i, however, is not randomly assigned, nor is it a pre-treatment covariate. Consequently, when we include M_i as a right-hand-side regressor as in equation (10.3), we are in a sense turning our experiment into an observational study and possibly introducing bias when we attempt to estimate b and d.

In order to see where bias creeps in, let's try to identify the parameters a, b, c, and d. If we assume that these parameters are the same for every subject (a special case in which the parameters represent not only average effects but constant effects for every subject), we can decompose the *total effect* of Z_i on Y_i into the *direct effect* that goes from Z_i to Y_i without passing through M_i, and the *indirect effect* that passes from Z_i to Y_i through M_i. In this model, the total effect of Z_i on Y_i is represented by

3 To be more precise, we could say that the assigned treatment Z_i affects the delivered treatment D_i, which in turn influences Y_i directly or indirectly through M_i. This formulation is correct but adds an extra layer of notation. For the moment, we assume that there are no problems of noncompliance (i.e., $Z_i = D_i$) until we consider encouragement designs in section 10.5 below.

BOX 10.1

This diagram depicts equations (10.1) and (10.3), illustrating a conventional mediation analysis in which an intervention Z exerts its influence on an outcome Y in two ways. The "direct" effect is the path from Z to Y; the parameter governing this relationship is d. The "indirect" path is from Z to M (governed by the parameter a) and from M to Y (governed by the parameter b). The disturbances e_1 and e_3 represent unobserved causes of M and Y, respectively.

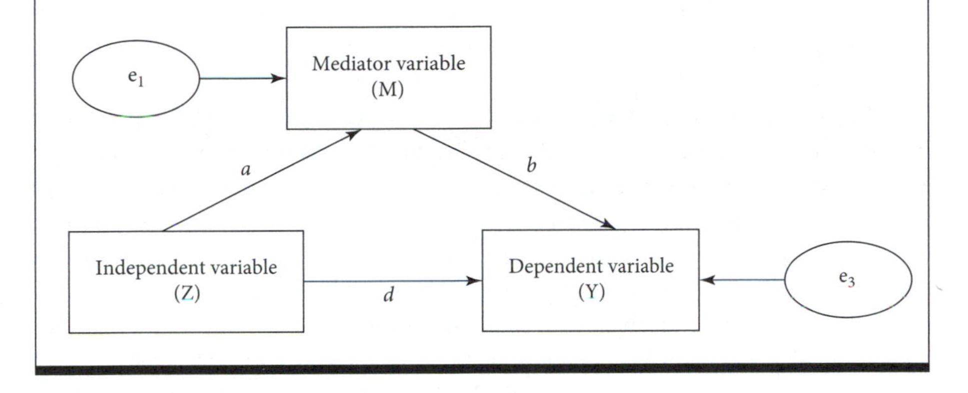

the parameter c in equation (10.2). The direct effect of Z_i on Y_i is the parameter d in equation (10.3). To see how c may be partitioned into direct and indirect effects, substitute equation (10.1) into equation (10.3):

$$Y_i = \alpha_3 + (d + ab)Z_i + (\alpha_1 + e_{1i})b + e_{3i}. \tag{10.4}$$

The total effect of Z_i on Y_i is $c = d + ab$. The direct effect of Z_i is d. The indirect or "mediated" effect is the product ab. The indirect effect, in other words, is the product of Z_i's influence on M_i and M_i's influence on Y_i.

When researchers speak of a causal chain they have in mind this product, where the parameter b dictates how much of Z_i's influence on M_i is transmitted from M_i to Y_i. It should be stressed, however, that the formula partitioning total effects into direct and indirect effects breaks down when coefficients vary from observation to observation. For subject i, let a_i represent the effect of Z_i on M_i, and let b_i represent the effect of M_i on Y_i. The expected value of $a_i b_i$ is not in general equal to $E[a_i]E[b_i]$. Instead, $E[a_i b_i] = E[a_i]E[b_i] + \text{Cov}(a_i b_i)$. When causal effects vary, we cannot infer the average indirect effect by estimating $E[a_i]$ and $E[b_i]$ and multiplying them together. The causal chain interpretation of the three-equation regression model in equations (10.1), (10.2), and (10.3) depends critically on the assumption that the regression parameters are the same for all subjects.

Suppose that, like most researchers who use regression to study mediation, we were to assume constant treatment effects in equations (10.1), (10.2), and (10.3). What happens when we estimate each of these three equations using regression? Applying regression to equation (10.1) poses no special problems. Because Z_i is randomly assigned, regression provides unbiased estimates of its effect on M_i. Estimating equation (10.2) poses no problems. Again, because Z_i is randomly assigned, regression provides unbiased estimates of c, the total effect of Z_i on Y_i. Equations (10.1) and (10.2) lend themselves to unbiased estimation because the experimental design ensures that the treatment is independent of the error terms. The problematic equation is (10.3), because the experimental design says nothing about the relationship between M_i and e_{3i}. The mediator M_i is not randomly assigned and could be systematically related to unmeasured causes of Y_i.

Thinking back to the study of reservations for female representatives in India, consider how bias could creep in. To simplify the discussion, we focus on just one of the mediating paths discussed above. Let Z_i represent randomly assigned reservations in 1997, Y_i the election of a female representative in the ensuing 2002 election, and M_i the number of women candidates running for office in 2002. The problem one confronts when inserting M_i into the regression model in equation (10.3) is that factors other than randomly assigned reservations cause women candidates to run for office. Female candidates may be more likely to run, for example, in constituencies where women are regarded in a more egalitarian light by the average voter. Suppose this unmeasured egalitarianism is what constitutes e_{1i}. This unmeasured disturbance is likely to be positively correlated with e_{3i}, the unmeasured factors that affect the election of a woman in 2002. More egalitarian districts tend to encourage women candidates to run *and* to elect women as representatives.

In order to see how the covariance between e_{1i} and e_{3i} leads to biased regression estimates, consider what the regression estimate of b would look like if we added subjects to the treatment and control groups until our experiment had an infinite number of subjects. Equation (10.5) expresses the limiting value of the regression estimator as the sum of the true parameter and a bias term:

$$\hat{b}_{N\to\infty} = b + \frac{\text{Cov}(e_{1i}, e_{3i})}{\text{Var}(e_{1i})}. \tag{10.5}$$

Equation (10.5) indicates that the estimator is biased when $\text{Cov}(e_{1i}, e_{3i}) \neq 0$. The same approach can be used to show that the OLS estimator of d is also biased when $\text{Cov}(e_{1i}, e_{3i}) \neq 0$:

$$\hat{d}_{N\to\infty} = d - a\frac{\text{Cov}(e_{1i}, e_{3i})}{\text{Var}(e_{1i})}. \tag{10.6}$$

The results in equations (10.5) and (10.6) are proven in the online appendix, but the basic principle may be familiar to readers with a background in regression analysis. If

M_i is related to the disturbance term e_{3i} adding M_i to the regression model will lead to biased estimates of the effect of M_i and biased estimates of any variables with which M_i is correlated. In this case, M_i is likely to be correlated with Z_i.

What can we infer about the direction of the bias? In our example and indeed in most examples, $\text{Cov}(e_{1i}, e_{3i})$ will be positive, which has two implications. First, according to equation (10.5), regression will tend to exaggerate the effect of M_i on Y_i. When interpreting the regression estimates, researchers will tend to overstate the extent to which a mediator transmits the causal influence of Z_i to Y_i. Second, applying equation (10.6) to the typical case in which $a > 0$, we see that regression will tend to underestimate d, the direct effect of Z_i on Y_i after controlling for M_i. This bias will lead researchers to conclude that Z_i exerts little causal influence on Y_i beyond what it transmits through M_i. The net effect of these two biases is to make the typical mediation analysis look "successful"—the mediator will seem to work, and the treatment will seem to have little effect on outcomes after controlling for the mediator.

Among researchers who apply regression to a randomly assigned Z_i and outcome variables M_i and Y_i, a common way of addressing concerns about specification error is to control for covariates other than M_i. Whether these controls are sufficient to eliminate the covariance between e_{1i} and e_{3i} is a matter of speculation, since these disturbance terms are unobserved. This type of argumentation is characteristic of nonexperimental data analysis but contrary to the spirit of experimental research, where assumptions are moored to experimental procedures wherever possible. The random assignment of Z_i says nothing about how e_{1i} and e_{3i} are related to each other or to background covariates.

In sum, the common practice of using regression to infer a causal chain hinges on two dubious assumptions. The first is constant treatment effects, without which the interpretation of ab as an indirect effect breaks down. The second assumption is that the disturbances e_{1i} and e_{3i} are unrelated. Without this assumption, regression will generate biased estimates when applied to equation (10.3). Neither assumption follows from an experimental design in which Z_i is randomly assigned.

10.2 Mediation Analysis from a Potential Outcomes Perspective

In order to appreciate the challenges that mediation analysis must overcome, it is helpful to characterize the data-generating process in much more general terms. Following the system of notation used in previous chapters, we can imagine a schedule of potential outcomes from which the observed values of M_i and Y_i are generated. Using potential outcomes, it is easy to come up with a brief illustration of a data-generating process that leads regression analysis astray.

TABLE 10.1
Example of potential outcomes for Y_i and M_i assuming no effect of M_i on Y_i

Share of the population	$M_i(1)$	$M_i(0)$	$Y_i(1)$	$Y_i(0)$	Z_i	M_i	Y_i	e_{1i}	e_{3i}
1/6	1	1	1	0	0	1	0	0	−2
1/6	1	1	1	0	1	1	1	−1	−2
1/6	3	0	3	2	0	0	2	−1	0
1/6	3	0	3	2	1	3	3	1	0
1/6	2	2	5	4	0	2	4	1	2
1/6	2	2	5	4	1	2	5	0	2

Consider a simple example, depicted in Table 10.1. For ease of exposition, we will assume that treatment assigned (Z_i) is the same as treatment delivered (D_i). The table presents four potential outcomes for an arbitrarily large sample. The potential outcome $M_i(1)$ is the value of the mediator if subject i were treated; $M_i(0)$ is the value if this subject were not treated. $Y_i(1)$ is the outcome if subject i were treated; $Y_i(0)$ is the outcome if this subject were not treated. In this contrived example, the potential outcomes $Y_i(z)$ respond only to the input z, and the mediator M_i plays no causal role whatsoever. Because M_i has no effect on Y_i, it cannot mediate the effect of Z_i. The assigned treatment (Z_i) determines which potential outcome we see, either $Y_i(0)$ and $M_i(0)$ or $Y_i(1)$ and $M_i(1)$. Comparison of the $Y_i(1)$ and $Y_i(0)$ columns in Table 10.1 reveals that the total effect of the treatment on Y_i is 1 for every subject in the sample: in every row, $Y_i(1) - Y_i(0) = 1$. In this example, M_i has no effect on Y_i; in the special case depicted in Table 10.1, $Y_i(z)$ responds only to the treatment and not the mediator. Thus, the true direct effect of the treatment (the parameter we earlier called d in equation 10.3) is 1, and the effect of M_i (which we earlier called b) is 0.

The next three columns of the table show what we would observe if we were to randomly assign half of the subjects to treatment and half to control. Z_i indicates the treatment that each subject receives. Y_i and M_i represent the observed value of the outcome and mediator. In order to link our discussion of potential outcomes to our earlier discussion of regression, the last two columns represent the disturbance terms if one were to model Y_i and M_i using linear equations. The variable e_{1i} represents unobserved factors that cause M_i; e_{3i} represents unobserved factors that cause Y_i.

We can re-write the columns in this table to follow the same format as the regression equations (10.1), (10.2), and (10.3), including the true intercepts and coefficients:

$$\begin{bmatrix} M_i \\ 1 \\ 1 \\ 0 \\ 3 \\ 2 \\ 2 \end{bmatrix} = 1 + 1 \begin{bmatrix} Z_i \\ 0 \\ 1 \\ 0 \\ 1 \\ 0 \\ 1 \end{bmatrix} + \begin{bmatrix} e_{1i} \\ 0 \\ -1 \\ -1 \\ 1 \\ 1 \\ 0 \end{bmatrix} \quad (10.7)$$

$$\begin{bmatrix} Y_i \\ 0 \\ 1 \\ 2 \\ 3 \\ 4 \\ 5 \end{bmatrix} = 2 + 1 \begin{bmatrix} Z_i \\ 0 \\ 1 \\ 0 \\ 1 \\ 0 \\ 1 \end{bmatrix} + \begin{bmatrix} e_{2i} \\ -2 \\ -2 \\ 0 \\ 0 \\ 2 \\ 2 \end{bmatrix} \quad (10.8)$$

$$\begin{bmatrix} Y_i \\ 0 \\ 1 \\ 2 \\ 3 \\ 4 \\ 5 \end{bmatrix} = 2 + 1 \begin{bmatrix} Z_i \\ 0 \\ 1 \\ 0 \\ 1 \\ 0 \\ 1 \end{bmatrix} + 0 \begin{bmatrix} M_i \\ 1 \\ 1 \\ 0 \\ 3 \\ 2 \\ 2 \end{bmatrix} + \begin{bmatrix} e_{3i} \\ -2 \\ -2 \\ 0 \\ 0 \\ 2 \\ 2 \end{bmatrix} \quad (10.9)$$

In this example, Z_i affects both M_i and Y_i, but there is no causal connection between M_i and Y_i. The true regression parameters are $a = 1, b = 0, c = 1, d = 1$. However, on average, when we apply regression to the *observed* data, we obtain:

$$\hat{M}_i = \hat{\alpha}_1 + \hat{a}Z_i = 1 + 1Z_i, \quad (10.10)$$

$$\hat{Y}_i = \hat{\alpha}_2 + \hat{c}Z_i = 2 + 1Z_i, \quad (10.11)$$

$$\hat{Y}_i = \hat{\alpha}_3 + \hat{d}Z_i + \hat{b}M_i = 1 + 0Z_i + 1M_i. \quad (10.12)$$

Because equations (10.10) and (10.11) include only the randomly assigned treatment Z_i, applying regression to these equations yields estimates that match the true values obtained from the full schedule of potential outcomes. Equation (10.12), however, produces biased estimates. In the true model, M_i has no effect on Y_i, and Z_i has a direct effect of 1. The estimates suggest just the opposite: Z_i has no direct effect on Y_i, and M_i has an effect of 1. Why are the estimates so severely biased? Because the unobserved factors that affect M_i are correlated with unobserved factors that affect Y_i. Regression presupposes that e_{1i} and e_{3i} are mean independent, which is to say that knowing the value of e_{1i} gives you no clues about the average value of e_{3i} and vice versa.

The covariance between e_{1i} and e_{3i} in this example, however, turns out to be 0.8, and the variance of e_{1i} is also 0.8. Applying equation (10.5), we obtain $\hat{b} = b + 1 = 1$. Applying equation (10.6), we obtain $\hat{d} = d - a(1) = 0$.

10.3 Why Experimental Analysis of Mediators Is Challenging

So regression analysis is fraught with problems when Z_i is randomly assigned but M_i is not. Naturally, the experimenter's response is to randomly assign M_i as well. This isn't a bad idea in principle, but before going down this path, we should first be clear about what we are up against when trying to estimate the effects of mediators. When researchers speak of the indirect causal influence on Y_i that Z_i transmits through M_i, they are addressing the following causal question: how would Y_i change if we were to hold Z_i constant while varying M_i by the amount it would change if Z_i were varied? Similarly, the direct effect of Z_i on Y_i controlling for M_i refers to this causal question: how would Y_i change if we were to vary Z_i while holding M_i constant at the value it would take on for a given value of Z_i? An experiment in which M_i is manipulated will not provide the answer to these questions, although it may come close. In order to see the mismatch between experimental results and what we are after when trying to estimate direct and indirect effects, let's take a closer look at the potential outcomes.

In order to represent direct and indirect effects using potential outcomes, we need to expand our system of notation. As before, define $M_i(0)$ as the potential value of M_i when $z = 0$; likewise, $M_i(1)$ is the potential value of M_i when $z = 1$. But now define $Y_i(m, z)$ to be the potential outcome of Y_i expressed when $M_i = m$ and $Z_i = z$. For example, the expression $Y(0, 1)$ refers to the potential outcome revealed when $M_i = 0$ and $Z_i = 1$. The expression $Y_i(M_i(1), 1)$ refers to the potential outcome expressed when Z_i equals 1 and when M_i takes on the potential outcome that occurs when $Z_i = 1$.

Sometimes the expanded potential outcome notation just re-expresses the old notation with more detail. For example, $Y_i(1) = Y_i(M_i(1), 1)$ because the potential outcome of Y_i when the treatment is administered is the same as the potential outcome of Y_i when the treatment is administered *and* M_i responds to the treatment. They are the same because $Y_i(1)$ incorporates all of the consequences that follow when the treatment is administered, and $M_i(1)$ is just one of these consequences. The same equality applies to $Y_i(0) = Y_i(M_i(0), 0)$. Using the $Y_i(m, z)$ notation, the total effect of Z_i on Y_i may be written $Y_i(M_i(1), 1) - Y_i(M_i(0), 0)$.

The direct effect of Z_i on Y_i controlling for M_i is more complicated. First, "the" direct effect does not necessarily have just a single definition. Rather, Z_i might exert a different effect on Y_i depending on the value of M_i. Second, when defining direct effects we must contend with the idea of a *complex potential outcome*, something that is purely imaginary. For example, $Y_i(M_i(0), 1)$ is the potential outcome expressed under two contradictory conditions: M_i takes on the potential outcome that occurs when $Z_i = 0$, yet $Z_i = 1$. This kind of potential outcome never occurs empirically. If $Z_i = 1$, we will observe $M_i(1)$. If $Z_i = 0$, we will observe $M_i(0)$. A complex potential outcome is based on a contrived situation in which Y_i responds to $Z_i = z$ and $M_i(1 - z)$.

Complex potential outcomes play an integral role in the definition of direct effects:

$$Y_i(M_i(0), 1) - Y_i(M_i(0), 0) \tag{10.13}$$

is the direct effect of Z_i on Y_i holding m constant at $M_i(0)$;

$$Y_i(M_i(1), 1) - Y_i(M_i(1), 0) \tag{10.14}$$

is the direct effect of Z_i on Y_i holding m constant at $M_i(1)$.

The direct effect in equation (10.13) describes the way in which Y_i responds to a change in Z_i holding m constant at $M_i(0)$. The first term in this equation is a complex potential outcome: it is the potential response of Y_i to the incongruous inputs $Z_i = 1$ and $M_i(0)$. The second direct effect in equation (10.14) also has an imaginary term, $Y_i(M_i(1), 0)$. Because only one of the two quantities that comprise each direct effect is empirical, estimating direct effects is impossible without additional assumptions.

The same holds for indirect effects. The indirect effect of Z_i on Y_i through M_i is the effect on Y_i of changing $M_i(0)$ to $M_i(1)$ while holding Z_i constant. Again, we confront a complex potential outcome for both definitions of an indirect effect:

$$Y_i(M_i(1), 1) - Y_i(M_i(0), 1) \tag{10.15}$$

is the indirect effect of $M_i(z)$ on Y_i holding Z_i constant at 1;

$$Y_i(M_i(1), 0) - Y_i(M_i(0), 0) \tag{10.16}$$

is the indirect effect of $M_i(z)$ on Y_i holding Z_i constant at 0.

Now we see why the study of mediation is so resistant to empirical investigation. Each of these equations involves one term that is fundamentally unobservable.

BOX 10.2

Complex Potential Outcomes

Potential outcomes are said to be complex when they respond to two or more inputs that cannot occur simultaneously. For example, potential outcomes of the form $Y_i(M_i(1 - z), z)$ respond to two inputs, $Z_i = z$ and $M_i(1 - z)$. When $Z_i = 0$, we observe the potential outcome $M_i(0)$, not $M_i(1)$. Yet the potential outcome $Y_i(M_i(1), 0)$ responds to the incongruous inputs $Z_i = 0$ and $M_i(1)$. Complex potential outcomes are inherently unobservable.

Even when we assume that both indirect effects are the same (i.e., the effect of M_i is the same regardless of the value of Z_i, which is another way of saying that there is no interaction between M_i and Z_i), we still cannot identify the indirect effect.

Bear in mind that we confront this formidable obstacle to inference even though this mediation exercise involves a much simpler setup than one ordinarily encounters in social science. Unlike Bhavnani's application, for example, we are here dealing with just a single mediator. The hypothetical mediator in our example is unusually well behaved: it is measured without error and unaffected by Y_i. The abstract example in this section represents the best-case scenario for social science applications, yet it reveals a fundamental limitation in what one can learn from an experiment that manipulates only Z_i—unless one is prepared to impose additional assumptions or consider special cases.

10.4 Ruling Out Mediators?

One special case in which the inference problems noted in the previous section can be circumvented occurs when the effect of Z_i on M_i is 0 for every observation. Under the sharp null hypothesis that $M_i(0) = M_i(1)$ for all subjects, the complex potential outcome $Y_i(M_i(0), 1)$ is the same as the observable potential outcome $Y_i(M_i(1), 1)$. Thus, the indirect effect defined in equation (10.15) is zero. Similarly, under the sharp null hypothesis of $Y_i(M_i(1), 0) = Y_i(M_i(0), 0)$, the indirect effect in equation (10.16) is zero.[4] The empirical implication is interesting: if we could demonstrate that the causal effect of Z_i on M_i is zero for all observations, we could also rule out M_i as a mediator. An experiment may be incapable of demonstrating the transmission of effects via a mediator, but it may indicate when mediation does *not* occur.

How can we establish that the effect of Z_i on M_i is zero for all observations? Recall from our discussion of heterogeneous treatment effects in Chapter 9 that estimating an average effect of zero does not necessarily imply that the effect of Z_i on M_i is zero for all observations. In the first place, our estimated ATE may be subject to sampling variability; when we speak of an estimate of zero, we have in mind a precisely estimated average effect that is close to zero, not merely a noisy estimate that is not significantly different from zero. Second, it is conceivable that positive effects among some observations offset negative effects among other observations, leading to an average effect of zero according to a test of the sharp null hypothesis of no effect. In order to rule out this possibility, we could stipulate that Z_i can only influence M_i

4 The same point can be made in relation to the regression model discussed in section 10.1. If $a_i = 0$ for all i, then even if the parameters b_i, c_i, and d_i in equations (10.2) and (10.3) vary across observations, the average total effect nevertheless equals the average direct effect: $E[a_i b_i + d_i] = 0E[b_i] + \text{Cov}(a_i, b_i) + E[d_i] = E[d_i]$, since the covariance between a constant and a variable is zero.

in one direction (e.g., reserving council seats for women cannot reduce the probability that women candidates run for office in the subsequent election). Alternatively, we could test the null hypothesis of homogeneous effects using the methods outlined in Chapter 9. The test statistic is the difference between the treatment and control groups' variances, and the p-value is calculated using randomization inference. Neither approach settles the question of heterogeneity. Our stipulations may be wrong, and our empirical investigation may lack adequate power to detect heterogeneity. That said, we do learn something useful about mediation when discovering an apparent lack of causal relationship between Z_i and proposed mediators; conversely, when a strong relationship between Z_i and M_i is detected, we know that M_i cannot be ruled out as a possible mediator.

Returning to the example of reserved seats for women legislators, the data suggest that certain mediators appear to be affected by the treatment. The percentage of wards in which at least one woman ran for office in the election following the reservation increased dramatically from 35.8% to 73.0%. On the other hand, the average turnout rate was very similar in the treatment group (41.6%) and control group (42.2%), suggesting that a surge in turnout may not be a mediator. In order to rule out this mediator more convincingly, additional analysis is required to establish that this zero effect is estimated precisely and that there is little indication of heterogeneous effects. We leave these additional analyses to the exercises.

10.5 What about Experiments That Manipulate the Mediator?

As mentioned earlier, one fundamental problem with regression-based mediation analysis stems from the fact that M_i is not independently manipulated through some sort of random intervention. In response to this problem, one might envision an experiment in which both Z_i (e.g., limes vs. no limes) and M_i (vitamin C vs. no vitamin C) are each manipulated randomly. Although this line of inquiry might be interesting in its own right, it does not identify the causal estimands in equations (10.15) and (10.16). Holding Z_i constant and varying M_i—for example, providing no limes but regulating diet so that some seafarers get a lime's worth of vitamin C—is similar in spirit to $Y_i(M_i(1),0) - Y_i(M_i(0),0)$, the indirect effect that Z_i transmits through M_i, holding Z_i constant at 0. But strictly speaking, $M_i(1)$ is the value of the mediator when subjects receive the treatment (limes). A lime's worth of vitamin C does not necessarily generate the same indirect effect on Y_i as a lime itself, due to the way that limes tend to be eaten or their nutrients absorbed. Moreover, providing vitamin tablets rather than limes might have other consequences for diet that could in turn affect Y_i. In other words, experiments that approximate changes in M_i may be instructive, but their ability to provide empirical estimates of complex potential outcomes inevitably involves additional assumptions.

Can one build a convincing case for mediation based on double experiments? In principle, the answer is yes. A series of studies that manipulate M_i in different ways may reveal that outcomes do not depend on the manner in which M_i is altered. This experimental strategy turns out to work in the case of vitamin C and scurvy. Limes cure scurvy, but so do equivalent doses of vitamin C from tablets, lemons, jicama, or broccoli.

In practice, experiments conducted in field settings rarely have the luxury of manipulating mediators in an exact manner. Thinking back to Bhavnani's list of possible mediators, it is difficult to come up with practical strategies for setting the number of female incumbents, voters' sense of whether it is appropriate or desirable to have women representatives, or the turnout rate in local elections. Frankly, most researchers would count themselves lucky if they were able to induce *any* noticeable change in these mediating variables. To the extent that one can move these mediators through some sort of intervention, the changes that are induced as a result are not necessarily equivalent to a reservation-induced change in terms of their effects on M_i or Y_i.

When researchers cannot set the level of M_i but rather intervene in the hopes of altering M_i in some way, the experiment amounts to the encouragement design that we discussed in Chapter 6. The experimenter introduces a random encouragement Z_i in an attempt to affect M_i, which under special conditions identifies the average treatment effect of M_i on Y_i among Compliers, those for whom M_i changes in the anticipated direction if and only if they are treated. The special conditions are excludability and monotonicity. Of particular concern in this context is excludability, which requires that the random encouragement has no effect on potential outcomes except insofar as it affects M_i. In other words, the encouragement must have no direct effect on Y_i. This assumption is much more restrictive than the framework with which we started the chapter, where both direct and indirect effects were modeled as free parameters. When we use an encouragement design and apply instrumental variables regression in order to estimate the CACE of M_i on Y_i, we are stipulating that $d = 0$.

Excludability is a strong assumption, particularly given the various mediators that may transmit Z_i's influence to Y_i. Reservations might affect the election of women not only by increasing the number of female incumbents but also by changing voters' attitudes and behavior. If our encouragement increases the number of female incumbents, it may also set in motion other mediators. In order to identify the effects of v mediators, at least v encouragements must be deployed in the experimental design, and the random encouragements must influence each mediator to some degree without affecting Y_i directly. Not easy.

In sum, researchers preparing to investigate mediation should appreciate what they are up against. In certain highly controlled settings, where mediators can be set to specific levels in ways that are assumed to have no direct influence on Y_i, researchers should endeavor to manipulate mediators in a variety of different ways. For example, different dosages of vitamin C could be administered via capsule, skin patch, or diet. If the apparent effect of M_i on Y_i is the same regardless of how change in M_i is induced, it may be reasonable to infer that the exclusion restriction is valid and that complex potential outcomes can be approximated by observable outcomes. In the more typical case in which researchers can only encourage changes in M_i, the challenges are greater. In the case of a single mediator, the encouragement must satisfy monotonicity and affect the mediator without otherwise affecting Y_i outcomes. For multiple mediators, multiple encouragements are required, and a variety of experimental encouragements may be necessary to make a convincing case that the mediators' effects on Y_i do not depend on how changes in the mediators were induced.

10.6 Implicit Mediation Analysis

Given the many practical challenges that arise when M_i is not manipulated directly or not manipulated at all, experimental researchers may wish to scale back their ambitions when investigating mediation. Rather than trace the full causal sequence that leads from Z_i to Y_i by measuring how each of the intervening variables transmits Z_i's influence, researchers might instead manipulate mediators implicitly by adding and subtracting elements of the treatment itself. The focus is no longer on how a Z_i-induced change in M_i influences outcomes but rather on the relative effectiveness of different classes of treatments whose attributes affect one or more mediators along the way. This form of mediation analysis is implicit in the sense that the researcher does not attempt to estimate the effects of observed changes in M_i; rather, the researcher posits the way in which various treatments affect one or more unobserved or implicit mediators.

To draw an example from the study of social welfare policy, several large-scale field experiments have evaluated the effects of conditional cash transfers. These programs involve government payments to poor families who agree to keep their children enrolled in school and bring them to health clinics to receive basic medical services. The experimental evidence suggests that these programs lead to improved educational outcomes for children in developing countries. Conditional cash transfer programs call to mind two causal pathways: cash and conditions. These programs might work because they provide cash subsidies to poor families, enabling them to invest in their children's education. Another possibility is that these programs work because they impose requirements on poor families that they must meet if they accept the cash subsidies. These two mediators have been investigated by randomly assigning families to one of three experimental groups: a control group that receives no subsidy or instructions from government, a group that receives cash without conditions, and a group that receives cash along with conditions. The experimental treatments are designed to highlight different mediating pathways, but the approach does not require the researcher to construct a comprehensive statistical model that includes measures of household expenditures and compliance with government requirements. If such measures exist, it is helpful to assess whether these intermediate outcomes change as expected by the experimental encouragements, but the analyst does not attempt to estimate the indirect effects of these mediators. A recent experiment using this design suggests that cash without conditions works about as well as cash with conditions.[5] The implication is that the key ingredient is household income, not the requirements that the government imposes on parents.

Three aspects of implicit mediation analysis make it attractive. First, it never strays from the unbiased statistical framework of comparing randomly assigned groups. Samples are not partitioned by post-treatment variables. Second, by adding and subtracting ingredients from an intervention, this approach lends itself to exploration and discovery of new treatments. Most social science research programs are in early stages of development, and scholars are still trying to test basic propositions about what works. Implicit mediation designs facilitate this process of exploration by providing clues about the active ingredients that cause a treatment to work especially well. Third, this line of research allows the researcher to gauge treatment effects on a wide array of outcome variables without necessarily committing to a set of claims about the precise causal sequence by which outcomes affect each other, such as the path diagram in Box 10.1. One may still measure "mediating" variables and examine whether the treatment has the expected influence on them. For example, one could assess how cash transfers affect parents' perceived level of economic security or their need for child labor as a source of income. These tests perform the same function as a *manipulation check* in laboratory experiments; they establish that the treatment influences intervening variables in an expected fashion.

Implicit mediation analysis relies to an important extent on theory. Every treatment involves a collection of ingredients that are subject to theoretical interpretation, sometimes more than one. Consider, for example, the four postcards listed in the appendix to this chapter. In August of 2006, 180,000 Michigan households were randomly divided into five groups: 100,000 households received no mail, and four sets of 20,000 households received one of the four postcards. Using public records, Gerber, Green, and Larimer measured voter turnout rates in each experimental group.[6] As is apparent from a close reading of the postcards, the treatments contain many ingredients, each of which is designed to influence the social costs that people may incur when failing to comply with the norm of voter participation. Here are the ingredients

5 Baird, McIntosh, and Özler 2009.
6 Gerber, Green, and Larimer 2008.

BOX 10.3

Definition: Manipulation Check

A manipulation check is a method for establishing the empirical relationship between the intended and actual treatments. Suppose a researcher seeks to assess whether classroom discussion improves student performance on end-of-year achievement tests. The intervention is intended to create a classroom environment that encourages discussion. In order to verify the link between the intended and actual treatments, the researcher might send observers to treatment and control classrooms, and instruct observers to record the proportion of each class period devoted to discussion. A manipulation check in this context is a statistical analysis confirming that, as expected, discussion was substantially more prevalent in treatment classrooms than in control classrooms.

we sought to introduce in each mailing. (The reader may decide whether we succeeded or inadvertently introduced other ingredients.) One ingredient is stating a widely shared social norm: the first experimental treatment was a mailing that urged recipients to do their civic duty. Another ingredient was surveillance: in addition to telling people to do their civic duty, we announced that they were part of an academic study, and that their participation in the upcoming election would be monitored (testing the Hawthorne effect, or the effect of being studied). A further ingredient is disclosure: a third treatment reported whether each voter in the household had voted in recent elections and promised to send an updated mailing to indicate whether ballots were cast in the upcoming election. A final treatment amplified the level of disclosure by reporting not only whether members of the household voted but also whether others on the block voted.

The results presented in Table 10.2 confirm that turnout rises as the social pressure to vote increases. The last two treatments are especially effective. Revealing the vote histories of those in the household and promising to update the histories with a follow-up mailing raises turnout by 4.8 percentage points, which is approximately ten times as effective as a conventional mailing that encourages voting. Revealing the vote histories of both the household and the neighbors increases turnout by 8.1 percentage points, which is an extraordinarily powerful effect even by comparison to other effective voter mobilization tactics, such as door-to-door canvassing.

This experiment provides important insights into some implicit mediators, even if the study in no way settles the question of how each of the specific components of social pressure combine to produce the observed effect. Both of the treatments that show recipients' official vote history produced large effects. The Neighbors treatment

TABLE 10.2

Illustration of implicit mediation: the effects of different forms of social pressure mail on voter turnout, 2006

	Experimental group				
	Control (not mailed)	Civic Duty (encouraged to vote)	Hawthorne (encouraged & monitored)	Self (encouraged, monitored, shown own past voting)	Neighbors (encouraged, monitored, shown own and others' past voting)
Percent voting	29.7%	31.5%	32.2%	34.5%	37.8%
N of individuals	191,243	38,218	38,204	38,218	38,201

Source: Gerber, Green, and Larimer 2008.

produces much larger effects. One interpretation is that the social costs of disclosing one's own vote history to others beyond the household impels people to vote. This is not the only interpretation, however. As so often happens in experimental science, two treatments of interest differ in more than one respect. It may be that what matters is not disclosure to the neighbors but rather what one infers about voting norms from seeing how many of the neighbors voted in prior elections. These and other nuances have been explored in more than a dozen subsequent experiments, as researchers address gaps in our design and extend the line of inquiry in new directions.[7] Some studies vary whether subjects are scolded about upholding voting norms; others vary the neighbors' apparent voting rates. The implicit mediation approach seems much more fruitful in this domain than a more traditional approach in which one attempts to predict voting using post-treatment variables such as voters' feelings of obligation to conform to norms of civic duty or concerns about what others will think of them if they fail to vote. Even an encouragement design seems hopeless here, given the difficulty of devising reliable ways of changing specific mediators without changing potential outcomes.

SUMMARY

When experimental researchers present evidence of a causal effect, invariably someone in the audience will ask the speaker to discuss the mediating factors that explain this effect. The stronger the experimental effect, the greater the audience's interest in mediators. When experimenters fail to offer evidence that explains how their inter-

7 See, for example, Mann 2010; Panagopoulos 2010; Davenport 2010; Aronow 2011.

vention's effect is transmitted, audiences have been known to grumble about "black box" experimentation.

One can scarcely fault scholars for expressing curiosity about the mechanisms by which an experimental treatment transmits its influence, especially when the treatment involves a package of ingredients. At the same time, impatience for answers reflects a failure to appreciate the challenge of rendering convincing evidence about mediation. This chapter has reviewed the strong statistical assumptions on which mediation analysis often rests. Even under the assumption that the mediators are measured perfectly and all causal effects are constant across subjects, one still must contend with the threat of bias when post-treatment variables are handled as though they were randomly assigned factors. Adding mediators as right-hand-side variables in regression models typically exaggerates the effect of the mediator and understates the direct effect of the treatment. Previous chapters have warned against attaching a causal interpretation to the apparent "effects" of covariates; here, the message is even stronger. Whereas the inclusion of pre-treatment variables as regressors does not jeopardize consistent estimation of the average treatment effect of the randomized intervention, the inclusion of post-treatment variables as regressors may well bias the estimation of the treatment's direct effect.

When mediators are manipulated experimentally, prospects for sound inference improve, but basic problems remain due to the impossibility of observing complex potential outcomes. In practice, researchers conducting field experiments rarely have the luxury of manipulating mediators directly, which means that they must rely on encouragement designs and their attendant assumptions. These designs are vulnerable to bias when exclusion restrictions are violated, as may occur when multiple mediators link cause and effect.

In light of these difficulties, this chapter has suggested two less ambitious lines of inquiry. One is to regard mediators as outcome variables, paying special attention to the question of when potential mediators may be dropped from consideration on the grounds that they appear to be unaffected by the randomly assigned treatment. Although you cannot prove the sharp null hypothesis of no effect, you can adduce evidence that suggests whether this conjecture is a reasonable approximation. This exercise may help winnow the number of plausible hypotheses about mediation.

A second suggestion is to vary the treatments in theoretically guided ways so as to manipulate mediators implicitly. By focusing only on experimental comparisons, this type of investigation tries to minimize the risk of bias, which haunts conventional mediation analysis. Implicit mediation analysis is a form of experimental investigation that has generated discoveries in the natural sciences. Causal mechanisms are posited, and a variety of experiments are conducted using different treatments or contexts in an effort to disable or augment these putative mechanisms. Like any theoretically guided investigation, implicit mediation analysis may lead to disagreements about the interpretation of experimental results, particularly when several mediators are thought to be affected by a set of treatments. The social pressure experiments described above illustrate how these disputes arise and how further experimentation gradually clarifies which mediators are operative. At best, implicit mediation analysis contributes to a more refined theoretical understanding of causal pathways. At worst, it generates experimental results indicating the total effects of various treatments, which may be interesting in their own right.

SUGGESTED READINGS

The use of post-treatment covariates is criticized by Gelman and Hill (2007). Robins and Greenland (1992) show how the causal chain interpretation of regression hinges on the constant effects assumption. Rubin (2005) critiques mediation analysis from a potential outcomes perspective, and our example in Table 10.1 is patterned after his. Imai, Keele, and Yamamoto (2010) suggest the use of sensitivity analysis to assess identifying assumptions. Spencer, Zanna, and Fong (2005) defend the use of experimental designs that try to manipulate mediators directly or implicitly.

EXERCISES: CHAPTER 10

1. Important concepts:
   (a) Suppose that equations (10.1), (10.2), and (10.3) depict the true causal process that generates outcomes. Referring to these equations, define the direct effect of Z_i on Y_i and the indirect effect that Z_i transmits through M_i to Y_i.
   (b) Explain why the equation *Total effect* = *Direct effect* + *Indirect effect* breaks down when the parameters of equations (10.1), (10.2), and (10.3) vary across subjects.
   (c) Suppose that the effect of M_i on Y_i varies from one subject to the next. Show that the indirect effect of Z_i on Y_i is zero when the treatment effect of Z_i on M_i is zero for all subjects.
   (d) Explain why the complex potential outcome $Y_i(M_i(0),1)$ defies empirical investigation.
   (e) Explain the distinction between the indirect effect that Z_i transmits to Y_i through M_i given in equations (10.15) and (10.16) and the causal effect of M_i, defined using $Y_i(m, z)$ notation as $Y_i(1, 0) - Y_i(0, 0)$ or $Y_i(1, 1) - Y_i(0, 1)$. (Hint: Look closely at how the mediator takes on its value).
2. When researchers use an encouragement design to study mediation, what assumptions must they make in order to satisfy the CACE Theorem from Chapter 6?
3. Consider the following schedule of potential outcomes for 12 observations. This table illustrates a special situation in which the disturbance e_{1i} is unrelated to the disturbance e_{3i}.
   (a) What is the average effect of Z_i on M_i?
   (b) Use yellow to highlight the cells in the table of potential outcomes to indicate which potential outcomes for Y_i correspond to $Y_i(M_i(0), 0)$. Use green to highlight the cells in the table of potential outcomes to indicate which potential outcomes for Y_i correspond to $Y_i(M_i(1), 1)$. Put an asterisk by the potential outcomes for Y_i in each row that correspond to the complex potential outcome $Y_i(M_i(0), 1)$. Put a pound sign by the potential outcomes for Y_i in each row that correspond to the complex potential outcome $Y_i(M_i(1), 0)$.

Observation	$Y_i(m = 0, z = 0)$	$Y_i(m = 0, z = 1)$	$Y_i(m = 1, z = 0)$	$Y_i(m = 1, z = 0)$	$M_i(z = 0)$	$M_i(z = 1)$
1	0	0	0	0	0	0
2	0	0	0	0	0	1
3	0	0	0	0	1	1
4	0	1	0	1	0	0
5	0	1	0	1	0	1
6	0	1	0	1	1	1
7	1	0	1	1	0	0
8	1	0	1	1	0	1
9	1	0	1	1	1	1
10	0	1	1	1	0	0
11	0	1	1	1	0	1
12	0	1	1	1	1	1

(c) What is the average total effect of Z_i on Y_i?

(d) What is the average direct effect of Z_i on Y_i holding M_i constant at $M_i(0)$? Hint: see equation (10.13).

(e) What is the average direct effect of Z_i on Y_i holding M_i constant at $M_i(1)$? Hint: see equation (10.14).

(f) What is the average indirect effect that Z_i transmits through M_i to Y_i when $Z_i = 1$? Hint: see equation (10.15).

(g) What is the average indirect effect that Z_i transmits through M_i to Y_i when $Z_i = 0$? Hint: see equation (10.16).

(h) In this example, does the total effect of Z_i equal the sum of its average direct and indirect effect?

(i) What is the average effect of M_i on Y_i when $Z_i = 0$?

(j) Suppose you were to randomly assign half of these observations to treatment ($Z_i = 1$) and the other half to control ($Z_i = 0$). If you were to regress Y_i on M_i and Z_i, you would obtain unbiased estimates of the average direct effect of Z_i on Y_i and the average effect of M_i on Y_i. (This fact may be verified using the R simulation at http://isps.research.yale.edu/FEDAI.) What special features of this schedule of potential outcomes allows for unbiased estimation?

(k) In order to estimate average indirect effect that Z_i transmits through M_i to Y_i, estimate the regressions in equations (10.1) and (10.3) and multiply the estimates of a and c together. Use the simulation to show that this estimator is unbiased when applied to this schedule of potential outcomes. Why does this estimator, which usually produces biased results, produce unbiased results in this example?

4. Earlier we indicated that in Bhavnani's experiment, the pathway between random reservations for women and voter turnout appears to be zero, suggesting that we may be able to rule out this mediator as a possible pathway.
   (a) With the replication dataset at http://isps.research.yale.edu/FEDAI, use randomization inference to test the sharp null hypothesis of no treatment effect on turnout in 2002 for any subject.
   (b) Following the steps described in Chapter 9, use randomization inference to test the null hypothesis that $\mathrm{Var}(\tau_i) = 0$.
   (c) It is tempting to include voter turnout in 1997 as a covariate when assessing the relationship between reservations and turnout in 2002, but is turnout in 1997 a pre-treatment covariate? Explain why or why not.
5. In most places in the United States, you can only vote if you are a registered voter. You become a registered voter by filling out a form and, in some cases, presenting identification and proof of residence. Consider a jurisdiction that requires and enforces voter registration. Imagine a voter registration experiment that takes the following form: unregistered citizens are approached at their homes with one of two randomly chosen messages. The treatment group is presented with voter registration forms along with an explanation of how to fill them out and return them to the local registrar of voters. The control group is presented with an encouragement to donate books to a local library and receives instructions about how to do so. Voter registration and voter turnout rates are compiled for each person who is contacted using either script. In the table below, Treatment = 1 if encouraged to register, 0 otherwise; Registered = 1 if registered, 0 otherwise; Voted = 1 if voted, 0 otherwise; and N is the number of observations).
   (a) Estimate the average effect of Treatment (Z_i) on Registered (M_i). Interpret the results.
   (b) Estimate the average total effect of treatment on voter turnout (Y_i).
   (c) Regress Y_i on X_i and M_i. What does this regression seem to indicate? List the assumptions necessary to ascribe a causal interpretation to the regression coefficient associated with M_i. Are these assumptions plausible in this case?
   (d) Suppose you were to assume that the treatment has no direct effect on turnout; its total effect is entirely mediated through registration. Under this assumption and monotonicity, what is the Complier average causal effect of registration on turnout?

Treatment	Registered	Voted	N
0	0	0	400
0	0	1	0
0	1	0	10
0	1	1	90
1	0	0	300
1	0	1	0
1	1	0	100
1	1	1	100

	No mail	Standard letter	Letter with threat	Letter with norms	Letter with threat & norms	Letter with appeal to fairness	Letter with threat & fairness
Payment of registration fee	1.58%*	8.62%	9.67%	8.23%	9.70%	8.19%	9.32%
Any response from recipient	N/A	43.09%	45.01%	40.70%	42.77%	38.82%	42.81%
N	2,586	6,858	6,694	6,825	6,960	6,920	6,750

*This figure assumes that 14.41% of the control group had undeliverable addresses, which is the same rate as the treatment groups. Note that unlike the treatment groups, the control group did not receive a letter or a prepaid return envelope.

6. Fellner, Sausgruber, and Traxler (2009) collaborated with an Austrian tax collection agency to examine the conditions under which people who own televisions pay the mandatory annual fee when requested to do so via an official letter from the agency.[8] The researchers randomly varied the content of the mailings so that it emphasized either (1) a threat of prosecution for tax evasion, (2) a fairness appeal to pay one's fair share rather than forcing others to bear one's tax burden, or (3) information stating the descriptive norm that 94% of households comply with this tax. These interventions seem to accentuate three mediators: fear of punishment, concern for fairness, and conformity with perceived norms. There are two outcome measures. One is whether the recipient responded to the request for an explanation for non-payment by mailing in a prepaid envelope. The other outcome, which is a subset of the first, is payment of the registration fee. The table above presents an excerpt of the results.
   (a) This experiment included two control groups, one that received no letter and another that received a standard letter. Explain how the use of two control groups aids the interpretation of the results.
   (b) Analyze the data using the statistical model of your choice, and assess the effectiveness of threats, assertion of norms, and appeals to fairness.
   (c) What light do these results shed on the question of why people respond (or fail to respond) to requests to pay taxes?
7. Several experimental studies conducted in North America and Europe have demonstrated that employers are less likely to reply to job applications from ethnic minorities than from nonminorities.
   (a) Propose at least two hypotheses about why this type of discrimination occurs.
   (b) Propose an experimental research design to test each of your hypotheses, and explain how your experiment helps identify the causal parameters of interest.
   (c) Create a hypothetical schedule of potential outcomes, and simulate the results of the experiment you proposed in part (b). Analyze and interpret the results.
8. Sometimes is it difficult and costly to conduct a long-term evaluation of policies or programs. For example, many states have instituted civics education requirements in high schools on

8 Fellner, Sausgruber, and Traxler 2011.

the grounds that this type of curriculum makes for a more knowledgeable and involved citizenry. However, it is often impossible to track students after they leave high school. Suppose you were asked to evaluate the impact of a recommended civics curriculum that is being considered by a state that currently does not have a civics requirement. You may randomly assign a large number of schools and students to different curricula, but you can only measure outcomes up to the point at which students leave school.

(a) Propose one or more mediating variables that you think explain why civics classes affect the attitudes and behaviors of students after they leave school.

(b) Propose a research design that would shed light on whether your hypothesized mediating variables are affected by civics classes.

(c) One problem with measuring short term outcomes is that effects may dissipate over time. Although your study cannot address this issue directly because long-term outcomes cannot be measured, suggest ways in which your design could at least shed some light on the rate at which effects decay over time.

9. Researchers who attempt to study mediation by adding or subtracting elements of the treatment confront the practical and conceptual challenge of altering treatments in ways that isolate the operation of a single causal ingredient. Carefully compare the four mailings from the Gerber et al. (2008) study, which are reproduced in the appendix to this chapter.

(a) Discuss the ways in which the treatments differ from one another.

(b) How might these differences affect the interpretation of Table 10.2?

(c) Suppose you were in charge of conducting one or more "manipulation checks" as part of this study. What sorts of manipulation checks would you propose, and why?

APPENDIX

A10.1 Treatment Postcards Mailed to Michigan Households

Note: The names listed on these sample mailings are fictitious.

30426-2 ||| || || || | XXX

For more information: (517) 351-1975
email: etov@grebner.com
Practical Political Consulting
P. O. Box 6249
East Lansing, MI 48826

PRSRT STD
U.S. Postage
PAID
Lansing. MI
Permit # 444

ECRLOT **C002
THE JONES FAMILY
9999 WILLIAMS RD
FLINT MI 48507

Dear Registered Voter:

DO YOUR CIVIC DUTY AND VOTE!

Why do so many people fail to vote? We've been talking about this problem for years, but it only seems to get worse.

The whole point of democracy is that citizens are active participants in government; that we have a voice in government. Your voice starts with your vote. On August 8, remember your rights and responsibilities as a citizen. Remember to vote.

DO YOUR CIVIC DUTY — VOTE!

(a) Civic Duty

30424-1 ||| || || ||

For more information: (517) 351-1975
email: etov@grebner.com
Practical Political Consulting
P. O. Box 6249
East Lansing, MI 48826

PRSRT STD
U.S. Postage
PAID
Lansing. MI
Permit # 444

ECRLOT **C001
THE SMITH FAMILY
9999 PARK LANE
FLINT MI 48507

Dear Registered Voter:

YOU ARE BEING STUDIED!

Why do so many people fail to vote? We've been talking about this problem for years, but it only seems to get worse.

This year, we're trying to figure out why people do or do not vote. We'll be studying voter turnout in the August 8 primary election.

Our analysis will be based on public records, so you will not be contacted again or disturbed in any way. Anything we learn about your voting or not voting will remain confidential and will not be disclosed to anyone else.

DO YOUR CIVIC DUTY — VOTE!

(b) Hawthorne

30422-4 ||| || || | ||

For more information: (517) 351-1975
email: etov@grebner.com
Practical Political Consulting
P. O. Box 6249
East Lansing, MI 48826

PRSRT STD
U.S. Postage
PAID
Lansing. MI
Permit # 444

ECRLOT **C050
THE WAYNE FAMILY
9999 OAK ST
FLINT MI 48507

Dear Registered Voter:

WHO VOTES IS PUBLIC INFORMATION!

Why do so many people fail to vote? We've been talking about the problem for years, but it only seems to get worse.

This year, we're taking a different approach. We are reminding people that who votes is a matter of public record.

The chart shows your name from the list of registered voters, showing past votes, as well as an empty box which we will fill in to show whether you vote in the August 8 primary election. We intend to mail you an updated chart when we have that information.

We will leave the box blank if you do not vote.

DO YOUR CIVIC DUTY — VOTE!

OAK	ST	AUG 04	Nov 04	Aug 06
9999	ROBERT WAYNE		Voted	———
9999	LAURA WAYNE	Voted	Voted	———

(c) Self

30423-3 ||| || || | |||

For more information: (517) 351-1975
email: etov@grebner.com
Practical Political Consulting
P. O. Box 6249
East Lansing, MI 48826

PRSRT STD
U.S. Postage
PAID
Lansing. MI
Permit # 444

ECRLOT **C050
THE JACKSON FAMILY
9999 MAPLE DR
FLINT MI 48507

Dear Registered Voter:

WHAT IF YOUR NEIGHBORS KNEW WHETHER YOU VOTED?

Why do so many people fail to vote? We've been talking about this problem for years, but it only seems to get worse. This year, we're taking a new approach. We're sending this mailing to you and your neighbors to publicize who does and does not vote.

This chart shows the names of some of your neighbors, showing which have voted in the past. After the August 8 election, we intend to mail an updated chart. You and your neighbors will all know who voted and who did not.

DO YOUR CIVIC DUTY — VOTE!

MAPLE DR	AUG 04	Nov 04	Aug 06
9995 JOSEPH JAMES SMITH	Voted	Voted	_____
9995 JENNIFER KAY SMITH		Voted	_____
9997 RICHARD B JACKSON		Voted	_____
9999 KATHY MARIE JACKSON		Voted	_____
9999 BRIAN JOSEPH JACKSON		Voted	_____
9991 JENNIFER KAY THOMPSON		Voted	_____
9991 BOB R THOMPSON		Voted	_____
9993 BILL S SMITH			_____
9989 WILLIAM LUKE CASPER		Voted	_____
9989 JENNIFER SUE CASPER		Voted	_____
9987 MARIA S JOHNSON	Voted	Voted	_____
9987 TOM JACK JACKSON	Voted	Voted	_____
9987 RICHARD TOM JOHNSON		Voted	_____
9985 ROSEMARY S SUE		Voted	_____
9985 KATHRYN L SUE		Voted	_____
9985 HOWARD BEN SUE		Voted	_____
9983 NATHAN CHAD BERG		Voted	_____
9983 CARRIE ANN BERG		Voted	_____
9981 EARL JOEL SMITH			_____
9979 DEBORAH KAY WAYNE		Voted	_____
9979 JOEL R WAYNE		Voted	_____

(d) Neighbors

CHAPTER 11

Integration of Research Findings

The long process of designing an experiment, gathering data, and analyzing results eventually culminates in a set of estimates. Now the question is what to make of them. In a narrow sense, interpreting the estimates is a matter of giving a clear description of the apparent treatment effects and the uncertainty surrounding them. But researchers are expected to address broader questions as well: How do the estimates square with other research? What theories are supported or called into question? What new hypotheses seem to be suggested? What do the results imply for programs or policies?

Connecting research findings to the broader world of ideas and interventions sparks the imagination. At the same time, the guesswork involved stands in juxtaposition to the cautious mindset of experimental inquiry. The experimental researcher turns to randomized trials in order to be on the firmest scientific footing, minimizing reliance on assumptions and conjectures. With the completion of the study, the researcher is expected to offer speculations that go well beyond the confines of the statistical results. Whereas the estimation of the sample average treatment effect rests on procedures that ensure unbiased inference, extrapolation to other treatments, settings, subjects, or outcomes introduces the risk of error. Extrapolation may also introduce *bias*; mistaken assumptions may lead one to systematically over- or underestimate the size of the effect in some other context.

One way to address the uncertainty associated with extrapolation is to conduct further experiments under varying conditions. This is a step in the right direction, but even something as simple as comparing results from two experiments requires some assumptions. Any two experiments differ along an unmanageably large number of dimensions. If the same experiment were repeated using samples drawn from the same population, perhaps the differences could be ascribed to random sampling variability alone. In practice, there will inevitably be something idiosyncratic about the populations from which the subjects were selected, the manner in which

What you will learn from this chapter:

1. The distinction between the sample and population average treatment effect.
2. How the standard error of the sample average treatment effect differs from the standard error of the population average treatment effect.
3. A Bayesian algorithm for updating prior beliefs given experimental evidence.
4. The assumptions underlying meta-analysis, a statistical technique for pooling experimental results.
5. Regression-based methods for modeling treatments that vary in intensity.
6. Methods for assessing the accuracy of a model's predictions, both in absolute terms and in relation to alternative models.

subjects were sampled, the particular way in which the treatment was administered, the context within which each experiment took place, and the procedures used to measure outcomes. Even if all of these elements were identical, the mere fact that two experiments took place at different times introduces the possibility that the underlying parameters that govern cause and effect may have changed. The situation becomes even more complex when one attempts to square one's own experimental results with published research findings. Eye-catching results are more likely to find their way into print than null findings, a phenomenon known as *publication bias*.[1] Your experimental results may differ from published findings because experiments described in research journals are not a random selection from the set of all experiments on a given topic.

Try as you may to be on secure empirical footing when making causal claims, you must invoke additional assumptions when drawing connections between empirical results or between experimental findings and broader theoretical claims. In an effort to make these assumptions explicit, researchers sometimes turn to statistical models when attempting to integrate research findings. A statistical model specifies the units to which a causal law applies and a function linking inputs to outcomes.[2] For example, a statistical model might stipulate that as the dosage of a

1 Because statistical significance is often regarded as a requirement for publication, one symptom of publication bias is an unusually large number of published p-values just below the 0.05 threshold (Gerber and Malhotra 2008a, 2008b). Another symptom is larger reported effects among studies with smaller samples; because smaller studies tend to have larger standard errors, their estimated effects need to be larger in order to achieve significance at the $p < 0.05$ level. Those conducting literature reviews often plot effect size against sample size in order to detect publication bias (Gerber, Green, and Nickerson 2001).

2 For a discussion of the technical requirements of a coherent statistical model, see McCullagh 2002 and accompanying responses from other authors.

drug increases, the expected physiologic effect increases proportionally. Twice the dosage leads to twice the effect. Or it might specify a quadratic function in order to allow for the possibility that excessive quantities of the drug have harmful effects. The key points to bear in mind when imposing a statistical model on experimental data are: (1) modeling assumptions, when correct, function like additional data, improving the precision with which causal effects are estimated; (2) modeling assumptions are potentially fallible, and uncertainty about the fallibility of modeling assumptions contributes to uncertainty in the statistical conclusions; and (3) in an effort to relax modeling assumptions, researchers may build more flexible models with additional parameters, but adding parameters tends to reduce statistical precision.

This chapter provides a brief introduction to two classes of models that are used to aid generalization. One is meta-analysis, a widely used method for pooling experimental research findings. The attraction of meta-analysis is that a series of small experiments may each be unable to speak to a hypothesis with precision, but when pooled together, these experiments may suggest a clear conclusion. The statistical model underlying meta-analysis makes some demanding assumptions, however. When one conducts a meta-analysis, one is, in effect, merging subjects from different experiments into a single dataset. Is it plausible to regard subjects from different studies as participants in the same grand experiment? Meta-analysis is used convincingly in research domains where subjects are drawn randomly from the same population or where sampling procedures are immaterial because treatment effects are reasonably assumed to be homogeneous across subjects. When these requirements are not met or when experimental procedures vary from one study to the next, meta-analysis may produce misleading results. Another class of models uses regression to extrapolate from one treatment to another. The most straightforward instance of model-guided extrapolation occurs when treatments vary in terms of dosage or intensity. A researcher who conducts an experiment to gauge the effect of one dosage may wish to generalize to higher or lower dosages, but an incorrect statistical model may lead to incorrect extrapolations. The statistical uncertainty that arises when one uses an experimental estimate to forecast what will happen in a different setting is a combination of sampling variability and modeling uncertainty; the latter is too often ignored.

Because generalization (with or without the aid of statistical models) involves guesswork, this chapter characterizes it in Bayesian terms. This approach defines probability with reference to a person's subjective assessment about the likelihood that an event will occur or that a claim is true. The Bayesian framework is especially useful when formalizing the way in which beliefs are updated as new information becomes available. Bayes' rule may be used to describe an idealized situation in which an experimenter launches a study with a hypothesis in mind and some sense of what outcomes are compatible or incompatible with it. When the experimental results become known, the experimenter uses the information to update his or her assessment of the hypothesis.[3]

In this chapter, we use a Bayesian framework to structure our discussion of three related topics. The first is the challenge of generalizing from the sample average treatment effect to the population average treatment effect. This section is designed to clarify the role of random sampling and to call attention to the difference between the standard errors associated with sample and population average treatment effects. The second topic is the challenge of drawing inferences from a series of experiments. We begin by considering this problem in abstract terms, describing a Bayesian model that formalizes how beliefs are updated as new evidence becomes available. The critical feature of this model is that it makes explicit allowance for the possibility that an experiment may generate a biased estimate of the population average treatment effect if subjects are not drawn randomly from the population. This model turns out to be quite general and also covers other sources of bias that potentially afflict experimental or observational research. With this analytic framework in place, we use the research literature on compliance with social norms to illustrate the concerns that arise when interpreting a series of replication studies involving the same treatment. The final section of the chapter looks at the topic of extrapolation when treatments vary in intensity. Statistical models are used to ascertain the underlying dose-response curve. Using an example of a suite of experiments in which different financial inducements were distributed in an effort to encourage Kenyans to purchase anti-malaria bed nets, we discuss the kinds of assumptions that come into play in the construction, cross-validation, and interpretation of dose-response models.

11.1 Estimation of Population Average Treatment Effects

When introducing the concept of sampling variability in Chapter 3, we conjured up an image of a hypothetical experiment that is conducted repeatedly under identical circumstances. The subjects' potential outcomes remain fixed; what changes is the way in which the subjects are allocated by chance to treatment and control groups. A *replication study* departs from this framework in a subtle but important way. Instead of re-running the experiment with the same set of subjects, a replication

3 Under this learning model, different researchers may still draw different inferences from a given set of experimental results if they approach the study with different initial beliefs about the claim being tested and have different assessments of whether the experimental design provides an informative test of the hypothesis. The researchers feed different inputs into Bayes' rule and emerge with different posterior beliefs.

study draws a new set of subjects and allocates them to treatment and control. The follow-up study may differ from the original experiment in several ways. First, the subjects and their potential outcomes change. Second, the treatment may change, or it may be administered in a different way. Third, the setting within which the treatment is deployed may change. Fourth, the criteria used to classify outcomes may change. Each of these elements may cause results to vary from one experiment to the next.

The most direct form of replication tries to hold the latter three factors constant while varying the subjects. If the subjects in both studies are selected at random from the same population, the second study in effect adds new subjects to the first experiment. When the treatment, context, and outcome measures are the same in both studies, sampling variability is the only remaining reason why the two experiments might generate different results. This time, however, sampling variability refers not to the set of outcomes that one observes across hypothetical experiments with the same subjects, but rather to the set of outcomes that one observes across a series of experiments each involving different samples from the same population.

Under random sampling, it is possible to provide a precise statistical characterization of this sample-to-sample variability. To see how random sampling works, consider the following example in which N experimental subjects are selected from a population of N^* people who donated more than \$100 to federal candidates. Suppose that N^* is very large in absolute terms and in relation to N, so that even if subjects were sampled with replacement, there is a negligible probability of selecting the same subject more than once. Under random sampling, N subjects are selected independently and with identical probability. The N subjects are then allocated randomly to treatment and control groups, with m subjects in treatment and $N - m$ in control. Replication of the study involves going back to the same list and drawing a new random sample of size N and again allocating m subjects to the treatment group and

BOX 11.1

Definition: Replication

When we use the term *replication,* we refer to an experiment that deploys the same design as an earlier study. A replication study is conducted under similar conditions, with similar subjects, treatments, and outcome measures.

The term replication is sometimes used to refer to an activity that is more accurately termed *verification*. Verification refers to an effort to reproduce reported statistical results using the original data. The purpose of verification is to detect clerical errors or determine the sensitivity of the reported results to modeling choices.

$N - m$ to the control group. When these conditions are met (independent random sampling from a fixed, large population), the standard error of the original experimental estimate of the ATE in the population of N^* units is:

$$\text{SE}(\widehat{\text{PATE}}) = \sqrt{\frac{\text{Var}(Y_i(1))}{m} + \frac{\text{Var}(Y_i(0))}{N - m}}. \tag{11.1}$$

Estimating this quantity empirically is straightforward. The sample variance among units assigned to the treatment group provides an estimate of $\text{Var}(Y_i(1))$, and the sample variance among units assigned to the control group provides an estimate of $\text{Var}(Y_i(0))$. Sampling variability of the estimated PATE diminishes as additional observations are added to the treatment and control groups, but the size of the population (N^*) plays no role. Because N^* is assumed to be very large in relation to N, the selection of subjects into the treatment group has no effect on the potential outcomes that are available to the control group. As a result, the covariance between $Y_i(0)$ and $Y_i(1)$ is absent from this formula.

Equation (11.1) differs from the variance formula given in equation (3.4), because in Chapter 3 we sought to estimate the ATE in the *sample*, whereas in this chapter we seek to estimate the ATE in the *population* from which our sample is drawn. Because sampling from a large population introduces an extra element of uncertainty, the standard error of the population ATE is greater than or equal to the standard error of the sample ATE. The two standard errors are approximately equal when treatment effects are assumed to be the same for all units (i.e., $\text{Var}(Y_i(0)) = \text{Var}(Y_i(1))$, and the correlation between $Y_i(0)$ and $Y_i(1)$ is 1.0), as would be the case under the sharp null hypothesis that the treatment effect is zero for every unit. (See exercise 11.4). When effects are equal for everyone, sampling from a larger population becomes inconsequential—no matter which subjects one selects, the sample's ATE is always the same.

Sampling subjects randomly from a large population is unusual in field experimental research. Typically the subjects are selected through *convenience sampling*, which is to say no particular sampling method other than the fact that the subjects were close at hand. For example, when conducting our 1998 study of voter mobilization, we selected a sample consisting of all registered voters in New Haven living in voting wards outside the Yale campus and residing at addresses where no more than two voters were registered. We did not draw a random sample of midsized towns or locations holding state and legislative elections; instead, we selected a proximal location where we could conduct and supervise a large canvassing campaign. When sampling procedures are ad hoc, it is difficult to quantify the uncertainty that arises when we extrapolate from a sample to a population. Convenience sampling introduces two possible sources of uncertainty that are not taken into account in equation (11.1), which assumes random sampling. The first is sampling bias. Subjects obtained through convenience sampling may have systematically larger or smaller ATEs than subjects in the population. Second, a convenience sample tends to be a clustered sample, drawing all of the subjects from a common location or period of time. Even if the experimental cluster were selected at random from the population of clusters, the subjects in the selected cluster may all share unobserved attributes that predict outcomes. The next section illustrates how departures from simple random sampling affect the manner in which inferences are drawn from a series of experiments.

11.2 A Bayesian Framework for Interpreting Research Findings

In this section, we present a framework for thinking about how beliefs are updated in the wake of evidence. The new terms in this section are *prior beliefs* and *posterior beliefs* (or "priors" and "posteriors" for short). Prior beliefs refer to one's subjective sense of where the truth lies in advance of seeing the experimental results; posteriors are one's subjective sense of where the truth lies after seeing the evidence. The truth we are trying to learn in this instance is the value of the ATE in a population. Beliefs may be characterized as distributions—our intuitions assign a certain probability to the claim that the true PATE lies, say, between 2 and 3. For any interval, we could discern our own beliefs by asking what odds we would take if we were to make a wager about the PATE's location. In theory, we could specify our beliefs about the location of the PATE over the entire number line, subject only to the constraint that the probability that the PATE is located somewhere between negative infinity and positive infinity is 1.0. For analytic convenience, we will assume that our prior beliefs are normally distributed. This assumption keeps the formulas below very simple. An example of a prior distribution is shown in Figure 11.1. This normal distribution has a mean of 0 and a standard deviation of 2. If these were your priors, you'd say there is a 50-50 chance that the PATE is greater than 0 and about a 68% chance that the PATE lies between −2 and 2.

Posterior beliefs are formed after seeing the experimental results. The posterior is a combination of prior beliefs and new information gleaned from the experiment. Again, in order to keep the algebra simple, we imagine that the experimental results are summarized by an estimate and a standard error, and that the sampling distribution of the estimate is normal, or bell-shaped. The normal distribution provides a reasonable approximation to the sampling distribution of any large experiment and many small experiments as well. (Refer back to Figure 4.1 for an example.) Figure 11.1 illustrates a normal sampling distribution when the estimate is 10 and the standard error is 1. Figure 11.1 also shows the resulting posterior distribution. In order to explain how one gets from a prior distribution to a posterior distribution using Bayes' rule, let's lay out the updating process a bit more formally.

BOX 11.2

Bayes' Rule for Discrete Outcomes

Bayes' rule allows us to relate two conditional probabilities, $P(H|E)$ and $P(E|H)$, that are the reverse of each other. When these two probability distributions are discrete, Bayes' rule takes the following simple form:

$$P(H|E) = \frac{P(E|H)P(H)}{P(E)} = \frac{P(E|H)P(H)}{P(E|H)P(H) + P(E|{\sim}H)(1 - P(H))}.$$

In order to bring this notation to life, imagine that we conduct a randomized experiment. The hypothesis is that the treatment has a positive effect. Before seeing the results, P(H) is our *prior* probability that the hypothesis is correct. Suppose the evidence (E) the experiment generates is a statistically significant treatment effect. We want to know the conditional probability that the hypothesis is true given the experimental result. $P(H|E)$ is the *posterior* probability that our hypothesis is correct given evidence (E) of a statistically significant effect. $P(E|H)$ is the conditional probability of finding a statistically significant treatment effect if our hypothesis is correct (i.e., the power of the test). The notation "~" indicates that a proposition is false. $P(E|{\sim}H)$ is therefore the probability of finding a significant treatment effect under the null hypothesis. If we supply values of P(H), $P(E|H)$, and $P(E|{\sim}H)$, we can calculate $P(H|E)$.

Suppose that before we conducted the experiment, we thought there was a 50-50 chance that the treatment works, so $P(H) = 0.50$. Suppose we conduct an experiment and obtain a statistically significant estimate. Given our prior beliefs and the data, how should we update our beliefs? Let's assume that $P(E|H)$, the power of our experiment, is 0.45. (See Chapter 3, section A3.1 on how to calculate the power of an experiment.) Since we are assessing statistical significance at the 0.05 level, we assume that $P(E|{\sim}H)$ is equal to 0.05. Plugging these numbers into Bayes' rule gives 0.90. We started out thinking that the treatment had a 50-50 chance of having a positive effect. Having seen the experimental result, we revise our prior beliefs. Our posterior view is that there is a 90% chance that the treatment has a positive effect.

Suppose you seek to estimate the population average treatment effect, $\bar{\tau}$, and suppose that in advance of gathering the data, you hold prior beliefs about the possible values of $\bar{\tau}$. As noted above, these prior beliefs are akin to a series of wagers: your beliefs enable you to guess the probability that $\bar{\tau}$ lies between any two numbers. Suppose that your prior beliefs about $\bar{\tau}$ are distributed normally with mean g and

FIGURE 11.1

Example of updating prior beliefs based on experimental results in order to form posterior beliefs

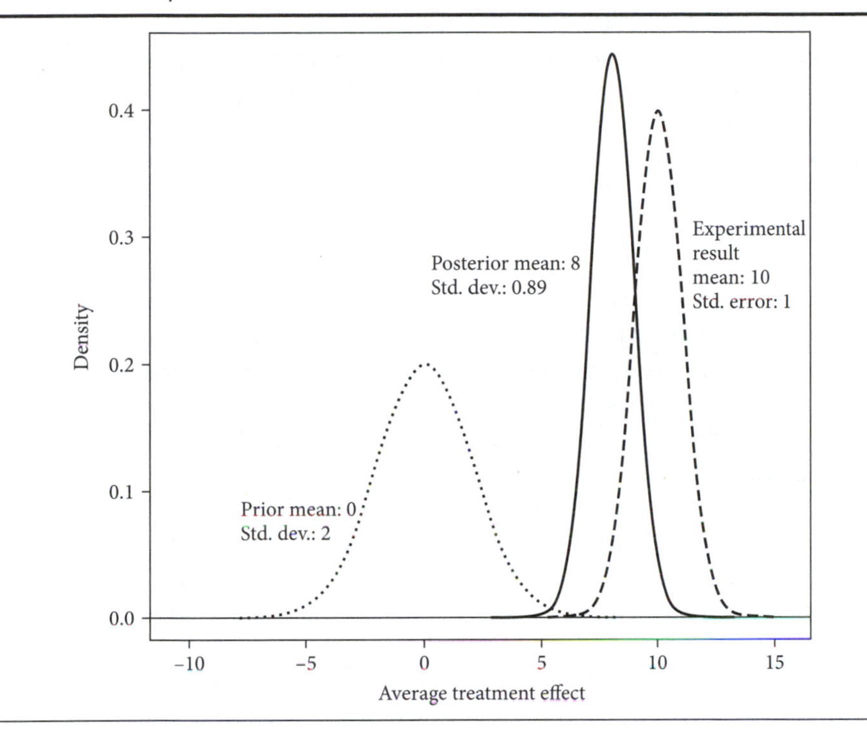

variance σ_g^2 (we use g for "guess"). The dispersion of your prior beliefs (σ_g^2) is of special importance. The smaller the value of σ_g^2, the more certain you are about the true parameter $\overline{\tau}$ in advance of seeing the data.

You now launch an experimental study. When the study is completed, you will in effect draw from the distribution of experimental outcomes (X_e) and observe the actual experimental result x_e. Before you examine the data, the *central limit theorem* leads you to expect that your estimator, X_e, will have a normal sampling distribution.[4] Given that $\overline{\tau} = t$ (the population average treatment effect equals the value t), X_e is normal with mean t and variance $\sigma_{x_e}^2$ if two conditions hold: (1) the experiment produces an unbiased estimate of the sample ATE (i.e., the treatments are assigned at random and non-interference and excludability hold); and (2) the sample of

4 The central limit theorem states that when observations are sampled independently from a large population, the sampling distribution of the sample mean converges to a normal distribution as sample size increases. With some fairly weak additional assumptions, this theorem can be extended to the difference-in-means estimator or the regression estimator.

experimental subjects is drawn randomly from a large population. If the subjects are not sampled at random from the population, the experiment will provide an unbiased estimate of the sample average treatment effect, but the estimate of the population average treatment effect may be biased.

In order to allow for possible sampling bias, we expand the model to include B, the random variable that denotes this bias. Suppose that your prior beliefs about B are distributed normally with mean β and variance σ_B^2. Again, smaller values of σ_B^2 indicate more precise prior knowledge about sampling bias. Infinite variance implies complete uncertainty. Further, we assume that priors about $\bar{\tau}$ and B are independent. This assumption makes intuitive sense: there is usually no reason to suppose ex ante that one can predict the sampling bias by knowing whether a causal parameter is large or small. Given that $\bar{\tau} = t$ (the true ATE equals t) and $B = b$ (the true bias equals b), we assume that the sampling distribution of X_e is normal with mean $t + b$ and variance $\sigma_{x_e}^2$. In other words, the study produces an estimate (x_e) that may be biased in the event that b is not equal to 0. In sum, our model of the research process assumes normal and independently distributed priors about the true effect and the bias.

The analytic results we now present describe how beliefs about the causal parameter $\bar{\tau}$ change as we learn the results of the experiment.[5] The more fruitful the research program, the more our posterior beliefs will differ from our prior beliefs. New data might give us a different posterior belief about the location of $\bar{\tau}$, or it might confirm our prior belief and reduce the variance (uncertainty) of these beliefs.

From the assumptions above, one can derive the posterior distribution of $\bar{\tau}$ given the experimental results.[6] The results are summarized in equations (11.2), (11.3), and (11.4). The posterior distribution of $\bar{\tau}$ is normally distributed with mean given by:

$$E(\bar{\tau} \mid X_e = x_e) = p_1 g + p_2(x_e - \beta) \tag{11.2}$$

and variance:

$$\sigma^2_{\bar{\tau}|x_e} = \frac{1}{\dfrac{1}{\sigma_g^2} + \dfrac{1}{\sigma_B^2 + \sigma_{x_e}^2}}, \tag{11.3}$$

where:

$$p_1 = \frac{\sigma^2_{\bar{\tau}|x_e}}{\sigma_g^2} = \frac{\sigma_B^2 + \sigma_{x_e}^2}{\sigma_g^2 + \sigma_B^2 + \sigma_{x_e}^2} \text{ and } p_2 = \frac{\sigma^2_{\bar{\tau}|x_e}}{\sigma_B^2 + \sigma_{x_e}^2} = \frac{\sigma_g^2}{\sigma_g^2 + \sigma_B^2 + \sigma_{x_e}^2} = 1 - p_1. \tag{11.4}$$

5 The results can also be used to describe how beliefs change after one examines an entire literature of experimental results. The precise sequence in which one examines the evidence does not affect our conclusions (see Gerber, Green, and Kaplan 2004 for a proof), but tracing this sequence does make the analytics more complicated. For purposes of exposition, therefore, we concentrate our attention on what happens as one moves from priors developed in advance of seeing the results to posterior views informed by all the evidence that one observes subsequently.

6 The proof may be found in Gerber, Green, and Kaplan 2004, which also derives the posterior distribution of B and the posterior correlation between $\bar{\tau}$ and B.

In other words, the posterior mean is a weighted average (since $p_1 + p_2 = 1$) of two terms: the prior expectation of the true ATE (g) and the experimental estimate corrected by the prior expectation of the bias ($x_e - \beta$). The biased estimate is re-centered to an unbiased estimate by subtracting off the prior expectation of the bias. Such re-centering is rarely, if ever, done in practice. It is also rare to see researchers make allowance for the extra uncertainty that follows when $\sigma_B^2 > 0$. In effect, when computing estimates and standard errors, researchers working with non-random samples implicitly assume that the sampling bias equals zero and that the uncertainty associated with this bias is also zero.

To get a feel for how prior beliefs are transformed into posterior beliefs using the weights p_1 and p_2, it is useful to consider several limiting cases. If before examining the data one were certain that the true effect were g, then $\sigma_g^2 = 0$, $p_1 = 1$, and $p_2 = 0$. In this case, one would ignore the data from the experiment and set $E(\overline{\tau} | X_e = x_e) = g$. In the less extreme case where one has some prior information about $\overline{\tau}$ such that $\sigma_g^2 < \infty$, p_2 nevertheless remains zero so long as one remains completely uninformed about sampling bias (i.e., $\sigma_B^2 = \infty$). In other words, in the absence of prior knowledge about bias, one accords zero weight to the experimental results. Note that this result holds even when the sample size of the study is so large that $\sigma_{x_e}^2$ is reduced to zero. For finite values of $\sigma_B^2 > 0$, the implication of equation (11.3) is that the statistical uncertainty associated with the experimental outcome ($\sigma_{x_e}^2$) understates the total uncertainty in the mind of the analyst, who also must contend with the possibility of bias.

This analytic framework turns out to be much more general than one might guess from the preceding discussion. The analytics are the same when we consider how we might combine evidence from an experiment and an observational study. Suppose our priors are based entirely on an unbiased experiment. We update these priors based on results from a possibly biased observational study. We apply the formula in equation (11.2): the estimate and sampling variance of the experiment provide the location and dispersion of our priors; the estimate, sampling variance, and uncertainty about bias provide the location and dispersion of the observational study that we use to update our priors. The implications of the preceding discussion still apply: unless our priors about the observational study's bias have finite variance, we accord zero weight to the observational evidence and rely solely on our (unbiased) experimental knowledge. This point holds even if the observational study involved a sample so large that its standard error appears to be zero. This example drives home the key point: a study's nominal standard errors may be quite misleading if the potential for bias is uncertain. Observational research may be accorded too much weight because the nominal standard errors that researchers examine ignore the uncertainty associated with bias.

A simple numerical example may help fix ideas about how the model works for the case in which priors are updated with experimental evidence. Suppose for the sake of illustration that your prior beliefs about the population ATE $\overline{\tau}$ were centered at $g = 10$ with a standard deviation of 5 (i.e., $\sigma_g^2 = 25$). Because your experimental subjects are not sampled randomly from the broader population, an extrapolation of the estimated sample ATE to the population ATE may be biased. Suppose that your prior beliefs about B are distributed normally with $\beta = 2$ and variance $\sigma_B^2 = 16$; in other words, you suspect that your sampling method causes your experimental estimates to overstate the population ATE. Your experiment produces an estimate of $x_e = 20$ with a standard error of 3 (i.e., $\sigma_{x_e}^2 = 9$). Applying the formulas given above, we find that the posterior distribution of $\bar{\tau}$ is normally distributed with mean given by:

$$E(\bar{\tau} | X_e = x_e) = p_1 g + p_2(x_e - \beta) = 14.0, \quad (11.5)$$

and variance:

$$\sigma_{\bar{\tau}|x_e}^2 = \frac{1}{\dfrac{1}{\sigma_g^2} + \dfrac{1}{\sigma_B^2 + \sigma_{x_e}^2}} = \frac{1}{\dfrac{1}{25} + \dfrac{1}{16 + 9}} = 12.5, \quad (11.6)$$

where:

$$p_1 = \frac{\sigma_{\bar{\tau}|x_e}^2}{\sigma_g^2} = \frac{12.5}{25} = 0.5 \text{ and } p_2 = \frac{\sigma_{\bar{\tau}|x_e}^2}{\sigma_B^2 + \sigma_{x_e}^2} = \frac{12.5}{16 + 9} = 0.5. \quad (11.7)$$

In short, your priors were 10, the experiment came in at 20, and your posterior ended up at 14, with a standard error of $\sqrt{12.5} = 3.54$.

In this example, uncertainty about bias has a profound effect on the posterior distribution, reducing the extent to which priors move in the direction of the experimental results. In the absence of this uncertainty ($\sigma_B^2 = 0$), the posterior mean would have been 15.9 instead of 14.0. Uncertainty about bias also results in greater posterior variance. If the researcher were certain about the bias, the posterior variance would have been 6.6 instead of 12.5. This example illustrates the core idea that posterior beliefs become less responsive to evidence as uncertainty about bias increases. The next section presents an example illustrating how assumptions about sampling bias affect the way in which one learns from a series of experiments.

11.3 Replication and Integration of Experimental Findings: An Example

Let's now turn to an example in order to see how factors that lead to cross-study variability play out in practice and apply the Bayesian framework for integrating experimental findings. In 2003, Goldstein, Cialdini, and Griskevicius conducted their "Room with a Viewpoint" experiment, which evaluated whether messages conveying "descriptive norms" about towel reuse caused guests of a mid-priced Tempe, Arizona hotel to alter their recycling behavior.[7] Hanging signs were placed in 190 hotel room

7 Goldstein, Cialdini, and Griskevicius 2008. Additional information about the studies was obtained through correspondence with Noah Goldstein.

bathrooms requesting that guests reuse their towels. One of two messages was randomly assigned to each room. In both experimental groups, the signs provided facts about the environmental benefits of recycling and instructed guests to participate in the hotel's towel reuse program by hanging towels on shower rods or towel racks. In the control condition, guests were told:

> HELP SAVE THE ENVIRONMENT.
>
> You can show your respect for nature and help save the environment by reusing your towels during your stay.

The descriptive norms treatment encouraged guests to:

> JOIN YOUR FELLOW GUESTS IN HELPING TO SAVE THE ENVIRONMENT.
>
> Almost 75% of guests who are asked to participate in our new resource savings program do help by using their towels more than once. You can join your fellow guests in this program to help save the environment by reusing your towels during your stay.

The outcome, which was measured by hotel housekeeping staff, was whether the sample of guests staying at least two nights in the hotel reused at least one towel. The authors report that 35.1% of the control group ($N = 222$) recycled as compared to 44.1% of the descriptive norms treatment group ($N = 211$). Inserting these estimates into equation (11.1), we find that this 9 percentage point effect has an estimated standard error of:

$$\widehat{SE}(\widehat{ATE}) = \sqrt{\frac{(0.441)(1 - 0.441)}{211} + \frac{(0.351)(1 - 0.351)}{222}} = 0.047, \quad (11.8)$$

or 4.7 percentage points, which implies $p < 0.05$ using a one-tailed test. This formula presupposes that whether an individual is selected into the sample is unrelated to that individual's potential outcomes. This assumption would be satisfied if each of the hotel patrons were selected independently and with equal probability from a large population of patrons.

The following year, these authors conducted a second experiment using the same experimental treatments (as well as others) in the same hotel. In terms of experimental procedure, the follow-up study is about as close as one can come to a pure replication: the same treatment, the same mode of administering the treatment, the same setting. In the second experiment, the standard message group recycled at a rate of 37.2% ($N = 277$) as compared to approximately 42.3% ($N = 334$) in the descriptive norms group. This difference of 5.1 percentage points has an estimated standard error of 4.0 percentage points, again using equation (11.1). Although the second study is more precise than the first, the second study alone would not lead us to reject the sharp null hypothesis of no effect if we used randomization inference to conduct a one-tailed test ($p > 0.05$). This standard error formula as applied to both studies, however, ignores non-quantifiable uncertainty due to the lack of random sampling. The samples of guests used in each study were selected by convenience. The authors do not compare the background attributes of the two groups of experimental subjects or provide any information about the seasons during which the two studies were conducted. With additional information about individual attributes or contextual variation, one could in theory reweight the replication sample to more closely resemble the original sample. Lacking additional information, we will have to impose assumptions if we want to combine the results from the two studies. Define the population as the hotel's guests for the period covering both studies. If we assume that customers were sampled independently and with equal probability from this population, and that customers in the second study were not exposed to the treatments in the first study (which might change the ATE in the second study), then the sole source of between-experiments variability is due to random sampling.

To summarize, the same team of investigators using the same treatments in the same hotel in two successive years produced estimates of 9.0 and 5.1 percentage points with estimated standard errors of 4.7 and 4.0 percentage points. What can we infer from the two studies, taken together? One rather superficial way to answer this question is the following: "The first study showed a statistically significant effect, but the second, larger study did not; because the initial results failed to replicate, the experimental effect cannot be considered robust." This interpretation has two weaknesses. Although it is true that the second experiment failed to produce a statistically significant estimate, this estimate is not a precisely estimated zero. If the only information that one had about the causal phenomenon in question were the results of the second experiment, one's best guess about the size of the effect would be 5.1 percentage points. Second, there is no reason to ignore the first experiment when forming a judgment about the claim addressed by the second experiment. If all one had to go on were these two studies, presumably one's best guess of the ATE would be somewhere between 5.1 and 9.0. When testing the null hypothesis of no effect, it seems reasonable to base the calculations on both studies combined in some way.

One such approach is to use the Bayesian framework described above. Recall that this framework asks two questions about each study: (1) Does it produce an unbiased estimate of the sample ATE? and (2) Are the observations sampled such that the experimental result provides an unbiased estimate of the population ATE? In regard to the first question, both experiments use random assignment and have no apparent problems of attrition, interference between observations, or asymmetries in the ways that outcomes are measured. On average, they should each recover the sample ATE. More questionable is the assumption that the samples were effectively independent random draws from the same population. Convenience sampling presents problems if we suspect that potential outcomes vary across contexts, in which case the samples are drawn from different populations, or that the two samples are not drawn randomly, in which case the two studies may share some of the same systematic biases.

Social science research articles rarely define the population explicitly. By characterizing norms as powerful behavioral cues that operate across a variety of different social settings, Goldstein, Cialdini, and Griskevicius seem to imply that the population should be defined broadly. Basically, anyone able to read and understand the implications of a message describing normative behavior should be susceptible to its influence. Alternatively, one could read this study as a program evaluation that informs hotels about the efficacy of different messages promoting recycling. Whether the population is defined as people who can read and comprehend an appeal based on descriptive norms or American hotel guests circa 2004, these two samples are skewed toward the patrons of mid-priced hotels in Tempe. Without additional evidence, we can only speculate about whether norm-based environmental appeals have unusually large or small effects among those lodging in this particular hotel during this time period. Below, we discuss some empirical efforts to assess whether these results generalize to other times and settings.

In order to get a feel for the statistical mechanics of how one pools results from different experiments, let's begin by assuming the best-case scenario: the subjects are sampled randomly from the same population. If the two samples are considered independent draws from the same population, the most efficient estimate of the treatment effect given both experimental results is calculated by pooling the results, weighting each by the inverse of its squared standard error. This method, which is known as *fixed effects meta-analysis*, parallels the formula for Bayesian updating. The researcher forms priors based on the results of the first experiment; these priors are updated based on the second experiment; and both samples are assumed to be unbiased ($E[B] = 0$) with no lingering uncertainty about the possibility of bias ($\sigma_B^2 = 0$). Rearranging equation (11.2) gives:

$$\widehat{\text{ATE}}_{pooled} = \frac{\frac{1}{\hat{\sigma}_1^2}}{\frac{1}{\hat{\sigma}_1^2} + \frac{1}{\hat{\sigma}_2^2}} \widehat{\text{ATE}}_1 + \frac{\frac{1}{\hat{\sigma}_2^2}}{\frac{1}{\hat{\sigma}_1^2} + \frac{1}{\hat{\sigma}_2^2}} \widehat{\text{ATE}}_2$$

$$= \frac{\frac{1}{0.047^2}}{\frac{1}{0.047^2} + \frac{1}{0.040^2}} 0.090 + \frac{\frac{1}{0.040^2}}{\frac{1}{0.047^2} + \frac{1}{0.040^2}} 0.051 = 0.067. \quad (11.9)$$

This weighted average of two experimental estimates accords greater weight to the study with a smaller standard error. In this case, the first experimental result is assigned a weight of 0.42, and the second is assigned a weight of 0.58. The result is a pooled estimate of 6.7 percentage points.

How does pooling affect the accuracy of the experimental estimates? Recall that the two experiments generated standard errors of 0.047 and 0.040. By comparison, the pooled estimate has an estimated standard error of:

$$\sqrt{\text{Var}(\widehat{\text{ATE}}_{pooled})} = \sqrt{\frac{1}{\frac{1}{\hat{\sigma}_1^2} + \frac{1}{\hat{\sigma}_2^2}}} = \sqrt{\frac{1}{\frac{1}{0.047^2} + \frac{1}{0.040^2}}} = 0.030. \quad (11.10)$$

BOX 11.3

Fixed Effects Meta-Analysis

Meta-analysis is a method for aggregating research findings. The data are aggregated in order to obtain a more precise estimate of the average treatment effect in the population from which each experimental sample is randomly selected.

By analyzing reported estimates and standard errors, meta-analysis approximates the results that one would obtain if the individual-level data from the various experiments were combined into a single dataset. Fixed effects meta-analysis amounts to a simple algorithm: weight each study's estimated ATE by its *precision*, which is defined as the inverse of its squared standard error, and divide by the sum of the precisions.

This formula illustrates an important principle about the accumulation of experimental evidence under random sampling. Any experiment that generates a finite standard error will help diminish the sampling variability of the pooled estimate. The standard error of the pooled estimate is therefore always smaller than the standard errors of each of the experiments whose results are pooled. In this case, the results of pooling are rather dramatic. When pooled, the experiments yield an estimate of 6.7 with a standard error of 3.0, which implies a one-tailed p-value of .013.

Having illustrated the mechanics of pooling experimental results, let's now return to the question of whether the Tempe studies are representative draws from a broader population. One attractive feature of this field experiment is that it lends itself to replication. The treatments are inexpensive to produce and deploy. Measurement of outcomes requires training and supervision but no additional cost. The study involves no risk to subjects, and their anonymity is easily protected. In an effort to test whether the Goldstein, Cialdini, and Griskevicius findings extend to other settings, Elizabeth Campbell conducted a replication in a pair of hotels near her hometown, one in Bedford, Indiana, and the other in La Grange, Kentucky.[8] Using the same hanging signs as in the original study, Campbell tracked towel usage in these two mid-priced hotels during the first quarter of 2009 and measured outcomes in the same manner as Goldstein, Cialdini, and Griskevicius, focusing only on guests staying at least two nights.

Before revealing the results of the experiment, let's first reflect on our priors. Campbell's research sites were selected based on convenience sampling, which again

8 Campbell 2009.

raises the question of what the underlying population is. Suppose we call the population American hotel guests in 2009. If we believe that differences in time and location are immaterial, we could use meta-analysis to pool the two Tempe studies with the two Campbell studies, simply by weighting each study by its *precision*, the inverse of its squared standard error (see equation 11.9). Suppose, on the other hand, we assume that region is immaterial (guests in all regions are equally affected by the treatment) but that in the years since 2004, the potential outcomes in the population of hotel guests may have changed. If we define the population to be hotel guests in 2009, the 2003–4 and 2009 studies may be combined using a formula that includes uncertainty about whether the earlier studies are potentially biased. The term bias in this instance does not imply that the Tempe studies are flawed. Rather, the concern is that if we seek to estimate the ATE in 2009, extrapolating based on the results of older studies may introduce "bias" even if these studies were conducted flawlessly.

As it turned out, Campbell observed reuse rates of 68.6% in the standard sign group ($N = 70$)and 68.8% in the descriptive norms group ($N = 80$) in Bedford; in LaGrange, the corresponding rates were 66.2% ($N = 68$) and 59.4% ($N = 64$). The estimated average treatment effects in the two sites are 0.2 with a standard error of 7.6 and −6.8 with a standard error of 8.5. In sum, in comparison to the Tempe studies, the average level of recycling in the control condition is higher, and the treatment effect is weaker.

Let's consider how two alternative meta-analyses play out. The first pools all four experiments as though they were independent random draws from the same population. Independence is a strong assumption. In effect, independence implies that obtaining an above-average estimate in the first Tempe study gives you no clues as to whether you are likely to get an above-average estimate from the second Tempe study. Positive dependence, on the other hand, will tend to make the two Tempe studies look more similar to each other than would be expected by chance.

When assembling the results of a research literature for purposes of meta-analysis, some rules of thumb are helpful when assessing whether independent sampling is a realistic assumption. For example, when studies contain several different treatment groups, the estimated effects for each of the treatments are correlated because they are each calculated with reference to the same control group. The same goes for experiments with multiple outcomes: the estimated effect of signs on recycling among hotel guests is not independent of the effect of signs on the guests' attitudes about the importance of environmental issues. Independence may also be jeopardized when (1) a single research group conducts multiple studies, (2) studies take place in similar settings, or (3) researchers are aware of each other's results. In the first case, research collaborators may share idiosyncratic ways of deploying treatments, measuring outcomes, or deciding which results to report. In the second case, studies conducted under similar circumstances may tend to produce more similar results than would be expected if they were independent draws from a larger population. In the third case, knowledge of how other experiments have come out may affect, perhaps unconsciously, how researchers analyze or report their experiments.

If we adopt the independence assumption, combining the four studies statistically is straightforward. When we assume independence, we can combine any number of studies in any order using the same formula. We can keep a running tally of all studies to date and then update the tally when a new study comes along. Applying equations (11.2) and (11.3) to Campbell's two studies gives a precision-weighted average of −2.9 with a standard error of 5.7. Earlier, we calculated the weighted average of the two Tempe studies to be 6.7 with a standard error of 3.0. Combining the two pairs of studies using the same formula yields a precision-weighted average of 4.6 with a standard error of 2.7. The Goldstein, Cialdini, and Griskevicius studies have a smaller standard error and therefore are accorded more weight than the Campbell studies, but the inclusion of the Campbell studies reduces the point estimate and its standard error.

For the sake of illustration, let's consider what happens when we make a different set of assumptions. An alternative approach considers the Campbell studies to be random draws from the population from the 2009 population of hotel guests and models the Tempe studies as potentially biased because the estimand is defined as the ATE among hotel guests in 2009. In effect, we have one unbiased estimate (from the pooled Campbell studies) and one possibly biased estimate (from the pooled Tempe studies). Suppose that prior to seeing the results our best guess about the Tempe studies is that they are unbiased ($\beta = 0$) but that the potential for bias exists. Suppose our priors about bias are normally distributed with a mean of 0 and a standard deviation of 3. This prior distribution implies that we believe there is a 1-in-6 chance that the bias is greater than 3. Under these assumptions, equation (11.2) produces a posterior mean of 3.2 with a standard error of 3.4. Of course, hunches about bias may vary. Yours may be different from ours. The point of the exercise is to show that the mere possibility of bias can have important consequences for meta-analysis. When we ignore the possibility of bias, the results suggest a strong effect that is distinguishable from zero ($p < 0.05$, one-tailed test); when the possibility of bias is introduced, the apparent effect is both weaker and statistically ambiguous.

The attraction of meta-analysis is that it provides a systematic method for summarizing research findings. It should be stressed, however, that meta-analysis invokes assumptions about sampling from well-defined populations.[9] The example above illustrates that the conclusions generated by meta-analysis may be sensitive to (1) how the population is defined, (2) whether the sampling procedure is assumed to be unbiased,

9 See Gleser and Olkin 2009 and Hedges 2009 for discussion of more complex meta-analysis models that can accommodate non-independent samples and other complications.

and (3) how much uncertainty surrounds the threat of bias.[10] In practice, meta-analyses often ignore the threat of bias, sometimes going so far as to include experimental and non-experimental studies in the same analysis without adjusting the weights accorded to potentially biased studies. As the examples in this section suggest, when the meta-analytic framework is expanded to allow for uncertainty about bias, the weight assigned to potentially biased evidence changes, sometimes dramatically.

The statistical assumptions underlying meta-analysis become stronger as we pool studies across substantive domains. In the example above, we limited our attention to the relatively tractable problem of pooling studies that all use the same treatment and the same standard for measuring outcomes. Matters become much more complex when we expand our purview to include all experimental studies that assess the effects of descriptive norms on willingness to engage in pro-social behavior, such as recycling. Clearly, reviewing this far-reaching literature is necessary if one is to assess the hypothesis that descriptive norms promote these behaviors. If a series of domain-specific meta-analyses were to show that the proclivity to pay taxes, contribute to charitable causes, or vote is unaffected by interventions that convey descriptive norms, the "Room with a Viewpoint" study would be cast in an altogether different light. Instead of revealing a basic proposition about human responsiveness to perceived social norms, the study would tell us something about the special conditions that cause norms-related interventions to work in this particular domain.

Meta-analysis is sometimes used to synthesize studies that use different treatments and outcomes. This all-encompassing approach is usually characterized as an attempt to extract an overarching population ATE from a diverse research literature. This type of meta-analysis not only invokes strong sampling assumptions; it also invokes strong measurement assumptions. Typically, meta-analyses of this kind skirt the issue of comparability by transforming the estimates from each experiment into a standardized effect. (Recall from Chapter 3 that a standardized effect is the estimated ATE divided by the standard deviation in the control group.) A meta-analysis based on standardized effects rests on the assumption that once the outcomes are rescaled in this way, each experiment is effectively a random draw from the same population. Given that the subjects, settings, and outcomes often differ across research domains, the random sampling assumption would be questionable even if the treatments were identical. But if treatments vary in intensity as well, results may vary across domains for reasons having to do with the strength of the treatment rather than sampling variability. The next section discusses some of the special modeling concerns that arise when analyzing treatments that vary in intensity.

10 This criticism does not imply that meta-analysis should be discarded in favor of narrative reviews. Critical evaluation of the assumptions that a meta-analysis invokes is possible because the assumptions are clear. It is much harder to have a meaningful argument about the conclusions derived from a narrative review because the procedures used to synthesize a collection of studies are seldom defined explicitly.

11.4 Treatments That Vary in Intensity: Extrapolation and Statistical Modeling

To this point, we have focused exclusively on binary treatments and defined a causal effect as the difference between a treated $Y_i(1)$ potential outcome and an untreated $Y_i(0)$ potential outcome. Some experiments, however, are designed to gauge the effects of treatments that vary in intensity. Rather than manipulate *whether* a subject is treated, they vary the *extent* of treatment. For example, medical experiments vary the dosage of experimental drugs that are administered to patients. Political mobilization experiments vary the number of mailings that subjects receive. Marketing experiments vary the number of times that an advertisement is aired in different regions.

Experimental research on incentives often involves treatments that vary in intensity. Dupas, for example, reports the results of an experiment in which Kenyan subjects were encouraged to purchase insecticide-treated bed nets designed to prevent malaria.[11] The treatment took the form of a voucher that lowered the price of the bed nets at a local store. Prices varied randomly across subjects, and Dupas measured outcomes by tracking which vouchers were actually redeemed.

This study represents an especially instructive example because Dupas varied the intensity of her treatments and replicated her experiment using different pricing schedules. The experiment was carried out in six different regions; within each region, vouchers varied in value, and the range of values presented to subjects differed from region to region. Thus, the data enable a researcher to explore two related questions about generalization: (1) Within a given region, how much does the apparent causal effect of changing a price from A to B help us to predict the effect of changing a price from C to D? and (2) Across regions, does the apparent effect of changing the price from A to B in one area help us to predict the consequences of similar price changes elsewhere?

The design and results of the bed nets experiment are depicted in Table 11.1. For each region and price (after subtracting the value of the voucher), the table reports the proportion of people who made a purchase. In Region 2, for instance, the bed net's prices were 40, 80, 120, or 200 shillings. The rate at which bed nets were purchased declines steadily as the price increases: 75.4% of the subjects who were offered a price of 40 shillings purchased a bed net, as compared to 17.0% of those offered a price of 200 shillings.

How might one represent this type of experiment in terms of potential outcomes? One option is to characterize each subject in Region 2 as having four potential outcomes, one of which is revealed when they are presented with a price of 40, 80, 120, or 200 shillings. This approach is nonparametric in the sense that it makes no assumptions about treatment intensity. There is no recognition of the fact that 80

11 Dupas 2012.

TABLE 11.1

Rates at which anti-malaria bed nets are purchased, by sales price (after subtracting the value of a randomly assigned voucher)

Price (in Kenyan Shillings)	% Purchased in Region 1	% Purchased in Region 2	% Purchased in Region 3	% Purchased in Region 4	% Purchased in Region 5	% Purchased in Region 6
0					96.9 (64)	98.1 (53)
40		75.4 (61)				
50			72.4 (58)	40.0 (35)		73.7 (19)
60					73.0 (37)	
70	55.2 (29)					
80		57.1 (70)				
90			55.0 (60)			
100	34.0 (47)			28.6 (49)		61.1 (18)
110					32.4 (37)	
120		28.1 (64)				
130	24.5 (49)					
140					37.9 (29)	
150			31.0 (58)	35.6 (45)		22.2 (18)
190	17.9 (28)					
200		17.0 (59)		10.3 (29)		
210			18.8 (48)			
250	6.7 (30)			7.7 (26)		

Note: Total number of households per group is in parentheses. The exchange rate at the time of this study was 65 shillings = $1.00.

is halfway between 40 and 120 or that 200 is greater than 120. In effect, this model of potential outcomes considers these to be four different treatments, each of which has a different label.

An alternative approach is to represent potential outcomes as a function of treatment intensity. Suppose, for purposes of illustration, that the potential outcomes were a subject's true underlying demand for a bed net rather than just a binary outcome of whether a purchase was made. Imagine a smooth function that translates price inputs into demand outputs. For example, one such function would be a linear relationship between price and demand. Since 80 is halfway between 40 and 120, the potential outcome for a price of 80 would be halfway between the potential outcomes for prices of 40 and 120: $Y_i(80) = (Y_i(40) + Y_i(120))/2$. A similar line of reasoning might lead us to stipulate that the potential outcome for a price of 200 is twice the average of the potential outcomes for 120 and 80. A linear function has the virtue of simplicity but is by no means the only modeling option. In principle, one may choose from a limitless supply of functions. One could, for example, stipulate that potential outcomes follow an inverted-V pattern, reaching their peak at $Y_i(120)$ but with $Y_i(40) = Y_i(200)$.

Theoretical intuitions help the analyst narrow the range of possible functions. If the menu of possibilities can be narrowed sufficiently, the data can help a researcher decide among them. For example, if the analyst settles on a linear model, regression can be used to find the intercept and slope that best fits the data. The empirical adequacy of the statistical model may be assessed by replicating the experiment using different levels of treatment intensity. Statistical models should always be viewed as provisional simplifications of reality. A successful modeling effort is one that produces theoretically guided predictions that are borne out in subsequent experiments.

In order to narrow the menu of possible models that may be applied to the bed net experiment, we start with some basic theoretical intuitions. In keeping with most studies of price and purchase behavior, we stipulate that demand diminishes as price increases. This stipulation rules out functions of price that are shaped like a parabola or sine wave. We also expect a smooth, continuous function. As two prices A and B get closer together, $Y_i(A)$ approaches $Y_i(B)$.

Rather than stipulate a particular diminishing function, we will let the data tell us whether demand declines with price at a constant rate (as assumed by a linear function) or more rapidly (as assumed by a logarithmic function) or something in between. Because Region 2 contains a large number of subjects divided evenly across experimental prices, we use it as our exploratory dataset, deriving predictions that will be tested on other regions, which present subjects with higher or lower prices. Plotting the data usually provides clues about the function that links price and purchase rates. The trick is to rescale the X and Y axes until the points fall along a line, because a line tells us that we can extrapolate from the purchase rates of any two of the four experimental groups in Region 2 to the purchase rates of the other two groups. Figure 11.2 shows the relationship between price and rates of purchase in Region 2 using two different methods for scaling prices. Figure 11.2a leaves price in its original units, and Figure 11.2b transforms the price scale into logarithms. Both plots show a reasonably linear pattern, but the logarithmic transformation is the more linear of the two, as gauged by the R-squared value when purchase rates are regressed on prices.

These plots nicely illustrate the advantages of modeling outcomes as a function of treatment intensity. Knowing $\hat{Y}_i(40)$ and $\hat{Y}_i(200)$ gives a fairly accurate sense of the location of $\hat{Y}_i(80)$ and $\hat{Y}_i(120)$. We observe $\hat{Y}_i(80) = 57.1\%$ and $\hat{Y}_i(120) = 28.1\%$ and the logarithmic model predicts 50.4% and 35.6%. The observed and predicted values do not coincide exactly, but the numbers are close enough to suggest that outcomes follow some sort of well-behaved function. Although prices such as 90 shillings or 150 shillings were never offered to subjects in this region, we have a reasonably good sense of how subjects would have answered had they been presented

FIGURE 11.2

Rates of actual and predicted purchase of anti-malaria bed nets in relation to price and the log of price

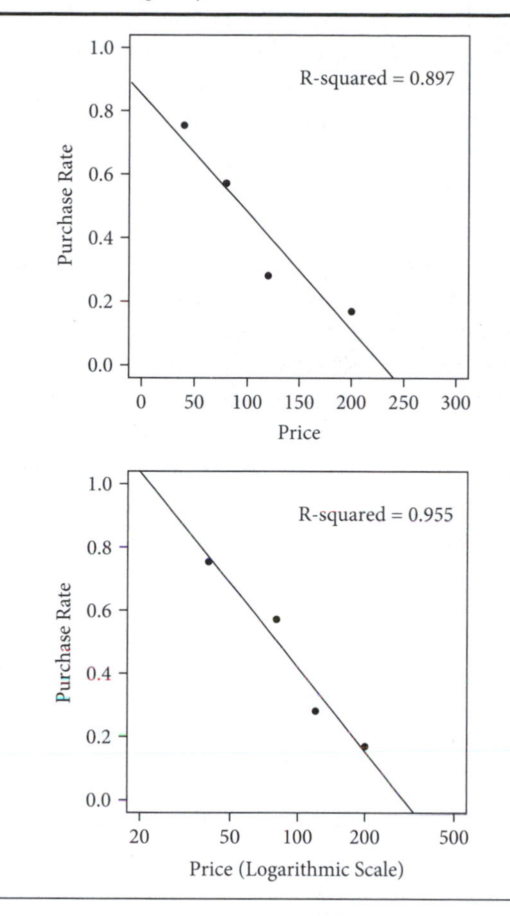

with these prices. We can also quantify our uncertainty about these hypothetical outcomes. Given a regression model in which the voucher price offered to subject i is denoted V_i:

$$Y_i = \beta_0 + \beta_1 \ln(V_i) + u_i = \beta_0 + \beta_1 V_i' + u_i, \tag{11.11}$$

we can estimate the standard error associated with the expected outcome for any hypothetical price. For example, suppose we wanted to calculate the standard error of the expected value of Y_i when V_i' (the log of V_i) is equal to v:

$$SE(\overline{Y} \mid V_i' = v) = \sigma_u \sqrt{\frac{1}{N} + \frac{(v - \overline{V}')^2}{\Sigma(V_i' - \overline{V}')^2}}, \tag{11.12}$$

where σ_u is the standard deviation of the disturbances. The term $(v - \overline{V}')^2$ is the squared difference between the hypothetical price and the average price, and $\Sigma(V_i' - \overline{V}')^2$ is the sum of the squared differences between the observed and average values of V_i'. This formula embodies an important principle about extrapolation. Prediction uncertainty is smallest when the hypothetical price is close to the average price in our experiment. As the hypothetical price (v) moves away from the average price that is actually offered $(\overline{V}')$, our forecasts become increasingly tenuous.[12]

One virtue of plotting the data and exploring hypothetical scenarios is that it helps detect flaws in one's statistical model. When we estimate equation (11.11), we obtain coefficients $\hat{\beta}_0 = 2.199$ and $\hat{\beta}_1 = -0.378$. Using these coefficients to predict purchase rates for a price of 20, we obtain predictions that are above 100%. And when this model is used to predict purchase rates for a price of 300, we obtain a purchase rate of less than 0%. These nonsensical predictions arise because our statistical model does not constrain outcomes to fall between 0% and 100%. There are many ways to impose this constraint, one of the most common being the log-odds transformation. Instead of predicting a proportion p, one predicts $\ln(p/(1 - p))$. This transformation allows the dependent variable to take on any value: the log-odds are negative when $p < 0.5$ and positive when $p > 0.5$. In order to move back and forth between the probability scale and the log-odds scale, note that if $\ln(p/(1 - p)) = \alpha$ it follows that:

$$p = \frac{e^{\alpha}}{1 + e^{\alpha}} = \frac{1}{1 + e^{-\alpha}}. \tag{11.13}$$

The logistic regression model expresses α as a regression equation, allowing log-odds to vary across subjects:[13]

$$\Pr[Y_i = 1] = \frac{1}{1 + e^{-(\beta_0 + \beta_1 \ln[V_i])}}. \tag{11.14}$$

One important feature of this transformation is that any given change in the log-odds scale may mean a large or small percentage point movement depending on the starting point. Starting at 50% and increasing to 73% moves one log-odds unit, but so does the move from 88.1% to 95.3%. Thus, extrapolations based on a logistic model may differ from extrapolations derived from a linear model.

When we adopt a logistic model, we must adjust the way we envision the data-generating process. In previous chapters, we imagined that each subject harbored potential outcomes that were revealed deterministically depending on the treatment. The logistic model implies that each person's potential outcomes are probabilities of

12 This principle would be accentuated further if this standard regression formula for prediction error took into account specification uncertainty, or the researcher's uncertainty about how to model the causal relationship between price and outcomes.

13 This regression model does not have an explicit disturbance term; instead, the random component of the model is reflected in the fact that the left hand side of the equation is a probability rather than a specific outcome.

purchase. The schedule of these probabilistic potential outcomes is determined by equation (11.14).[14] For example, the parameters of this model may imply that a subject has a 75% probability of purchase if offered a price of 100 shillings and a 50% probability if offered a price of 200 shillings. Unfortunately, we cannot observe these probabilities. Instead we observe the result of a coin flip that comes up heads with a specified probability, in this case either 75% or 50%. The subjects in this experiment do not literally flip coins; the idea of coin flipping is a metaphor for the host of unobserved factors that cause someone with a given probability to make a bed net purchase or not.

In order to get a sense of how to model price effects within the logistic framework, we graph the relationship between price and the log-odds of purchase. Figure 11.3a plots the log-odds of purchase against price, while Figure 11.3b plots it against the log of price. Again, we see that both relationships are reasonably linear, but the latter plot is the more linear of the two. The model relating the log-odds of purchase and the log of price fits the *observed* data about as well as equation (11.11); the advantage of the log-odds transformation is that it never produces inadmissible forecasts for very low or very high prices.

Having settled on the model in equation (11.14) based on our exploration of Region 2, we now assess its adequacy more rigorously using in-sample and out-of-sample tests. The model in equation (11.14) features two parameters, an intercept and a slope. Recall that in Region 2, there are four experimental groups. Thus, we have 4 goups − 2 parameters = 2 degrees of freedom, which is to say two more pieces of information than we need in order to estimate the model. This surplus information allows us to assess the fit between the model's empirical predictions and what we actually observe. If the fit is poor, the surplus information also allows us to specify a more complex model with one or two additional parameters.

Adding extra parameters to the model is a mixed blessing. On the one hand, additional parameters improve the fit between the model's predictions and our actual results; when the number of parameters equals the number of experimental groups, the fit will always be perfect. On the other hand, when we add extra parameters, we run the risk of "over-fitting" the random idiosyncrasies of our experimental results, jeopardizing our ability to make accurate forecasts about the consequences of setting prices at levels other than what we observe. For example, in Region 4 there is a strange upward bump in demand at 150 shillings. When we apply the model that we developed based on an analysis of Region 2 to all of the regions, we could add parameters to account for this wayward result, but a better modeling strategy might be to ignore this anomaly, treating it as a fluke of sampling variability rather than a reflection of the true causal law that links price to demand. In general, try to make

14 An equivalent individual-level model posits a linear regression model like equation (4.7) where the disturbance term is drawn from a logistic distribution (which is like a normal distribution but with fatter tails). The dependent variable in this model is the unobserved propensity to make a purchase. When the right hand side of this model is greater than zero, one observes an actual purchase; otherwise, no purchase.

FIGURE 11.3

Log adds of actual and predicted purchase of anti-malaria bed nets in relation to price and the log of price

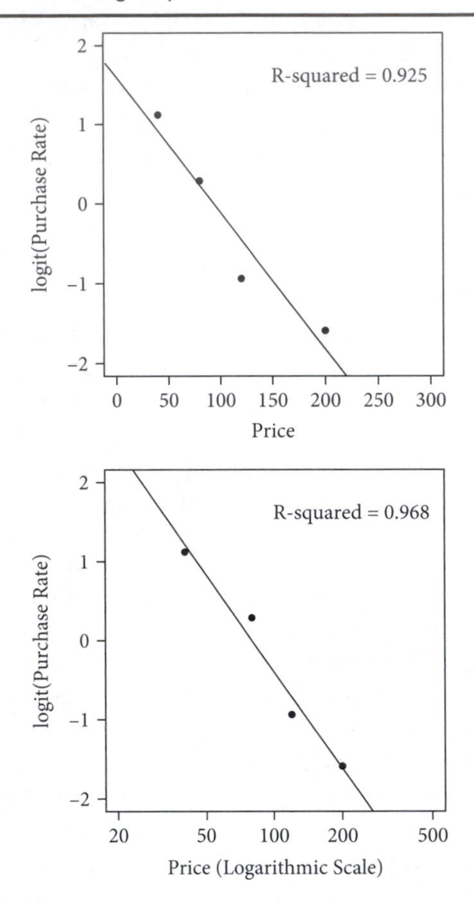

Note: The Y-axis indicates the logit, or log-odds of purchase. For example, the log-odds of a purchase rate of 75.4% is $\ln[0.754/(1 - 0.754)] = 1.12$.

do with parsimonious model specifications, adding parameters only when anomalies challenge the adequacy of the model over a series of experiments.

Table 11.2 reports the estimated slopes and intercepts when logistic regression models are applied to the data from each region. (An equivalent approach is to estimate a single regression that includes indicator variables for region and interactions between each region and the log of price.) The slope estimates suggest that apart from Region 4, subjects in the different regions respond similarly to price changes. A more parsimonious way to model purchase behavior in all six regions would be to assume that price has the same effect everywhere:

$$\Pr[Y_i = 1] = \frac{1}{1 + e^{-(\beta_0+\beta_1 R_{1i}+\beta_2 R_{2i}+\beta_3 R_{3i}+\beta_4 R_{4i}+\beta_5 R_{5i}+\beta_6 \ln[V_i])}}, \quad (11.15)$$

where the R_{ji} are indicator variables marking each region and V_i denotes the bed net price. In effect, this model presupposes that the regions in our study are sampled from a larger population of regions for which we want to learn the ATE.[15] The estimated effect of log price is given in the last row of Table 11.2. The estimate for β_6 of -1.623 implies that the log-odds of purchase decline by 1.623 for each unit increase in the log of price. For example, doubling the price means increasing the log of price by $\ln[2] \approx 0.693$, which means a decrease of $(0.69)(1.623) \approx 1.125$ in the log-odds of purchase. In order to convert this effect into percentage-point terms, consider a person whose probability of making a purchase is 50%. The log-odds of 50% is $\ln(0.5/(1 - 0.5)) = 0$. Reducing the log-odds of purchase by 1.125 would lower this probability to $1/(1 + e^{1.125}) = 0.245$.

How should we assess the empirical adequacy of the logistic regression models presented in Table 11.2? In order to gauge the fit of a model, we can compare actual purchase rates to the purchase rates that are predicted by the estimated regression equation. If predicted rates diverge markedly from actual rates, the specification we posited is called into question. Table 11.3 shows the expected and actual purchases

TABLE 11.2

Logistic regressions of bed net purchases on price, by region

		Rate of purchase modeled as a function of the log of price	
Region	N	Intercept	Slope
1	183	8.531 (2.275)	−1.978 (0.480)
2	254	7.873 (1.233)	−1.791 (0.272)
3	224	7.692 (1.361)	−1.694 (0.292)
4	184	3.323 (1.496)	−0.912 (0.316)
5	103	9.177 (2.767)	−2.029 (0.605)
6	55	8.784 (3.119)	−1.921 (0.683)
All	1,003	Not shown	−1.623 (0.149)

Note: Estimated standard errors in parentheses. Regressions for Regions 5 and 6 exclude prices of zero because the log of zero is undefined. The statistical model used to estimate the slope when all regions are combined is given in equation (11.15).

15 See Humphreys 2009 on the distinction between estimation using dummy variables for blocks and inverse probability weights. The latter is suited to unbiased estimation of sample average treatment effects, whereas the former is designed to efficiently estimate a population ATE under random sampling.

TABLE 11.3

Actual and predicted responses to price for two nested logistic regression models

		Observed frequencies		MODEL I: Different slopes for each region			MODEL II: One slope for all regions		
Region	Price	Purchases	Non-purchases	Predicted purchases	Predicted non-purchases	χ^2	Predicted purchases	Predicted non-purchases	χ^2
1	70	16	13	15.44	13.56	0.04	14.18	14.82	0.46
1	100	16	31	16.92	30.08	0.08	16.41	30.59	0.02
1	130	12	37	12.29	36.71	0.01	12.71	36.29	0.05
1	190	5	23	3.82	24.18	0.42	4.46	23.54	0.08
1	250	2	28	2.52	27.48	0.12	3.24	26.76	0.53
2	40	46	15	47.58	13.42	0.24	46.10	14.90	0.00
2	80	40	30	35.43	34.57	1.19	35.08	34.92	1.38
2	120	18	46	21.21	42.79	0.73	21.90	42.10	1.06
2	200	10	49	9.78	49.22	0.01	10.92	48.08	0.09
3	50	42	16	43.14	14.86	0.12	42.57	15.43	0.03
3	90	33	27	31.05	28.95	0.25	30.90	29.10	0.29
3	150	18	40	18.04	39.96	0.00	18.37	39.63	0.01

3	210	9	39	9.76	38.24	0.08	10.16	37.84	0.17
4	50	14	21	15.36	19.64	0.21	20.02	14.98	4.23
4	100	14	35	14.39	34.61	0.01	14.83	34.17	0.07
4	150	16	29	10.04	34.96	4.56	8.25	36.75	8.90
4	200	3	26	5.25	23.75	1.17	3.58	25.42	0.11
4	250	2	24	3.97	22.03	1.15	2.32	23.68	0.05
5	0	62	2	N/A	N/A				
5	60	27	10	26.08	10.92	0.11	24.59	12.41	0.71
5	110	12	25	15.22	21.78	1.16	15.74	21.26	1.55
5	140	11	18	8.70	20.30	0.87	9.67	19.33	0.27
6	0	52	1	N/A	N/A				
6	50	14	5	14.84	4.16	0.22	14.20	4.80	0.01
6	100	11	7	8.73	9.27	1.15	8.82	9.18	1.06
6	150	4	14	5.43	12.57	0.54	5.98	12.02	0.98
					Total χ^2, 12 d.f.	14.44 $p = .27$		Total χ^2, 17 d.f.	22.10 $p = .18$

and non-purchases for each experimental group. The expected frequencies in the columns labeled Model I and Model II illustrate the consequences of different modeling decisions. One model applies equation (11.14) to each region separately, thus allowing each region to respond to price at different rates. An alternative model pools the regions together using equation (11.15), using indicator variables for region but applying a single price parameter to all regions.

Although it is useful to visually inspect the discrepancies between actual and predicted outcomes, a more rigorous assessment of goodness-of-fit employs a chi-square test, the results of which are reported at the foot of the table. This test is calculated as follows: within each experimental group and outcome category, square the difference between the actual number of observations and the predicted number, and divide the result by the predicted number of observations. Table 11.3 shows this calculation for each of the experimental groups, which allows the analyst to gain some insight into where mispredictions occur. For example, in the first row of Table 11.3, we see that Region 1 produced 16 purchases and 13 non-purchases when the asking price was 70 shillings. The model assuming a single price effect across all regions generates predictions of 14.18 and 14.82, respectively. The contribution to the overall chi-square in this row is $(16 - 14.18)^2/14.18 + (13 - 14.82)^2/14.82 = 0.46$.

In the Dupas study, there are 24 experimental groups in all (not counting the two experimental groups that were offered prices of zero), which means that there are 24 sets of purchasers and 24 sets of non-purchasers. In order to calculate the overall chi-square, we add up the 48 statistics. The p-value for the chi-square depends on the degrees of freedom. We start with 24 degrees of freedom, one for each experimental group. The model with distinct slopes for each region consumes a total of 12 degrees of freedom (six intercepts and six slopes). The chi-square test is based on the remaining $24 - 12 = 12$ degrees of freedom. When we constrain all of the slopes to be the same, the model consumes 7 degrees of freedom (six intercepts and one slope), leaving $24 - 7 = 17$ degrees of freedom.[16]

Because the null hypothesis is that the data were generated by the stipulated model, a significant chi-square test would indicate that the model fits poorly. By this criterion, both models seem to perform adequately, with p-values above 0.05.[17] In other words, we cannot reject the null hypothesis that the experimental data were generated by the models we posited. This conclusion does not prove that either model in fact generated the data; it simply means that the data are not at odds with each model's predictions. The deviations between what we observe and what the models predict are sufficiently small that they may be attributed to sampling variability.

16 A more exacting test would exclude Region 2 from the estimation on the grounds that data from this region were used to formulate the regression model.

17 We use a conventional chi-square test here rather than one based on randomization inference, as we are no longer working within a standard potential outcomes framework.

How do we choose between these competing models? The likelihood ratio test may be used to compare the goodness-of-fit of two *nested* models. Recall from Chapter 9 that two models are said to be nested when one model can be written as a special case of the other. The model that applies one slope to all regions is a special case of a model that applies a different slope to each region. When comparing nested models, one asks whether the loss of degrees of freedom is rewarded with a significant improvement in model fit. The null hypothesis is that the data were generated by the more parsimonious model, which posits a single price effect across all regions. In order to test the null hypothesis, compare the log-likelihood statistic (a measure of goodness-of-fit that is conventionally reported by logistic regression routines) from the null model (−585.04) with the log-likelihood statistic from the alternative model (−581.79). Twice the absolute difference (6.50) is distributed chi-square with degrees of freedom equal to the number of additional parameters in the alternative model (in this case 5). A significant test statistic rejects the null model in favor of the model with more parameters. In our example, the *p*-value is 0.26, indicating that the null hypothesis of a single price effect cannot be rejected. In other words, the model that allows for a different price effect in each of the six regions is not a significant improvement over the more parsimonious model that assumes the same price effect everywhere. The usual presumption in favor of parsimonious models leads us to prefer the model with a single slope.

It should be stressed that although the model that we settled on shows a reasonably good fit to the data, it is not necessarily the true model. Other models that were not examined may fit the data as well or better. Moreover, the model we selected may fail to account for additional evidence that we have yet to consider. Indeed, the limitations of this particular model become evident when we calculate the predicted purchase probabilities when the offered price is zero, a complication we leave for the exercises. That said, this model is arguably a big step up from an analysis that treats the experimental groups as unordered categories and fits 24 parameters to the 24 observed purchase rates. The latter approach exactly reproduces the experimental results but cannot speak to the question of whether behavioral regularities exist across prices or regions.

SUMMARY

The central theme of this chapter is the challenge of extrapolating from samples to populations. We began by calling attention to the subtle but important difference in the meaning and estimation of sampling variability as applied to the estimated sample ATE and population ATE. Extrapolating from sample to population introduces additional uncertainty. The Bayesian learning model presented in this chapter calls attention to two sources of uncertainty. One source of uncertainty is random sampling variability; another is the threat of bias due to convenience sampling, publication bias, model misspecification, and other systematic sources of error.

Both sources of uncertainty must be borne in mind when integrating research findings. This chapter has introduced basic principles of meta-analysis while at the same time cautioning readers that models that ignore systematic error may produce misleading conclusions. Meta-analysis is a statistical model designed to estimate a population average treatment effect using a collection of experimental studies with similar treatments, comparable outcome measures, and subjects sampled randomly from the same population. Strong and often untenable statistical assumptions must be invoked when meta-analysis is used to integrate research findings from disparate experiments or a mix of experiments and observational studies.

Statistical models may be used to systematize the process of extrapolation. One class of experiments that is well suited to statistical modeling involves treatments that vary along a single dimension, such as quantity or duration. Formulating a statistical model presents a trade-off between parsimony and predictive accuracy. We describe one approach, which is to use a portion of the experimental data for exploratory purposes, developing a model that appears to fit the observed outcomes and makes sensible predictions. This model is then evaluated using additional experimental evidence, with special attention paid to whether the parameters of interest appear to generalize from one subset of the data to the other.

Connecting research findings to one another and to broader theoretical propositions is a difficult undertaking. The social world is complex and continually changing. Experimental research can contribute a great deal to the assessment of causal effects for different subjects, treatments, and contexts, but modeling assumptions play an irreducible role in any attempt to integrate research findings, extract theoretical propositions, or formulate policy recommendations. Recalling equation (11.12), bear in mind that the farther you stray from the center of your data, the more uncertain your claims will be. In a broader sense, the farther you extrapolate from the evidence at hand, the more dependent you become on possibly fallible assumptions.

SUGGESTED READINGS

When analyzing experimental or observational data, social scientists routinely apply statistical models that invoke assumptions about the distributions from which disturbances are drawn and the functions that link interventions and outcomes. Wooldridge (2010) and Freedman (2005) discuss regression, non-linear models, and nested hypothesis tests. An accessible introduction to Bayes' rule and its application in statistics is Bolstad (2007). Hedges and Olkin (1985) provide an introduction to meta-analysis as a statistical method, and the edited volume by Cooper, Hedges, and Valentine (2009) discusses a variety of challenges that confront systematic literature reviews. Cole, and Stuart (2010) illustrate the statistical issues that arise when one attempts to generalize from a randomized trial to a target population. Gerber, Green, and Kaplan (2004) present a Bayesian framework for thinking about how to combine priors, unbiased evidence, and potentially biased evidence.

EXERCISES: CHAPTER 11

1. Important concepts:
   (a) Explain the distinction between a sample average treatment effect and a population average treatment effect. Why might a researcher be primarily interested in one rather than the other?
   (b) What is a meta-analysis? Why is meta-analysis a better way to summarize research findings than comparing the number of studies that show significant estimated treatment effects to the number of studies that show insignificant estimated treatment effects?
   (c) Using equations (11.2), (11.3), and (11.4), provide a hypothetical example to illustrate how uncertainty about the possibility of bias affects the way in which prior beliefs are updated in light of new evidence.
   (d) What does it mean to conduct a hypothesis test that compares two "nested" models?
2. Identify a research literature where an experiment was replicated at least once. Carefully consider the manner in which subjects and contexts were selected, as well as the methods used to administer treatments and measure outcomes. If publication bias is a concern, note that, too. Based on your close reading of these studies, assess whether the estimated ATE from each study can be said to constitute an independent random sample from some larger population.
3. Suppose one were to sample N subjects at random from a population of N^* people. An experiment is performed whereby m of the N subjects are assigned to receive a treatment, and the remaining $N - m$ are assigned to the control group. Suppose that sometime after the treatment is administered, outcomes are measured for all N^* people.
   (a) Suppose one estimates the population ATE by comparing the mean outcome among the m subjects in the treatment group to the mean outcome among the $N^* - m$ subjects who were not assigned to the treatment. Is this estimator unbiased?
   (b) Would the appropriate standard error of this difference-in-means estimator be equation (11.1), equation (3.4), or neither?
4. Suppose that $N = 2m$. Using equations (3.4) and (11.1), show that SE(SATE) $\approx$ SE(PATE) when the treatment effect is constant across subjects. Hint: In this case, $\text{Var}(Y_i(0)) = \text{Var}(Y_i(1)) = \text{Cov}(Y_i(0), Y_i(1))$.
5. Using the Bayesian updating equations, show algebraically how the priors represented in Figure 11.1 combine with the experimental results depicted in order to form a posterior distribution with a mean of 8 and a standard deviation of 0.89.
6. Prior to a 2006 primary election, Gerber, Green, and Larimer sent a sample of registered voters in Michigan an encouragement to vote that disclosed whether each person registered to vote at that address had voted in previous elections.[18] (The self mailing is described in Chapter 10.) A similar mailing was sent to Michigan voters in 2007, prior to municipal elections in small cities and towns,[19] and to Illinois voters in 2009, prior to a special congressional election.[20] For comparability, we restrict each of the samples to the set of households containing just one registered voter.

18 Gerber, Green, and Larimer 2008.
19 Gerber, Green, and Larimer 2010.
20 Sinclair, McConnell, and Green 2012.

Study	Number in the control group	Number voting in the control group	Number in the treatment group	Number voting in the treatment group
2006 Michigan	26,481	8,755	5,310	2,123
2007 Michigan	348,277	88,960	12,391	3,791
2009 Illinois	15,676	2,600	9,326	1,936

(a) Estimate the ATE for each study.

(b) Estimate the standard error for each study. Use the standard errors (squared) to calculate the precision of each study.

(c) Assuming that these three samples are random draws from the same population, calculate a precision-weighted average of the three studies. (Hint: weight each estimate by the inverse of its squared standard error.)

(d) Show that this estimate is identical to what one obtains by using the Bayesian updating formula recursively: use the results from the 2006 study as your priors, and update them based on the 2007 study to form a posterior mean and posterior variance based on equations (11.2) and (11.3); then update this posterior using the results from the 2009 study.

(e) Use equation (11.3) to estimate the variance of the precision-weighted average. Take the square root of the variance in order to obtain the standard error. In order to estimate the 95% confidence interval, use the following procedure, which is based on a large-sample approximation. Obtain the lower bound of the interval by subtracting 1.96 times the standard error from the precision-weighted average; obtain the upper bound of the interval by adding 1.96 times the standard error to the precision-weighted average.

(f) Explain why the confidence interval formed in part (e) is likely to understate the true amount of uncertainty associated with the estimate of the population ATE.

7. According to the logistic regression coefficients reported in Table 11.2, the intercept in Region 1 is 8.531 and the slope is −1.978. Based on these numbers, what proportion of those offered a price of 100 shillings is expected to buy a bed net? How does this compare to the actual rate of purchases at this price?

8. Table 11.3 presents observed and predicted values for each of the experimental sites in the Kenya bed nets study. Data for this experiment may be found at http://isps.research.yale.edu/FEDAI.

   (a) Verify that mispredictions mostly occur in Region 4 by calculating this region's contribution to the total chi-square statistics for each model.

   (b) Re-estimate the two logistic regression models presented in Table 11.3, this time excluding Region 4. Reproduce Table 11.3.

   (c) Although excluding Region 4 from the analysis improves the apparent fit of both models, why might it be considered a questionable practice to exclude Region 4?

9. Because the log transformation of price is undefined when price is zero, we excluded the zero price condition from the analysis of bed net purchases in Tables 11.2 and 11.3.

   (a) If we exclude zero prices from our experimental analysis, will our estimate of the causal effect of price be biased?

(b) Suppose we reasoned that a nominal price of zero nevertheless involves some transaction cost, as villagers have to make the effort to redeem their vouchers. For a given subject, we may model the probability of making a purchase as:

$$Pr[Y_i = 1] = \frac{1}{1 + e^{-(\beta_0 + \beta_1 \ln[V_i + \gamma])}},$$

where γ represents the transaction cost of redeeming the voucher. In order to estimate γ, insert a positive value of γ, and use logistic regression to estimate the revised model; note the value of the log-likelihood for this model. Repeat this exercise for different values of γ. Obtain the "maximum likelihood estimate" of γ by finding the value of γ that maximizes the log-likelihood.

(c) What is the substantive interpretation of the maximum likelihood estimates of γ and β_1? (Note that the standard errors using this method understate the true sampling variability because they are conditional on a particular choice of γ. Ignore the reported standard errors, and just interpret the estimates.)

10. This chapter discussed the modeling issues that arise when randomly assigned treatments vary in intensity or duration. The table below considers a different case, where subjects are randomly assigned treatments but then choose to take different dosages. In this experiment, subjects were randomly assigned to receive one of two pre-recorded phone messages.[21] The treatment script encouraged people to vote and revealed whether members of the household voted in the past two elections. The control script encouraged people to recycle. Both calls were made the day before the election. Both calls were answered at similar rates. The table below presents voting rates for households that answered the phone call. Voting rates are broken down by how long the person answering the phone took to hang up after the recorded message began.

	Duration of call before respondent disconnected				
Treatment group	**1-10 seconds**	**11-20 seconds**	**21-30 seconds**	**31-40 seconds**	**Total**
Call encouraged voting	16.6 (187)	17.4 (784)	19.7 (983)	24.3 (2,032)	21.4 (3,986)
Call encouraged recycling	17.5 (143)	18.3 (619)	18.9 (1,132)	19.8 (2,012)	19.2 (3,906)

Entries are percent voting, with *N*s in parentheses. The sample is restricted to one-voter households. Both scripts were approximately 35 seconds long.

(a) Focusing only on households that answered the phone, estimate the apparent average effect of assignment to the script that encouraged voting?

(b) Does this table provide convincing evidence that "the longer a person listens to a recorded message that encourages voting, the more effective that message will be in terms of boosting voter turnout"? Why or why not?

21 Gerber et al. 2010.

CHAPTER 12

Instructive Examples of Experimental Design

Crafting an informative experiment is a highly customized undertaking. Every project requires investigators to devise treatments, allocate subjects, and measure outcomes in ways that will speak convincingly to a specific research question. Familiarity with the principles covered in previous chapters is an important starting point; the next step is to learn to design studies so as to minimize reliance on untestable assumptions, address threats to inference, and deflect potential problems of implementation.

One good way to learn to link abstract principles to real-world application is to read instructive examples of field experimental research. In this chapter we discuss a series of thought-provoking field experiments in order to illustrate the kinds of design challenges that researchers routinely encounter. In each section, we present a simplified account of an actual experiment that highlights valuable lessons that can be applied to other research questions. (For the sake of brevity, we skip over many important aspects of these experiments, but we encourage readers to examine the original articles.) For each application, we formalize the core hypotheses using potential outcomes, show the conditions under which causal parameters are identified, and work from abstract models to experimental design and analysis.

We begin with examples showing how experimental design may be informed by theoretical ideas and debates. In one study, the experimental treatments are cleverly designed to distinguish between two closely related theories, while in another, the allocation of subjects to treatments is guided by conjectures about which subjects are most likely to respond to the intervention. A third example focuses on outcome measures, showing how failure to consider the full range of subjects' responses to an intervention can lead to misleading interpretations. We next illustrate how using a diverse array of treatments makes an experiment more informative and robust. Sometimes researchers tailor treatments to match the circumstances of the subjects who receive them; we discuss the implications of this design for analysis and

What you will learn from this chapter:

1. How experiments may be designed to test competing theoretical interpretations.
2. How sampling can facilitate estimation of average treatment effects within subgroups.
3. The advantages of factorial designs.
4. The inference problems that arise when the treatments are tailored to subjects' characteristics.
5. The complications that arise when a treatment involves several components, some of which encounter noncompliance.
6. How to address instances in which some subjects have missing outcomes, missing background data, or missing information concerning their compliance with the assigned treatment.

interpretation. In order to highlight the importance of exclusion restrictions, we consider an instance where the treatment consists of many components, some of which encounter noncompliance. Finally, we discuss an instructive example in which some subjects have missing outcomes, missing background data, or missing information concerning their compliance with the assigned treatment. In each case, we consider ways in which one might adjust the design and analysis in order to address these complications.

12.1 Using Experimental Design to Distinguish between Competing Theories

Should health products like bed nets and water disinfectant be given away in developing countries free of charge, or should they be sold? Those who argue for free distribution say that charging denies access to the poorest households. Others counter that products that are distributed for free are perceived to have little value and go unused. This argument against free distribution is based on the hypothesis that paying produces a *sunk-cost* effect; the very act of having paid makes failure to use the product psychologically costly. The notion that distributing health products at little or no cost could reduce usage is intriguing and, if true, important for policy design. Well-meaning interventions that lower the price of a widely purchased health product might inadvertently diminsh its usage. Beyond its implications for public health programs, the claim that paying for something causes greater demand is part of a broader line of research emphasizing how consumer behavior is shaped by psychological factors. It would be useful to know if there is anything to this striking claim. Unfortunately, the sunk-cost hypothesis is hard to test.

The psychology of sunk cost implies that the more a consumer pays for a product, the more they will tend to use it. It is tempting to test this hypothesis by performing an experiment that randomly assigns households to groups that are offered the product for free or at various prices. Researchers could then measure the difference in usage among those who acquire the product for free versus those who decide to pay for it. A problem with this approach is that more than one theory predicts that, conditional on acquisition, those who pay a higher price may use the product more than those who get the product for free. The *screening* hypothesis says that a low-income household's willingness to pay for a health product is, all else equal, correlated with the benefit that the household expects to receive from acquiring the product. Average usage rates for those who purchase at different prices are expected to increase with price because higher prices discourage those consumers who see less benefit from using the product. So there is a problem interpreting evidence that higher price leads to greater usage: both the sunk-cost and the screening hypotheses make the same prediction, namely that among those who acquire a product, the higher the price, the higher the rate of usage.

Ashraf, Berry, and Shapiro set out to design an experiment to disentangle sunk-cost and screening effects.[1] Their study gauges the effect of price on Zambian households' usage of a water disinfectant, Clorin. In order to understand the identification strategy behind their experiment, let's consider each household's potential outcomes. Define $Q_i(x)$ as the quantity of Clorin usage that occurs if Clorin is purchased at price x. Define $Y_i(x)$ as the decision whether to purchase Clorin at price x; $Y_i(x) = 1$ when a purchase occurs at price x, and $Y_i(x) = 0$ if no purchase occurs.

Suppose that households were randomly assigned to be offered Clorin at two prices, p and p', where $p' > p$. The researcher observes which households buy Clorin at each price and measures their usage. We could use the average usage levels among those who make purchases at each price to estimate two quantities, $E[Q_i(p') \mid Y_i(p') = 1]$ and $E[Q_i(p) \mid Y_i(p) = 1]$. Unfortunately, this information is not enough to disentangle the screening and the sunk-cost effects. The expected difference in average usage rates can be written:

$$E[Q_i(p') \mid Y_i(p') = 1] - E[Q_i(p) \mid Y_i(p) = 1]$$

$$= \overbrace{E[Q_i(p') \mid Y_i(p') = 1] - E[Q_i(p) \mid Y_i(p') = 1]}^{\text{Sunk-cost effect}}$$

$$+ \overbrace{E[Q_i(p) \mid Y_i(p') = 1] - E[Q_i(p) \mid Y_i(p) = 1]}^{\text{Screening effect}}. \qquad (12.1)$$

1 Ashraf, Berry, and Shapiro 2010.

by subtracting and adding $E[Q_i(p) \mid Y_i(p') = 1]$. The first right-hand-side term is the usage generated by the sunk-cost effect. This term is equal to the change in average usage for a subgroup of households (those who purchase when the price is p') at two different prices p' and p. For this subgroup, the difference in usage is attributable to the effect of price alone on these households' desire to use the product. According to the sunk-cost hypothesis, this effect will be positive. The second right-hand-side term represents the screening effect. It shows how average usage levels for a given price (p) differ for two sets of households: the set of households that would agree to purchase the Clorin at the higher price and the set that would agree to purchase at the lower price. The sunk-cost effect plays no role in the second term because both groups' usage $Q_i(p)$ is based on the same price. According to the screening hypothesis, the households that are willing to purchase at a higher price are more eager to use the product, and therefore the screening term should be positive. As previously noted but now formalized, both theories contribute a positive term to the sum. Therefore, if we observe greater usage among those who purchase at the high versus low price, we cannot distinguish between these two interpretations.

The decomposition in equation (12.1), however, suggests a solution to this identification problem. The key is to recognize that price serves two distinct roles: it determines whether a purchase occurs and how much usage occurs given a purchase. When the experimenter asks whether a household will purchase Clorin at a given price p, the response reveals $Y_i(p)$. When the experimenter charges the household a price for the good, the price at which the product changes hands determines $Q_i(p)$. Call the price to which the households respond initially the *offer price*, and call the price for which the product is sold the *transaction price*. Recognizing that price plays these two roles suggests how to estimate the two effects: make the offer price different from the transaction price.

In a field setting, Ashraf, Berry, and Shapiro distinguish offer and transaction prices by first asking each household if it wishes to purchase Clorin at a stated price p' and then, after receiving a response, announcing to those who agree to purchase that they are eligible for a discount and may purchase Clorin at price p. Households are randomly assigned an initial stated price and a discounted price, allowing the researchers to estimate three of the four quantities in Table 12.1. (No attempt is made to estimate $E[Q_i(p') \mid Y_i(p) = 1]$, for $p' > p$. In other words, the researchers never raise the price after the initial offer; fortunately, this quantity is not needed to distinguish the sunk-cost effect from the screening effect.) The screening hypothesis can be tested by assessing whether usage increases when the offer price rises, holding constant the transaction price. The sunk-cost hypothesis can be tested by assessing whether usage increases when the transaction price rises, holding constant the offer price.

The exclusion restriction plays a crucial role in this design. In particular, the model presented above assumes that potential outcomes for usage, $Q_i(x)$, are a function of just one input, transaction price. This assumption allows one to test the

TABLE 12.1

Expected potential outcomes under varying offer and transaction prices

		Transaction price	
		p'	p
Offer price	p'	$E[Q_i(p') \mid Y_i(p') = 1]$	$E[Q_i(p) \mid Y_i(p') = 1]$
	p	$E[Q_i(p') \mid Y_i(p) = 1]$	$E[Q_i(p) \mid Y_i(p) = 1]$

Note: $Y_i(x)$ refers to whether the household agrees to purchase Clorin when initially offered a price of x. $Q_i(x)$ refers to the quantity of Clorin used given a transaction price of x. The two experimental prices are p and p', where $p' > p$. The shaded cell reflects the fact that Ashraf et al. do not attempt to estimate $E[Q_i(p') \mid Y_i(p) = 1]$; for obvious reasons, the surprise transaction price is never increased beyond the original offer price.

sunk-cost hypothesis by comparing the usage levels of those who accept a price of p' and pay an actual price of p' to those who accept a price p' but pay a surprise discount price p. Suppose, however, that consumers are motivated by sunk-cost reasoning but, when thinking back on how much they paid for Clorin, remember only the price they were offered initially. The data would give the misleading impression that no sunk-cost effect exists. Fortunately, this scenario has testable implications: one could ask respondents who received different randomly assigned prices to recall what they paid for Clorin. Concern about the exclusion restriction can also inspire researchers to be proactive in their research design; the authors took steps to reaffirm the transaction price with customers in order to help increase its cognitive salience.

The Clorin experiment is an elegant example of a general design principle. If a causal proposition follows from more than one theory, randomly assign different components of the treatment in order to differentiate the mechanisms suggested by each theory. This principle may be used to inform theoretical debates across an array of substantive domains, a point that we emphasize in the exercises to this chapter.

12.2 Oversampling Subjects Based on Their Anticipated Response to Treatment

An important feature of tax compliance is that certain sources of income are much easier to conceal than others. For example, American taxpayers with business and farm income have much more flexibility in what they report to tax collection agencies than do citizens with only wage income, which is reported by employers directly to the government. Taxpayers whose tax returns in the previous year suggest that their income is primarily from business or farming are labelled *high-opportunity* taxpayers.

In the early 1990s, the Minnesota Department of Revenue sponsored an innovative series of experiments in an effort to discover ways of increasing compliance with tax laws, especially among high-opportunity taxpayers. In one of these experiments, Slemrod, Blumenthal, and Christian (2001) assess the effects of an official letter warning taxpayers of a possible audit. Among the outcomes considered is the income that subjects subsequently report in relation to the income they reported in the year prior to the intervention. The hypothesis is straightforward: the change in income reported by households that receive a warning letter will be greater, on average, than the change in income reported by households that do not receive a letter.

We use the Slemrod et al. study to illustrate how researchers may adjust the way in which subjects are sampled and allocated to experimental groups when the primary research aim is to learn about conditional average treatment effects, or average treatment effects within subgroups. Table 12.2 shows the distribution of income and opportunity among Minnesota households. The design question is how to sample subjects from the six combinations of income and opportunity. In this study, sampling is constrained by the fact that auditing and tracking tax returns is costly in terms of staff time. Audits must be conducted for treatment households (which receive the letter), and subsequent tax returns must be tracked closely for households in both the treatment and control groups. As discussed in Chapter 3, researchers seeking a precise estimate of the ATE would ordinarily allocate a similar number of subjects to each experimental condition, but concerns about non-interference pose another constraint: if a large proportion of the state's taxpayers were told to expect an audit, the effects of this treatment might spill over to the control group either through

TABLE 12.2

The population of Minnesota households partitioned by income and opportunity to misreport income to Department of Revenue

	Low opportunity	High opportunity	Total
Low income	449,017	2,120	451,137
Medium income	1,290,233	50,920	1,341,153
High income	52,093	8,456	60,549
Total	1,791,343	61,496	1,852,839

"High opportunity" describes taxpayers who have sources of income that are relatively easy to conceal from tax collection agencies.

conversations or media coverage. In this experiment, the researchers deployed only 1,537 treatments, or approximately one letter for every 1,200 households in the state.[2]

For an experiment with a given size treatment group, which types of households should receive letters? The simplest design is to select a random sample of the Minnesota population. If this procedure were followed, then the number of subjects assigned to treatment from each of the six population groups would be expected to follow the pattern described in Table 12.3. Leaving aside the possibility that the variance of the outcome measure may differ from one taxpayer category to the next, this sampling strategy is an efficient way to estimate the average treatment effect among all Minnesota households. Notice, however, that fully two-thirds of the treatments are applied to low-opportunity middle-income households. Only 51 households in the high-opportunity category receive letters, and just 7 of these households are high income. High-income and high-opportunity taxpayers are potentially an immense source of additional tax revenue. Sampling with equal probability from the population of households would not enable the Department of Revenue to assess treatment effects within these especially important subgroups.

The actual sampling scheme used in this experiment reflects the value of obtaining enough data to estimate treatment effects among high-income and high-opportunity households. Table 12.4 shows the number of households assigned to treatment in each of the six population segments. The authors' decision to deviate from equal-probability sampling increased the number of high-opportunity treatments tenfold and tripled the number of treatments directed at high-income households. Because the standard error of the mean varies in proportion to the square root of the sample size, lowering the number of subjects in the middle-income low-opportunity group raises the standard error of its estimated mean by about 40%.

TABLE 12.3

Expected number of subjects assigned to treatment when the population of Minnesota households is sampled with equal probability

	Low opportunity	High opportunity	Total
Low income	373	2	375
Medium income	1,070	42	1,112
High income	43	7	50
Total	1,486	51	1,537

2 The actual number of treatments is 1,714, but we exclude those who were subsequently dropped from the analysis for, among other things, not filing a tax return. We take up the issue of attrition later in the chapter.

TABLE 12.4

Actual number of subjects assigned to treatment in the Minnesota tax compliance experiment

	Low opportunity	High opportunity	Total
Low income	381	52	433
Medium income	520	397	917
High income	107	80	187
Total	1,008	529	1,537

On the other hand, the tenfold increase in the high-income high-opportunity group lowers the standard error of its estimated mean by about 70%.

In principle, the control group could have comprised all untreated households. However, measuring outcomes requires close examination of income records. Because of resource constraints, only 20,831 households were selected into the control group. The control group is nevertheless much larger than the treatment group, and its composition of taxpayer types more closely matches the population, although the pattern of oversampling is a muted version of the pattern for the treatment group. Table 12.5 compiles the distribution of all households across the six taxpayer types, the distribution of taxpayer types by treatment and control groups, and the estimated ATEs within each type.

Recall from Chapters 3 and 4 that under this sampling scheme one cannot obtain unbiased estimates of the overall ATE by comparing average outcomes in the treatment and control groups because the probability of assignment to treatment varies across the six categories of taxpayers. (A naive comparison of unweighted means ignores the fact that high-income households comprise 12.2% of the treatment group but only 4.5% of the control group.) The proper way to estimate the population average treatment effect is to construct a weighted average of the estimated ATEs for each taxpayer category, where the weights are determined by the share of the Minnesota population that falls in each category.[3] For example, the estimated ATE in

3 An algebraically identical way to adjust the data to estimate the population average treatment effect is to reweight the observations. For each experimental group calculate:

$$\text{weighted average} = \Sigma w_i Y_i / \Sigma w_i,$$

where w_i is the number of subjects in unit i's block divided by the number of subjects in unit i's block who are assigned to that experimental group. The estimate of the PATE is the weighted average for the treatment group minus the weighted average for the control group. In the Slemrod et al. experiment, there are six blocks. To see how to form weights for one of the blocks, consider subjects in the low-income, low-opportunity block: for each of the 381 subjects in the treatment group, $w_i = 449{,}017/381$, and for the 4,829 control group subjects, $w_i = 449{,}017/4{,}829$.

TABLE 12.5

Distribution of treatment and control group subjects and experimental outcomes, by household type

		Proportion of taxpayer population within each household type	Proportion of treatment group within each household type	Proportion of control group within each household type	Avg. change in reported income, treatment group	Avg. change in reported income, control group	Estimated ATE
Low income	Low opportunity	0.242	0.248	0.232	1,609	1,490	119
	High opportunity	0.001	0.034	0.006	6,502	3,204	3,298
Medium income	Low opportunity	0.696	0.338	0.655	960	497	463
	High opportunity	0.027	0.258	0.063	3,546	1,539	2,007
High income	Low opportunity	0.028	0.070	0.033	−18,721	−2,659	−16,062
	High opportunity	0.005	0.052	0.012	−33,513	12,150	−45,663
	Total N	1,852,839	1,537	20,831			

Change in reported income is difference between income reported in the post-intervention tax year and income reported in the pre-intervention tax year.

the low-income low-opportunity group (\$1,609 − \$1,490 = \$119) would receive a weight of 0.242 because 449,017 of the 1,852,839 Minnesota households fall into this group. Applying population weights to each group's estimated ATE gives:

$$\begin{aligned} &0.242 \cdot (\$1{,}609 - \$1{,}490) + 0.001 \cdot (\$6{,}502 - \$3{,}204) + 0.696 \cdot (\$960 - \$497) \\ &\quad + 0.027 \cdot (\$3{,}546 - \$1{,}539) + 0.028 \cdot (-\$18{,}721 - (-\$2{,}659)) \\ &\quad + 0.005 \cdot (-\$33{,}513 - \$12{,}150) \approx -\$278. \end{aligned} \tag{12.2}$$

Surprisingly, this estimate indicates that the letters led to a slight *decrease* in tax revenue, although the confidence interval (−\$1,187, \$631) suggests that one cannot rule out the possibility that the letters had no average effect. Using a somewhat different outcome measure, the authors find a statistically significant increase ($p < 0.01$) in the proportion of taxpayers reporting more income in 1994 than in 1993, suggesting that the letter did cause some taxpayers to change their behavior.

The strength of the authors' sampling strategy is that it provides an opportunity to investigate whether and how treatment effects vary by subgroup. As Table 12.5 shows, estimated treatment effects vary markedly across the six taxpayer categories, with low- and middle-income taxpayers reporting more and high-income taxpayers reporting less. Especially striking is the behaviour of high-income high-opportunity taxpayers, whose reported income appears to decline by \$45,663 in the wake of the intervention. This finding was unanticipated—the Department of Revenue expected their letters would increase reported income in this especially lucrative group of taxpayers—and illustrates yet again that rigorous scientific evaluations sometimes challenge the beliefs of seasoned professionals. For our purposes, Table 12.5 provides an opportunity to walk through the analysis and interpretation of conditional average treatment effects.

Recall from Chapter 9 that subgroup analysis may be used to address several related research questions. The first is whether the pattern of results for each subgroup leads us to reject the null hypothesis that all subgroups share the same average treatment effect. Although the data from this experiment are not publicly available, the authors do report the proportion of each experimental group that increased its reported income, and this information enables us to perform a hypothesis test. Our test statistic is the F-test comparing two nested regressions in which increased reported income is the outcome variable. The first regression allows treatment effects to vary by subgroup. The restricted regression assumes that treatment effects are the same across subgroups. Comparing the F statistic for the actual experiment to the sampling distribution of this statistic under the null hypothesis that the treatment effect for all subjects is the same as the estimated ATE, we obtain a p-value of 0.07. This borderline result provides some equivocal evidence of treatment effect heterogeneity.

A second question is whether any of the subgroups show convincing evidence of a positive treatment effect. Here a one-sided test reflects the goals of the study, namely to evaluate an intervention thought to increase tax revenue. One must apply a correction for multiple comparisons, since there are six opportunities for a significant positive result to arise by chance. (The probability of at least one false positive test is $1 - 0.95^6 = 26\%$.) The correction to the usual 0.05 significance level places the critical p-value for each hypothesis test at $0.05/6 = 0.008$. As it turns out, none of the six hypothesis tests are significantly positive, regardless of whether a Bonferroni correction is applied. Finally, one might ask whether any of the subgroups show a treatment effect that is significantly different from zero, regardless of sign. Recall that for one of the subgroups, the high-income high-opportunity taxpayers, the estimates indicate a large and statistically significant *negative* effect of the warning letter. For this subgroup, the authors report a treatment effect of −45,663 with a standard error of 17,394. If we assume that the sampling distribution is approximately normal, this ratio of estimate to standard error implies a p-value of 0.004, which leads us to reject the null hypothesis of no treatment effect even after applying the Bonferroni correction. The authors are appropriately cautious about this unexpected result; they consider substantive explanations for the observed pattern (e.g., taxpayers in this subgroup were prompted to prepare their tax returns more carefully in the wake of the treatment) but do not find much in the data to support these subsidiary hypotheses. Another reason for caution has to do with the way in which this p-value is calculated. Throughout this book, we have been hesitant to rely on normal approximations when calculating p-values, and this example illustrates a case in which a normal approximation may produce misleading results. The distribution of year-to-year changes in income among high-income households may have a few very large positive or negative values; as a result, the difference-in-means estimator may have a non-normal sampling distribution unless the number of subjects in each experimental group is very large. Randomization inference would be particularly useful here, as it provides exact p-values for a test of the sharp null hypothesis of no effect regardless of sample size or the distribution of outcomes.

This study provides an instructive example of how research objectives guide the sampling and allocation of subjects. When the average treatment effect within certain subgroups is of special interest, experimental designs may be adapted so that the probability of treatment varies from one subgroup to the next. Special care must be exercised when analyzing these experiments: weights must be applied if the aim is to generate unbiased estimates of the population average treatment effect, and the problem of multiple comparisons must be addressed if the aim is to test whether treatment effects are statistically significant within particular subgroups.

12.3 Comprehensive Measurement of Outcomes

Researchers may draw misleading conclusions about the effects of an experimental intervention when they confine their attention to a narrow set of outcomes. For example, suppose you were interested in the effect of a treatment that encourages civic participation, an outcome category that might encompass a range of behaviors including voting, reading local newspapers, discussing public affairs with neighbors, and showing up at town meetings. What if you only measure whether the subject shows up at the next town meeting? In order to interpret your estimated ATE as the average effect of your intervention on civic participation, you will have to assume that the behavior you measure is representative of the civic behaviors that go unmeasured. This assumption may be risky. Suppose that subjects exposed to your treatment are indeed more likely to show up at the town meeting. Your intervention seems effective. But imagine that your intervention actually has two effects: it makes subjects more likely to attend town meetings but less likely to vote in local elections. Now the overall effect on civic participation is ambiguous. The same point holds if the treatment causes subjects to attend the next town meeting but to skip the meeting after that, a pattern known as intertemporal substitution. The broader design principle: measure an array of conceptually related outcomes over time, so that conclusions do not rest on the assumption that all outcomes move in the same direction in the wake of an intervention.

A field experiment by Simester, Hu, Brynjolfsson, and Anderson illustrates the importance of this design principle.[4] Customers of a clothing retailer were assigned to two groups, low and high advertising. Over an eight-month period, the low-advertising group received 12 catalogues in the mail and the high-advertising group received 17. The central research question is which advertising strategy is more profitable. The results show quite convincingly that the high-advertising condition raised catalogue sales during the treatment period, defined as the advertising period and the subsequent four months. The strong and statistically significant estimates seem to imply that the retailer should ramp up its advertising. It turns out, however, that the short-run results are misleading; although more advertising caused a short-run boost in catalogue sales, it also caused a decline in catalogue sales during the eight months following the treatment period. Moreover, customers who received extra catalogues made fewer purchases through the company's Internet site. Although the short-term results look promising, a comprehensive assessment of outcomes shows that sending extra catalogues actually loses money.

This example illustrates how experiments that measure outcomes too narrowly can run afoul of the excludability assumption. The authors set out to study the profitability of an advertising strategy rather than the short-run sales that occurred through a single purchase channel. Call $Y_i(d)$ each subject's total purchases under treatment ($d = 1$) or control ($d = 0$), and divide these purchases into short-term $Y_i^S(d)$ and long-term $Y_i^L(d)$ purchases, such that $Y_i(d) = Y_i^S(d) + Y_i^L(d)$. If total purchases are the true outcomes of interest but we compare the average $Y_i^S(1)$ to the average $Y_i^S(0)$, the expected value of our estimate is off by the expected difference between $Y_i^L(1)$

4 Simester, Hu, Brynjolfsson, and Anderson 2009.

and $Y_i^L(0)$, and our estimator will be biased. The excludability assumption is violated because experimental assignment affects outcomes (total sales) through a pathway (long-term sales) that is ignored in our analysis.

Similar concerns arise whenever there is an imperfect match between the outcome as a theoretical construct and the outcome that is actually measured. For reasons of feasibility or convenience, researchers frequently measure outcomes that are early or incomplete manifestations of the true outcomes of interest. In medical studies, for example, treatments designed to prevent heart attacks are sometimes evaluated according to whether blood tests reveal a decline in cholesterol, a *proxy outcome* or *surrogate outcome* that is correlated with heart attacks. Cholesterol is easy to measure in the short run, whereas heart attacks occur infrequently and necessitate the use of large samples or extended periods of observation. However, the use of cholesterol as a proxy variable introduces a new substantive assumption about the relationship between cholesterol and heart attacks, thereby opening the door to a possible violation of excludability. Recall from section 2.7.1 that when errors of measurement vary systematically by experimental condition, a comparison of treatment and control group means may no longer generate unbiased estimates of the ATE. In this example, a treatment that has no effect on cholesterol might nevertheless reduce the risk of heart attack; in this case, the use of cholesterol as a proxy outcome would overstate adverse health outcomes in the treatment group relative to the control group. In the social sciences, where proxy variables and short-run outcomes are frequently used, their adequacy should be carefully evaluated and an active topic of empirical investigation.

12.4 Factorial Design and Special Cases of Non-Interference

Bertrand and Mullainathan's study of racial discrimination in the labor market illustrates two important insights, the value of administering a wide array of treatments and more than one treatment to each subject.[5] The authors' primary research objective is to assess whether a job applicant's race affects whether a firm grants the applicant an interview. To restate the research objective using potential outcomes notation, let each firm be a subject, and let $Y_i = 1$ if firm i grants an interview and $Y_i = 0$ otherwise. For a given resume, each firm has two potential outcomes, $Y_i(d)$, where $d = 1$ when a resume is from a person with a stereotypically white name and $d = 0$ when a resume is from a person with a stereotypically black name. The treatment effect is 1 if the firm grants an interview to the applicant who submits the white

5 Bertrand and Mullainathan 2004.

resume but not the applicant who submits the black resume; 0 if the firm grants an interview to neither applicant or both applicants; and −1 if the firm grants an interview to the applicant submitting the black resume but not the applicant submitting the white resume. The average treatment effect represents the degree to which firms discriminate in favor of white applicants.

In order to estimate this ATE, the authors use a type of field experiment known as an *audit study*. In a typical application of this design, research confederates apply for loans, make retail purchases, inquire about the availability of housing, or ask for assistance.[6] The researcher measures how loan officers, landlords, or public officials respond. The experimental manipulation involves varying some attribute of the confederate, such as race, gender, or accent, while holding constant other attributes, such as age, attire, or how a request is worded. For example, in order to study rental housing discrimination, a researcher might send white and black confederates with similar financial and employment status to try to rent an apartment. The effect of race is measured by the difference in the landlord's responses to the black and white renters.

The traditional audit design has some drawbacks. First, because people vary across so many dimensions, it is challenging to implement an experiment in which all else is equal except for the applicants' race. Second, the confederates know they are participating in an audit study (indeed, they are often extensively coached), and this knowledge may have subtle effects on how they behave. Unfortunately, the confederates' behavior in the field is often hard to observe, which makes it hard to assess this concern empirically.

Bertrand and Mullainathan devise an audit study that avoids these potential problems. By emailing resumes to firms that have posted help-wanted ads, the researchers eliminate the need for confederates. In order to signal each applicant's race while holding all other aspects of the applicant's qualifications constant, fictitious applicants are randomly assigned putatively white or black names. Using population records and a pilot survey, the authors determine which first and last names lead a typical person to correctly infer the race of an applicant. The use of email instead of confederates eliminates extraneous variation in how the treatments are administered. Employers see resumes that are identical apart from the applicant's race.

The authors extend the basic audit design by varying resume quality in order to assess whether the effect of race changes depending on an applicant's qualifications. If employers are reluctant to hire blacks because they perceive them to be unqualified, resumes that demonstrate that the applicant is highly qualified should overcome this stereotype. Varying applicant quality has another advantage: the researcher can calibrate the race effect by describing it in relation to the effect of resume quality. For example, the experiment may indicate that attaching a white rather than black name

6 Fix and Struyk 1993; Choi, Ondrich, and Yinger 2005; Pager 2007; Butler and Broockman 2011.

to a resume is, on average, equivalent to adding five years of job experience to the resume.

Recall from Chapter 9 that a factorial design is an experiment that randomly allocates subjects to all possible combinations of treatments. The factorial design in this study involves two factors, applicant race (black or white) and resume quality (high or low).[7] Each of the four experimental groups contains approximately the same number of subjects; because the two factors are, by design, uncorrelated, interpretation of the results is straightforward. Table 12.6 shows the rate at which firms receiving each type of resume granted the applicant an interview. From this table, one can estimate four interesting quantities: the average effect of race regardless of resume quality ($9.65 - 6.45 = 3.20$); the average effect of high resume quality regardless of race ($8.75 - 7.35 = 1.40$); the average effect of high resume quality for white applicants ($10.79 - 8.50 = 2.29$); and the average effect of high resume quality for black applicants ($6.70 - 6.19 = 0.51$). Evidently, white applicants are called more frequently than black applicants, and the return to resume quality appears to be larger for white applicants than for black applicants. Another, equally valid way to interpret the table is to say that the effect of race is especially large when resume quality is high. This treatment-by-treatment interaction falls short of statistical significance ($p = 0.24$) but remains noteworthy nevertheless. Had the authors not employed a two-factor design, they would have been unable to detect the fact that increasing resume quality is of relatively little value for black applicants.

TABLE 12.6

Interview invitation rates, by race and quality of resume

	Overall	Low quality	High quality	High-low (p-value)
White names	9.65 [2,435]	8.50 [1,212]	10.79 [1,223]	2.29 (0.06)
Black names	6.45 [2,435]	6.19 [1,212]	6.70 [1,223]	0.51 (0.61)
Overall		7.35 [2,424]	8.75 [2,446]	1.40 (0.07)
White–Black (p-value)	3.20 (0.00)			

Note: Entries are percentage of resumes of each type that prompt a call or email from the prospective employer to schedule an interview. This table is adapted from Bertrand and Mullainathan 2004 Table 4, panel A. Following the authors, all p-values are based on two-tailed tests, although the case could be made for one-tailed tests given the expectation that white names and high quality resumes will garner more interviews.

7 For ease of exposition, we ignore other factors in the experiment, such as the sex or address of the applicant, and other nuances, such as the fact that quality is signaled through a variety of different attributes, such as job experience or education.

The researchers also diversify the treatments in a way that makes results less dependent on idiosyncratic design decisions. Bertrand and Mullainathan could have selected one white name and one black name when deploying their experimental treatments, but what if their selected names happened to coincide with a name in a pop song or a newly released movie? (We are aware of an audit study that used only one name that, by some freakish chance, became prominent in a national news story while the experiment was in the field.) The solution is familiar to those who work with investments: diversify. Instead of using one name, the authors assembled dozens of acceptable names and randomly assigned one to each resume. (This design also enabled the researchers to verify that all putatively white or black names generate the same treatment effects.) Diversification also guided the selection of job skills, education, experience, language background, and technical training that made for high- or low-quality applications. Although diversification requires the experimenter to formulate and manage a large number of randomly varying treatments, it reduces the risk that the experiment will be undermined by a single unfortunate choice.

Factorial design subtly changes the interpretation one attaches to the difference-in-means estimator. The difference in interview rates for white and black applicants represents the estimated effect of race weighted by the share of resumes of different quality. In this experiment, approximately the same number of high- and low-quality resumes were sent out, and so the estimand is the average treatment effect of race given a collection of resumes that are half high-quality and half low-quality. A researcher could easily reweight the data to reflect any desired mix of resumes—for example, a pool of resumes that were 75% high quality and 25% low quality. A drawback of factorial design is that it may reduce the precision with which a researcher can estimate the average effect of one specific combination of treatments. For example, if the primary aim were to estimate the effects of race given low-quality resumes, allocating resumes to the high-quality condition would reduce the number of observations available to estimate the parameter of interest, thereby increasing sampling variability. In this experiment, the researchers' aims were broader, and they used a factorial design to explore a wide array of interactions.

The Bertrand and Mullainathan study also illustrates how powerful the non-interference assumption can be. For now, ignore the factorial design and assume that the researchers had used a single resume from either a black or white applicant. An interesting feature of this experiment is that rather than send each firm either the black or white version of the resume, the researchers send each firm *both* black and white versions. In other words, the design enables the researchers to observe both $Y_i(0)$ and $Y_i(1)$ for each firm. The researcher can now calculate (rather than estimate) whatever quantity is needed from the complete schedule of potential outcomes! Racial discrimination can be calculated at the firm level without uncertainty.

Something seems wrong. Without randomly assigning resumes (all firms get the same pair of resumes) we nevertheless obtain a bounty of information about potential outcomes. Ordinarily, the discussion of potential outcomes starts with the stipulation that the researcher cannot observe both $Y_i(0)$ and $Y_i(1)$. The reason we can do so here is that we have, rightly or wrongly, introduced a strong non-interference assumption: no matter whether the firm is sent one resume or four, its potential outcomes $Y_i(d)$ remain the same.

The broader insight is that by invoking a non-interference assumption, a researcher may obtain a full schedule of potential outcomes using a within-subject design (see Chapter 8). Ordinarily, a researcher randomly assigns each subject to just one experimental condition on the grounds that non-interference will be violated if more than one treatment were applied. The plausibility of non-interference in this experiment seems to hinge on the number of resumes sent to each firm. Implicitly, we envision a human resources officer sifting through email applications. Perhaps non-interference would be a reasonable assumption if a few experimental emails were mixed in with a large number of other resumes, but what if several dozen experimental emails comprised the majority of the applicant pool?[8]

In practice, of course, the researchers did not send the same resumes with different names; they sent four resumes with different names and randomly varying profiles. As a result, the responses of each firm do not reveal $Y_i(0)$ and $Y_i(1)$ but rather a set of potential outcomes that are defined over a range of attributes other than race. For this reason, the experiment does not reveal treatment effects for each firm.[9] A firm that seems to discriminate because it interviews two white applicants and no black applicants could conceivably have based its choices on the particular qualifications listed in each resume.

Although the design does not yield a full schedule of potential outcomes for each firm, under the non-interference assumption, it does provide a great deal of added statistical power. Rather than compare responses across a heterogeneous collection of firms, the authors reduce statistical uncertainty by looking for evidence of discrimination within each firm. Discrimination is assessed by counting up the number of firms that seem to favor white applicants, black applicants, or neither. The authors find that 111 of the 1,323 firms grant the white applicants more interviews than the black applicants; 46 firms granted the black applicants more interviews than the white applicants; and the remainder granted the same number to each group.[10] The null hypothesis is that interviews are granted in a race-neutral way, in which case

8 In principle, the non-interference assumption can be tested. Assuming that it is possible to apply more than one treatment at the same time, the researcher can compare the effect of treatment B when it is the only treatment versus when it is applied along with treatment A. If subjects are randomly assigned to these alternatives, significant differences between the estimated effect of treatment B when it is or is not accompanied by treatment A is evidence of interference.

9 Another complication is the possibility that firms have several human resources staff reviewing resumes. If chance plays a role in determining which staff member reviews each resume and staff have different potential outcomes, the observed outcomes will only partially reveal a firm's average potential outcomes.

10 Overall, interviews were rare in this sample; 1,103 firms granted no interviews to any experimental applicants.

the probability that a firm grants more interviews to whites is the same as the probability that a firm grants more interviews to blacks. If this hypothesis were true, the probability of observing a percentage as high as $111/(111 + 46) = 70.7\%$ is essentially zero. The null hypothesis that firms treat white and black applicants identically is therefore rejected. Because we earlier rejected this null hypothesis when examining Table 12.6, this result does not show off the statistical advantages of controlling for firm-level heterogeneity. The results are more striking when we revisit the null hypothesis of no resume quality effect. Recall that across all resumes, high-quality applicants were 1.40 percentage points more likely to receive an invitation for an interview, which implies a two-tailed p-value of 0.07. When looking at firm-level patterns, we see that 99 of the 1,323 firms granted the high-quality applicant a higher number of interviews than the lower-quality applicant, whereas the lower-quality applicant received a larger number of interviews in just 66 instances. The two-tailed probability of obtaining an imbalance this large is 0.01.[11] In sum, the Bertrand and Mullainathan study illustrates the many benefits of factorial design: a wider range of research hypotheses may be tested; diversification of treatments guards against unanticipated problems; and, in special cases, varying treatments facilitates the unobtrusive application of multiple treatments per subject.

12.5 Design and Analysis of Experiments In Which Treatments Vary with Subjects' Characteristics

A growing number of field experiments assess the impact of informing citizens about the performance of their elected representatives. This research agenda is motivated by the concern that without this information, citizens will unwittingly vote for corrupt or incompetent candidates. Experiments have studied the effects of information in a variety of settings. Banerjee et al. study the effects of distributing scorecards printed in newspapers to Indian slum dwellers; Ferraz and Finan study a naturally-occurring experiment in which random audits of Brazilian municipalities were publicized by local media; Humphreys and Weinstein provide Ugandans with evidence about their legislators' behavior.[12] In each study, subjects receive different treatments depending on how their representatives have performed. These studies are leading examples of a broader class of experiments that in some way tailor treatments to the subjects who receive them.

11 This method also offers a more precise assessment of the interaction between resume quality and race. Among black applicants, high-quality resumes are favored 52 times, and low-quality applicants are favored 45 times. Among white applicants, high-quality resumes are favored 78 times, and low-quality resumes are favored 49 times. This interaction remains statistically insignificant (two-tailed $p = 0.13$), but the confidence interval surrounding the interaction effect is now narrower.

12 Banerjee, Kumar, Pande and Su 2010; Ferraz and Finan 2008; Humphreys and Weinstein 2010.

To appreciate the special issues of interpretation that arise when treatments vary systematically across subjects, consider an experiment by Chong, De La O, Karlan, and Wantchekon that assesses the effect of providing voters with information about the level of corruption in local government.[13] In Mexico, the federal government conducts periodic audits of municipal finances. Some municipalities receive high marks (they spend authorized funds appropriately and without accounting irregularities symptomatic of corruption), and others do not. Simplifying the design of this study for ease of exposition, suppose that in each municipality precincts are randomly assigned to one of two experimental groups. Control precincts receive a flyer shortly before the election with information about when and where to vote; treatment precincts receive a flyer with this information plus a brief summary of the federal audit's findings concerning the municipality within which the precinct is located. The outcome is the share of the vote that each precinct casts for the incumbent's party.

In order to highlight the subtle identification problem presented by this design, let's take a closer look at the potential outcomes. Suppose that there were just two audit results: honest (no discoveries of serious improprieties) and corrupt (many examples of official misdeeds). Table 12.7 presents a hypothetical schedule of potential outcomes (incumbent vote shares) for four precincts. Two precincts are located in a municipality with an audit judgment of corrupt, and two are located in a municipality deemed honest. The audit result is a pre-treatment covariate that is denoted X_i in the table. The potential outcome $Y_i(0)$ indicates the incumbent's vote share if the precinct receives the control flyer. Modeling treated potential outcomes is a bit more complicated. Let $Y_i(honest)$ denote the outcome when the information in the flyer says the municipal government is honest, and let $Y_i(corrupt)$ denote the outcome when the flyer declares the municipal government to be corrupt. Suppose the design is blocked on municipality such that one precinct from each municipality is

TABLE 12.7

Potential outcomes from hypothetical leafleting experiment in which precincts receive information about the corruption of their municipal government

	$Y_i(honest)$	$Y_i(corrupt)$	$Y_i(X_i)$	$Y_i(0)$	X_i
1	60	45	60	60	Honest
2	50	35	50	50	Honest
3	60	30	30	40	Corrupt
4	70	50	50	50	Corrupt
Average	60	40	47.5	50	

13 Chong, De La O, Karlan, and Wantchekon 2011.

randomly assigned to treatment. When assigned to the treatment group, subjects 1 and 2 will reveal to the researcher the value of the potential outcome $Y_i(honest)$, while subjects 3 and 4 will reveal the potential outcome $Y_i(corrupt)$. Because only truthful flyers are distributed, the researcher can never observe $Y_i(corrupt)$ for precincts in the honest municipality or $Y_i(honest)$ for precincts in the corrupt municipality. What the researcher does observe can be characterized as $Y_i(X_i)$, the potential outcome that occurs when the flyer displays the audit result indicated by municipality type X_i.[14]

The complete schedule of potential outcomes displayed in Table 12.7 can be used to calculate the true average treatment effects. First, define the treatment as a flyer that conveys an auditor's findings, which, depending on the municipality, will report honesty or corruption. The potential outcomes that are observed when a precinct is treated are reported in the column titled $Y_i(X_i)$. The average treated potential outcome is 47.5, and the control average is 50. Therefore, the ATE of receiving a flyer with some type of auditor's report is -2.5. Next, what is the ATE of telling voters that their municipal government is honest? The average potential outcome when voters are informed that their government is honest is 60, which is 10 points higher than the average value of $Y_i(0)$. Finally, the average of $Y_i(corrupt)$ is 40, which is 10 points lower than the average value of $Y_i(0)$. In sum, flyers reporting honesty have an ATE of 10, and flyers reporting corruption have an ATE of -10.

Next, consider which potential outcomes the experiment reveals and therefore which quantities the experiment can be used to estimate. Each of the precincts has the same chance of being randomized into the control group; therefore, the expected value of $Y_i(0)$ is 50. Similarly, all four precincts have an equal chance of being randomly assigned to receive some type of auditor's report, so the expected value of $Y_i(X_i)$ is 47.5. The difference between the treatment and control groups is therefore an unbiased estimator of the average treatment effect of receiving some type of auditor's report.

Estimation becomes trickier when we try to learn about the average effect of each version of the treatment. Because only precincts 1 and 2 are ever shown the honest version of the treatment, and each of these subjects has a 50% chance of being in the treatment group, the expected value of the potential outcomes among those shown the honest treatment is $0.5 \cdot 60 + 0.5 \cdot 50 = 55$. By similar reasoning, the expected value of incumbent vote share for those subjects randomized to see the corrupt version of the treatment is $0.5 \cdot 30 + 0.5 \cdot 50 = 40$.

When analyzing data from this experiment, one temptation would be to evaluate the effect of telling people the government is honest versus corrupt by comparing the average vote share of those precincts shown the honest version to the average

14 We use the notation $Y_i(X_i)$ for simplicity even though it violates our usual notation conventions. A more exacting notation might be $Y_i(1, X_i = x)$, where the first argument refers to treatment status, and the second argument describes the background attribute that determines which potential outcome is triggered by the distribution of truthful leaflets.

vote share of precincts shown the corrupt version. In expectation, this method would tell you that flyers declaring the government to be honest rather than corrupt boost incumbent vote share by $55 - 40 = 15$ percentage points.[15] This comparison, however, is prone to bias and in this example does not recover the true ATE of $60 - 40 = 20$. This flawed estimation approach conflates the effect of the treatment and preexisting differences in the types of precincts that get each of the treatment variations. Although assignment to treatment versus control is random, which *type* of treatment a precinct receives is not randomly assigned. Instead, the treatment subjects receive depends on their attributes, and these attributes may be correlated with their potential outcomes. Looking back at the potential outcomes in Table 12.7, the precincts in the honest municipality tend to be more supportive of the incumbent.

The guiding principle when using experiments to estimate average treatment effects is to compare only groups formed by random assignment. The average treatment effect of each type of leaflet among *all* precincts cannot be estimated here because precincts were ineligible to receive certain treatments. However, because random assignment does occur within municipality type, it is possible to obtain unbiased estimates of the effect of the honest version of the treatment for precincts in the honest municipality. We can also assess the effect of the corrupt treatment for precincts in the corrupt municipality. According to Table 12.7, the ATE of the honest treatment in the honest municipality is 0, and the effect of the corrupt treatment in the corrupt municipality is -5.

It is important to avoid one final pitfall. The expected result from this simulated experiment seems to suggest that "telling people that an administration is honest has no effect (ATE $= 0$), but telling them it is corrupt causes them to vote against the incumbent's party (ATE $= -5$)." Taken at face value, this erroneous interpretation might prompt us to speculate about the value or credibility of positive news versus negative news. Before we go too far down this path, remember that this experiment cannot tell us about the effects of informing voters in the honest municipality that their government is corrupt, nor does it tell us how voters in the corrupt municipality would react if the auditor declared their government to be honest. What we have before us is a pair of distinct experiments, each involving different treatments and different subjects.

Let's take a closer look at the assumptions that would allow us to use this design to identify the difference in average treatment effects, defined as the average effect of the honest flyer minus the average effect of the corrupt flyer:

$$E[Y_i(honest) - Y_i(0)] - E[Y_i(corrupt) - Y_i(0)]. \tag{12.3}$$

15 Similarly, you could compare the honest treatment vote shares to the control average outcomes and conclude that declaring the government to be honest has an effect of $55 - 50 = 5$, while declaring the government to be corrupt has an effect of $40 - 50 = -10$. This is also a bad idea.

Unfortunately, the experiment we have described does not allow us to estimate either quantity. Instead, we observe the difference between the effect of the honest flyer in honest municipalities and the effect of the corrupt flyer in corrupt municipalities. In expectation, we observe:

$$E[(Y_i(honest) - Y_i(0)) \mid X_i = honest] - E[(Y_i(corrupt) - Y_i(0)) \mid X_i = corrupt]. \tag{12.4}$$

Under what conditions does the observed difference in treatment effects provide an unbiased estimate of equation (12.3), the average difference in treatment effects? In order to see the connection between equations (12.3) and (12.4), let $E[honest] = 1 - E[corrupt]$ be the expected proportion of subjects in honest municipalities, and rewrite the ATE of the honest flyer across all subjects as a weighted average of the ATEs for subjects in honest and corrupt municipalities:

$$\begin{aligned} E[Y_i(honest) - Y_i(0)] &= E[honest] \cdot E[(Y_i(honest) - Y_i(0)) \mid X_i = honest] \\ &\quad + (1 - E[honest]) \cdot E[(Y_i(honest) - Y_i(0)) \mid X_i = corrupt]. \end{aligned} \tag{12.5}$$

Because $E[corrupt] = 1 - E[honest]$, we can rearrange equation (12.5) so that it may be more easily compared to equation (12.4):

$$\begin{aligned} E[Y_i(honest) - Y_i(0)] &= E[(Y_i(honest) - Y_i(0)) \mid X_i = honest] + (E[corrupt]) \cdot \\ &\quad \{E[(Y_i(honest) - Y_i(0)) \mid X_i = corrupt] - E[(Y_i(honest) - Y_i(0)) \mid X_i = honest]\}, \end{aligned}$$

or

$$\begin{aligned} E[Y_i(honest) - Y_i(0)) \mid X_i = honest] &= E[Y_i(honest) - Y_i(0)] - (E[corrupt]) \cdot \\ &\quad \{E[(Y_i(honest) - Y_i(0)) \mid X_i = corrupt] - E[(Y_i(honest) - Y_i(0)) \mid X_i = honest]\} \end{aligned} \tag{12.6}$$

Equation (12.6) tells us that the ATE for precincts within honest municipalities equals the ATE among all municipalities when the last term in the equation is zero, a situation that arises when treatment effects are the same in corrupt and honest municipalities.

Similarly, the ATE of the corrupt flyer may be written as a weighted average of the ATEs for subjects in honest and corrupt municipalities.

$$\begin{aligned} E[Y_i(corrupt) - Y_i(0)] &= E[corrupt] \cdot E[(Y_i(corrupt) - Y_i(0)) \mid X_i = corrupt] + \\ &\quad (1 - E[corrupt]) \cdot E[(Y_i(corrupt) - Y_i(0)) \mid X_i = honest], \end{aligned} \tag{12.7}$$

which implies that:

$$\begin{aligned} E[(Y_i(corrupt) - Y_i(0)) \mid X_i = corrupt] &= E[Y_i(corrupt) - Y_i(0)] - (E[honest]) \cdot \\ &\quad \{E[(Y_i(corrupt) - Y_i(0)) \mid X_i = honest] - E[(Y_i(corrupt) - Y_i(0)) \mid X_i = corrupt]\}. \end{aligned} \tag{12.8}$$

When does the estimation approach in equation (12.4) yield unbiased estimates? Substituting (12.6) and (12.7) into (12.4), we see that the bias disappears when

$$E[corrupt] \cdot \{E[(Y_i(honest) - Y_i(0)) \mid X_i = corrupt] - E[(Y_i(honest) - Y_i(0)) \mid X_i = honest]\} = E[honest] \cdot \{E[(Y_i(corrupt) - Y_i(0)) \mid X_i = honest] - E[(Y_i(corrupt) - Y_i(0)) \mid X_i = corrupt]\}. \quad (12.9)$$

One situation in which equation (12.9) holds occurs when the average treatment effects of both the honest and corrupt treatments are the same for subjects in honest and corrupt municipalities; under that scenario, both sides of the equation equal zero. The assumption of equal average effects runs counter to a number of seemingly plausible hypotheses (e.g., the treatment reduced support for the incumbent party in corrupt municipalities but had no effect on voting in honest municipalities, where the information on the flyer merely restated what citizens already thought about the integrity of their municipal government. The general point is that unless one is prepared to defend the assumption implied by equation (12.9), the interpretation of any observed contrast in treatment effects across municipalities is ambiguous: the two leaflets' messages may have different effects, or the apparent interaction may reflect the differences between the subjects in the two distinct experiments.

Equation (12.9) amounts to a stronger assumption than researchers are ordinarily willing to make, but there is no need for this substantive assumption if one employs an alternative experimental design: randomly assign all subjects to all treatments rather than only the treatments that are tailored to their circumstances. In other words, flyers would inform voters in each precinct that their municipal government was found to be corrupt or honest, without regard to what federal auditors in fact reported. This design is impractical in this instance because authors sought to distribute only truthful information.

A large and important class of experiments tailors treatments to subjects. In experimental studies of political representation, sometimes voters are told about the behavior of their representatives; sometimes representatives are told about the opinions of their constituents.[16] There are sound reasons to tailor treatments to the subjects who receive them—researchers may seek to avoid disseminating false information or information that recipients may dismiss as implausible. In addition, reasonable arguments can be made about why the conditions of equation (12.9) hold in a given application. Researchers might point out that untreated potential outcomes appear to be similar across municipalities, that subjects in different municipalities exhibited similar patterns of behavior in the past, or that auditors' evaluations are prone to error such that similar municipalities are often given different ratings. Our point is not to challenge

16 Butler and Nickerson 2011.

the design or inferences drawn from any particular leafleting experiment but rather to underscore the fact that design decisions that reflect practical constraints may have important implications for analysis and interpretation.

12.6 Design and Analysis of Experiments In Which Failure to Receive Treatment Has a Causal Effect

Criminal justice systems in democracies tend to place enormous emphasis on the rights of the accused. Failure to respect these rights can in some cases lead to the dismissal of charges despite overwhelming evidence of guilt. In contrast, victims of crime receive relatively little attention, although effects of the crime on the victim can be severe and long-lasting. Reformers have proposed an alternative policy model known as restorative justice, which attempts to repair some of the harm done to the victim while at the same time assisting the offender toward rehabilitation. One program inspired by the restorative justice movement involves face-to-face meetings between the offenders and their victims. During these 90-minute meetings, which are arranged and moderated by police officers, the offender discusses the crime, offers victims and their family an apology, and signs an "outcome agreement" that summarizes "promises by the offender to undertake either reparations to the victim or community, or rehabilitation of themselves, or both."[17]

Several field experiments have investigated whether these offender-victim conferences benefit crime victims, where benefits are defined to include reduced symptoms of post-traumatic stress and a lessened desire for revenge.[18] These experiments vary in terms of who the offenders are and how they are encouraged to participate, but the basic design involves randomly assigning an offender-victim pair to either the regular criminal justice track (control) or the victim conference track (treatment). To measure the effect of the restorative justice intervention, researchers survey victims in both the treatment and control groups.

Translating this intervention into potential outcomes is complicated by the fact that the treatment consists of several ingredients: a meeting, an apology, a signed agreement, and subsequent adherence to this agreement. In any given attempt to administer restorative justice, one or more of these elements may be lacking. For example, Sherman and his colleagues report that offenders or victims sometimes failed to show up for the scheduled meetings, so that in the end 16% of victims in the treatment group did not receive an apology; even when conferences came off as

17 Sherman et al. 2005, p.379.
18 Ibid.

planned, roughly 25% of the offenders failed to abide by their agreements.[19] The intricacy of the treatment makes it an instructive example. We have one random assignment but multiple sources of noncompliance.

In order to see the implications of noncompliance in this context, let's start with a stripped-down model of potential outcomes that focuses solely on the causal effect of receiving an apology from the offender. Drawing on the framework presented in Chapters 5 and 6, let the potential outcome that results from a given treatment and experimental assignment be $Y_i(z, d)$, where $z = 1$ when an apology is assigned ($z = 0$ otherwise), and $d = 1$ when an apology is received ($d = 0$ otherwise).

Suppose that an experiment on restorative justice confronts one-sided noncompliance because some of the offenders who are assigned to apologize ($z = 1$) fail to show up for their conference with victims. As a result, offenders fail to offer an apology to victims in the treatment group ($d = 0$). Noncompliance is one-sided in the sense that no offenders in the control group offer a spontaneous apology.[20] When confronted by noncompliance, an experimental researcher may estimate two kinds of causal effects. The intent-to-treat effect is the average effect of assignment to the conference: $E[Y_i(z = 1, d(z = 1))] - E[Y_i(z = 0, d(z = 0))]$. The ITT is important for program evaluation because it represents the expected benefit of the program for a randomly selected victim. An ITT may be low because the assigned treatment is ineffectual or because few subjects received the treatment. Because the authors seek to evaluate the effectiveness of the restorative justice program considered as a whole, they focus on the ITT. Assuming random assignment, non-interference, and symetrical measurement procedures, unbiased estimates of the ITT are obtained by comparing average outcomes in the assigned treatment and control groups.

Another estimand is the average treatment effect for Compliers, the victims who receive an apology if and only if they are assigned to the treatment group. Identifying this causal effect requires additional assumptions. Recall from our discussion of one-sided noncompliance in Chapter 5 that a key assumption is the exclusion restriction, or the assumption that random assignment (z) has no effect on potential outcomes except insofar as it affects whether one receives treatment. This assumption implies that $Y_i(z = 1, d = 0) = Y_i(z = 0, d = 0)$. In words, the potential outcome if an offender is assigned to apologize but fails to do so is the same as the potential outcome when no apology is assigned or received. This assumption is likely to be violated in this context because victims become outraged when an offender fails to show up to a session and deliver an apology.[21] In order to see how identification of the ATE

19 These figures are based on results reported on p.379 of Sherman et al. 2005 and in Figure 1.

20 In fact, the authors' interviews with victims did reveal that some who were assigned to the control group had received apologies. We nevertheless assume one-sided noncompliance in order to keep the example simple.

21 Sherman et al. 2005, p.386.

among Compliers hinges on the exclusion restriction, write out the expanded set of potential outcomes for each subject i:

> $Y_i(0, 0) = Y_i(z = 0, d = 0) =$ outcome when the subject is assigned to the control group and receives no apology,
>
> $Y_i(1, 0) = Y_i(z = 1, d = 0) =$ outcome when the subject is assigned to the restorative justice group but receives no apology, and
>
> $Y_i(1, 1) = Y_i(z = 1, d = 1) =$ outcome when the subject is assigned to the restorative justice group and receives an apology.

Because we are assuming one-sided noncompliance, we can divide the subjects into two groups, Compliers (for whom $d_i(1) = 1$) and Never-Takers (for whom $d_i(1) = 0$).

The average outcome among subjects assigned to the treatment group, in expectation, equals:

$$E[(Y_i(1,1))\,|\,D_i(1) = 1] \cdot E[D_i(1)] + E[(Y_i(1,0))\,|\,D_i(1) = 0] \cdot (1 - E[D_i(1)]). \tag{12.10}$$

Similarly, the expected outcome in the control group equals:

$$E[(Y_i(0,0))\,|\,D_i(1) = 1] \cdot E[D_i(1)] + E[(Y_i(0,0))\,|\,D_i(1) = 0] \cdot (1 - E[D_i(1)]. \tag{12.11}$$

The difference between the treatment and control group averages is an unbiased estimate of the ITT, which is a weighted average of the effect of successfully delivered apologies and undelivered apologies:

$$\overbrace{\{E[(Y_i(1,1))\,|\,D_i(1) = 1] - E[(Y_i(0,0))\,|\,D_i(1) = 1]\}}^{\text{Average effect of an apology among Compliers}} E[D_i(1)] + \\ \overbrace{\{E[(Y_i(1,0))\,|\,D_i(1) = 0] - E[(Y_i(0,0))\,|\,D_i(1) = 0]\}}^{\text{Average effect of a no-show among Never-Takers}}(1 - E[D_i(1)]). \tag{12.12}$$

In order to estimate the Complier average causal effect using our experimental data, we would ordinarily divide our estimate of the ITT in equation (12.12) by our estimate of $E[D_i(1)]$. Suppose that our sample were large enough so that we knew the exact value of $E[D_i(1)]$ and the ITT in equation (12.12). The estimated CACE would then be:

$$\frac{\text{ITT}}{E[D_i(1)]} = \overbrace{\{E[(Y_i(1,1))\,|\,D_i(1) = 1] - E[(Y_i(0,0))\,|\,D_i(1) = 1]\}}^{\text{Complier average causal effect}} + \\ \frac{\overbrace{\{E[(Y_i(1,0))\,|\,D_i(1) = 0] - E[(Y_i(0,0))\,|\,D_i(1) = 0]\}(1 - E[D_i(1)])}^{\text{Bias}}}{E[D_i(1)]}. \tag{12.13}$$

In other words, the instrumental variables estimator introduced in Chapter 5 will not recover the CACE unless certain assumptions are imposed. The most important of these assumptions is excludability. Under excludability, potential outcomes in the wake of a no-show are the same as the potential outcomes when no apology is scheduled, and the second term equals zero since $Y_i(1,0) = Y_i(0,0)$. In that case, dividing the ITT by $E[D_i(1)]$ yields $E[(Y_i(1,1) - Y_i(0,0)) | D_i(1) = 1]$, the Complier average causal effect.[22]

When this exclusion restriction is rejected as implausible, the experimental results no longer lend themselves to clear interpretation. The core problem is that, if the exclusion restriction fails, the treatment in effect comes in two varieties, an apology and a no-show. A single randomization leaves us short of the information needed to assess the causal effect of either apologies or no-shows. How might this identification problem be remedied?

Some design modifications might allow estimation of the average effect of receiving an apology. The most straightforward solution would be to tighten the administration of the program so that offenders no longer miss meetings or neglect to apologize when meetings take place. This change in procedures in fact occurred as this experimental literature developed; the authors report that noncompliance problems were eventually eliminated by working with offenders who were in police custody.[23] Another design solution is to expand the number of experimental conditions from two to three in order to randomly vary the extent to which offenders were encouraged to conform to the restorative justice protocol. This design strategy uses the average outcomes in the control group, the standard encouragement group, and the strong encouragement group to identify the average treatment effect of an apology or a no-show. The identification strategy underlying this design presupposes three different types of subjects: Compliers (those who show up when encouraged in any way), Reluctant-Compliers (those who show up only when strongly encouraged), and Never-Takers. The key identifying assumptions in this model are that the average treatment effect of a no-show is the same regardless of whether the offender is a Reluctant-Complier or Never-Taker and that the average treatment effect of an apology is the same regardless of whether it comes from a Complier or Reluctant-Complier. We leave the formal proof as an exercise. The bottom line is that an augmented experimental design coupled with supplementary modeling assumptions allows a researcher to overcome a violation of the exclusion restriction and identify the distinct effects of apologies and no-shows. An even more elaborate design (with its own accompanying modeling assumptions) would be needed if one sought to tease apart the distinct effects of having the offender abide by the outcome agreement.

---

22 The CACE may also be written $E[(Y_i(1) - Y_i(0)) | D_i(1) = 1]$, since assigned ($z$) and actual ($d$) treatments are the same for Compliers.

23 Sherman et al. 2005, p. 386.

When an experimental intervention involves several components (e.g., conferences, apologies, pledges), each involving noncompliance, care must be taken to define the causal estimands and assess whether they can be identified given an experimental design and a set of statistical assumptions. Before one attempts to estimate the causal influence of one or more ingredients, write out a model of potential outcomes, describe subgroups that may be defined in terms of their patterns of compliance with assigned treatments, define the estimand(s), and determine which causal parameters are identified. Quite often, this exercise yields helpful design insights. The exercise can also be valuable when analyzing naturally-occurring experiments, despite the fact that the design is a fait accompli. Random allocations, such as the Vietnam draft lottery, change the probability that subjects receive the treatment of interest, such as military service, which in turn may affect subsequent economic, health, or political outcomes. But lotteries may also change potential outcomes through other channels. Just as assignment to restorative justice might have an effect on victims' potential outcomes apart from whether they receive an apology, receiving a low draft lottery number might affect potential outcomes for reasons other than military service. The same point holds if the treatment in the draft lottery study were defined not as military service but rather as participation in military combat. Careful attention to modeling assumptions helps the researcher keep track of the distinction between statistical results and causal interpretations.

12.7 Addressing Complications Posed by Missing Data

Many important experiments encounter problems of attrition, or missing outcomes for some subjects. Data may also be missing for variables other than outcomes. For example, researchers may be unable to gather covariate data for some subjects. Experimenters who use encouragement designs and therefore experience two-sided noncompliance (see Chapter 6) may find it impossible to measure whether every subject has been exposed to the treatment. In this section, we discuss the implications of missing outcome, covariate, and treatment data.

An instructive example is the series of experiments conducted by Howell and Peterson on the effects of school vouchers on student achievement, the lessons from which have shaped the design and analysis of subsequent experimental evaluations.[24] In these experiments, private school vouchers worth approximately $2,000 per year (in 2010 dollars) were distributed at random to a subset of voucher applicants. In the New York City voucher experiment, approximately 1,000 vouchers were randomly

24 The Howell and Peterson studies are described in their 2002 book and discussed in a 2003 exchange with Krueger and Zhu. Other methodological commentary may be found in Barnard, Frangakis, Hill, and Rubin 2003. Wolf et al. (2010) build on some of the key methodological lessons of the Howell and Peterson experiment.

distributed among 5,000 applicants. The 1,000 winners comprise the treatment group, and 1,000 students randomly selected from the non-winners serve as controls. For both the treatment and control groups, educational achievement was measured by standardized tests administered annually one, two, and three years after the vouchers were awarded.

Attrition occurred because some students did not show up to take the tests, which were administered outside of school. Outcome measures were missing for about 25% of the sample following the first school year, and attrition increased further in years two and three. The precise causes of attrition are unknown. Some families may have failed to show up due to indifference or incidental scheduling conflicts. Others may have failed to show up for reasons related to their treatment status—families that lost the lottery may have become disgruntled and uncooperative, and families that won the lottery may have failed to show up because of dissatisfaction with their new schools. Whatever the reasons, Table 12.8 indicates that attrition was consistently higher among subjects in the control group.

As explained in Chapter 7, when missingness is related to potential outcomes, attrition can lead to biased estimates of the ATE. Although a number of strategies have been proposed for analyzing experiments that encounter attrition, none of them is especially attractive, as they each impose strong assumptions. In this application, the strong assumptions have to do with the potential outcomes of students who are missing from the treatment and control groups. If the missing subjects were effectively a random sample of all subjects, attrition would be innocuous (except for the precision lost due to reduced sample size). Unfortunately, that does not seem to be the case here, as the missing in both treatment and control groups are, on average, poorer and have lower baseline test scores.[25] A fallback assumption is that missingness is

TABLE 12.8

Attrition rates in the school voucher experiment, by race

	All students		African American students	
	Control	Treatment	Control	Treatment
Baseline test	12.1	10.8	12.3	10.7
One-year follow-up	24.6	19.4	26.1	21.3
Two-year follow-up	39.8	31.1	44.3	33.0
Three-year follow-up	35.5	31.1	38.9	34.6
N	1,010	1,080	422	488

Source: Howell and Peterson 2002. Figures exclude students who entered kindergarten in the first year of the study. Attrition is defined as not taking the math test.

25 Howell and Peterson 2002, p. 645.

independent of potential outcomes conditional on background attributes, such as baseline scores. This assumption cannot be tested directly but is suggested by the fact that in the Howell and Peterson study background attributes such as baseline test scores bear approximately the same relationship to missingness in both treatment and control groups. This pattern suggests that missingness could be innocuous, but one never knows because similarity on observed attributes does not guarantee similarity in terms of potential outcomes.

What can be learned if we make no assumptions about the relationship between missingness and potential outcomes? Let's examine the extreme value bounds for the treatment effect estimates. To simplify the example, we focus on one reported treatment effect, the effect of one year of private school for African American students. The authors report that private school is associated with a 3.35 increase in national percentile ranks on the standardized test (the standard error of the estimate is 1.48).[26] They also encounter substantial rates of attrition; Table 12.8 reports that 78.7% of the treatment group and 73.9% of the control group took a one-year follow-up test. In order to keep the calculations simple, suppose that attrition rates are the same (25%) in both treatment and control groups. The maximum test score is 100, and the minimum score is 0, so the maximum treatment effect in the missing group is 100 and the minimum effect is -100. Therefore the upper bound is $(3.35 \cdot 0.75) + (100 \cdot 0.25) = 27.5$, and the lower bound is $(3.35 \cdot 0.75) + (-100 \cdot 0.25) = -22.5$. The width of these bounds is 50, which dwarfs the estimated ATE of 3.35. Without imposing assumptions, the bounds span a huge range, extending from a large positive effect to a large negative effect.

Can we narrow these bounds without making strong substantive assumptions? One possibility is to reduce the number of families in the treatment group and use the savings to pay for more intensive follow-up efforts among those who do not show up for the standardized test. The details of optimal design of experiments with attrition and enhanced follow-up are found in Chapter 7; here we provide some rudimentary calculations to sketch out the benefits of allocating resources to a follow-up round of outcomes assessment.

Let's compare alternative experimental designs that may be implemented for the same budget. In keeping with the design of the Howell and Peterson study, suppose we start with 2,000 subjects and allocate 1,000 subjects to the treatment group and 1,000 to the control group. Each subject assigned to the treatment group is offered a voucher worth $2,000, and all those offered the voucher use it. Control group observations are costless. Consider the following possible modification of the original experimental design: reduce the number of subjects by 20% to 1,600 by reducing the treatment and control groups by 200 each, and redirect the money to obtaining

26 To account for noncompliance, Howell and Peterson estimate a two-stage least squares regression using actual attendance in a private versus public school in the main regression and using the treatment group assignment as an instrumental variable. To focus on the key issue at hand we ignore noncompliance in our discussion.

a higher rate of reported outcome measures. To evaluate the consequences of this change in design, we must posit a relationship between spending and improvements in outcome recovery rates. When using standard outcome measurement (which produced a 25% attrition rate), we expect 1,200 of the 1,600 subjects to have measured outcomes. The reduced size of the treatment group frees up $200 \cdot \$2{,}000 = \$400{,}000$ in voucher funds, which can now be redirected to efforts to obtain test scores from the approximately 400 subjects who did not attend the testing session. Imagine that spending $2,000 per family on invitations to testing sessions, scheduling accommodations, and financial incentives will succeed in obtaining test scores from 90% of those who fail to respond to the standard efforts. The budget allows you to randomly select 200 of the 400 missing students for the intensive follow-up.

Compared to the baseline design, the modified design has fewer subjects but much less of an attrition problem. We evaluate the baseline and alternative design using two criteria. First, what is the standard error of the estimated ATE under the original and modified design? The original design includes $2{,}000 \cdot 0.75 = 1{,}500$ tested subjects, while the modified design includes $1{,}600 \cdot 0.75 + 200 \cdot 0.90 = 1{,}380$ tested subjects, which means that if attrition really were unrelated to potential outcomes, the modified design would result in standard errors that were approximately $\sqrt{1500/1380} = 1.08$ times the standard errors from the original design, or 8% larger. Second, since we do not trust the assumption that missingness is effectively random, we compare extreme value bounds. The extreme value bounds for the modified design turn out to be dramatically smaller. Assume that half of the follow-up respondents come from the treatment group and half from the control group, and that the estimated ATE among them is $\hat{F}$. The new extreme value bounds are:

$$\textit{Upper bound} = 3.35 \cdot 0.75 + 0.25 \cdot (0.90\hat{F} + 0.10 \cdot 100) = 5.0 + 0.225\hat{F}, \quad (12.14)$$

$$\textit{Lower bound} = 3.35 \cdot 0.75 + 0.25 \cdot (0.90\hat{F} - 0.10 \cdot 100) = 0.225\hat{F}. \quad (12.15)$$

This small change in design reduces the width of the extreme value bounds from 50 to 5, illustrating how a follow-up sampling design may dramatically reduce the statistical uncertainty associated with attrition. In this example, we have assumed that the intensive follow-up does not affect students' math ability or the way in which ability is translated into math scores. When implementing a follow-up data collection effort, researchers must take care not to violate this key assumption.

The voucher experiment presents another complication: some subjects are missing data on pre-treatment covariates. Students entering kindergarten, unlike subjects in higher grades, did not take a baseline academic proficiency test prior to the start of the experiment. As a result, the kindergarten cohort has missing baseline test scores. How should the analyst address the issue of missing covariate data? In order to simplify the discussion, let's assume that the only missing data problem is one of missing covariates, not missing outcomes. Under this scenario, a researcher can still obtain unbiased estimates of the ATE without covariate data. The estimates may be less precise than they would be if covariate data were available, but they remain valuable nonetheless.

One way to approach the data is to divide subjects into two groups, those with missing baseline scores and those without missing scores. An unbiased estimate of the treatment effect among subjects with baseline test scores is the change in average test scores for the treatment group minus the change in average test scores for the control group. For the group with missing baseline data, an unbiased treatment effect estimate is the difference in the average test scores for the treatment group minus the average control group test scores. In order to obtain an estimate of the sample average treatment effect, weight the change-score estimate by the proportion of subjects who have baseline data, and the missing-data estimate by the proportion of subjects who are missing baseline data. If one is prepared to assume that treatment effects are the same in the two subsamples because treatment effects are constant in the population, the best way to obtain precise estimates of the treatment effect is to weight each estimate by its precision, or the inverse of its squared standard error (see Chapter 11). The use of baseline information reduces the standard error and therefore increases the weight given to students for which baseline test scores are available.

Finally, let's briefly consider the problem of missing data as it relates to noncompliance. Suppose that researchers have difficulty measuring who attends private school. Imagine that researchers know that 80% of the treatment subjects attend private schools (because the researchers write the voucher checks to the schools), but school attendance in the control group is less clear. Researchers suspect that 5% of the control group attends private school, but the true figure could be higher. How does mismeasurement affect the estimate of the ATE among Compliers, those who attend private school if and only if offered a voucher? Recall from Chapter 6 that the estimator of the CACE is the estimated ITT divided by the observed difference in treatment rates between treatment and control groups. If the researchers' hunch is right and 5% of the control group attends private school, the denominator is $0.80 - 0.05 = 0.75$. If private school attendance in the control group were actually 15%, the denominator would shrink to $0.80 - 0.15 = 0.65$, which means that the estimated CACE will rise. The larger point is that measurement error in assessments of compliance works in a somewhat counterintuitive way. When researchers paint a rosy picture of compliance with assigned treatments, their estimates may understate the true effect of the treatment on Compliers.

SUMMARY

This chapter has used seven thought-provoking field experiments to show how abstract design principles play out in real-world applications. Much more could be said about these exemplary studies, and readers are encouraged to delve into these and other suggested readings. Your own repertoire of design ideas will expand as you become familiar with the strategies that other scholars have used.

In addition to the specific lessons conveyed by particular experiments, this chapter has stressed the importance of reading and conducting research with an eye to underlying assumptions. Writing out a model of potential outcomes helps to clarify the estimand. Showing the conditions under which an experiment may help estimate a parameter of interest highlights crucial substantive assumptions and may suggest ways to improve the experimental design. When examining the details of the randomization, the experimental conditions, and the outcome measures, consider ways that core assumptions (random assignment, excludability, and non-interference) may be jeopardized. When analyzing data and interpreting results, think about whether the procedures used to handle such complications as noncompliance, attrition, or heterogeneous treatment effects are susceptible to bias. If you suspect that these procedures are deficient, reflect on whether their limitations can be evaluated or overcome by an alternative research design.

Experiments are sometimes heralded as the gold standard of causal inference because random assignment facilitates unbiased inference. Random assignment is without question an enormous asset, but experiments are properly judged along many other dimensions as well. One can appreciate outstanding experimental research while at the same time reflecting on debatable assumptions, unanswered questions, and possible design improvements. Experiments are not so much exemplars against which all other methods should be judged as a cumulative enterprise in which new designs are continually proposed to address potential flaws and objections. Rather than characterize experiments as a gold standard, a more apt metaphor would be gold prospecting, which is slow and laborious but when conditions are right, gradually extracts flecks of gold from tons of sediment.

SUGGESTED READINGS

The number of social science articles and research reports that deploy experimental designs in field settings is large and growing rapidly. In order to locate experimental research in your field of interest, you might consult one of the following sources. *The Digest of Social Experiments*, 3rd ed. (Greenberg and Shroder 2004), which contains brief summaries of 240 social experiments, has been succeeded by *Randomized Social Experiments eJournal*, also edited by Greenberg and Shroder, which links to hundreds of on-line manuscripts. See http://www.ssrn.com/update/ern/ern_random-social-experiments.html. Another useful source is the *Annual Reviews* series. For example, in the *Annual Review of Economics*, Kremer and Holla (2009) review experimental interventions designed to improve education in the developing world; in the *Annual Review of Psychology*, Paluck and Green (2009) review experimental interventions designed to reduce prejudice. See http://www.annualreviews.org/search/advanced for a searchable database of articles. Google Scholar (www.scholar.google.com) is a useful free resource, and Web of Knowledge (www.webofknowledge.com) is a subscription database. Both allow users to locate academic journal articles and the articles that they reference.

EXERCISES: CHAPTER 12

1. Stewart Page performed an audit study to measure the extent to which gay people encounter discrimination in the rental housing market.[27] Answer the following questions, which direct your attention to specific page numbers in the original article.
   (a) Who are the subjects in this experiment (p. 33)?
   (b) What is the treatment (pp. 33–34)?
   (c) One criticism of audit studies is that in addition to differing with respect to the intended treatment (in this case, sexual orientation of the renter), the treatment and control group also differ in other ways that might be related to the outcome variable. What is the technical name for the assumption that audit studies may violate?
   (d) Suppose that the experiment used one male caller to make calls that mentioned sexual orientation and another male caller to make calls that did not. How would this procedure affect your interpretation of the apparent degree of discrimination against gay men?
   (e) Take a careful look at the treatment and control scripts, and consider some ways that the treatment and control conditions might differ in addition to transmitting information about the potential renter's sexual orientation. Are the scripts the same length? Do both scripts seem similar in terms of tone and style? How might the incidental differences between scripts affect the generalizations that can be drawn from this study?
   (f) How might you design an experiment to eliminate some or all of these incidental differences between scripts?
   (g) Based on the description on pages 33–34, how are subjects assigned to the treatment groups? What is the implication if random assignment was not used?
2. Over the past several decades, trust in government has declined. Among the possible culprits is the rise of confrontational TV news shows, which are thought by some to produce citizen disgust and disengagement. An influential study by Mutz and Reeves investigated the effects of uncivil political discourse by scripting and producing two versions of a candidate debate.[28] Subjects were randomly assigned to be shown either the uncivil (treatment) or civil (control) debate. After viewing the treatment or control video, subjects were asked about their level of trust in government.
   (a) Who are the experimental subjects in the first Mutz and Reeves experiment (p.4)?
   (b) Let the variable X_i categorize subjects according to whether they regularly watch political television shows ($X_i = 1$) or not ($X_i = 0$). Let the conditional average treatment effects be denoted $E[(Y_i(1) - Y_i(0)) | X_i = 1])$ and $E[(Y_i(1) - Y_i(0)) | X_i = 0])$. Does your intuition suggest that these CATEs will be similar or different? Why?
   (c) Write the expression for the average treatment effect as a weighted average of the CATEs of those who do and do not watch political TV shows.
   (d) The researchers estimated the average treatment effect and found the uncivil video reduced trust in government. Suppose that only 5% of the general public watches

27 Page 1998.
28 Mutz and Reeves 2005.

shows that convey this treatment. To what extent does the experiment support a claim that exposure to uncivil political programs caused a decline in trust in government among the general public?

(e) Critics of cable TV shows argue that the programs should be encouraged to be more civil. Can the estimated ATE be used to predict the effect of increasing the civility of cable shows on the overall public level of trust in government?

(f) Suppose that a company which tracks television viewers provides you a list of three million potential subjects, along with data on their TV viewing habits. How would you select the subjects for a follow-up experiment if you were interested in estimating how trust in government would change if political TV programs were to become more civil?

(g) The researchers also measure whether aggressive shows are more engaging to audiences. They use multiple outcome measures: a survey item response and a physiological measure, galvanic skin response (see pp. 10–11 for a discussion). What is the rationale for using the physiological measure? What potential problem with survey response is it designed to address?

3. In an experiment designed to evaluate the effects of political institutions, Olken randomly assigned 49 villages in Indonesia to alternative political processes for selecting development projects.[29] Some villages were assigned to the status quo selection procedure (village meetings with low attendance), while others were assigned to use an innovative method of direct elections (a village-wide plebiscite). Consistent with expectations, participation in the plebiscite was 20 times greater than attendance at the village meetings. Olken examines the new procedure's effect on which projects are selected and how the villagers feel about the selection process. He finds that there are minimal changes in which projects are selected. However, a survey after the project selection found that the villagers who were assigned to the plebiscite reported much greater satisfaction with the project selection process, and were significantly more likely to view the selection as fair, and the project as useful and in accordance with their own and the people's wishes.

(a) One part of this experiment focuses on whether the treatment influences which projects villages select. These results are reported in Figure 1, and the study is described on pp. 244–247. Describe the experimental subjects. What units are assigned to treatment versus control? What is the treatment?

(b) Suppose that in Indonesia, the plebiscite method is rare, but the village meeting is very common. How would this affect your interpretation of the findings?

(c) The level of satisfaction is measured by survey responses. From the description on p. 250, can you tell who conducted the surveys and whether the interviewers were blinded as to the respondents' assignment to treatment or control? Why might survey measures of satisfaction be susceptible to bias?

(d) There is no indication that the treatment and control villages had contact with each other. Imagine, however, that people regularly communicated across village lines. What assumption might be violated by this interaction? Discuss how cross-village

29 Olken 2010.

communication might affect treatment effect estimates. What design or measurement strategy might address possible concerns?

(e) Olken concludes that, consistent with the views of many democratic theorists, participation in political decision making can substantially increase satisfaction with the political process and political legitimacy. Does the experiment provide convincing evidence for this general proposition? What are some of the limitations noted by Olken (see pp. 265–266)? What additional limitations does the experiment have? How might you address these concerns in a future experiment?

(f) It is often claimed that short-term effects may diminish over time, but the short-run outcome measurements nevertheless reliably indicate the direction, if not the magnitude, of the long-term effects. However, if an institutional change is thought to be a durable feature of the political world, leaders and voters may change their behavior and the way they compete for power. Speculate on why the long-term effects of the plebiscite on satisfaction with the decision process might be negative despite the initial positive response.

4. In section 12.5, we considered a hypothetical experiment in which leaflets were distributed to publicize an audit that declared local government to be honest or corrupt. Suppose another experiment of this kind were conducted in 40 municipalities, half of which are honest and half corrupt. Half of the honest municipalities are randomly assigned to receive leaflets publicizing the auditor's finding of honesty, and half of the corrupt municipalities are randomly assigned to receive leaflets publicizing the auditor's report of corruption. Outcomes are the incumbent mayor's vote share in an upcoming election. The data from the experiment are used to estimate the following regression:

$$Voteshare_i = \beta_0 + \beta_1 Leaflet_i + \beta_2 Honest_i + \beta_3 Leaflet_i \times Honest_i + u_i,$$

where $Voteshare_i$ is the incumbent's vote share (from 0 to 100 percent), $Leaflet_i$ is scored 1 if the municipality is randomly treated with a leaflet (0 otherwise), $Honest_i$ is scored 1 if the municipality receives an audit rating declaring it to be honest (0 if it was declared corrupt), and u_i is the disturbance term. Suppose the regression estimates (and estimated standard errors in parentheses) are as follows: $\hat{\beta}_0 = 30\ (4)$, $\hat{\beta}_0 = -15\ (5)$, $\hat{\beta}_2 = 25\ (5)$, $\hat{\beta}_3 = 35\ (7)$. Interpret the results, taking care not to assume that the average treatment effect of leaflets announcing the honest rating in honest municipalities is the same as the average treatment effect of leaflets announcing the honest rating in corrupt municipalities. (Hint: Use the regression coefficients to figure out what the regression results would be if honest and corrupt municipalities were analyzed separately. See section 9.4).

5. The Simester et al. study showed how incomplete outcome measurement can lead to erroneous conclusions. On that note, suppose researchers are concerned with the health consequences of what people eat and how much they weigh. Consider an experiment designed to measure the effect of a proposal to help people diet. Subjects are invited to a dinner and are randomly given regular-sized or slightly larger than regular-sized plates. Hidden cameras record how much people eat, and the researchers find that those given larger plates eat substantially more food than those assigned small plates. A statistical test shows that

the apparent treatment effect is far greater than one would expect by chance. The authors conclude that a minor adjustment, reducing plate size, will help people lose weight.

(a) How convincing is the evidence regarding the effect of plate size on what people eat and how much they weigh?

(b) What design and measurement improvements do you suggest?

6. As noted in section 12.1, experiments are sometimes motivated by a desire to test two rival explanations for an empirical regularity. Each of the three examples below features a clash between competing explanations. For each topic, propose an experiment that would, in principle, shed light on the causal influence of each explanation. Assume that you have a very large budget and a good working relationship with governments and other organizations that might implement your experiments.

(a) Does imprisonment reduce crime because convicts have fewer opportunities to break the law, or does imprisonment deter crime by teaching prisoners about the penalties they face if they re-offend?

(b) Do employers in the United States discriminate against black job applicants because they believe them to be less economically productive than whites, or do employers discriminate against black job applicants because they harbor negative attitudes toward black people in general?

(c) Does face-to-face communication with voters before Election Day raise voter turnout because it reminds people about an upcoming election that they might otherwise forget, or because it conveys the importance of the choices that will be presented to voters?

7. In the Slemrod et al. experiment, measuring the outcome variables involved some effort and cost to match names and state tax return records. Outcome measurements were obtained for only a randomly selected portion of the households available to serve as control group observations.

(a) Suppose that additional resources were made available to the researchers, and they gathered outcomes for randomly selected taxpayers who were not selected for treatment. (Assume that this was the only thing they could spend the money on.) How would including these additional observations in the control group affect the properties of the weighted difference-in-means estimator? Is it still unbiased? How does its standard error change?

(b) Records are sometimes lost over time. Suppose that before the second round of outcome measurement were launched, some taxpayer records went missing. What additional assumption is necessary for the combined old and new control group outcome measurements to be an unbiased estimate of the same estimand as the old outcome measurements?

8. According to social psychologists, performance on standardized tests may be affected by seemingly minor contextual features, such as the instructions read to those about to take a test and the similarity between the test-taker and other students taking the test at the same time. This literature implies that subtle asymmetries across treatment and control in how outcomes are measured may have a material effect on test scores. Suppose you were designing an experiment similar to the voucher experiment described in section 12.6.

Instead of bringing students to a common testing center for testing, you have decided to use the standardized tests that students ordinarily take in their own schools.

(a) What are some important potential sources of asymmetry in outcome measurement? Consider among other things how the test is administered, who proctors the test, who grades the test, the mixture of students in the room for a testing session, and whether the administration and grading is blinded to the subject's group status.

(b) How would you design your study to reduce bias due to asymmetric outcome measurement of the treatment and control subjects?

(c) Suppose you want to investigate the impact of the measurement asymmetries you discuss in part (a). Describe an experimental design to estimate the effect of the measurement asymmetries.

9. As pointed out in section 12.4, sending resumes via email seems to have several advantages over typical face-to-face audit studies of racial discrimination. However, an email treatment is a more subtle method of communicating race than a face-to-face meeting. What if some employers do not notice the name on the job application or incorrectly guess the race of the applicant? For simplicity, assume that each human resource officer either concludes that the applicant is black or white. Suppose that when sent any white resume, a human resources officer has an 80% chance of surmising that it is from a white applicant. When sent any black resume, a human resources officer has a 90% chance of surmising that it is from a black applicant. Suppose that making a mistaken classification of a white resume is independent of making a misclassification of a black resume. Recall from Table 12.6 that 9.65% of the white resumes received callbacks, as opposed to 6.45% of the black resumes.

   (a) For definitional purposes, consider assignment to the white resume to be assignment to treatment, and consider assignment to the black resume to be assignment to control. To show how misclassification is analogous to noncompliance, use the classification system in Chapter 6 to describe the four types of subjects: what proportion of subjects are Compliers, Never-Takers, Always-Takers, and Defiers?

   (b) What is the $\widehat{ITT_D}$ in this case?

   (c) What assumption(s) are needed to interpret the ratio of ITT/ITT_D as the Complier average causal effect? Suppose that when analyzing the data in Table 12.6, you assumed that these assumptions were satisfied; what would be your estimate of the CACE?

   (d) Does the rate of noncompliance have any bearing on the statistical significance of the relationship between race and interviews that the authors report in Table 12.6?

   (e) What steps do Bertrand and Mullainathan take to reduce the rate of misclassification? Do they measure the rate of misclassification? What methods might you use to measure misclassification rates? What are some strengths and weaknesses of your proposal?

10. One limitation of the restorative justice experiment described in section 12.6 is that one cannot identify the distinct effects of an apology or a no-show; instead, one can only estimate the effects of a treatment that is a combination of the two. Suppose that in

order to identify the ATE of an apology as well as the ATE of a no-show, you assigned subjects randomly to one of three experimental groups: a control group, a standard encouragement group, and a strong encouragement group. The identification proof posits three different types of subjects: Compliers (those who show up when encouraged in any way), Reluctant-Compliers (those who show up only when strongly encouraged), and Never-Takers.

(a) Write the expected outcome in the control group as a weighted average of the expected outcomes among Compliers, Reluctant-Compliers, and Never-Takers.

(b) Write the expected outcome in the standard encouragement group as a weighted average of the expected outcomes among Compliers, Reluctant-Compliers, and Never-Takers. Your model should acknowledge that Compliers will offer an apology, but Reluctant-Compliers and Never-Takers will be no-shows.

(c) Write the expected outcome in the strong encouragement group as a weighted average of the expected outcomes among Compliers, Reluctant-Compliers, and Never-Takers. Your model should acknowledge that Compliers and Reluctant-Compliers will offer an apology, but Never-Takers will be no-shows.

(d) Explain why the experimental design allows us to estimate the shares of the three types of subjects.

(e) Notice that in the three equations (a), (b), and (c) there are four parameters: the ATE of a no-show among Never-Takers, the ATE of a no-show among Reluctant-Compliers, the ATE of an apology among Compliers, and the ATE of an apology among Reluctant-Compliers. No matter how you manipulate the three equations, you cannot solve for each of the four parameters. In other words, with more unknown parameters than equations, you cannot identify either of the apology effects or either of the no-show effects. Suppose you assume instead that the ATE of a no-show is the same regardless of whether a Reluctant-Complier or Never-Taker is at fault and that the ATE of an apology is the same regardless of whether it comes from a Complier or Reluctant-Complier. Now you have reduced the number of unknowns to just two parameters. Revise your equations (a), (b), and (c) to reflect this assumption, and show that it allows you to identify the apology effect and the no-show effect.

11. One reason for concern about attrition in the school voucher experiment described in section 12.7 was that, after the first year, the attrition rate was greater in the control group than the treatment group. Intuitively, the problem with comparing the treatment and control group outcomes is that the post-attrition control group is no longer the counterfactual for the post-attrition treatment group in its untreated state. The trimming bounds described in Chapter 7 attempt to extract from the post-attrition treatment group (which has a larger percentage of the randomly assigned group reporting) a subset of subjects who can be compared to the control group and used to bound the treatment effect. The dataset for this exercise at http://isps.research.yale.edu/FEDAI contains subjects of any race in the Howell and Peterson study who took a baseline math test. The outcome measure (Y_i) is the change in math scores that occurred between the baseline test and the test that was taken after the first year of the study.

(a) What percentage of the control group is missing outcome data? What percentage of the treatment group is missing outcome data?

(b) Among students with non-missing outcome data, what are the average outcomes for the control group and treatment group?
(c) What is the distribution of outcomes for the treatment group? What is the range of outcomes? What outcomes correspond to the 5%, 10%, 15%, 25%, 50%, 75%, 85%, 90%, and 95% percentiles?
(d) To trim the top portion of the treatment group distribution, what value of Y_i is the 93.6 percentile of the treatment group? (The value 93.6 is the control group reporting rate divided by the treatment group reporting rate.)
(e) What is the average value of the treatment group observations that are less than the 93.6 percentile value? Call this average treatment effect L_B. Confirm that the percentage of the original treatment group that remains is equal to the percentage of the control group with outcome data.
(f) Subtract the control group average from L_B.
(g) To trim the bottom portion of the treatment group distribution, what treatment group outcome corresponds to the 6.4 percentile? (The value 6.4 is calculated by subtracting 93.6 from 100.)
(h) What is the average value of the treatment group observations that are greater than the 6.4 percentile? Call this average treatment U_B. Confirm that the percentage of the original treatment group that remains after trimming is equal to the percentage of the control group with outcome data.
(i) Subtract the control group average from U_B.
(j) The lower and upper bounds that you calculated in parts (f) and (i) are designed to bound an ATE for a particular subgroup. Describe this subgroup.

12. In private school voucher studies, treatment group observations are much more expensive than control observations. Assume the experiment is free from attrition and noncompliance. Suppose that the researchers have a fixed budget of $2M, each treatment group observation costs $2,000, and each control observation costs $200. The table below shows four possible ways to use the budget to form treatment and control groups.

	Option 1	Option 2	Option 3	Option 4
Treatment	950	750	600	900
Control	500	2500	4000	1000

Let the standard deviation of outcomes in the treatment group and the control group be the same, and equal to s.

(a) Estimate the standard error for the difference-in-means estimator using the formula in equation (3.6), letting the number of observations assigned to treatment be n_t and the number of observations assigned to control be n_c. The standard error may be written:

$$s\sqrt{\frac{1}{n_c} + \frac{1}{n_t}}.$$

In the table, which allocation of subjects to treatment and control produces the most precise estimate?

There is a general method for minimizing the standard error subject to a budget constraint. Suppose the cost per observation in the control and treatment groups are p_c and p_t, respectively, and both groups have the same standard deviation. To minimize the standard error of the difference-in-means, assign subjects to groups in proportion to the square root of the cost ratio. The following questions illustrate the derivation behind this idea.

(b) If n_t is the number of subjects you assign to the treatment group, how much money is spent on the treatment group?

(c) If n_c is the number of subjects you assign to the control group, how much money is spent on the control group?

(d) Express the budget B as the total spent on the treatment group and control group.

Set up a constrained maximation problem by defining the Lagrangian equation (Dixit 1990):

$$L(q, n_c, n_t) = s\sqrt{\frac{1}{n_c} + \frac{1}{n_t}} - q(B - n_t p_t - n_c p_c).$$

Take the partial derivative of L with respect to n_c, n_t, and q, and set each of the partial derivatives equal to zero. (If your calculus is rusty, use an online calculator to take derivatives.) The values of n_c and n_t that satisfy these conditions minimize the standard error subject to the budget constraint.

(e) Set the partial derivative with respect to n_t equal to the partial derivative with respect to n_c. Manipulate the resulting equation to show that

$$\frac{p_c}{p_t} = \left(\frac{n_t}{n_c}\right)^2.$$

From this result it follows that the ratio of the size of the treatment group to the size of the control group is equal to the inverse of the square root of the ratio of the costs of each type of observation. Thus, if a treatment group observation costs 10 times as much as a control group observation, the standard error minimizing division of resources places $\sqrt{10} \approx 3.2$ times as many observations in the control group.

(f) When the cost of treatment and control group observations is the same, what is the appropriate way to allocate the budget to n_t and n_c?

CHAPTER 13

Writing a Proposal, Research Report, and Journal Article

In order to make an enduring contribution to knowledge, you must do more than conduct research; you must also record and communicate pertinent information so that others can understand and appreciate your contribution. Pertinent information erodes quickly. A few months after an experiment concludes, the researchers who conducted it will have difficulty remembering key details about the recruitment of subjects, the treatments, the random assignment, and the outcome measures. A few years later, no one will be quite sure who has the data or in what format.[1] The time, effort, and resources that went into the research will be squandered unless you take pains to carefully document your experiment's design, implementation, and results.

Writing not only preserves research findings, it enables and inspires others to build on these findings in a methodical way. Clear documentation of experimental findings and procedures facilitates replication. Writing that calls attention to unresolved puzzles and research opportunities attracts scholarly interest and guides research down an efficient path.

This chapter provides a checklist for authors as they wend their way through three stages of experimental research. The first stage involves a research proposal, sometimes occasioned by the need to secure funding for an experimental project. By describing the aims of the study and how it will be conducted, the proposal helps sharpen your experimental design and provides an opportunity to explain how you will analyze the data once they are gathered. As noted in Chapters 4 and 9, ex ante plans help bolster the credibility of the eventual data analysis by limiting the role of discretion. The next phase is the preliminary research report, an extended set of lab notes and statistical findings that contains the raw materials from which the research article will be crafted. Much of the research report can and should be written prior to

1 When assembling the material for this book, we contacted dozens of authors in the hopes of using their data for examples and exercises. Scholars were cooperative, but rarely were they able to retrieve data from studies that were more than five years old, except in instances where the data had previously been archived in a public repository.

What you will learn from this chapter:

1. How to write a research proposal describing an experiment's aims and procedures.
2. How to assemble a research report summarizing the details of how the experiment was executed and providing an inventory of statistical results.
3. How to compose a research article that situates the experiment within a broader literature and methodically describes the method and results.
4. How to preserve the results in a manner that allows you and other researchers to reconstruct the statistical findings.

completing the experiment. The basic principle is to "write as you go," so that important details do not fade from memory. Building on the reporting standards that currently guide medical research, we recommend a list of items to include in the final product, a social science research article. These recommendations are more stringent and burdensome than the current reporting requirements of social science journals, but they represent a set of standards that we ourselves aspire to meet when conducting research. Many of the reporting practices we recommend are inspired by deficiencies we see in our own work.

We have written this chapter with the writer in mind, but our checklists are meant to guide the reader as well. One of the main aims of an experimental write-up is to allay readers' concerns about nonstatistical sources of error, such as flawed procedures or hidden biases. Readers need to see certain crucial pieces of information in order to make sense of the results and the uncertainty surrounding them. Authors who neglect to provide key details create needless uncertainty. Although unadorned and methodical presentation of pertinent information robs experimental research articles of some artistic merit, the scientific advantages are clear.

13.1 Writing the Proposal

We should say at the outset that we debated whether to write this section in the future tense or the past tense. In a perfect world, everything would be laid out neatly before the experiment gets underway. In practice, last-minute adjustments occur. This section is written in the past tense as though the experiment had been launched a few minutes ago, and the research team has run back to their computers to record the final details of the design.

1. Spell Out the Research Hypothesis. A useful first step in any experimental project is to describe, in a single sentence, the causal parameter that you intend to estimate. This exercise will force you to specify the treatment, subjects, and context. For example, "This experiment gauges the extent to which state legislators are less likely to respond to requests from constituents with Hispanic surnames than to requests from constituents with Anglo surnames." A clear description of the estimand will help you and your readers assess whether the excludability assumption is satisfied by your experimental design. Are there reasons to suspect that the assigned treatment (z_i) influences outcomes for reasons other than the treatment that is actually delivered (d_i)? In this example, is there anything about the intervention or the assessment of outcomes that might lead you to suspect that assignment influences outcomes for reasons other than constituents' surnames? Are the coders of legislators' responses to constituent mail blind to the subjects' assigned experimental condition?

Next, explain whether (and why) you expect the causal parameter to have a particular sign or magnitude. You might, for example, expect a negative effect on the grounds that legislators and their staffs tend to regard Hispanics less favorably than Anglos or that they place less electoral value on Hispanic constituents, who are perceived to be less likely to vote or donate money to political campaigns. Or you might remain agnostic, pointing out that while negative effects are possible, it may be that legislators are more responsive to Hispanics because sending a brief letter is an inexpensive way to burnish their standing in the Hispanic community. Laying out your prior beliefs about the parameter will help guide hypothesis testing later on. For example, if you expect that Hispanics will receive fewer responses than Anglos, the relevant hypothesis test will be one-sided. Observing a negative effect will lead you to reject the null hypothesis of no negative effect. On the other hand, if you expect ethnic surnames to have some sort of causal effect but you cannot say which way it will go, the hypothesis test you conduct later on will be two-sided. Either large positive or large negative effects will cause you to reject the null hypothesis of no effect.

If you anticipate heterogeneous treatment effects, indicate which subgroup(s) you expect to show particularly large or small effects. Explain the grounds for this hypothesis. For example, you might anticipate that low-achieving students will be most helped by an intervention that allocates more teaching aides to elementary school classrooms on the grounds that low-achieving children benefit most from adult attention and supervision. Specifying subgroup interactions in advance addresses statistical concerns that arise when researchers sift through the data ex post in search of interactions. As explained in Chapter 9, the number of planned comparisons determines the Bonferroni correction for multiple comparisons.

Specifying the research hypothesis serves a number of purposes. It forces researchers to be clear about what is being tested and what the outcomes are. Explaining the reasoning behind your hypotheses helps readers to understand what is at stake theoretically and allows them to assess the correspondence between the aims of the experiment and the actual implementation. Laying out your priors about the direction of the ATE in advance structures the ways in which hypothesis tests will be conducted. It also allows you, after the results come in, to make credible claims about how your subjective beliefs have been updated in light of the new evidence (see Chapter 11).

2. Describe the Treatment in Detail. Without a clear understanding of what the treatment is, a reader has no way of interpreting the causal effect that your experiment purports to estimate. Take, for example, door-to-door canvassing prior to an election. For starters, you might say, "Canvassers knocked on doors and encouraged people to vote." We need more detail. Were canvassers instructed to follow a script? If so, provide it. Did you watch the canvassers deliver the script? If so, characterize the gist, tone, and length of the conversations that you observed. A clear description of the treatment helps the reader interpret the results and allows other researchers to replicate your experiment.

Describing the treatment also means describing the circumstances in which it was deployed. Who administered the treatment? Who were the canvassers—were they from the local area? Given the canvassers' age, education, gender, and ethnicity, how are they likely to be perceived by the people whom they canvass? When did the canvassing occur—a few days before the election or a few months? What kind of election is it? What issues or candidates were at stake? How much campaign communication were voters in the study likely to receive in addition to the experimental canvass? As explained in Chapter 11, your description of the context helps researchers to situate your study in the experimental literature.

It is also important to describe the "treatment" that the control group receives. Sometimes that answer is "nothing." Those who are not canvassed by a political campaign go about their daily lives with one less five-minute interruption. Sometimes the situation is more complicated. Suppose, in an education experiment, you were to assign students in the treatment group to receive an additional class in mathematics. In order to make room for this math class in their schedules, students will have to forgo one class in some other subject. In order to understand the net treatment effect, we need to know how the course schedule of the treatment group differs from the course schedule of the control group.

3. Describe the Criteria by Which Subjects Were Included in the Experiment. How did subjects come to be included in your experiment? What criteria determined which units were eligible for random assignment? For example, the experimental observations may consist of "television markets in Texas other than Dallas and Houston" or "undergraduates at an Ivy League college who, when pre-registering for courses, expressed a preference for one of eight freshman seminars." Earlier, you mentioned the subjects when stating your hypothesis in general terms. This is the place to spell out the details. Sample restrictions are an inevitable part of experimentation. State as clearly as possible what the restrictions were. If subjects were sampled in some systematic way from a larger population, describe the population and sampling method.

Sometimes there are certain areas or people that for practical reasons *must* be treated. For example, freshmen apply to seminars and are placed on waiting lists. Students are admitted to the classes on a random basis—unless the instructor and prospective student had some prior relationship, in which case the instructor may insist that the student be admitted off of the wait-list. In principle, you could ignore the instructor's wishes, with possibly unpredictable consequences for the experiment. (If the instructor denounces the random assignment in a way that makes students aware that a study is being conducted, the study may lose its capacity to generate results that generalize to settings in which assignment is unobtrusive.) Or you could allow some students to cross over from the assigned control group to the treatment group when the instructor allows them special admission to the course, but this approach diminishes the power of the study for reasons explained in Chapters 5 and 6. A better approach is to confer with instructors in advance of the randomization, giving them a chance to discreetly designate a must-admit list. These must-treat observations are not considered subjects, as they are not randomly allocated. Your description should underscore the fact that certain units were excluded from the randomization, at the very least to remind you to exclude these observations when assembling the experimental results.

Another reason to be explicit about sample restrictions is so that your readers can understand the context in which the experimental subjects received the treatment. The freshman seminars, for example, are populated not by randomly assigned students from the wait-list but by a combination of wait-listed students and students with a preexisting relationship with the instructor. The mixed composition of the classroom may subtly change the nature of the treatment. In effect, the freshman seminar as a treatment consists of the classroom dynamics generated by a mixture of the two groups of students.

4. Explain How Subjects Were Randomly Assigned to Experimental Groups. Describe the random assignment procedure in detail. For example, "In order to allocate subjects to treatment and control groups, the following procedure was used. Subjects were listed in a spreadsheet and assigned a random number. The 1,000 rows of this spreadsheet were then sorted by this random number in ascending order. The first 400 subjects in the spreadsheet were deemed the control group, and the remainder was deemed the treatment group." For purposes of later verification, it is always a good idea to leave the random numbers in the dataset. If you used statistical software to draw a random sample, cut-and-paste the lines of computer code into your research proposal. Where possible, include a random number seed in your program so that the pattern of random assignments is reproducible.

Because random assignments may be subverted by researchers who seek a particular experimental outcome or who prefer that certain subjects receive the treatment, the official CONSORT reporting standards for medical experiments request that the researchers indicate who conducted the randomization.[2] Indeed, best practices in medical research specify that random assignment be conducted by a

2 Schulz, Altman, and Moher 2010.

person not connected to the study. Given the time constraints that field experimenters work under, a reasonable best practice is to conduct the random assignment using a random number seed so that the procedure is automated and the results are reproducible.

A related concern has to do with throwing out "bad randomizations," allocations that produce treatment and control groups whose background attributes differ in some way. Discarding bad randomizations in order to improve precision is an acceptable practice (assuming that the experimenter is not trying to finagle the randomization so that specific subjects end up in the treatment group), but it does require the experimenter to take special precautions when estimating the ATE and conducting hypothesis tests. When estimating the ATE, a researcher may need to weight the data if screening caused subjects to have different probabilities of assignment to treatment (see Box 4.5). When applying randomization inference to experimental data that were allocated based on some screening criterion, a researcher should consider only the set of possible randomizations that would have been allowed by the screening procedure. Randomization inference therefore requires the researcher to specify the exact criteria used to reject certain random allocations and accept others.

Under simple random assignment or complete random assignment, each unit is assigned to treatment and control with equal probability (see Box 2.5). If you used simple or complete random assignment, the description of the randomization process should say so. The two most common departures from simple random assignment are blocked assignment and clustered assignment (see Chapters 3 and 4). Blocked assignment means that observations are first divided into distinct strata, and subjects within each stratum are allocated randomly to treatment and control. The probability of assignment to treatment is constant within each block but may vary across blocks. For example, one might conduct a blocked randomization in which men and women are randomly assigned in different pools: among men, two-thirds of subjects are randomly assigned to the treatment group; among women, half of subjects are randomly assigned to the treatment group. In effect, each block is its own experiment. Clustered assignment means that subjects are assigned not as individual units but rather as groups of units. For example, rather than assign individual students to treatment and control, an education researcher might assign classrooms or grades or entire schools. Clearly describing the clustered randomization procedure is extremely important, as subtle details may have important implications for the estimation of the ATE and confidence intervals. As explained in Chapters 3 and 4, failure to properly account for blocking or clustering can lead to biased estimates and standard errors.

5. Summarize the Experimental Design. Having discussed the treatments and process of allocating subjects to experimental conditions, provide a table or figure that indicates how many observations were allocated to each condition. In the case of block randomized designs, provide a table for each of the blocks; a scatterplot describing the relationship between block size (X-axis) and proportion of subjects assigned to the treatment condition (Y-axis) may be used if the number of blocks is very large. If the design involves cluster randomization, provide a table indicating the number of clusters in each experimental condition as well as the number of individual subjects in each condition. Describe the distribution of subjects per cluster: how many subjects does the average cluster comprise, and what is the standard deviation of this number? This information may also be displayed using an individual values plot: the X-axis indicates each cluster's experimental assignment, and the Y-axis displays the number of subjects per cluster. You may want to link this description to a discussion of your estimation approach. Recall from Chapters 3 and 4 that average cluster size may affect sampling variability, and variation in cluster size may undermine the unbiasedness of difference-in-means estimation.

6. Check the Soundness of the Randomization Procedure. Random assignment should, in expectation, create treatment and control groups that have similar background characteristics, or *covariate balance.* Whether you conducted your own randomization or relied on a government agency to conduct a lottery, it is useful to verify the integrity of the randomization procedure by assessing covariate balance. As explained in Chapter 4, a high degree of covariate balance does not prove that treatments were assigned randomly; conversely, poor covariate balance can occur even when treatments are in fact assigned randomly. The point of this exercise is to assess whether the degree of covariate balance is in line with what one would expect to see given the use of random assignment. Typically, this exercise reveals that covariates are balanced, a finding that helps assure readers that there is one less nonstatistical source of error to worry about. If the observed randomization turns out to be extremely unlikely given randomization, researchers should take a closer look at the randomization steps to see whether there is a flaw in the procedure or some misunderstanding about the way in which random assignment was carried out. If substantively large imbalances are discovered but the randomization procedure appears sound, one could discard this bad randomization and re-randomize. If the experiment is already underway, the results should later be presented both with and without controls for covariates so that readers will have a sense of whether the estimation approach has a material effect on the statistical results.

When assessing imbalance, try to gauge the magnitude of the imbalance and the probability of obtaining the observed level of imbalance by chance. The first question is addressed through a presentation of descriptive statistics. If you used simple random assignment or clustered random assignment, present a table of means that compares the treatment and control groups' background attributes. If the background variables are continuous, compare the standard deviations as well. If you used block random assignment, compare weighted means, where the weights are the inverse of the probability of selection into the observed treatment condition.

Randomization inference may be used to assess whether the degree of covariate balance is statistically unexpected. If you used simple or complete random assignment to form two experimental groups, use regression to predict treatment assignment based on the covariates at your disposal.[3] In order to calculate the p-value of the F-statistic that is reported by the regression routine, generate this statistic for a large number of simulated randomizations using the methods described in Chapter 4. Under the null hypothesis of random assignment, the p-value tells you the probability of obtaining the degree of covariate imbalance at least as great as what you in fact observed in your sample. If you used blocked random assignment and probabilities of treatment assignment vary by block, estimate a weighted regression of treatment assignment on covariates, weighting each observation by the inverse of its probability of being assigned to that condition, and record the F-statistic.[4] In order to find the p-value of this statistic, use your blocking procedure to simulate a large number of possible random assignments, and form the sampling distribution of the F-statistic. If your experimental groups were formed using clustered assignment, form the sampling distribution of the F-statistic by using your clustered assignment procedure to simulate the sampling distribution.

What should an experimenter do if imbalance is detected? Suppose the average value of a prognostic indicator were higher in the treatment than the control group and a test of the null hypothesis of covariate balance indicates that $p < 0.001$. One possibility is that the researcher obtained an unlucky draw, but another possibility is that there is some flaw in the assignment procedure, perhaps due to a programming error or confusion over how observations were randomly assigned. A common mistake is to conduct a test of balance that ignores the use of blocked or clustered assignment. Carefully review your random assignment procedure and the way in which you conducted the test of imbalance. If you find that random assignment did not in fact occur, you will have to redo the randomization and re-launch the experiment.

7. Describe the Outcome Measures. Describe in detail each of the outcomes and the manner in which each is measured. For example: "Voter turnout was measured using official records obtained from the Secretary of State. Those subjects appearing on the list of voters were scored a 1; all others were scored a 0." If you are measuring outcomes indirectly using a set of indicators, explain why it is reasonable to suppose that your measures tap into the underlying construct of interest. For example, if you are measuring "political knowledge" using a ten-question multiple-choice test, describe the items, their correlations with one another, and their correlations with background factors that are expected to predict political knowledge.

---

3 Use multinomial logistic regression if you have three or more treatment groups, and conduct randomization inference on the reported log-likelihood statistic.

4 Weighted regression should also be used when blocking is not used but subjects are assigned to treatment with varying probabilities.

Detail is particularly important when describing attrition, or missing outcome data. Describe how many subjects in each experimental condition have missing outcomes. (See the flow diagram example in Box 13.3.) Indicate whether the relationship between covariates and missingness is similar across experimental conditions. If follow-up efforts are made to gather outcome data for subjects whose outcomes were missing during the initial round of data collection (see Chapter 7), describe the sampling and measurement procedures.

When describing your outcome measures, indicate whether special procedures have been put into place in order to minimize measurement asymmetries. If your outcomes are measured via surveys, indicate whether the interviewers are blind to the respondents' experimental condition. Is there anything about the interviewers, the way questions are worded, or the context in which the survey is administered that might encourage respondents in the treatment group to give distinctive responses? For example, are the survey interviewers connected in any way to the administration of the treatment itself?

If your experiment involves noncompliance, define compliance with the assigned treatment. Describe in detail how compliance is measured. Receiving the treatment is an "outcome" and requires its own set of tables when the results of the experiment become available.

8. Describe How You Plan to Analyze the Data. In previous chapters, we discussed the advantage of spelling out a plan in advance. In a nutshell, analytic plans help limit the scope of discretion and are therefore especially helpful in situations where the experimental results turn out to be ambiguous. (Many researchers, ourselves included, have failed to specify plans in advance and later wished they had done so.) When following a plan, the analyst cannot pick and choose results depending on whether they "look good" or generate statistically significant estimates. Commitment to a plan therefore helps make results more credible.

Because the reasoning behind the statistical analysis is laid out in the proposal, analytic plans can be as austere as a few lines of computer code that will be used to generate statistical tables once the outcomes become available. A simple analytic plan would indicate how you will (1) display graphically the distribution of outcomes for each experimental group; (2) compute average outcomes and standard deviations for each experimental group; (3) compute the average treatment effect and its standard error; (4) use regression to estimate the ATE after controlling for a specific list of covariates; (5) use randomization inference to test the sharp null hypothesis of no treatment effect; (6) when working with continuous outcome measures, compare variances across experimental groups and/or use the nonparametric bounds described in Chapter 9 to get a sense of whether the data suggest heterogeneous effects; and (7) test possible interactions between the treatment and specified covariates. One nice benefit of having programmed your plans in advance is that you will be ready to go once the outcome data become available.

As the experiment unfolds, unforeseen implementation problems may undo or complicate your plan.[5] For example, failure to apply the treatment as planned to all members of the treatment group may require additional analysis that gauges the average treatment effect among Compliers (see Chapters 5 and 6). Or failure to observe outcomes in the treatment and control groups may necessitate additional analyses that assess whether the patterns of attrition are different for treatment and control groups (see Chapter 7).

Unforeseen statistical results are a different matter. After the experimental outcomes become available, researchers are encouraged to explore post hoc interactions under the general admonition to "know thy data," but any discoveries that emerge from this exploration should be interpreted with caution in light of the multiple-comparisons problem described in Chapter 9. Keep track of the number of comparisons you make and apply a Bonferroni correction when conducting hypothesis tests.

9. Archive Your Data and Experimental Materials. Finally, while you wait for your experimental results to come in, create a physical or electronic archive of your treatment materials—your scripts, recorded messages, mailings, and so forth. Be sure to collect whatever materials you need from the people who carried out the treatments. For example, if you are conducting a voter mobilization study using door-to-door canvassers, gather the lists showing who was contacted before these materials disappear. If you are studying the effects of classroom instruction, make a copy of the handouts, readings, lesson plans, and quizzes for both treatment and control classrooms.

10. Register Your Experiment. In the social sciences, few experiments are registered in any formal way. In the medical sciences, registration of randomized trials at sites such as ClinicalTrials.gov or ISRCTN.org has become mandatory. Registration is a simple process; it essentially amounts to submitting your research proposal so that it becomes part of a permanent public record.

If one has already crafted a research proposal, why is registration necessary? Think about it from the standpoint of the reader of a research literature. If experiments that produce splashy findings are more likely to find their way into print, a survey of the published literature may lead to biased inferences. In order to correct for publication bias, one must survey the unpublished literature as well. A registration system makes it possible to locate studies that for whatever reason never found their way into publication. Registration also helps keep track of all of the pieces of an

---

5 Sometimes plans contain built-in contingencies. For example, so-called *adaptive experimental designs* allow the researcher to analyze the data from an ongoing experiment to determine whether the results look sufficiently promising to warrant additional data collection. For a discussion of the statistical complications posed by this design and a field experimental application involving car inspections, see Schneider 2007.

experiment, some of which may be published and others not. Registration makes just as much sense in the social sciences as the medical sciences, and it is only a matter of time before scholarly norms about registration develop in disciplines such as psychology, economics, political science, sociology, education, and criminology.

13.2 Writing the Research Report

The waiting is over: the data have been checked for accuracy, and the experimental results are in. The research team is poring over the output from the planned statistical analysis. The function of the research report is to present a full array of results, not just those deemed to be sufficiently interesting for the research article. These results test the hypotheses laid out in the proposal, sometimes in multiple ways so that the researcher may assess the robustness of the results. Later, when the research article is written, it will present findings more economically, referencing the more extensive presentation of results contained in the research report. Sometimes the research report becomes an online appendix to the published article, available to those who visit the publisher's website.

The research report need not have a clear thesis. It may be a sprawling document with few stylistic virtues. Its main function is to provide reference material, largely in the form of narratives, tables, and figures. Due to the importance of the research report as an archive of experimental results, special care should be taken to ensure that tables and figures are clearly labeled and contain essential descriptive information. Here are some suggestions for presenting information in tables, whether in the research report or published article:

1. A heading that indicates what the outcome is and the groups within which the outcomes are calculated. For example, "Voter turnout rates, by experimental condition."
2. A clear indication of what the numbers in each row and column represent. For example, "Entries are means, with standard deviations in parentheses."
3. Indication of what the *N*s are within each experimental group.
4. If the table reports an estimate of a parameter such as the average treatment effect, it should also report the statistical uncertainty associated with that estimate in the form of a standard error or confidence interval.
5. Indication of how outcome variables, treatments, or covariates are defined or coded. For example, a description of the variables in a regression equation might read as follows: "The outcome, voter turnout, is scored 1 for voters and 0 for nonvoters. The treatment is scored 0, 1, 2, or 3, depending on the number of mailings that were sent to each voter. The covariate *Past Votes* is scored 0 to 8, reflecting the number of votes that each subject cast in eight prior elections."

Although the research report is a discursive compendium of statistical results, it is best to arrange the tables starting with the simplest and most transparent. Follow the dictum "crosstabs before regressions": present a comparison of means across experimental groups before proceeding to more opaque forms of analysis, such as regression.

Whether used in a preliminary report or published article, the main virtue of figures is their ability to illustrate the distribution of outcomes. Here are some suggested types of figures and accompanying annotations. An *individual values plot* (see Box 13.1) is a useful way to give readers a sense of the distribution of outcomes in each experimental group, which in turn has implications for sampling variability (Chapter 3) and treatment effect heterogeneity (Chapter 9). Annotate the graph so that readers can glance at it and infer the mean and standard deviation for each experimental group.

Another useful graph is the *added variable plot* (see Box 13.2), which shows the relationship between the treatment and the outcome after controlling for covariates. The added variable plot is generated by plotting two sets of residuals. The residuals of Y_i are obtained by regressing Y_i on the covariates. The residuals of the treatment are obtained by regressing d_i on the covariates. The resulting plot shows the relationship between the portions of d_i and Y_i that are orthogonal to the covariates. In essence, the added variable plot depicts the process by which a linear regression estimates the effect of the treatment after controlling for covariates.

One of the most important figures in any research report tracks the flow of observations from sample definition to experimental assignment to groups for which outcomes are available. If attrition occurs, it should be described in the figure and discussed in detail in the research report. See Box 13.3 for an example of this type of flow chart, which, under the CONSORT standards, is required when describing medical trials.

The research report is essentially an update of the proposal, with additional information about outcomes. When experiments involve noncompliance, tables should describe the relationship between the assigned treatments and the realized treatments. If there is ambiguity about how to measure compliance with the treatment, the research report should show an array of tables that invoke different measurement standards. The same goes for attrition. If there is ambiguity about how to measure the dependent variable or how to impute a missing outcome, present tables that invoke different measurement assumptions.

Another valuable function of the research report is to provide "placebo tests" of various sorts, showing that the treatment has no effect where no effect is expected. For example, treatments should not affect outcomes measured prior to the experiment. Treatments are also expected to influence outcomes in some domains and not others. Encouraging someone to recycle should have negligible effects on whether they turn out to vote. These kinds of null findings might be deemed too banal to

BOX 13.1

Example of an Individual Values Plot

The individual values plot is used to display the results from experiments with continuous outcome measures. The individual values plot is essentially a scatterplot, but the X-axis is "jittered" by a small amount so that points sharing identical X and Y values become noticeable.

For example, Titiunik (2010) studied the effects of randomly assigned term length among Texas state senators. Some senators were assigned two-year terms, others four-year terms. The outcome measure is the number of bills introduced during a legislative session. The plot shows the dispersion of the points in the two experimental groups. A small amount of jitter enables the reader to discern observations that would otherwise be on top of each other. The horizontal bars indicate the means of each group. The graph seems to suggest that senators with four-year terms have higher means and more widely varying outcomes.

Distributions under treatment and control

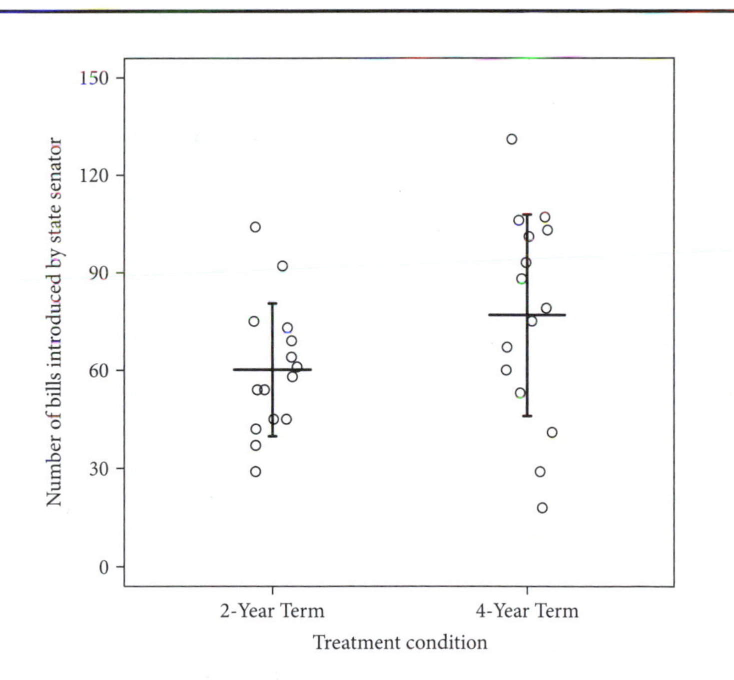

Horizontal bars represent the means of each experimental group. Vertical bars indicate one standard deviation above and below each mean.

BOX 13.2

Example of an Added Variable Plot

An added variable plot is an individual values plot that controls for covariates. Regression is used to purge Y_i and d_i of the covariance they share with the covariates. Y_i is regressed on the covariates, and residuals e_{Yi} are generated for each observation. d_i is regressed on the covariates, and residuals e_{di} are generated for each observation. A scatterplot is generated, with e_{di} on the X-axis and e_{Yi} on the Y-axis. For example, using the Texas state senate data described above (Titiunik 2010), we may create an added variable plot showing the relationship between term length and number of bills introduced, controlling for the senator's party, the Democratic vote share in the senator's district in the previous presidential election, and the senator's own vote share in the previous election.

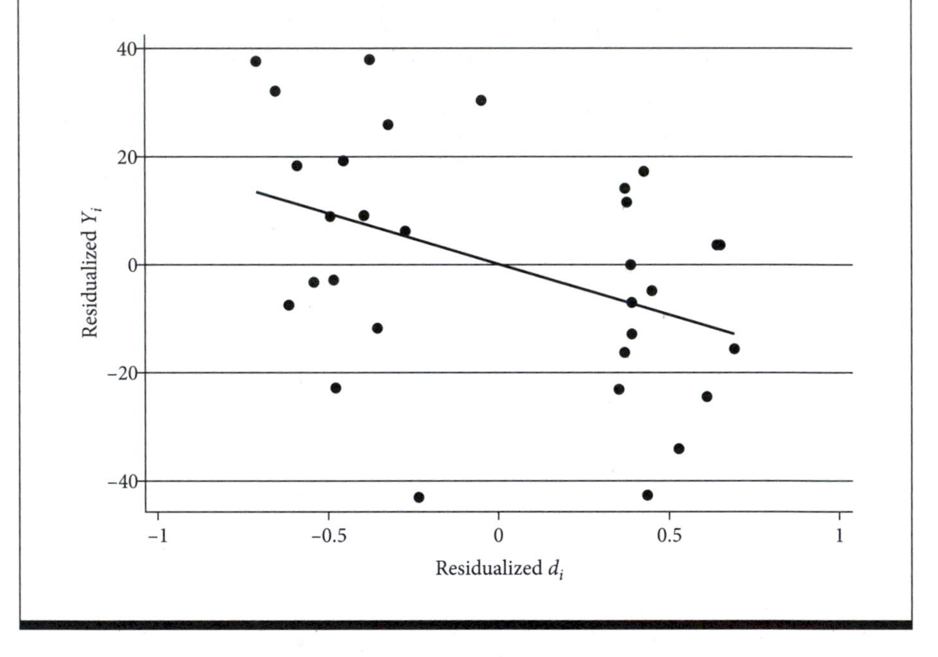

be presented in a research article, but by establishing that the experimental results follow the predicted pattern when various correlated outcomes are considered, this exercise bolsters the credibility of the results. If subjects encouraged to vote or recycle *both* had higher turnout than the control group, we might suspect that random assignment placed an unusual set of subjects into the control group or that something about the administration of the treatments went awry.

BOX 13.3

Illustration of a Flow Diagram Tracking the Observations over the Course of the Experiment

A flow diagram helps document sampling, allocation, delivery of treatment, and attrition. A helpful example may be found in Cotterill et al.'s (2009, p. 407) field experiment on the effectiveness of door-to-door canvassing on street-level recycling rates. Their flow diagram shows that they started with 209 streets, deleted some problematic streets, randomly allocated the remaining 194, and successfully canvassed 61% of the houses on streets in the treatment group. No attrition occurred in this cluster-randomized experiment.

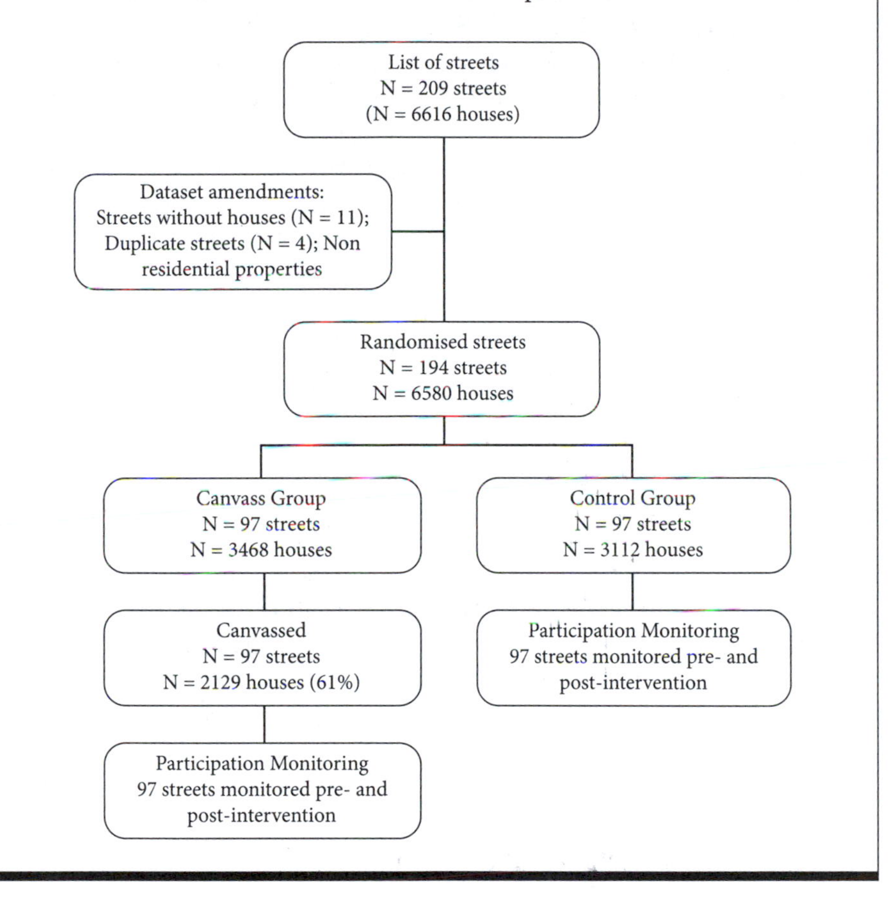

13.3 Writing the Journal Article

The format of social science research articles is constantly evolving, but the most common format remains the 7,500-word essay. The precise specifications for this essay vary, but the typical article has the following elements:

1. An introduction that poses the research question and places it in the context of prior studies and ongoing theoretical debates.
2. A signposting paragraph that indicates how the essay is organized and foreshadows the findings.
3. A more structured discussion of the literature that identifies a promising research opportunity and/or gap in existing knowledge.
4. Presentation of research hypotheses.
5. Explanation of the identification strategy, perhaps accompanied by a statistical model that indicates how observable quantities will be used to estimate parameters of interest.
6. A summary of the experimental design, describing the random allocation procedure, treatment, outcomes, and context.
7. A presentation of experimental results, perhaps accompanied by robustness checks and subgroup analysis.
8. A discussion of the substantive and methodological implications, with suggestions for future lines of investigation.
9. A list of references to related scholarly work and supporting documents.
10. Tables and figures.

The space allotted to each of these ingredients depends on the academic journal for which the essay is written. Some journals encourage authors to offload technical material to unpublished supporting documents, usually available online. Many general interest journals instruct authors to focus on the most theoretically engaging aspects of their research, which often means lengthy introductions designed to convey the importance of the topic and novelty of the research. More specialized journals also encourage authors to "think big" but tend to be more forgiving when it comes to focused research projects that make an incremental empirical, theoretical, or methodological contribution. Both general interest and special interest articles tend to be similar in length. Rarely in social science does one encounter the brief technical research articles that are common in the natural sciences. When laying out the minimum reporting standards below, we realize that journal editors may choose to divert some of this required information to supporting materials rather than information supplied in the published version of the article. Supplementary materials are fine so long as interested readers can reliably find this information years later.

Below is an inventory of what we suggest researchers should report when describing the experimental design, statistical procedures, and results in the body of a research article. The list summarizes the information required to understand the experimental results, to assess the statistical uncertainty surrounding the estimates, and to evaluate non-statistical sources of uncertainty. Elements of the research proposal and report are designed to answer the queries below, so it should not be surprising that many of the elements repeat the recommendations described earlier in this chapter.

1. Describe the estimand, and explain how the experimental design identifies the parameter of interest. When appropriate, address threats to the excludability and non-interference assumptions.
2. Describe the treatment, the manner in which it was administered, and the context in which it was administered.
3. Describe the experimental subjects. Explain the criteria used to determine which participants were eligible for inclusion in the study.
4. Describe the procedure by which units (or clusters of units) were randomly assigned to experimental groups.
5. Summarize the experimental design, indicating the number of subjects assigned to each experimental condition.
    a. If noncompliance occurred, provide a table showing the relationship between the assigned and realized treatments.
    b. If attrition occurred, provide a table indicating rates of attrition for each assigned treatment condition and, within each assigned condition, showing the relationship between attrition and covariates that are believed to predict outcomes.
6. Provide a table describing covariate balance across experimental groups. Conduct a hypothesis test to assess whether the observed degree of imbalance is within the expected range.
7. Describe the outcome measure or measures and when they were gathered. Indicate whether those gathering outcome measures were blind to the subjects' treatment status.
8. Report the average outcome in each experimental condition (and within each block if the number of blocks is small; otherwise, report weighted averages, using inverse probability weights). For non-binary outcomes, report the standard deviation in each experimental condition, and present individual values plots (using weighted data for block-randomized experiments and other randomized designs that assign subjects to treatment with varying probabilities.)
9. Estimate average treatment effects based on the comparisons specified in your research proposal. Report the estimates and accompanying standard errors (or confidence intervals) both with and without covariates.

10. Conduct the hypothesis tests laid out in your research proposal. Report the test statistics and accompanying p-values.
11. Assess whether the interaction hypotheses described in your proposal are borne out by the data. If you conduct several hypothesis tests when testing treatment-by-treatment interactions or treatment-by-covariate interactions, use a Bonferroni correction.
12. Assess any post hoc interaction hypotheses. (Hypotheses of this sort may be suggested by reviewers when they evaluate your manuscript for publication.)
13. Disclose the sources of funding for the study and any potential conflicts of interest that might possibly affect the judgment of the researchers involved.
14. If human subjects regulations apply to your study, indicate whether it was approved by an institutional review board. (See Appendix A.)

The body of the research article is summarized in the title, abstract, and keywords. Titles of social science articles play an important role in bibliographic searches, which in turn are important for comprehensive literature reviews. For that reason, try to include the terms "experiment" or "field experiment" in your title.

13.4 Archiving Data

When properly crafted, a research article provides an informative and permanent record of the experimental protocol and results. The final step in the research process is to preserve the data and research materials for future use. Archiving is required by an increasing number of funding agencies and academic journals; even when optional, archiving is a sound scientific practice that may work to your benefit in the long run. Regardless of whether you plan to share your data with others, you probably will have occasion to use them again yourself. You may write a follow-up paper that revisits the data with a new research question, or you may teach a class in which the data are used as an example.

Researchers seeking to archive their experimental data and materials confront four main preservation issues: format, media, documentation, and curatorship.

1. Format: Technological change is so rapid that anything stored in a proprietary digital format is likely to be unreadable in twenty years. The more generic the format, the better. In addition to saving your data in their current form, save them as a comma-delimited ASCII file. If your dataset is modest in size, print out a copy as a backup in case the digital version is lost.
2. Media: Making a stash of CD-ROM discs might suffice for five to ten years, but you'll probably misplace them or discover that they have become unreadable with the passage of time. Hard drives deteriorate. Cloud storage depends on the long-term survival of the service providers that sell storage space. A better solution is

to deposit your data and materials in a dedicated archive that is responsible for maintaining the integrity of the data files and materials (see "Curatorship" below).

3. Documentation: Datasets must be documented with so-called *metadata* that explain how to interpret each of the variables. Without metadata that label each variable and category, the data are unintelligible. Another important form of documentation is the series of commands that reproduce the results presented in your report or article (which should be part of the digital archive). This type of program or script enables researchers to reconstruct the details of your coding decisions and statistical procedures.
4. Curatorship: Personal archiving seems like a poor strategy over the long haul; even well-intentioned researchers let their experimental data and accompanying metadata molder away. A better idea is to put your research materials into an archive that is maintained by an academic institution, scholarly organization, or research journal as part of its core mission.

BOX 13.4

Example of a Planning Document and Research Report

An instructive example of how a planning document evolves into a research report may be found at http://isps.research.yale.edu/FEDAI. The experiment evaluated the effects of a conditional cash transfer program called Family Rewards, which sought to improve health, education, and employment outcomes among low-income families in New York City. Families were given cash rewards, for example, if they attended parent-teacher conferences, obtained a library card, received preventative health screening, or maintained full-time employment.

The planning document situates the proposed experiment in the existing experimental and theoretical literature; introduces the incentive scheme and other aspects of the intervention; describes the sampling, recruitment, and random assignment procedures; and lists the outcomes of interest and explains how they will be measured over a five-year period using a combination of administrative records, surveys, and field observation. Finally, the document explains in general terms how the data will be analyzed (but stops short of specifying the subgroups to be examined).

The interim report (Miller, Riccio, and Smith 2009) summarizes the results one year into the intervention, and provides an extensive list of tables comparing treatment and control outcomes. The interim report also functions as a planning document insofar as it describes the subgroups to be analyzed when more subjects are enrolled in the experiment and longer-term outcomes become available.

Preservation and documentation require time and effort, and scholars are often too busy with the next experiment to properly archive the one they just completed. One way to lower the costs of archiving at the end is to make documentation and preservation part of your research procedures at each phase of the project. For example, before sending your manuscript off to a journal for review, prepare a dataset, program, and accompanying materials so that anyone can reproduce and interpret your results.

SUMMARY

One of the challenges of conducting an experiment is maintaining careful records. Experiments, especially when conducted in field settings, can involve a fair amount of last-minute frenzy. It is easy to lose track of information or forget important design decisions. This chapter has stressed the importance of documenting key details about the experimental intervention, the assembly and assignment of subjects, the measurement of outcomes and covariates, and the context in which the experiment occurred.

Documentation not only helps researchers organize and retain key details, it also serves as a commitment device. A planning document automates much of the analysis and, in so doing, limits the role of discretion. The researcher still retains a fair amount of latitude when deciding which results to present in the final research article, but because adherence to the planning document bolsters the credibility of the results, the researcher has an incentive to present the planned analysis. The planning document potentially serves another valuable purpose. By encouraging researchers to think carefully about what they will do when the results come in, the process of composing a research plan may help to improve the experimental design.

This chapter has also emphasized the enduring value of providing readers with key details about the experiment's design, implementation, and analysis. In the social sciences, there seems to be little consensus about what information should be conveyed in an experimental write-up. In preparing this book, we sifted through hundreds of journal articles and frequently had difficulty understanding exactly how experiments were conducted or analyzed. (Even our own articles are sometimes lacking in this regard.) The primary aim of this chapter is to help improve and standardize the presentation of experimental designs and results so that scientific contributions are more easily understood and evaluated. Another important aim is to preserve scientific contributions by archiving data and the accompanying documentation that makes them intelligible. Again, in the course of assembling examples for this book, we were struck by how few field experiments are publicly accessible in well-documented form. Sadly, the data from many important field experiments seem to have disappeared because they were never archived. Through clever design and careful oversight, these field experiments overcame a series of nettlesome financial, ethical, and administrative challenges, only to vanish years later when researchers' aging computers became inoperative.

SUGGESTED READINGS

The suggested reporting standards described here are patterned after the Consolidated Standards of Reporting Trials (CONSORT) (Altman et al. 2001) that are currently enforced by leading medical journals. See Boutron, John, and Torgerson (2010) for a description of these standards and an attempt to apply them to social science research. See the websites of professional associations in the social sciences for new reporting and replication standards. The book's website links to recommendations from the Research Standards Committee from the APSA Experimental Research Section. For a description of how to document and archive data, see ICPSR's *Guide to Social Science Data Preparation and Archiving: Best Practice throughout the Data Life Cycle* (2009).[6] For some helpful tips regarding graphics and presentation of causal relationships, see Gelman and Hill (2007), pp. 167–181, 551–563.

EXERCISES: CHAPTER 13

1. Middleton and Rogers report the results of an experiment in which ballot guides were mailed to randomly assigned precincts in Oregon prior to the 2008 November election.[7] The guides were designed to encourage voters to support certain ballot measures and oppose others. Load the example dataset from http://isps.research.yale.edu/FEDAI. The dataset contains election results for 65 precincts, each of which contains approximately 550 voters. The outcome measure is the number of net votes won by the sponsors of the guide across the four ballot measures that they endorsed or opposed. The treatment is scored 0 or 1, depending on whether the precinct was assigned to receive ballot guides. A prognostic covariate is the average share of the vote cast for Democratic candidates in 2006.
   (a) Estimate the average treatment effect, and illustrate the relationship between treatment and outcomes graphically using an individual values plot.
   (b) Interpret the graph in part (a).
   (c) Use randomization inference to test whether the apparent difference-in-means could have occurred by chance under the sharp null hypothesis of no treatment effect for any precinct. Interpret the results.
   (d) Suppose it were the case that when randomly assigning precincts, the authors used the following screening procedure: no random allocation was acceptable unless the average 2006 Democratic support score in the treatment group was within 0.5 percentage points of the average 2006 Democratic support score in the control group. Do all subjects have the same probability of being assigned to the treatment group? If not, re-estimate the ATE, weighting the data as described in Box 4.5. Redo your

6 Available at http://www.icpsr.umich.edu/files/ICPSR/access/dataprep.pdf.
7 Middleton and Rogers 2010.

hypothesis test in part (c) subject to this restriction on the randomization. Interpret the results.

2. Select a published article that presents the design and analysis of a field experiment. Based on the publication and any supplementary materials provided by the authors, try to fill in as much of the reporting checklist for research articles as you can. What pieces of information, if any, went unreported? Does the failure to address one or more items on the checklist affect the confidence that you place in the results they report?
3. Conduct your own randomized experiment, based on one of the suggested topics in Appendix B.
   (a) Compose a planning document.
   (b) Take an online research ethics course, and obtain your certification to conduct human subjects research. Obtain approval for your study from the institutional review board at your college or university.
   (c) Conduct a small pilot study to work out any problems in administering the treatment or measuring outcomes.
   (d) Conduct the experiment. Construct a data file and supporting metadata.
   (e) Compose a research report.

APPENDIX A

Protection of Human Subjects

This appendix discusses the responsibilities one shoulders when conducting studies that involve human subjects. We review federal regulations that researchers at American universities should be aware of when designing and implementing field experiments. We suggest four general guidelines to help researchers keep their experimental designs within the zone of permissible research.

A.1 Regulatory Guidelines

Even if field experiments were not experiments, they would be subject to regulatory oversight insofar as they involve research on human beings. Federal regulations in the United States govern research, defined as "systematic investigation, including research development, testing and evaluation, designed to develop or contribute to generalizable knowledge."[1] These regulations require institutions that receive federal funds, such as universities, to set up an institutional review board (IRB) to evaluate all proposed research involving human subjects, defined as living individuals from whom a researcher obtains data or private information "through intervention or interaction."[2] In a nutshell, IRBs assess risks to subjects, safeguard confidential information, and, when proposed research involves obtaining private information or putting participants at risk, ensure that subjects are given an explanation and an opportunity to withdraw from the research. Special scrutiny is given to proposals to study vulnerable populations, such as pregnant women, prisoners, children, or those unable to give informed consent.[3] On the other hand, broad exemptions apply to research involving public officials, surveys, educational testing, taste testing, or observation

1 Federal Policy 45 CFR 46.102(d).

2 Federal Policy 45 CFR 46.102(f).

3 These regulations may be found at http://ohsr.od.nih.gov/guidelines.

of public behavior.[4] Even when researchers believe that their research is exempt from IRB review, they are still required to submit a description of their proposed research to the IRB, which then determines whether an exemption is warranted.

The manner in which codes protecting human subjects are applied to social science experiments remains in flux, and researchers frequently complain that regulations are interpreted differently by different IRBs.[5] Some areas of human subjects regulation are more ambiguous than others. Protection of private information seems fairly clear. Researchers are prohibited from divulging any confidential information they obtain, and there is a strong presumption against revealing the identities of study participants, even when their behavior is publicly observable. More ambiguity concerns the operational definition of what constitutes harm to human subjects. Federal guidelines define harm in relation to what is termed the *minimal risk threshold:* "*Minimal risk* means that the probability and magnitude of harm or discomfort anticipated in the research are not greater in and of themselves than those ordinarily encountered in daily life or during the performance of routine physical or psychological examinations or tests."[6] IRBs are required to give special scrutiny to proposed research that involves more than minimal risk, but assessing risk is sometimes fraught with uncertainty. We return to this point below.

Another confusing aspect of the regulatory system concerns its jurisdiction. Because "research" is defined to exclude interventions that do not involve "systematic investigation . . . designed to develop or contribute to generalizable knowledge," it seems that some interventions that are both harmful and directed at vulnerable populations may fall outside the regulatory framework, even when they are deployed by scholars at universities. In other words, the same intervention that would be regulated if administered to half of the subject pool as part of an experiment would be unregulated if administered to all of the subject pool for reasons other than research.

Similarly, those outside federally funded entities are essentially unregulated, even though their research results may find their way into scholarly hands. So-called reality shows unabashedly assign contestants randomly to outlandish experimental conditions. The TV show *Survivor* determined by spin of a wheel which contestants would be required to drink a "smoothie" containing ingredients such as rotting octopus, while the show *Fear Factor Live* used the "Wheel of Fear" to randomly instruct certain contestants to lie down in a bed of cockroaches.[7] These types of experiments are conducted without IRB approval but enter the public domain and sometimes become the basis of academic publication.[8] The principle that public data fall outside

4 Federal Policy 45 CFR 46.101(b).
5 Yanow and Schwartz-Shea 2008.
6 Federal Policy 45 CFR 46.102(i).
7 Burnett 2003; Sandler 2007, p. 160.
8 See, for example, Gertner 1993, Metrick 1995, and Post et al. 2008.

the purview of IRB review applies to a broad array of naturally-occurring experiments such as military induction, tax audits, jury selection, and other lotteries.

Researchers who conduct field experiments sometimes find themselves in the murky area between regulated and unregulated data collection. Suppose a nongovernmental organization orchestrates a randomized intervention (outside the purview of IRBs) and hires university researchers to analyze the data, which are not in the public domain. If university researchers were conducting the intervention, they would be required to obtain IRB approval before launching the study, but in this case, the intervention and data collection have already taken place, and IRBs are typically unwilling to review research projects retrospectively. Consult with your IRB to find out its policy regarding these borderline cases.

A.2 Guidelines for Keeping Field Experiments within Regulatory Boundaries

Given the complex and evolving regulatory environment within which field experiments operate, researchers need to confer with their local IRBs in order to understand what the rules are and how they are likely to be applied. What follows is a set of general guidelines designed to help you stay within regulatory boundaries.

1. Avoid Assigning Subjects to Experimental Conditions That You Expect Will Harm Them. Expectations play a critical role here. Expectations should be informed by credible scientific evidence about the treatments in question. Prior to conducting an experiment, there may be good reason for skepticism about whether, on average, the treatment is beneficial or harmful. Indeed, the history of science is replete with examples of medical treatments that were initially believed to be beneficial to subjects but were demonstrated to be useless or harmful.

Uncertainty about cause and effect tends to be even greater in social and policy research. Consider, for instance, experiments designed to test whether driver training for high school students reduces the number of accidents and driving infractions. Intuition suggests that the control group incurs risks because students who go untrained will become accident-prone drivers. The implication seems to be that the experiment puts subjects in the control group in a risky situation for the sake of research, but the randomized experiments that have addressed this question paint a more complex picture. Evidently, the treatment group's risks are at least as great because driver training seems to encourage students to get their licenses at an earlier age.[9] Ironically, risk assessment was made possible by a randomized experiment.

9 Roberts and Kwan 2001; Vernick et al. 1999.

Evaluating threats to subjects becomes more complex as we leave domains in which harms are tangible (e.g., impaired health, loss of material resources) and consider outcomes about which people have differing opinions. Suppose experimental interventions encourage subjects to change their childrearing practices, attend a political demonstration, or adopt a new set of religious beliefs. How should we define and assess harm in these cases? Should the assessment be confined to tangible harms such as personal injury or financial loss? There are no easy answers, and federal regulations, which were drafted with biomedical research in mind, offer little guidance.

Setting aside the question of what constitutes harm, the "avoid inflicting harm" principle has the following implication for experimental design. Whenever possible, induce what you believe to be helpful treatments. If you believe that SAT prep classes are beneficial and want to study their effects on SAT performance, randomly encourage those who would not take the prep class to take it. Do not randomly encourage subjects who would otherwise take the course not to do so.[10]

A corollary to this principle is that one should look for ways of mitigating losses to subjects who may be harmed if they do not receive the treatment.[11] The stepped-wedge design (see Chapter 8) is a useful way to achieve a balance between random assignment and treatment of all subjects. The design presumes that treatments may be administered at different points in time and assigns a certain proportion of the subject pool to an early treatment and the remainder to a late treatment. When outcomes are measured for all subjects prior to the second phase of treatment, the late treatment group serves as the control for the early treatment group.

2. Exposing Subjects to Significant Risk of Harm Requires Their Informed Consent. If you plan to implement experiments that expose subjects to more than minimal risks, IRBs will typically require you to disclose these risks in ways that subjects can understand. The requirement that subjects understand the ramifications of risk and voluntarily enter into an agreement usually rules out participation by minors, prisoners, and mentally challenged individuals. Subjects must actively consent, and they may withdraw from participation in the treatment (or control) if they later change their minds.

When subjects are exposed to minimal risk and these subjects do not come from vulnerable populations, IRBs may waive the informed consent requirement.[12] An acceptable rationale for dispensing with informed consent is that "the research could

10 Notice that these two encouragement designs focus on different sets of Compliers and therefore involve different causal estimands.

11 Another corollary is that you should try to not reduce the number of beneficiaries (Humphreys 2011). This principle would be violated, for example, if an NGO plans to give 1,000 villagers $100, and you suggest giving a random sample of 500 villagers $200 and 500 villagers nothing. A design that would satisfy the no harm principle would be to select 2,000 villagers and give half $100.

12 Federal Policy 45 CFR 46.116(d).

not be practicably carried out without the waiver."[13] If you plan to carry out an experiment that depends on unobtrusive measurement, you may have to make the case to your IRB that your study cannot be practicably carried out if subjects (especially subjects in the control group) are made aware of their participation.

3. *Take Precautions to Protect Anonymity and Confidentiality.* It is rare for the goals of scientific inquiry to impel a researcher to disclose the identities of subjects or any confidential information that subjects may have provided to researchers. Confidential information should be stored securely and not redistributed without protecting the subjects' anonymity. On occasion, a researcher may wish to merge the records of the experimental database with another confidential database, and doing so requires sharing a list containing subjects' identifying information. The question is how to do so in a way that meets with IRB approval. This situation may be handled by identifying a trustworthy third party, perhaps representatives from the IRB itself, to serve as a go-between that merges both datasets. Confidentiality could be further protected if the intermediary were to aggregate the data slightly, perhaps generating a dataset where subjects are randomly grouped into clusters of 10 subjects apiece, or to cloak the actual dataset in a much larger dataset consisting of fictitious data.

4. *Confer with Your IRB as You Plan Your Research.* Before designing and implementing experiments, researchers should familiarize themselves with the regulations that apply in their home country and the countries in which they plan to conduct research. Discuss your project ideas with the IRB chair or staff in order to get a sense of the concerns they are likely to have when evaluating your proposal. Be prepared to explain your research design and how it compares to similar designs that have been used by other scholars and approved by IRBs in the past. By discussing project ideas with IRB representatives early in the design process, you leave yourself plenty of opportunity to make adjustments that will satisfy the IRB and your research aims.

13 Federal Policy 45 CFR 46.116(d)3.

APPENDIX B

Suggested Field Experiments for Class Projects

In the course of illustrating concepts and techniques, we have made frequent reference to thought-provoking studies that use experimental methods to understand politics, education, markets, crime, and a host of other topics. You can learn a great deal from these and other exemplars, but the only way to truly solidify your understanding is to design, implement, and analyze an experiment of your own.

This appendix is designed to guide you through the process of conducting your own study. We walk through a hypothetical project and present a list of experiments that may be conducted inexpensively and with minimal risk to subjects. Even if none of the projects mentioned below match your substantive interests, you may nevertheless find it instructive to think about the design and implementation challenges they each present.

B.1 Crafting Your Own Experiment

Some researchers decide what to study based on their ongoing substantive interests; others look for experimental opportunities wherever they arise. For readers who are new to field experimentation and are looking for an inexpensive class project that can be completed over the course of a few weeks, it is usually best to conduct a practice experiment in a domain where a research opportunity presents itself. Very well; where do research opportunities present themselves?

Everyday conversation is one source of ideas. The next time you find yourself in the company of friends or relatives, ask them about their jobs. What do they do? What do they think are some of the more effective ways of getting results? What kinds of activities or strategies seem to them to be ineffective?

Suppose you strike up a conversation with a realtor. It's not uncommon to hear a realtor advance the hypothesis that a sale is more likely to occur if you bake bread when prospective buyers tour your home. The conjecture sounds plausible enough to be interesting, and a quick search of the Internet confirms that baking advice is frequently dispensed to home sellers. The intervention seems to involve minimal risk to subjects, since the bread-baking treatment is currently a recommended practice, and pleasant odors are something that a homebuyer might ordinarily encounter when shopping for a home. Is the hypothesis worth studying experimentally? Would anyone care if your rigorous experiment found that the effect were zero? Would anyone care if, on the other hand, you found that the effect were a 5% reduction in the average time that a house were on the market before selling? The answer to both questions seems to be yes. If baking in fact has negligible effects, realtors could redirect sellers' energies elsewhere; more broadly, it suggests that extraneous factors like aroma have a trivial effect on buyers' economic valuations. On the other hand, if aroma were in fact to produce an appreciable effect, the financial implications could be immense, and the general phenomenon suggests something theoretically interesting about the rationality of economic decisions.

How would you go about orchestrating a study of baking and home sales? The first step is to see whether an analogous experiment has already been conducted. If so, it might be helpful to use it as a template for your own research or a springboard for design improvements. The next step is to have a follow-up conversation with this realtor (or any others you know) in order to find out whether an experiment is feasible and, if so, under what conditions. It may be that sellers who still live in their homes may be amenable to a random encouragement. (See Chapter 6 on encouragement designs.) You may have to orchestrate the encouragement yourself, and you may even have to supply the necessary pans and ingredients if you want a large share of Compliers. Another option is to work with a sample of houses that sellers no longer live in; in the treatment condition, you bake prior to the realtor's open house and remain out of sight. Using either design, this experiment requires spare time, transportation, bread-making ingredients, a rudimentary knowledge of baking, and, if you are an academic researcher, approval of your IRB. But most of all, it requires coordination with the realtor, who after a few days will probably regret that the subject of the experiment ever came up. You will need to muster all of your charm in order to maintain good relations with the realtor and others in the realty office, who play a critical role in coordinating your treatments and furnishing outcome data (sales date and price) for the homes in the treatment and control conditions. If you become adept at baking bread, delivering some fresh loaves to their office might buoy their spirits. Continue the experiment for a set period or when a pre-specified number of homes have been shown in treatment or control conditions. Take note of a subtle statistical issue: don't end the study based on a decision rule that says "stop once statistical significance has been achieved." This stopping rule leads to biased significance tests, and you'll need to adjust your significance test results to take this bias into account.[1]

1 See Jennison and Turnbull 2000.

The broader point is to keep your eyes open for testable propositions. Everywhere you turn, people advance hypotheses. They speculate about whether eating candy makes children hyperactive, whether banner ads on Web pages tend to be obnoxious because obnoxious ads work best, or whether amorousness is related to the temperature at which the dorm rooms are heated. Exercise your experimental imagination by trying to think of research designs that would enable you to test these conjectures or others you encounter. What will you randomize? Who or what will be your subjects? How will you measure outcomes? What kinds of implementation problems can you expect—violations of the exclusion restriction, noncompliance, attrition, interference—and how might your experiment be designed to minimize their ill effects? What kind of statistical power do you expect, given the sample size and covariates at your disposal?

Another important source of experimental inspiration is the recurrent dilemma. Our jobs, neighbors, families, carpools, community groups, schools, shopping centers, and governments continually present us with problems of allocation and enforcement. If you are in a position to decide which products should be stocked near the checkout counter or how to get post-operative patients to show up for physical therapy or how to advertise the elementary school's fundraiser, you have already overcome one of the most important obstacles to conducting an experiment. Some of the most remarkable experiments ever conducted in field settings occurred because practitioners in charge of allocating resources were eager to answer a causal question.

When choosing among experimental opportunities, lean toward projects that seem to involve fewer measurement headaches. In some cases, governments or survey organizations routinely gather outcome data; you need only furnish the experimental intervention. Sometimes extensive background information is available as well, which allows you to take advantage of covariates in your design and analysis. Also attractive are instances where you can easily verify that the treatment has been administered as intended.

Finally, when selecting a project, follow your intuitions about what interventions are likely to have a powerful impact. Although there is nothing wrong with conducting a solid experiment that debunks a hypothesis and convincingly demonstrates an average treatment effect of zero, you are likely to have more fun and more easily gain the cooperation of others if your intervention generates useful effects.

B.2 Suggested Experimental Topics for Practicum Exercises

In order to get you thinking about possible experimental projects, we have culled from recent studies a list of ideas that meet two criteria: they appear to pose minimal risk to subjects and can be conducted on a very small budget. Of course, vigilance is required to ensure that unforeseen risks and costs are kept in check, and it is always a good idea to conduct a small pilot test in order to detect unforeseen implementation problems. Several project ideas involve interventions that ordinarily arise on or near college campuses, but the ideas are easily adapted to other settings as well. Prior to conducting a pilot test or implementing an experiment growing out of these or other ideas, researchers (including students) who work in regulated settings such as universities or hospitals should obtain approval from the relevant IRB (see Appendix A).

Helping Behavior and Discrimination. Experiments may be designed to assess the conditions under which requests for assistance elicit cooperation. One set of conditions concerns attributes of the person or persons requesting assistance: their ethnicity, social class, accent, dress, stature, religious jewelry, political buttons, etc. Some attributes, such as dress and social class, may be easily manipulated. (See King and Ahmad for an example of an experiment in which a confederate changed from secular dress to traditional Muslim dress.)[2] In order to vary attributes such as height or skin tone, you will need multiple confederates who vary with respect to the treatment but are otherwise similar in appearance. The request for help should be tailored to the research setting. Possibilities include requests for directions, suggested restaurants of a given description, information about closing times, and so forth. One can also manipulate the costliness of the request by asking for more or less detailed instructions. A variety of outcomes may be measured: the length and quality of the response, whether additional unsolicited assistance was offered, and nonverbal behaviors. When designing an experiment along these lines, keep in mind several design challenges. You will need to define the unit of randomization and think carefully about the randomization procedure: which subjects will be eligible to be in your sample, and how will you allocate them to experimental conditions? Precautions must be taken to ensure that nothing about the subject affects the probability of assignment to an experimental condition. You may encounter some people who refuse to listen to your requests; this behavior may be in response to the treatment and should be classified as an outcome; take care not to exclude these observations from the dataset. Space your requests geographically and temporally and record outcomes discreetly so that subjects remain unaware that an experiment is taking place. Perhaps the biggest challenge is switching between treatments while maintaining the same standards of measurement when assessing outcomes. These experiments are not blinded; the subjects' treatment status is known to those who measure outcomes. The concern is that even if subjects respond to different treatments in the same way on average, confederates may expect to find a treatment effect and code subjects' responses in accordance with this expectation. A stronger research protocol would station observers unfamiliar with the hypothesis in locations where they can measure outcomes accurately but unobtrusively.

2 King and Ahmad 2010. See also Gabriel and Banse 2006.

Modeling Social Norms and Contagious Behavior. Experiments may be used to assess whether behaving in accordance with a social norm increases the chances that others will do likewise.[3] The behaviors in question might include picking up litter, holding doors in busy entranceways, donating money to homeless people, wiping feet when entering a building, obeying pedestrian street lights, switching off a cell phone when entering a library, washing hands before leaving a restroom, giving up one's seat on a subway to an elderly person, or returning one's shopping cart in a supermarket parking lot. Again, the challenges include defining the random assignment procedure and achieving adequate temporal and spatial separation between interventions. This experiment also requires careful planning to ensure that outcomes are measured unobtrusively and in a manner that preserves symmetry between treatment and control conditions.

Attractiveness and Social Cues. Certain kinds of social psychology experiments lend themselves to adaptation to field settings. For example, Argo, Dahl, and Morales find that when an unusually attractive shopper (confederate) handles a piece of clothing, other shoppers (subjects) find that piece of clothing especially desirable.[4] This kind of hypothesis may be adapted to field settings by having confederates of varying levels of attractiveness each handle a series of randomly rotating items. For example, if there are two confederates, in store 1, confederate 1 handles blouse A and jeans B, while confederate 2 handles blouse C and jeans D. In store 2, they reverse which merchandise they handle. Another confederate measures how many customers handle the same objects within ten minutes and how many purchases occur. The challenge in implementing this design is to do so unobtrusively and in ample numbers of stores. Attractiveness-related interventions (as well as interventions related to other attributes, such as social class or political ideology) may be deployed in other contexts where people make choices, such as speed-dating.

Recruitment for Collective Action and Charitable Activity. Experiments may be designed to assess how different types of messages affect the success with which people are recruited to participate in activities such as environmental cleanups, political rallies, and charitable activities (e.g., donating time at community centers or making non-monetary donations such as books or eyeglasses).[5] This line of research is similar to requests for help but usually involves a more extensive description of an organized activity and demands considerable and perhaps sustained participation by the subject. Ideally, this type of study involves no deception: the organized activity is real, and subjects' participation is valued. One challenge of this type of study is interference between units: subjects in the treatment group may, in turn, recruit subjects in the control

3 See, for example, Kallgren, Reno, and Cialdini 2000. Also of interest are Mazerolle, Roehl, and Kadleck 1998 and Harcourt and Ludwig 2006.

4 Argo, Dahl, and Morales 2008.

5 For an experiment involving book donations, see John et al. 2011.

group. One might consider directing randomly assigned recruitment messages to different clusters of individuals (e.g., each block or dorm or floor gets its own message), trying to keep cluster size as small as possible while preventing interference between units. Another challenge is keeping track of who is who in each experimental group so that behavioral outcomes (e.g., number of volunteered hours) can be measured.

Program Participation. Institutions often have policies in place that encourage but do not require public participation. Examples include recycling of household waste, conservation of electricity during peak usage hours, and home insulation.[6] Experiments may be used to assess different messaging strategies for increasing compliance with these policy objectives. Possibilities include publicizing the financial incentives that governments or utility companies create to encourage compliance or providing feedback about how one's household, block, dorm room, or other unit is performing in terms of meeting a social objective. Research challenges include defining units of randomization, implementing an intervention that achieves high rates of compliance and minimal spillover across treatment conditions, and obtaining outcome data, either by working in collaboration with utilities or monitoring outcomes such as recycling through direct observation.

Enforcement of Laws and Norms. If researchers are able to work collaboratively with police or traffic enforcement, a variety of low-risk experiments may be conducted at little or no cost. A simple experiment involves parking a police car (which may be unoccupied, so that it does not involve the redeployment of police patrols) near a busy intersection with four-way stop signs.[7] If the intersection in question is selected randomly from a sample of similar intersections, researchers may measure the effect of the treatment on the likelihood that vehicles come to a complete stop. The challenge of this study is to measure outcomes in the treatment and control sites simultaneously. Intersections should be adequately separated so that slowing drivers down at a treated intersection has no appreciable effect on how they drive at a control intersection. Similar kinds of approaches may be applied to the assiduousness with which parking enforcement officers monitor certain areas during an experimental period (say, one week) on compliance with parking regulations the following week.

Tipping at Restaurants. A surprisingly large research literature illustrates how tipping may be used as an outcome measure. Several experiments vary the manner in which waitstaff present restaurant bills to customers.[8] Possible treatments include a handwritten note from the server, complimentary candies, drawings that include smiley faces, and patriotic symbols, such as flag stickers. When implementing this

6 Examples include Cotterill et al. 2009. Study 2 of Fisher and Ackerman 1998 may also serve as a useful design template. For a study conducted in collaboration with a utility company, see Ayres, Raseman, and Shih 2009.

7 For an example of enforcement-related experimentation, see Vaa 1997. This type of research dates back to the 1930s; see Moore and Callahan 1943.

8 Rind and Bordia 1996; Rind and Strohmetz 1999; Strohmetz et al. 2002.

type of experiment, take care to ensure that servers are blind to the experimental condition until the point at which the bill is ready to be delivered to the subjects. Bills should be presented in ways that prevent customers from comparing treatments. This type of experiment lends itself to investigation of heterogeneous treatment effects, as researchers can test whether treatments interact with, for example, server attributes or the size of the bill.

Tutoring and Short-term Education Programs for Adults. In order to avoid the complications that arise when using minors as subjects, this line of experiments should focus on adult learners.[9] In order to minimize risk, the instructional programs should not involve potentially hazardous activities, such as the operation of machinery. Short courses in languages, citizenship, or software are examples of the kinds of education programs that allow for experiments that can be completed over a few weeks. Treatments might include variations in teaching style, learning aids, drilling, discussion, and so forth. Outcomes might be the final exam grade or a specially designed test that can be used in a variety of different classes. Either way, researchers must try to minimize attrition. In order to maintain symmetry, the person doing the grading should be blind to treatment assignment. Another practical challenge is dividing subjects into treatment and control groups. Ideally, one would assign individuals to classes with different treatments; as a practical matter, it may be necessary to assign existing classes to different treatments, in which case clustering should be taken into account when analyzing the results.

Advertising. Events ranging from artistic performances to tag sales rely on advertising to get the word out. If one can form a collaborative relationship with groups that routinely advertise upcoming events, one may be able to craft an experiment to assess the effects of advertising on turnout. Because it is often difficult to track who shows up at these events, the easiest designs are those that vary the volume of advertising and measure the number of people who show up. For example, if a group ordinarily places 30 posters in the nearby area, consider a design in which some events are advertised with 30 posters and others with 60 posters. If you provide the extra labor necessary to expand the volume of advertising, the group gets extra publicity, and you get an experiment. The main challenge of this study is accumulating ample numbers of advertised events so that reliable comparisons can be made between treatment and control attendance, while at the same time separating treatments geographically and temporally so as to minimize spillover.

Discrimination in Housing and Job Markets. None of the experiments mentioned above involves deception. The closest we came was the experiment involving helping behavior, where a confederate asked for directions. That barely qualifies as deception, since the question is still a question, even if it is asked of many people. In all

9 An example of a learning study conducted on adults may be found in Karpicke and Blunt 2011. For examples of analogous research designs involving children, see James-Burdumy, Dynarski, and Deke 2008.

of the other experiments, the interventions and behavioral outcomes involve actual activities, such as tag sales or recycling efforts. Deception is commonly used in studies of market discrimination. Bertrand and Mullainathan, for example, sent fictitious resumes in response to job ads, varying, among other things, the putative race of the applicant.[10] Page had putatively homosexual or heterosexual callers inquire about whether rental apartments were still available.[11] These research paradigms may be implemented inexpensively and in ways that impose minor costs on employers or landlords. Care should be taken to distribute resumes or inquiries lightly so as not to jeopardize the non-interference assumption (see Chapter 12) or impose undue burdens on subjects. If deception is deemed unacceptable, an alternative is to use an actual applicant who goes by more than one name, and the names have different ethnic or gender connotations.

The experiments described above by no means exhaust the supply of inexpensive projects that seem to pose minimal risks to subjects. We suggest them because they can be completed expeditiously without requiring special expertise in a particular substantive domain. Any energetic, friendly, and well-organized person with basic research skills can, for example, conduct a rigorous study assessing whether tips increase when restaurant patrons are presented with a bill containing a smiley face. The researcher will have to find cooperative waitstaff, train them, and closely supervise the implementation of the intervention. Persistence is another helpful trait, as setbacks are common in field research. Fortunately, setbacks serve a pedagogic purpose. It takes practice to draw the connection between mishaps in the field and chapters in this book. For example, if waitstaff fail to follow the random assignment when annotating some of the bills, they will not instruct you to address noncompliance using the methods discussed in Chapters 5 and 6. With practice, you will learn to diagnose problems and adjust your design, analysis, and interpretation accordingly.

10 Bertrand and Mullainathan 2004.

11 Page 1998.

REFERENCES

Abadie, Alberto. 2003. "Semiparametric Instrumental Variable Estimation of Treatment Response Models." *Journal of Econometrics* 113: 231–63.

Abbring, Jaap H., and James J. Heckman. 2007. "Econometric Evaluation of Social Programs, Part III: Distributional Treatment Effects, Dynamic Treatment Effects and Dynamic Discrete Choice, and General Equilibrium Policy Evaluation." In *Handbook of Econometrics, Volume 6B*, James J. Heckman and Edward Leamer, eds. Amsterdam: Elsevier Science.

Allison, Paul D. 2002. *Missing Data*. Thousand Oaks, CA: Sage.

Allison, Paul D., and Robert M. Hauser. 1991. "Reducing Bias in Estimates of Linear Models by Remeasurement of a Random Subsample." *Sociological Methods and Research* 19(4): 466–92.

Amberson, James, B.T. McMahon, and Max Pinner. 1931. "A Clinical Trial of Sanocrysin in Pulmonary Tuberculosis." *American Review of Tuberculosis* 24: 401–35.

Angrist, Joshua D. 1990. "Lifetime Earnings and the Vietnam Era Draft Lottery: Evidence from Social Security Administrative Records." *American Economic Review* 80: 313–36.

Angrist, Joshua D. 1991. "The Draft Lottery and Voluntary Enlistment in the Vietnam Era." *Journal of the American Statistical Association* 86: 584–95.

Angrist, Joshua D. 2006. "Instrumental Variables Methods in Experimental Criminological Research: What, Why and How." *Journal of Experimental Criminology* 2: 23–44.

Angrist, Joshua D., Eric Bettinger, and Michael Kremer. 2006. "Long-Term Educational Consequences of Secondary School Vouchers: Evidence from Administrative Records in Colombia." *American Economic Review* 96: 847–62.

Angrist, Joshua D., Stacey Chen, and Brigham Frandsen. 2009. "Did Vietnam Veterans Get Sicker in the 1990s? The Complicated Effects of Military Service on Self-Reported Health." NBER Working Paper 14781.

Angrist, Joshua D., and Guido W. Imbens. 1994. "Identification and Estimation of Local Average Treatment Effects." *Econometrica* 62: 467–76.

Angrist, Joshua D., Guido W. Imbens, and Donald B. Rubin. 1996. "Identification of Causal Effects Using Instrumental Variables." *Journal of the American Statistical Association* 91: 444–55.

Angrist, Joshua D., and Victor Lavy. 2009. "The Effects of High Stakes High School Achievement Awards: Evidence from a Randomized Trial." *American Economic Review* 99: 1384–1414.

Angrist, Joshua D., and Jörn-Steffen Pischke. 2009. *Mostly Harmless Econometrics: An Empiricist's Companion*. Princeton: Princeton University Press.

Arceneaux, Kevin. 2005. "Using Cluster Randomized Field Experiments to Study Voting Behavior." *The Annals of the American Academy of Political and Social Science* 601: 169–79.

Argo, Jennifer, Darren Dahl, and Andrea Morales. 2008. "Positive Consumer Contagion: Responses to Attractive Others in a Retail Context." *Journal of Marketing Research* 45: 690–701.

Aronow, Peter M. 2011. "A General Method for Detecting Interference Between Units in Randomized Experiments." *Sociological Methods and Research*, forthcoming.

Ashraf, Nava, James Berry, and Jesse M. Shapiro. 2010. "Can Higher Prices Stimulate Product Use? Evidence from a Field Experiment in Zambia." *American Economic Review* 100: 2383–2413.

Ayres, Ian, Sophie Raseman, and Alice Shih. 2009. "Evidence from Two Large Field Experiments that Peer Comparison Feedback Can Reduce Residential Energy Use." NBER Working Paper 15386.

Bagues, Manuel, and Berta Esteve-Volart. 2011. "Politicians' Luck of the Draw: Evidence from the Spanish Chrismas Lottery." FEDEA Working Paper 2011-01.

Baird, Sarah, Craig McIntosh, and Berk Özler. 2009. "Desigining Cost-Effective Cash Transfer Programs to Boost Schooling among Young Women in Sub-Saharan Africa." *The Quarterly Journal of Economics* 126: 1709–53.

Balke, Alexander, and Judea Pearl. 1993. "Nonparametric Bounds on Causal Effects From Partial Compliance Data." Technical Report R-199, University of California, Los Angeles.

Banerjee, Abhijit, Selvan Kumar, Rohini Pande, and Felix Su. 2010. "Do Informed Voters Make Better Choices? Experimental Evidence from Urban India." Unpublished manuscript.

Barnard, John, Constantine E. Frangakis, Jennifer L. Hill, and Donald B. Rubin. 2003. "Principal Stratification Approach in Broken Randomized Experiments: A Case Study of School Choice Vouchers in New York City." *Journal of the American Statistical Association* 98(462): 299–323.

Baron, Reuben M., and David A. Kenny. 1986. "The Moderator-Mediator Variable Distinction in Social Psychological Research: Conceptual, Strategic, and Statistical Considerations." *Journal of Personality and Social Psychology* 51: 1173–82.

Beaman, Lori, Raghabendra Chattopadhyay, Esther Duflo, Rohini Pande, and Petia Topalova. 2009. "Powerful Women: Does Exposure Reduce Bias?" *Quarterly Journal of Economics* 124: 1497–1540.

Beath, Andrew, Fotini Christia, and Ruben Enikolopov. 2012. "Winning Hearts and Mindsthrough Development: Evidence from a Field Experiment in Afghanistan." Massachusetts Institute of Technology Working Paper No. 2011-14.

Bergan, Daniel. 2009. "Does Grassroots Lobbying Work?" *American Politics Research* 37: 327–52.

Bertrand, Marianne, and Sendhil Mullainathan. 2004. "Are Emily and Greg More Employable than Lakisha and Jamil? A Field Experiment on Labor Market Discrimination." *The American Economic Review* 94: 991–1013.

Bhattacharya, Jay, Azeem M. Shaikh, and Edward Vytlacil. 2008. "Treatment Effect Bounds under Monotonicity Assumptions: An Application to Swan-Ganz Catheterization." *American Economic Review: Papers & Proceedings 2008* 98(2): 351–56.

Bhavnani, Rikhil R. 2009. "Do Electoral Quotas Work after They Are Withdrawn? Evidence from a Natural Experiment in India." *American Political Science Review* 103: 23–35.

Bitler, Marianne P., Jonah B. Gelbach, and Hilary W. Hoynes. 2006. "What Mean Impacts Miss: Distributional Effects of Welfare Reform Experiments." *The American Economic Review* 96: 988–1012.

Bolstad, William M. 2007. *Introduction to Bayesian Statistics*. New York: Wiley-Interscience.

Boruch, Robert F. 2005. *Place Randomized Trials: Experimental Tests of Public Policy*. Thousand Oaks, CA: Sage.

Boutron, Isabelle, Peter John, and David J. Torgerson. 2010. "Reporting Methodological Items in Randomized Experiments in Political Science." *The Annals of the American Academy of Political and Social Science* 628: 112–31.

Box, Joan. 1980. "R.A. Fisher and the Design of Experiments, 1922–1926." *American Statistician* 34: 1–7.

Boylston, Arthur W. 2008. "Did Clinical Science Begin in 1767?" *JLL Bulletin: Commentaries on the History of Treatment Evaluation,* www.jameslindlibrary.org.

Braga, Anthony A., and Brenda J. Bond. 2008. "Policing Crime and Disorder Hot Spots: A Randomized Controlled Trial." *Criminology* 46: 577–607.

Brownell, Kelly D., Albert J. Stunkard, and Janet Michelle Albaum. 1980. "Evaluation and Modification of Exercise Patterns in the Natural Environment." *American Journal of Psychiatry* 137: 1540–45.

Bruhn, Miriam, and David McKenzie. 2009. "In Pursuit of Balance: Randomization in Practice in Development Field Experiments." *American Economic Journal: Applied Economics* 1: 200–32.

Bullock, John G., Donald P. Green, and Shang E. Ha. 2010. "Yes, but What's the Mechanism? (Don't Expect an Easy Answer)." *Journal of Personality and Social Psychology* 98: 550–58.

Burnett, Mark (Dir.). 2003. "Everyone's Hero." *Survivor Pearl Islands: The Complete Seventh Season*. CBS. DVD.

Butler, Daniel M., and David E. Broockman. 2011. "Do Politicians Racially Discriminate against Constituents? A Field Experiment on State Legislators." *American Journal of Political Science* 55: 463–77.

Butler, Daniel M., and David W. Nickerson. 2011. "Can Learning Constituency Opinion Affect How Legislators Vote? Results from a Field Experiment." *Quarterly Journal of Political Science* 6(1): 55–83.

Byar, David P. 1985. "Assessing Apparent Treatment-Covariate Interactions in Randomized Clinical Trials." *Statistics in Medicine* 4: 255–63.

Camerer, Colin F. 1998. "Can Asset Markets Be Manipulated? A Field Experiment with Racetrack Betting." *Journal of Political Economy* 106(3): 457–82.

Camerer, Colin F. 2011. "The Promise and Success of Lab-Field Generalizability in Experimental Economics: A Critical Reply to Levitt and List." SSRN paper accessed at http://ssrn.com/abstract=1977749.

Campbell, Elizabeth. 2009. "Room with Another Viewpoint? Environmental Messaging and Changes in Energy-Use Behavior." Yale University Working Paper.

Canning, Ray R. 1956. "Does an Honor System Reduce Class-Room Cheating? An Experimental Answer." *The Journal of Experimental Education* 24: 291–96.

Card, David, Stefano DellaVigna, and Ulrike Malmendier. 2011. "The Role of Theory in Field Experiments." *Journal of Economic Perspectives* 25: 39–62.

Cardiac Arrhythmia Suppression Trial II Investigators. 1992. "Effect of the Antiarrhythmic Agent Moricizine on Survival after Myocardial Infarction." *The New England Journal of Medicine* 327(4): 227–33.

Carrell, Scott E., and James E. West. 2010. "Does Professor Quality Matter? Evidence from Random Assignment of Students to Professors." *The Journal of Political Economy* 118(3): 409–32.

Caughey, Devin M., and Jasjeet S. Sekhon. 2011. "Elections and the Regression Discontinuity Design: Lessons from Close U.S. House Races, 1942–2008." *Political Analysis* 19: 385–408.

Chalmers, Iain. 2001. "Comparing Like with Like: Some Historical Milestones in the Evolution of Methods to Create Unbiased Comparison Groups in Therapeutic Experiments." *International Journal of Epidemiology* 30: 1156.

Chang, Jae Bong, Jayson Lusk, and Bailey Norwood. 2009. "How Closely Do Hypothetical Surveys and Laboratory Experiments Predict Field Behavior?" *American Journal of Agricultural Economics* 91: 518–34.

Chattopadhyay, Raghabendra, and Esther Duflo. 2004. "Women as Policy Makers: Evidence from a Randomized Policy Experiment in India." *Econometrica* 72: 1409–43.

Chauchard, Simon. 2010. "Can the Experience of Political Power by a Member of a Stigmatized Group Change the Nature of Day-to-Day Interpersonal Relations? Evidence from Rural India." Manuscript, Dartmouth University.

Chin, Michelle, Jon Bond, and Nehemia Geva. 2000. "A Foot in the Door: An Experimental Study of PAC and Constituency Effect on Access." *The Journal of Politics* 62: 534–49.

Choi, Ondrich, and Yinger 2005. "Do Rental Agents Discriminate against Minority Customers? Evidence from the 2000 Housing Discrimination Study." *Journal of Housing Economics* 14: 1–26.

Chong, Alberto, Ana L. De La O, Dean Karlan, and Leonard Wantchekon. 2011. "Looking Beyond the Incumbent: The Effects of Exposing Corruption on Electoral Outcomes." NBER Working Paper 17679.

Clingingsmith, David, Asim Ijaz Khwaja, and Michael Kremer. 2009. "Estimating the Impact of the Hajj: Religion and Tolerance in Islam's Global Gathering." *Quarterly Journal of Economics* 124: 1133–70.

Clinton, Joshua D., and John S. Lapinski. 2004. "'Targeted' Advertising and Voter Turnout: An Experimental Study of the 2000 Presidential Election." *The Journal of Politics* 66: 69–96.

Cochran, William G. 1976. "Early Development of Techniques in Comparative Experimentation." In *On the History of Statistics and Probability*, ed. Donald Owen. New York: Dekker.

Cochran, William G. 1977. *Sampling Techniques* 3rd Edition. New York: Wiley.

Cochran, William G., and Gertrude Cox. 1957. *Experimental Designs*. New York: Wiley.

Cohen, Jacob. 1988. *Statistical Power Analysis for the Behavioral Sciences.* 2nd Edition. Hillsdale, NJ: Erlbaum Associates.

Cole, Stephen R. and Elizabeth A. Stuart. 2010. "Generalizing Evidence from Randomized Clinical Trials to Target Populations: The ACTG-320 Trial." *American Journal of Epidemiology* 172: 107–15.

Cooper, Harris, Larry Hedges, and Jeff Valentine, eds. 2009. *The Handbook of Research Synthesis and Meta-Analysis*. New York: Russell Sage Foundation.

Cotterill, Sarah, Peter John, Hanhua Liu, and Hisako Nomura. 2009. "Mobilizing Citizen Effort to Enhance Environmental Outcomes: A Randomized Controlled Trial of a Door-to-Door Recycling Campaign." *Journal of Environmental Management* 91: 403–10.

Cox, D. R., and N. Reid. 2000. *The Theory of the Design of Experiments*. Boca Raton: Chapman & Hall/CRC.

Cox, David R. 1958. *Planning of Experiments*. New York: Wiley.

Crump, Richard K., V. Joseph Hotz, Guido W. Imbens, and Oscar A. Mitnik. 2008. "Nonparametric Tests for Treatment Effect Heterogeneity." *The Review of Economics and Statistics* 90: 389–405.

Davenport, Tiffany C. 2010. "Public Accountability and Political Participation: Effects of a Face-to-Face Feedback Intervention on Voter Turnout of Public Housing Residents." *Political Behavior* 32: 337–68.

Davenport, Tiffany C. 2011. "The Draft and the Ballot Box: The Effect of Conscription Risk on American Political Behavior: 1969–1972." PhD Dissertation, Yale University.

Davenport, Tiffany C., Alan S. Gerber, Donald P. Green, Christopher W. Larimer, Christopher B. Mann, and Costas Panagopoulos. 2010. "The Enduring Effects of Social Pressure: Tracking Campaign Experiments over a Series of Elections." *Political Behavior* 32: 423–30.

Davis, Douglas, and Charles Holt. 1993. *Experimental Economics*. Princeton, NJ: Princeton University Press.

De Paola, Maria. 2009. "Does Teacher Quality Affect Student Performance? Evidence from an Italian University." *Bulletin of Economic Research* 61: 353–77.

Dean, Angela M., and Daniel Voss. 1999. *Design and Analysis of Experiments*. New York: Springer.

Dehejia, Rajeev H. 2005a. "Program Evaluation as a Decision Problem." *Journal of Econometrics* 125: 141–73.

Dehejia, Rajeev H. 2005b. "Practical Propensity Score Matching: A Reply to Smith and Todd." *Journal of Econometrics* 125: 355–64.

Dehejia, Rajeev H., and Sadek Wahba. 1999. "Causal Effects in Nonexperimental Studies: Reevaluating the Evaluation of Training Programs." *Journal of the American Statistical Association* 94: 1053–62.

Devoto, Florencia, Esther Duflo, Pascaline Dupas, William Pariente, and Vincent Pons. 2011. "Happiness on Tap: Piped Water Adoption in Urban Morocco." *American Economic Journal*, forthcoming.

DiNardo, John, Justin McCrary, and Lisa Sanbonmatsu. 2006. "Constructive Proposals for Dealing with Attrition: An Empirical Example." University of Michigan Working Paper.

Ditlmann, Ruth, and Paul Lagunes. 2010. "Differential Treatment of Latinos Changes over the Course of Social Interactions." Yale University Working Paper.

Dixit, Avinash K. 1990. *Optimization in Economic Theory*. New York: Oxford University Press.

Djebbari, Habiba, and Jeffrey Smith. 2008. "Heterogenous Impacts in PROGRESA." *Journal of Econometrics* 145: 64–80.

Dobkin, Carlos, and Reza Shabani. 2009. "The Health Effects of Military Service: Evidence from the Vietnam Draft." *Economic Inquiry* 47: 69–80.

Doleac, Jennifer L., and Luke C. D. Stein. 2010. "The Visible Hand: Race and Online Market Outcomes." Stanford University Working Paper.

Duflo, Esther, Rema Hanna, and Stephen Ryan. 2010. "Incentives Work: Getting Teachers to Come to School." Massachusetts Institute of Technology Working Paper.

Dugard, Pat, Portia File, and Jonathan Todman. 2012. *Single-case and Small-n Experimental Designs: A Practical Guide to Randomization Tests.* New York: Routledge.

Dunning, Thad. *Natural Experiments in the Social Sciences.* New York: Cambridge University Press, forthcoming.

Dupas, Pascaline. 2012. "Short-Run Subsidies and Long-Run Adoption of New Health Products: Evidence from a Field Experiment." Stanford University Working Paper.

Eden, Thomas, and Ronald A. Fisher. 1927. "Studies in Crop Variation, IV: The Experimental Determination of the Value of Top Dressings with Cereals." *Journal of Agricultural Science* 17: 548–62 and Collected Papers, 2, no. 57.

Edgington, Eugene S., and Patrick Onghena. 2007. *Randomization Tests.* 4th Edition. London: Chapman & Hall.

Eldersveld, Samuel. 1956. "Experimental Propaganda Techniques and Voting Behavior." *The American Political Science Review* 50: 154–65.

Erikson, Robert S., and Laura Stoker. 2011. "Caught in the Draft: The Effects of Vietnam Draft Lottery Status on Political Attitudes." *American Political Science Review* 105: 221–37.

Ernst, Michael D. 2009. "Teaching Inference for Randomized Experiments." *Journal of Statistics Education* 17(1), www.amstat.org/publications/jse/v17n1/ernst.html.

Fellner, Gerlinde, Rupert Sausgruber, and Christian Traxler. "Testing Enforcement Strategies in the Field: Threat, Moral Appeal and Social Information." *Journal of the European Economic Association*, forthcoming.

Ferraz, Claudio, and Frederico Finan. 2008. "Exposing Corrupt Politicians: The Effects of Brazil's Publicly Released Audits on Electoral Outcomes." *Quarterly Journal of Economics* 123: 703–45.

Fieldhouse, Edward, David Cutts, Paul Widdop, Peter John, and Rod Ling. 2010. "Do Impersonal Mobilisation Methods Work? Evidence from a Get-Out-the-Vote Experiment from the 2009 English European Elections." Paper prepared for the Midwest Political Science Association National Conference, Chicago.

Finkelstein, Amy, Sarah Taubman, Bill Wright, Mira Bernstein, Jonathan Gruber, Joseph P. Newhouse, Heidi Allen, Katherine Baicker, and The Oregon Health Study Group. 2011. "The Oregon Health Insurance Experiment: Evidence from the First Year." NBER Working Paper No. 17190.

Fisher, Robert J., and David Ackerman. 1998. "The Effects of Recognition and Group Need on Volunteerism: A Social Norm Perspective." Journal of Consumer Research. 25: 262–75.

Fisher, Ronald A. 1935. *The Design of Experiments.* London: Oliver and Boyd.

Fix, Michael, and Raymond J. Struyk. 1993. *Clear and Convincing Evidence: Measurement of Discrimination in America.* Washington, DC: Urban Institute Press.

Flemming, Gregory, and Kimberly Parker. 1998. "Possible Consequences of Non-Response for Pre-Election Surveys: Race and Reluctant Respondents." Report for the Pew Research Center for the People and the Press.

Forsetlund, Louise, Iain Chalmers, and Arild Bjørndala. 2007. "When Was Random Allocation First Used to Generate Comparison Groups in Experiments to Assess the Effects of Social Interventions?" *Economics of Innovation and New Technology* 16(5): 371–84.

Freedman, David. 2005. *Statistical Models: Theory and Practice*. Cambridge: Cambridge University Press.

Freedman, David. 2008. "On Regression Adjustments to Experimental Data." *Advances in Applied Mathematics* 40: 180–93.

Freedman, David, Robert Pisani, and Roger Purves. 1998. *Statistics*. 3rd Edition. New York: W. W. Norton.

Freedman, David, Robert Pisani, and Roger Purves. 2007. *Statistics*. 4th Edition. New York: W. W. Norton.

Fried, Brian, Paul Lagunes, and Atheendar Venkataramani. 2010. "Corruption and Inequality at the Crossroad: A Multimethod Study of Bribery and Discrimination in Latin America." *Latin American Research Review* 45: 76–97.

Fryer, Jr., Roland G. 2011. "Financial Incentives and Student Achievement: Evidence from Randomized Trials." *Quarterly Journal of Economics* 126(4): 1755–98.

Gabriel, Ute, and Rainer Banse. 2006. "Helping Behavior as a Subtle Measure of Discrimination Against Lesbians and Gay Men: German Data and a Comparison Across Countries." Journal of Applied Social Psychology 36: 690–707.

Galiani, Sebastian, Martin Rossi, and Ernesto Schargodsky. 2010. "Conscription and Crime: Evidence from the Argentine Draft Lottery." FEEM Working Paper 55.2010.

Gelman, Andrew, and Jennifer Hill. 2007. *Data Analysis Using Regression and Multilevel/Hierarchical Models*. New York: Cambridge University Press.

Gerber, Alan S., James G. Gimpel, Donald P. Green, and Daron R. Shaw. 2011. "How Large and Long-lasting Are the Persuasive Effects of Televised Campaign Ads? Results from a Randomized Field Experiment." *American Political Science Review* 105: 135–50.

Gerber, Alan, and Donald Green. 2000. "The Effects of Canvassing, Telephone Calls, and Direct Mail on Voter Turnout: A Field Experiment." *American Political Science Review* 94: 653–63.

Gerber, Alan, and Donald Green. 2005. "Do Phone Calls Increase Turnout? An Update." *The Annals of The American Academy of Political and Social Science* 601: 142–54.

Gerber, Alan, Donald Green, and Edward Kaplan. 2004. "The Illusion of Learning from Observational Research." In *Problems and Methods in the Study of Politics*, ed. Ian Shapiro, Rogers Smith, and Tarek Massoud. New York: Cambridge University Press.

Gerber, Alan, Donald Green, Edward Kaplan, and Holger Kern. 2010. "Baseline, Placebo, and Treatment: Efficient Estimation for Three-Group Experiments." *Political Analysis* 18: 297–315.

Gerber, Alan S., Donald P. Green, Holger Kern, and Chris Blattman. 2011. "Addressing Missing Outcome Data in Randomized Experiments: A Design-Based Approach." Paper presented at the 69th meeting of the Midwest Political Science Association.

Gerber, Alan, Donald Green, and Christopher Larimer. 2008. "Social Pressure and Voter Turnout: Evidence from a Large-Scale Field Experiment." *American Political Science Review* 102: 33–48.

Gerber, Alan, Donald Green, and Christopher Larimer. 2010. "An Experiment Testing the Relative Effectiveness of Encouraging Voter Participation by Inducing Feelings of Pride or Shame." *Political Behavior* 32: 409–22.

Gerber, Alan S., Donald P. Green, and David Nickerson. 2001. "Testing for Publication Bias in Political Science." *Political Analysis* 9: 385-92.

Gerber, Alan, Donald Green, and Ron Shachar. 2003. "Voting May Be Habit Forming: Evidence from a Randomized Field Experiment." *American Journal of Political Science* 47: 540–50.

Gerber, Alan, Gregory Huber, David Doherty, Conor Dowling, and Shang Ha. 2010. "Personality and Political Attitudes: Relationships across Issue Domains and Political Contexts." *American Political Science Review* 104: 111–33.

Gerber, Alan, and Neil Malhotra. 2008a. "Do Statistical Reporting Standards Affect What Is Published? Publication Bias in Two Leading Political Science Journals." *Quarterly Journal of Political Science* 3: 313–26.

Gerber, Alan, and Neil Malhotra. 2008b. "Publication Incentives and Empirical Research: Do Reporting Standards Distort the Published Results?" *Sociological Methods and Research* 37: 3–30.

Gertner, Robert. 1993. "Game Shows and Economic Behavior: Risk Taking on 'Card Sharks.'" *Quarterly Journal of Economics* 108: 507–21.

Gibson, John, David McKenzie, and Steven Stillman. 2011. "The Impacts of International Migration on Remaining Household Members: Omnibus Results from a Migration Lottery Program." *The Review of Economics and Statistics* 92: 1297–1318.

Glass, Gene. 1976. "Primary, Secondary, and Meta-Analysis of Research." *Educational Researcher* 5: 3–8.

Gleser, Leon, and Ingram Olkin. 2009. "Stochastically Dependent Effect Sizes." In *The Handbook of Research Synthesis and Meta-Analysis*, ed. Harris Cooper, Larry Hedges, and Jeff Valentine. New York: Russell Sage Foundation.

Gneezy, Uri, Ernan Haruvy, and Hadas Yafe. 2004. "The Inefficiency of Splitting the Bill." *The Economic Journal* 114: 265–80.

Goldstein, Noah J., Robert B. Cialdini, and Vladas Griskevicius. 2008. "A Room with a Viewpoint: Using Social Norms to Motivate Environmental Conservation in Hotels." *Journal of Consumer Research* 35: 472–82.

Gosnell, Harold. 1927. *Getting Out the Vote: An Experiment in the Stimulation of Voting*. Chicago: The University of Chicago Press.

Graham, John W. 2009. "Missing Data Analysis: Making It Work in the Real World." *Annual Review of Psychology* 60: 549–76.

Green, Donald P., and Peter M. Aronow. 2011. "Analyzing Experimental Data Using Regression: When Is Bias a Practical Concern?" Yale University Working Paper.

Green, Donald P., and Alan S. Gerber. 2002. "The Downstream Benefits of Experimentation." *Political Analysis* 10: 394–402.

Green, Donald P., and Alan Gerber. 2003. "The Underprovision of Experiments in Political Science." *The Annals of the American Academy of Political and Social Science* 589: 94–112.

Green, Donald P., and Alan S. Gerber. 2008. *Get Out the Vote! How to Increase Voter Turnout.* 2nd Edition. Washington, DC: Brookings Institution Press.

Green, Donald P., Shang E. Ha, and John G. Bullock. 2010. "Enough Already about 'Black Box' Experiments: Studying Mediation Is More Difficult Than Most Scholars Suppose." *Annals of the American Academy of Political and Social Sciences* 628: 200–08.

Green, Donald P., and Holger L. Kern. 2011. "Modeling Heterogeneous Treatment Effects in Survey Experiments with Bayesian Additive Regression Trees." Yale University Working Paper.

Green, Donald P., Terence Y. Leong, Holger L. Kern, Alan S. Gerber, and Christopher W. Larimer. 2009. "Testing the Accuracy of Regression Discontinuity Analysis Using Experimental Benchmarks." *Political Analysis* 17: 400–17.

Green, Donald P., and Vavreck, Lynn. 2008. "Analysis of Cluster-Randomized Experiments: A Comparison of Alternative Estimation Approaches." *Political Analysis* 16: 138–52.

Green, Donald P., and Daniel Winik. 2010. "Using Random Judge Assignments to Estimate the Effects of Incarceration and Probation on Recidivism among Drug Offenders." *Criminology* 48: 357–87.

Greenberg, David, and Mark Shroder. 2004. *The Digest of Social Experiments. 3rd Edition.* Washington, D.C.: Urban Institute Press.

Grimmer, Justin, Eitan Hersh, Brian Feinstein, and Daniel Carpenter. 2011. "Are Close Elections Random?" Stanford University Working Paper.

Grose, Christian R. 2009. "A Field Experiment of Participatory Shirking among Legislators: Pressuring Representatives to Show Up for Work." Paper presented at the 2009 meeting of the American Political Science Association.

Gruber, Jonathan. 2006. "The Role of Consumer Copayments for Health Care: Lessons from the RAND Health Insurance Experiment and Beyond." Report for the Kaiser Family Foundation.

Guala, Francesco. 2005. *The Methodology of Experimental Economics.* New York: Cambridge University Press.

Guan, Mei, and Donald Green. 2006. "Non-Coercive Mobilization in State-Controlled Elections: An Experimental Study in Beijing." *Comparative Political Studies* 39: 1175–93.

Gueron, Judith M. 2002. "The Politics of Random Assignment: Implementing Studies and Impacting Policy." In *Evidence Matters: Randomized Trials in Education Research*, ed. Frederick Mosteller and Robert F. Boruch. Washington, DC: Brookings Institution Press.

Habyarimana, James, Macartan Humphreys, Daniel Posner, and Jeremy Weinstein. 2007. "Why Does Ethnic Diversity Undermine Public Goods Provision?" *American Political Science Review* 101: 709–25.

Hacking, Ian. 1990. *The Taming of Chance.* New York: Cambridge University Press.

Hansen, Ben B., and Jake Bowers. 2009. "Attributing Effects to a Cluster-Randomized Get-Out-the-Vote Campaign." *Journal of the American Statistical Association*, 104(487): 873–85.

Harcourt, Bernard E., and Jens Ludwig. 2006. "Broken Windows: New Evidence from New York City and a Five-City Social Experiment." *The University of Chicago Law Review* 73: 271–320.

Hastings, Justine, Thomas Kane, and Douglas Staiger. 2006. "Gender and Performance: Evidence from School Assignment by Randomized Lottery." *American Economic Review* 96: 232–36.

Hastings, Justine, Thomas Kane, Douglas Staiger, and Jeffrey Weinstein. 2007. "The Effect of Randomized School Admissions on Voter Participation." *Journal of Public Economics* 91: 915.

Hastings, Justine, and Jeffrey Weinstein. 2008. "Information, School Choice, and Academic Achievement: Evidence from Two Experiments." *Quarterly Journal of Economics* 123: 1373–1414.

Heckman, James J. 1979. "Sample Selection Bias as a Specification Error," *Econometrica* 47: 153–61.

Heckman, James J., and Jeffrey A. Smith. 1995. "Assessing the Case for Social Experiments." *Journal of Economic Perspectives* 9(2): 85–110.

Heckman, James J., Jeffrey A. Smith, and Nancy Clements. 1997. "Making the Most Out of Programme Evaluations and Social Experiments: Accounting for Heterogeneity in Programme Impacts." *The Review of Economic Studies* 64: 487–535.

Hedges, Larry, and Ingram Olkin. 1985. *Statistical Methods for Meta-Analysis*. Orlando: Academic Press.

Hedges, Larry. 2009. "Improving Generalizations from Social Experiments." Northwestern University Working Paper.

Hewitt, Catherine E., and David J. Torgerson. 2006. "Is Restricted Randomisation Necessary?" *British Medical Journal* 32: 1506–8.

Hill, Austin. 1951. "The Clinical Trial." *British Medical Bulletin* 7: 278–82.

Hill, Austin. 1952. "The Clinical Trial." *The New England Journal of Medicine* 247: 113–19.

Hill, Russell A., and Robert A. Barton. 2005. "Red Enhances Human Performance in Contests." *Nature* 435: 293.

Holland, Paul W. 1986. "Statistics and Causal Inference." *Journal of the American Statistical Association*. 81: 945–60.

Hong, Guanglei, and Stephen W. Raudenbush. 2006. "Evaluating Kindergarten Retention Policy: A Case Study of Causal Inference for Multilevel Observational Data." *Journal of the American Statistical Association* 101: 901–10.

Hough, Leslie. 2010. "Experimenting with an N of 1." Yale University Working Paper.

Howell, William G., and Paul E. Peterson, eds. 2002. *The Education Gap*. Washington, DC: Brookings Institution Press.

Hróbjartsson, Asbjørn, Peter C. Gøtzsche, and Christian Gluud. 1998. "The Controlled Clinical Trial Turns 100 Years: Fibiger's Trial of Serum Treatment of Diphtheria." *British Medical Journal* 317: 1243–45.

Hudgens, Michael G., and M. Elizabeth Halloran. 2008. "Toward Causal Inference With Interference." *Journal of the American Statistical Association* 103: 832–42.

Hughes, R. E. 1975. "James Lind and the cure of scurvy: an experimental approach." *Medical History* 19(4): 342–51.

Humphreys, Macartan. 2009. "Bounds on Least Squares Estimates of Causal Effects in the Presence of Heterogeneous Assignment Probabilities." Columbia University Working Paper.

Humphreys, Macartan, and Jeremy Weinstein. 2010. "Policing Politicians: Citizen Empowerment and Political Accountability in Uganda." Unpublished manuscript.

Hussey, Michael A., and James P. Hughes. 2007. "Design and Analysis of Stepped Wedge Cluster Randomized Trials." *Contemporary Clinical Trials* 28: 182–91.

Hyde, Susan. 2010. "Experimenting in Democracy Promotion: International Observers and the 2004 Presidential Elections in Indonesia." *Perspectives on Politics* 8: 511–27.

Imai, Kosuke. 2008. "Variance Identification and Efficiency Analysis in Randomized Experiments under the Matched-Pair Design." *Statistics in Medicine* 27: 4857–73.

Imai, Kosuke, Luke Keele, and Teppei Yamamoto. 2010. "Identification, Inference and Sensitivity Analysis for Causal Mediation Effects." *Statistical Science* 25: 51–71.

Imai, Kosuke, and Aaron Strauss. 2011. "Estimation of Heterogeneous Treatment Effects from Randomized Experiments, with Application to the Optimal Planning of the Get-Out-the-Vote Campaign." *Political Analysis* 19: 1–19.

Imbens, Guido W. 2011. "Experimental Design for Unit and Cluster Randomized Trials." Paper presented at the International Initiative for Impact Evaluation, 3ie.

Imbens, Guido W., and Thomas Lemieux. 2008. "Regression Discontinuity Designs: A Guide to Practice." *Journal of Econometrics* 142: 615–35.

Inter-university Consortium for Political and Social Research (ICPSR). 2009. *Guide to Social Science Data Preparation and Archiving: Best Practice throughout the Data Life Cycle.* 4th Edition. Ann Arbor: ICPSR.

James-Burdumy, Susanne, Mark Dynarski, and John Deke. 2008. "After-School Program Effects on Behavior: Results from the 21st Century Community Learning Centers Program National Evaluation." *Economic Inquiry* 46: 13–18.

Jennison, Christopher, and Bruce Turnbull. 2000. *Group Sequential Methods with Applications to Clinical Trials.* London: Chapman & Hall.

Jerit, Jennifer, Jason Barabus, and Scott Clifford. 2012. "Comparing Treatment Effects in Parallel Experiments." *Journal of Politics*, forthcoming.

John, Peter, Sarah Cotterill, Liz Richardson, Alice Moseley, Graham Smith, Gerry Stoker, and Corinne Wales. 2011. *Nudge, Nudge, Think, Think: Using Experiments to Change Civic Behaviour.* London: Bloomsbury.

Kagel, John, and Alvin Roth. 1995. *The Handbook of Experimental Economics.* Princeton, NJ: Princeton University Press.

Kallgren, Carl, Raymond Reno, and Robert Cialdini. 2000. "A Focus Theory of Normative Conduct: When Norms Do and Do Not Affect Behavior." *Personality and Social Psychology Bulletin* 26: 1002–12.

Karlan, Dean. 2005. "Using Experimental Economics to Measure Social Capital and Predict Financial Decisions." *American Economic Review* 95: 1688–99.

Karlan, Dean, and John List. 2007. "Does Price Matter in Charitable Giving? Evidence from a Large-Scale Natural Field Experiment." *The American Economic Review* 97: 1774.

Karpicke, Jeffrey D., and Janell R. Blunt. 2011. "Retrieval Practice Produces More Learning Than Elaborative Studying with Concept Mapping." *Science* 331: 772–75.

Keizer, Kees, Siegwart Lindenberg, and Linda Steg. 2008. "The Spreading of Disorder." *Science* 322: 1681–85.

Kempthorne, Oscar. 1979. *The Design and Analysis of Experiments.* Huntington, NY: R.E. Krieger.

King, Eden, and Afra Ahmad. 2010. "An Experimental Field Study of Interpersonal Discrimination Toward Muslim Job Applicants." *Personnel Psychology* 63: 881–906.

Kling, Jeffrey. 2006. "Incarceration Length, Employment, and Earnings." *American Economic Review* 96: 863–76.

Kling, Jeffrey R., Jeffrey B. Liebman, and Lawrence F. Katz. 2007. "Experimental Analysis of Neighborhood Effects." *Econometrica* 75: 83–119.

Kohn, Paul M., Reginald G. Smart, and Alan C. Ogborne. 1984. "Effects of Two Kinds of Alcohol Advertising on Subsequent Consumption." *Journal of Advertising* 13: 34–40, 48.

Kremer, Michael, and Alaka Holla. 2009. "Improving Education in the Developing World: What Have We Learned from Randomized Evaluations?" *Annual Review of Economics* 1: 513–42.

Kremer, Michael, Edward Miguel, and Rebecca Thornton. 2009. "Incentives to Learn." *The Review of Economics and Statistics* 91: 437–56.

Krueger, Alan B., and Diane M. Whitmore. 2001. "The Effect of Attending a Small Class in the Early Grades on College-Test Taking and Middle School Test Results: Evidence from Project STAR." *The Economic Journal* 111: 1–28.

Krueger, Alan B., and Pei Zhu. 2004a. "Another Look at the New York City School Voucher Experiment." *American Behavioral Scientist* 47(5): 658–98.

Krueger, Alan B., and Pei Zhu. 2004b. "Inefficiency, Subsample Selection Bias, and Nonrobustness: A Response to Paul E. Peterson and William G. Howell." *American Behavioral Scientist* 47(5): 718–27.

Kuehl, Robert O. 1999. *Design of Experiments: Statistical Principles of Research Design and Analysis*. Pacific Grove: Duxbury-Thomson Learning.

Lalive, Rafael, and Alejandra Cattaneo. 2009. "Social Interactions and Schooling Decisions." *The Review of Economics and Statistics* 91: 457–77.

LaLonde, Robert J. 1986. "Evaluating the Econometric Evaluations of Training Programs with Experimental Data." *The American Economic Review* 76: 604–20.

Lee, David. 2005. "Training, Wages, and Sample Selection: Estimating Sharp Bounds on Treatment Effects." NBER Working Paper 11721.

Lee, David S. 2008. "Randomized Experiments from Non-random Selection in U.S. House Elections." *Journal of Econometrics* 142: 675–97.

Lin, Winston. 2010. "Agnostic Notes on Regression Adjustments to Experimental Data: Reexamining Freedman's Critique." UC Berkeley Working Paper.

Lindo, Jason, and Charles Stoecker. 2010. "Drawn into Violence: Evidence on 'What Makes a Criminal' from the Vietnam Draft Lotteries." IZA Discussion Paper 5172.

List, John. 2004. "Neoclassical Theory versus Prospect Theory: Evidence from the Marketplace." *Econometrica* 72: 615–25.

Little, Roderick, and Donald Rubin. 1987. *Statistical Analysis with Missing Data*. New York: John Wiley.

Little, Roderick, and Donald Rubin. 2002. *Statistical Analysis with Missing Data*. 2nd Edition. New York: John Wiley.

Lock, Kari Frazer. 2011. "Rerandomization to Improve Covariate Balance in Randomized Experiments." PhD Dissertation, Harvard University.

Loewen, Peter, Royce Koop, Jamie Settle, and James Fowler. 2010. "A Natural Experiment in Proposal Power and Electoral Success." University of Toronto at Mississauga Working Paper.

Lohr, Sharon L. 2010. *Sampling: Design and Analysis*. Boston: Brooks/Cole.

Ludwig, Jens, Jeffrey R. Kling, and Sendhil Mullainathan. 2011. "Mechanism Experiments and Policy Evaluations." NBER Working Paper 17062.

Lyall, Jason. 2009. "Does Indiscriminate Violence Incite Insurgent Attacks?" *Journal of Conflict Resolution* 53: 331–62.

Mann, Christopher B. 2010. "Is There Backlash to Social Pressure? A Large-Scale Field Experiment on Voter Mobilization." *Political Behavior* 32: 387–407.

Manski, Charles F. 1989. "Anatomy of the Selection Problem." *The Journal of Human Resources* 24: 343–60.

Manski, Charles F. 1995. *Identification Problems in the Social Sciences*. Cambridge: Harvard University Press.

Manski, Charles F. 1997. "Monotone Treatment Response." *Econometrica* 65: 1311–34.

Manski, Charles F. 2007. *Identification for Prediction and Decision*. Cambridge: Harvard University Press.

Manski, Charles F. 2012. "Identification of Treatment Response with Social Interactions." *The Econometrics Journal*, forthcoming.

Mazerolle, Lorraine Green, James F. Price, and Jan Roehl. 2000. "Civil Remedies and Drug Control: A Randomized Field Trial in Oakland, California." *Evaluation Review* 24: 212–41.

Mazerolle, Lorraine Green, Jan Roehl, and Colleen Kadleck. 1998. "Controlling Social Disorder Using Social Remedies: Results from a Randomized Field Experiment in Oakland, California." *Crime Prevention Studies* 9: 141–59.

McCullagh, Peter. 2002. "What Is a Statistical Model?" *The Annals of Statistics* 30: 1225–1310.

Merrill, Ray. 2010. *Introduction to Epidemiology*. 5th Edition. Toronto: Jones and Bartlett Publishers.

Metrick, Andrew. 1995. "A Natural Experiment in 'Jeopardy!'" *American Economic Review* 85: 240–53.

Michalopoulos, Charles. 2005. "Precedents and Prospects for Social Experiments." In *Learning More from Social Experiments: Evolving Analytic Approaches*, ed. Howard S. Bloom. New York: Russell Sage Foundation.

Middleton, Joel A., and Peter M. Aronow. 2011. "Unbiased Estimation of the Average Treatment Effect in Cluster-Randomized Experiments." Yale University Working Paper.

Middleton, Joel, and Todd Rogers. 2010. "Defend Oregon's Voter Guide Program." Report for the Analyst Institute.

Miller, Cynthia, James Riccio, and Jared Smith. 2009. "A Preliminary Look at Early Educational Results of the Opportunity NYC–Family Rewards Program." MDRC Research Note.

Moore, Ryan. 2010. "blockTools: Blocking, Assignment, and Diagnosing Interference in Randomized Experiments." Version 0.5-2.

Moore, Underhill, and Charles C. Callahan. 1943. "Law and Learning Theory: A Study in Legal Control." *Yale Law Journal* 53: 1–136.

Morgan, Stephen L., and Christopher Winship. 2007. *Counterfactuals and Causal Inference: Methods and Principles for Social Research*. New York: Cambridge University Press.

Mullainathan, Sendhil, Ebonya Washington, and Julia Azari. 2010. "The Impact of Electoral Debate on Public Opinions: An Experimental Investigation of the 2005 New York City Mayoral Election." In *Political Representation*, ed. Ian Shapiro, Susan Stokes, Elizabeth Wood, and Alexander Kirshner. New York: Cambridge University Press.

Muralidharan, Karthik, and Venkatesh Sundararaman. 2011. "Teacher Performance Pay: Experimental Evidence from India." *Journal of Political Economy* 119: 39–77.

Murray, David M. 1998. *Design and Analysis of Group-Randomized Trials*. New York: Oxford University Press.

Mutz, Diana C., and Byron Reeves. 2005. "The New Videomalaise: Effects of Televised Incivility on Political Trust." *American Political Science Review* 99(1): 1–15.

Newhouse, Joseph. 1989. "A Health Insurance Experiment." In *Statistics: A Guide to the Unknown*, ed. Judith Tanur, Frederick Mosteller, William Kruskal, et al. Belmont, CA: Wadsworth.

Newhouse, Joseph. 1993. *Free for All? Lessons from the RAND Health Insurance Experiment*. Cambridge, MA: Harvard University Press.

Neyman, Jerzy. 1990 (1923). "On the Application of Probability Theory to Agricultural Experiments: Essay on Principles, Section 9." *Statistical Science* 5: 465–80.

Nickerson, David W. 2005. "Scalable Protocols Offer Efficient Design for Field Experiments." *Political Analysis* 13: 233–25.

Nickerson, David W. 2007. "Quality Is Job One: Professional and Volunteer Voter Mobilization Calls." *American Journal of Political Science* 51: 269–82.

Nickerson, David W. 2008. "Is Voting Contagious? Evidence from Two Field Experiments." *American Political Science Review* 102: 49–57.

Novak, Kenneth, Jennifer Hartman, Alexander Holsinger, and Michael Turner. 1999. "The Effects of Aggressive Policing of Disorder on Serious Crime." Policing: An International *Journal of Police Strategies and Management* 22: 171–94.

Nowell, Clifford, and Doug Laufer. 1997. "Undergraduate Student Cheating in the Fields of Business and Economics." *The Journal of Economic Education* 28: 3–12.

O'Brien, David J., and Valeri V. Patsiorkovski. 1999. "Russian Village Household Panel Surveys, 1995–1997." Inter-university Consortium for Political and Social Research (ICPSR) [Distributor] V1 [Version].

Olken, Benjamin. 2007. "Monitoring Corruption: Evidence from a Field Experiment in Indonesia." *Journal of Political Economy* 115: 200–49.

Olken, Benjamin A. 2010. "Direct Democracy and Local Public Goods: Evidence from a Field Experiment in Indonesia." *American Political Science Review* 104(2): 243–67.

Olson, Mancur. 1965. *The Logic of Collective Action*. Cambridge: Harvard University Press.

Orr, Larry L. 1999. *Social Experiments: Evaluating Public Programs with Experimental Methods*. Thousand Oaks, CA: Sage Publications.

Page, Stewart. 1998. "Accepting the Gay Person: Rental Accommodation in the Community." *Journal of Homosexuality* 36: 31–39.

Pager, Devah. 2007. "Is Racial Discrimination a Thing of the Past?" *The Academy Blog: Annals of the American Academy of Political and Social Science*.

Pager, Devah, Bruce Western, and Bart Bonikowski. 2009. "Discrimination in a Low-Wage Labor Market: A Field Experiment." *American Sociological Review* 74: 777–99.

Paluck, Elizabeth. 2009. "Reducing Intergroup Prejudice and Conflict Using the Media: A Field Experiment in Rwanda." Journal of Personality and Social Psychology 96: 574–87.

Paluck, Elizabeth L., and Donald P. Green. 2009. "Prejudice Reduction: What Works? A Critical Look at Evidence from the Field and the Laboratory." *Annual Review of Psychology* 60: 339–67.

Panagopoulos, Costas. 2009. "Partisan and Nonpartisan Message Content and Voter Mobilization: Field Experimental Evidence." *Political Research Quarterly* 62: 70–76.

Panagopoulos, Costas. 2010. "Affect, Social Pressure and Prosocial Motivation: Field Experimental Evidence of the Mobilizing Effects of Pride, Shame and Publicizing Voting Behavior." *Political Behavior* 32: 369–86.

Panagopoulos, Costas. 2011. "Timing Is Everything? Primacy and Recency Effects in Voter Mobilization Campaigns." *Political Behavior* 33: 79–93.

Peisakhin, Leonid, and Paul Pinto. 2010. "Is Transparency an Effective Anti-Corruption Strategy? Evidence from a Field Experiment in India." *Regulation and Governance* 4: 261–80.

Peterson, Paul E., and William G. Howell. 2004. "Efficiency, Bias, and Classification Schemes: A Response to Alan B. Krueger and Pei Zhu." *American Behavioral Scientist* 47(5): 699–718.

Petterson-Lidbom, Per. 2004. "Does the Size of the Legislature Affect the Size of Government? Evidence from Two Natural Experiments." Discussion Papers 350, Government Institute for Economic Research (VATT).

Pocock, Stuart J., Susan E. Assmann, Laura E. Enos, and Linda E. Kasten. 2002. "Subgroup Analysis, Covariate Adjustment and Baseline Comparisons in Clinical Trial Reporting: Current Practice and Problems." *Statistics in Medicine* 21: 2917–30.

Post, Thierry, Martin van den Assem, Guido Baltussen, and Richard Thaler. 2008. "Deal or No Deal? Decision Making under Risk in a Large-Payoff Game Show." *American Economic Review* 98: 38–71.

Ratcliffe, Jerry H., Travis Taniguchi, Elizabeth R. Groff, and Jennifer D. Wood. 2011. "The Philadelphia Foot Patrol Experiment: A Randomized Controlled Trial of Police Patrol Effectiveness in Violent Crime Hotspots." *Criminology* 49: 795–831.

Rind, Bruce, and Prashant Bordia. 1996. "Effect on Restaurant Tipping of Male and Female Servers Drawing a Happy, Smiling Face on the Backs of Customers' Checks." *Journal of Applied Social Psychology* 26: 218–25.

Rind, Bruce, and David Strohmetz. 1999. "Effect of Restaurant Tipping of a Helpful Message Written on the Back of Customers' Checks." *Journal of Applied Social Psychology* 29: 139–44.

Roberts, Ian, and Irene Kwan. 2001. "School-Based Driver Education for the Prevention of Traffic Crashes." *Cochrane Database of Systematic Reviews* 3: CD003201.

Robins, James M., and Sander Greenland. 1992. "Identifiability and Exchangeability for Direct and Indirect Effects." *Epidemiology* 3: 143–55.

Robins, Philip K. 1985. "A Comparison of the Labor Supply Findings from the Four Negative Income Tax Experiments." *The Journal of Human Resources* 20: 567–82.

Rondeau, Daniel, and John List. 2008. "Matching and Challenge Gifts to Charity: Evidence from Laboratory and Natural Field Experiments." *Experimental Economics* 11: 253–67.

Rosen, Jeff. 2010. "The Effects of Race and Grammar Quality on the Responsiveness of American State Legislators: A Field Experiment." Unpublished manuscript.

Rosenbaum, Paul R. 1984. "The Consequences of Adjustment for a Concomitant Variable That Has Been Affected by the Treatment." *Journal of the Royal Statistical Society, Series A (General)* 147: 656–66.

Rosenbaum, Paul R. 2002. *Observational Studies*. 2nd Edition. New York: Springer.

Rosenbaum, Paul R. 2007. "Interference Between Units in Randomized Experiments." *Journal of the American Statistical Association* 102: 191–200.

Rosenbaum, Paul R. 2010. *Design of Observational Studies*. New York: Springer-Verlag.

Rosenberger, William F., and John M. Lachin. 2002. *Randomization in Clinical Trials: Theory and Practice*. New York: Wiley.

Rothwell, Peter M. 2005. "Subgroup Analysis in Randomised Controlled Trials: Importance, Indications, and Interpretation." *The Lancet* 365: 176–86.

Rowe, Candy, Julie M. Harris, and S. Craig Roberts. 2005. "Seeing Red? Putting Sportswear in Context." *Nature* 437: E10.

Rubin, Donald B. 1980. "Randomization Analysis of Experimental Data: The Fisher Randomization Test Comment." *Journal of the American Statistical Association* 75: 591–93.

Rubin, Donald B. 1986. "Statistics and Causal Inference: Comment: Which Ifs Have Causal Answers." *Journal of the American Statistical Association* 81: 961–62.

Rubin, Donald B. 1990. "Formal Mode of Statistical Inference for Causal Effects." *Journal of Statistical Planning and Inference* 25: 279–92.

Rubin, Donald B. 2001. "Comment on 'Surprises from Self-Experimentation: Sleep, Mood and Weight.'" *Chance* 14: 16–17.

Rubin, Donald B. 2005. "Causal Inference Using Potential Outcomes: Design, Modeling, Decisions." *Journal of the American Statistical Association* 100: 322–31.

Rubin, Donald B. 2008. "For Objective Causal Inference, Design Trumps Analysis." *The Annals of Applied Statistics* 2: 808–40.

Sacerdote, Bruce. 2001. "Peer Effects with Random Assignment: Results for Dartmouth Roommates." *The Quarterly Journal of Economics* 116(2): 681–704.

Salsburg, David. 2001. *The Lady Tasting Tea: How Statistics Revolutionized Science in the Twentieth Century*. New York: W.H. Freeman.

Samii, Cyrus, and Peter M. Aronow. 2012. "On Equivalencies Between Design-Based and Regression-Based Variance Estimators for Randomized Experiments." *Statistics and Probability Letters* 82(2): 365–70.

Sanbonmatsu, Lisa, Jeffrey R. Kling, Greg J. Duncan, and Jeanne Brooks-Gunn. 2006. "Neighborhoods and Academic Achievement: Results from the Moving to Opportunity Experiment." *The Journal of Human Resources* 41: 649–91.

Sandler, Corey. 2007. "Econoguide Disneyland Resort, Universal Studios Hollywood, and Other Major Southern California Attractions Including Disney's California Adventure." Springfield: Globe Pequot.

Schneider, Henry S. 2007. "Empirical Studies of the Effects of Information Asymmetry." PhD Dissertation, Yale University.

Schneider, Henry S. 2009. "Agency Problems and Reputation in Expert Services: Evidence from Auto Repair." *Journal of Industrial Economics*, forthcoming.

Schochet, Peter Z. 2010. "Is Regression Adjustment Supported by the Neyman Model for Causal Inference?" *Journal of Statistical Planning and Inference* 140: 246–59.

Schulz, Kenneth F., Douglas G. Altman, and David Moher. 2010. "CONSORT 2010 Statement: Updated Guidelines for Reporting Parallel Group Randomized Trials." *Annals of Internal Medicine* 152: 1–7.

Shadish, William R., Thomas D. Cook, and Donald T. Campbell. 2002. *Experimental and Quasi-Experimental Designs for Generalized Causal Inference*. Boston: Houghton Mifflin.

Shaffer, Juliet. 1995. "Multiple Hypothesis Testing." *Annual Review of Psychology* 46: 561–84.

Sherman, Lawrence W., Dennis Rogan, Timothy Edwards, Rachel Whipple, Dennis Shreve, Daniel Witcher, William Trimble, The Street Narcotics Unit, Robert Velke, Mark Blumberg, Anne Beatty, and Carol Bridgeforth. 1995. "Deterrent Effects of Police Raids on Crack Houses: A Randomized Controlled Experiment." *Justice Quarterly* 12: 755–81.

Sherman, Lawrence W., Janell Schmidt, Dennis Rogan, Douglas Smith, Patrick Gartin, Ellen Cohn, Dean Collins, and Anthony Bacich. 1992. "The Variable Effects of Arrest on Criminal Careers: The Milwaukee Domestic Violence Experiment." *Journal of Criminal Law and Criminology* 83: 137–69.

Sherman, Lawrence W., Heather Strang, Caroline Angel, Daniel Woods, Geoffrey C. Barnes, Sarah Bennett, and Nova Inkpen. 2005. "Effects of Face-to-Face Restorative Justice on Victims of Crime in Four Randomized, Controlled Trials." *Journal of Experimental Criminology* 1: 367–95.

Sherman, Lawrence W., and David Weisburd. 1995. "General Deterrent Effects of Police Patrol in Crime 'Hot Spots': A Randomized, Controlled Trial." *Justice Quarterly*, 12(4): 625–48.

Simester, Duncan, Yu (Jeffrey) Hu, Erik Brynjolfsson, and Eric T. Anderson. 2009. "Dynamics of Retail Advertising: Evidence from a Field Experiment." *Economic Inquiry* 47: 482–99.

Sinclair, Betsy, Margaret McConnell, and Donald P. Green. 2012. "Detecting Spillover Effects: Design and Analysis of Multi-Level Experiments." *American Journal of Political Science* forthcoming.

Slemrod, Joel, Marsha Blumenthal, and Charles Christian. 2001. "Taxpayer Response to an Increased Probability of Audit: Evidence from a Controlled Experiment in Minnesota." *Journal of Public Economics* 79: 455–83.

Smith, Gordon C. S., and Jill P. Pell. 2003. "Parachute Use to Prevent Death and Major Trauma Related to Gravitational Challenge: Systematic Review of Randomised Controlled Trials." *British Medical Journal* 327: 1459–61.

Smith, Jeffrey A., and Petra E. Todd. 2005a. "Does Matching Overcome LaLonde's Critique of Nonexperimental Estimators?" *Journal of Econometrics* 125: 305–53.

Smith, Jeffrey A., and Petra E. Todd. 2005b. "Rejoinder." *Journal of Econometrics* 125: 365–75.

Sobel, Michael E. 2006. "What Do Randomized Studies of Housing Mobility Demonstrate? Causal Inference in the Face of Interference." *Journal of the American Statistical Association* 101: 1398–1407.

Sondheimer, Rachel Milstein, and Donald P. Green. 2009. "Using Experiments to Estimate the Effects of Education on Voter Turnout." *American Journal of Political Science* 54: 174–89.

Spencer, Steven J., Mark P. Zanna, and Geoffrey T. Fong. 2005. "Establishing a Causal Chain: Why Experiments Are Often More Effective Than Mediational Analyses in Examining Psychological Processes." *Journal of Personality and Social Psychology* 89: 845–51.

Stock, James H., and Mark W. Watson. 2007. *Introduction to Econometrics.* 2nd Edition. Boston: Pearson/Addison Wesley.

Strohmetz, David, Bruce Rind, Reed Fisher, and Michael Lynn. 2002. "Sweetening the Till: The Use of Candy to Increase Restaurant Tipping." *Journal of Applied Social Psychology* 32: 300–09.

Tanur, Judith, Frederick Mosteller, and William Kruskal, et al., eds. 1989. *Statistics: A Guide to the Unknown.* Belmont, CA: Wadsworth.

Titiunik, Rocío. 2010. "Drawing Your Senator from a Jar: Term Length and Legislative Behavior." University of Michigan Working Paper.

Torgerson, David J., and Carole J. Torgerson. 2008. *Designing Randomised Trials in Health, Education and the Social Sciences: An Introduction.* Basingstoke: Palgrave Macmillan.

U.S. Department of Health and Human Services, Administration for Children and Families. 2010. "Head Start Impact Study. Final Report." Washington, DC: Office of Planning, Research and Evaluation.

Vaa, Truls. 1997. "Increased Police Enforcement: Effects on Speed." *Accident Analysis and Prevention.* 29: 373–85.

van Helmont, Jean-Baptiste. 1662. Oriatrike, or Physick Refined: *The Common Errors Therein Refuted and the Whole are Reformed and Rectified.* London: Lodowick-Loyd.

Vernick, Jon S., Guohua Li, Susanne Ogaitis, Ellen J. MacKenzie, Susan P. Baker, and Andrea C. Gielen. 1999. "Effects of High School Driver Education on Motor Vehicle Crashes, Violations, and Licensure." *American Journal of Preventive Medicine* 16: 40–46.

Waldfogel, Joel. 1991. "Aggregate Inter-Judge Disparity in Federal Sentencing: Evidence from Three Districts (D.Ct., S.D.N.Y., N.D.Cal.)." *Federal Sentencing Reporter* 4: 151–54.

Weisburd, David, and Lorraine Green. 1995. "Policing Drug Hot Spots: The Jersey City Drug Market Analysis Experiment." *Justice Quarterly* 12(4): 711–35.

Wilson, James Q., and George L. Kelling. 1982. "Broken Windows: The Police and Neighborhood Safety." *The Atlantic* 127: 29–38.

Wolf, Patrick J., Babette Gutmann, Michael Puma, Brian Kisida, Lou Rizzo, Nada Eissa, and Matthew Carr. 2010. *Evaluation of the DC Opportunity Scholarship Program: Final Report.* U.S. Department of Education, Institute for Education Sciences, National Center for Education Evaluation and Regional Assistance, NCEE 2010-4018. Washington, DC: U.S. Government Printing Office.

Wooldridge, Jeffrey M. 2002. *Econometric Analysis of Cross Section and Panel Data.* Cambridge, MA: MIT Press.

Wooldridge, Jeffrey M. 2010. *Econometric Analysis of Cross Section and Panel Data.* 2nd Edition. Cambridge, MA: MIT Press.

Yanow, Dvora, and Peregrine Schwartz-Shea. 2008. "Reforming Institutional Review Board Policy: Issues in Implementation and Field Research." *PS, Political Science and Politics* 41: 483–94.

Zinovyeva, Natalia, and Manuel Bagues. 2010. "Does Gender Matter for Academic Promotion? Evidence from a Randomized Natural Experiment." FEDEA Working Paper 2010–15.

INDEX

Note: Material in figures or tables is indicated by italic page numbers. Footnotes are indicated by *n* after the page number.